Introduction to Microwave Imaging

This self-contained, introductory text will enable readers to easily understand the fundamentals of microwave and radar image generation. Written with the complete novice in mind, and including an easy-to-follow introduction to electromagnetic scattering theory, it covers key topics such as forward models of scattering for interpreting S-parameter and time-dependent voltage data, S-parameters and their analytical sensitivity formulas, basic methods for real-time image reconstruction using frequency sweep and pulsed radar signals, and metrics for evaluating system performance. Numerous application examples and practical tutorial exercises provided throughout allow for quick understanding of key concepts, and sample MATLAB codes implementing key reconstruction algorithms accompany the book online.

This one-stop resource is ideal for graduate students taking introductory courses in microwave imaging, as well as researchers and industry professionals wanting to learn the fundamentals of the field.

Natalia K. Nikolova is a Professor in the Department of Electrical and Computer Engineering at McMaster University and a Canada Research Chair in High-Frequency Electromagnetics. She is a Fellow of both the IEEE and the Canadian Academy of Engineering, as well as a former IEEE Distinguished Microwave Lecturer.

EuMA High Frequency Technologies Series

Series Editor
Peter Russer, Technical University of Munich

Homayoun Nikookar, *Wavelet Radio*
Thomas Zwick, Werner Wiesbeck, Jens Timmermann, and Grzegorz Adamiuk (Eds.), *Ultra-Wideband RF System Engineering*
Er-Ping Li and Hong-Son Chu, *Plasmonic Nanoelectronics and Sensing*
Luca Roselli (Ed.), *Green RFID Systems*
Vesna Crnojević-Bengin, *Advances in Multi-Band Microstrip Filters*
Natalia K. Nikolova, *Introduction to Microwave Imaging*

Forthcoming

Peter Russer, Johannes Russer, Uwe Siart, and Andreas Cangellaris, *Interference and Noise in Electromagnetics*
Maurizio Bozzi, Apostolos Georgiadis, and Ke Wu, *Substrate Integrated Waveguides*
Luca Pierantoni and Fabio Coccetti, *Radiofrequency Nanoelectronics Engineering*
Alexander Yarovoy, Introduction to *UWB Wireless Technology and Applications*
Philippe Ferrari, Rolf Jakoby, Onur Karabey, and Gustavo Rehder, *Reconfigurable Circuits and Technologies for Smart Millimeter-Wave Systems*
Luca Roselli, *Enabling Technologies for the Internet of Things*
Nuno Borges Carvalho, Alessandra Costanzo, Apostolos Georgiadis, and Franco Mastri, *Wireless Power Transfer*
Costas Sarris and Andrew Austin, *Uncertainty Quantification in Computational Electromagnetics*

Introduction to Microwave Imaging

NATALIA K. NIKOLOVA
McMaster University, Ontario

CAMBRIDGE
UNIVERSITY PRESS

CAMBRIDGE
UNIVERSITY PRESS

University Printing House, Cambridge CB2 8BS, United Kingdom

One Liberty Plaza, 20th Floor, New York, NY 10006, USA

477 Williamstown Road, Port Melbourne, VIC 3207, Australia

4843/24, 2nd Floor, Ansari Road, Daryaganj, Delhi - 110002, India

79 Anson Road, #06-04/06, Singapore 079906

Cambridge University Press is part of the University of Cambridge.

It furthers the University's mission by disseminating knowledge in the pursuit of education, learning and research at the highest international levels of excellence.

www.cambridge.org
Information on this title: www.cambridge.org/9781107085565

First published 2017

Printed in the United States of America by Sheridan Books, Inc.

A catalogue record for this publication is available from the British Library

ISBN 978-1-107-08556-5 Hardback

To Michael and Mihail

"Natalia Nikolova has created an outstanding, self-contained resource for students, researchers, and practitioners in the field of microwave imaging. The clarity, depth, and breadth of this masterful treatise are hallmarks of her exceptional talents as a distinguished researcher and educator. She leads the reader through a compelling landscape of field theory, computational electromagnetics, and technology to the cutting edge of microwave imaging. This book is a must read for students and experts alike."

Wolfgang J. R. Hoefer, University of Victoria

"In this reference book, Natalia Nikolova has opportunely and comprehensively gathered together different microwave image reconstruction algorithms as resulting from various processing options of Maxwell's equations, in frequency or time domains.

As such, it constitutes an extremely useful toolbox with practical operating guidance enabling all those, firstcomers, students or experienced researchers and engineers, faced with microwave imaging applications to select the best-suited algorithm for their test cases, implement it on a computer and, finally, evaluate its performance according to different image quality criteria."

Jean-Charles Bolomey, University Paris-Sud XI

Contents

Preface

Microwave imaging refers to applications and technology that exploit electromagnetic radiation of frequencies anywhere between several hundred mega-hertz (MHz) and several hundred giga-hertz (GHz). Depending on the frequency, this radiation penetrates many optically opaque media (e.g., fog, foliage, soil, wood, brick, concrete, ceramics, plastics, clothing, etc). It has been used for decades in marine and flight radar, remote sensing, weather radar, reconnaissance and target tracking, and underground surveillance. These applications are referred to as *long range*, indicating that the distance between the imaged target and the radar system is much greater than any of the following three measures: (i) the target's size, (ii) the size of the antenna or the sensor array, and (iii) the wavelength.

More recently, microwave and millimeter-wave systems have been deployed in a variety of *short-range* applications where the inspected scene is at a distance comparable to or smaller than at least one of the three measures listed earlier. These include security and surveillance systems for concealed weapon detection and through-the-wall imaging. For instance, the whole body scanners for airplane passenger screening utilize millimeter waves. Microwave systems are also used in nondestructive testing and inspection for structural integrity. The imaging of body parts and tissues has attracted much attention and is the focus of research throughout the world.

Short-range microwave imaging is a vibrant field of research and development that is projected to grow exponentially in the coming years. The main driving force behind this growth is the unprecedented miniaturization of the radio-frequency, microwave, and millimeter-wave technologies. The size and the cost of the hardware needed to perform microwave sensing have dramatically decreased with the advent of the radio-on-a-chip (RoC), the on-chip software-defined radios (SDRs), and the single-chip radars operating well into the 70-GHz bands. Coherent signal measurements with amplitude and phase information are now possible with electronic circuits occupying areas less than a square centimeter and with a price that ranges anywhere between tens and several hundred dollars. Multiplexing hundreds, even thousands, of microwave sensors through digital control is within reach, providing the ability to bypass expensive and bulky RF switching and power-distribution networks. As a result of these advancements, large arrays of microwave sensors are now affordable. Microwave imaging has arrived at a stage of technological development that enables its ubiquitous deployment.

This book is intended for the beginner in microwave imaging. Currently, many graduate students, researchers, and engineers are venturing into this promising field of study.

However, despite the large body of research publications in this field, introductory-level material is lacking. Moreover, due to the interdisciplinary nature of the subject, research publications are scattered throughout journals and the collections of the physics, biomedical, and engineering societies. This book's goal is to provide a one-stop introductory resource. As such, its focus is on the fundamentals of scattering theory and on the methods of real-time imaging, which are known as direct or linear methods. The material is supported with numerous solved problems and examples. Most of these are analytical or employ simple numerical computations so that the reader can appreciate the underlying concepts.

Chapter 1 introduces the scalar model of wave scattering. This model is fundamental to the methods of pulsed-radar imaging, which almost exclusively rely on the scalar approximation of the electromagnetic field. Microwave tomography is also based on the two-dimensional scalar model of scattering. Moreover, understanding the scalar theory is a necessary step toward understanding the vector theory of electromagnetic scattering, which is based on Maxwell's equations. Last but not least, the scalar scattering models are fundamental in acoustic imaging, which is akin to microwave imaging and which has inspired many of the microwave inversion concepts and strategies. Special attention is paid to the concepts of Green's functions and fundamental solutions, which are considered in the cases of the wave equation, the damped-wave equation, and the Helmholtz equation in parallel. This extensive subject has been condensed into about 90 pages by focusing on key concepts only. Many illustrations and examples support the most difficult concepts such as adjoint solutions, causal versus acausal solutions, back-propagation, and surface equivalence. The chapter ends with a detailed discussion of the Born and the Rytov scalar models of scattering.

Chapter 2 presents the basics of the vectorial model of electromagnetic scattering. The concept of Green's dyadic (or Green's tensor) is introduced in analogy with Green's function in the scalar models. The electromagnetic surface equivalence principle is discussed in the framework of radiation and scattering. Special attention is paid to the simplifying approximations of the vectorial model that are widely used in microwave imaging.

Chapter 3 introduces the concept of scattering parameters (or S-parameters), which are the most common type of responses measured in imaging experiments. Response sensitivities are discussed in the context of the S-parameters. Sensitivity formulas for both constitutive parameters and shape parameters are given along with analytical examples that illustrate their meaning and implementation. Sensitivity analysis is an essential part of electromagnetic modeling. It plays a critical role in imaging since it provides the Fréchet derivative to optimization-based and sensitivity-based reconstruction methods. Chapter 3 also introduces the data equation of electromagnetic scattering in the form of S-parameter responses. This equation is central to the developments in subsequent chapters because it serves as the forward model, on which the direct inversion methods are based.

Chapter 4 discusses linear inversion methods that solve the inverse problem in real space. This is in contrast to Chapter 5, which introduces the Fourier-space linear inversion methods. Chapter 4 presents qualitative inversion methods such as sensitivity maps, which operate on frequency-domain data, as well as synthetic focusing, which operates on time-domain (or pulsed) data. Time-reversal and 3D deconvolution methods are briefly discussed. Quantitative linear inversion is also introduced through the method of scattered-power maps.

Chapter 5 discusses in detail the methods of microwave holography, both qualitative and quantitative. Microwave holography is akin to synthetic aperture radar (SAR); thus, this subject is important for further SAR studies. The principle of diffraction tomography is also presented, and its close relationship to microwave holography is shown. This chapter also focuses on the important subject of spatial and frequency sampling rates, which are central to the design of imaging experiments.

Chapter 6 is dedicated to the performance metrics in imaging, such as spatial resolution, contrast sensitivity, system dynamic range, and signal-to-noise ratio. The sources of noise and uncertainty in measurements are discussed along with methods to assess experimentally the data quality produced by the particular acquisition hardware. These metrics are important as guidelines in setting the design specifications of the measurement system. They are also useful as a means of comparing these systems.

The concluding Chapter 7 prepares the reader for further study in advanced methods and applications of microwave imaging, namely, the nonlinear iterative reconstruction approaches. These are classified and are reviewed briefly. The linear (or direct) reconstruction approaches introduced in this text are limited to weak scattering, and this limitation can be overcome by nonlinear reconstruction. The linear approaches, however, are the platform on which understanding of the advanced methods rests. They also serve as linearized inversion modules within the framework of many iterative techniques. This latter role is discussed in the case of the Born iterative method and the distorted Born iterative method.

It is my hope that this book will elevate the novice to the level necessary to understand the literature on the subject. Most importantly, I hope that it will help the reader in making a first step toward innovation. The next frontier of wireless technology is in imaging, detection, and sensing. Perhaps the reader will join the new generation of scientists and engineers who will advance this frontier.

Acknowledgments

This book emerged from the efforts of many people whose support I must acknowledge. My graduate students spent many hours reading drafts of the book chapters and suggesting examples and aids to understanding. Their input made this book easier to read and better suited for the younger generation. I would like to extend special thanks to Prof. Reza K. Amineh of the New York Institute of Technology (NYIT) for having read the whole manuscript and for providing great insight and many valuable suggestions. I would also like to acknowledge the invaluable help and patience of the staff at Cambridge University Press.

Throughout the work on this book, I had the unwavering support and patience of my husband, Michael Bearder, which I deeply appreciate.

Notation

$\mathbf{r}$	position vector
(x, y, z)	position in rectangular coordinates
(ρ, ϕ, z)	position in cylindrical coordinates
(r, θ, ϕ)	position in spherical coordinates
$\mathbf{E}$	electric field phasor $(\mathrm{V \cdot m^{-1}})$
$\mathbf{H}$	magnetic field phasor $(\mathrm{A \cdot m^{-1}})$
$\mathbf{D}$	electric flux density phasor $(\mathrm{C \cdot m^{-2}})$
$\mathbf{B}$	magnetic flux density phasor $(\mathrm{Wb \cdot m^{-2}})$
$\mathcal{E}$	time-dependent electric field $(\mathrm{V \cdot m^{-1}})$
$\mathcal{H}$	time-dependent magnetic field $(\mathrm{A \cdot m^{-1}})$
$\mathcal{D}$	time-dependent electric flux density $(\mathrm{C \cdot m^{-2}})$
$\mathcal{B}$	time-dependent magnetic flux density $(\mathrm{Wb \cdot m^{-2}})$
Φ	electric scalar potential phasor (V)
φ	time-dependent electric scalar potential (V)
$\mathbf{J}$	electric current density phasor $(\mathrm{A \cdot m^{-2}})$
$\mathcal{J}$	time-dependent electric current density $(\mathrm{A \cdot m^{-2}})$
$\mathbf{J}_{\mathrm{s}}$	electric surface current density phasor $(\mathrm{A \cdot m^{-1}})$
$\mathcal{J}_{\mathrm{s}}$	time-dependent electric surface current density $(\mathrm{A \cdot m^{-1}})$
$\mathbf{J}_{\mathrm{i}}$	impressed electric current density phasor (source) $(\mathrm{A \cdot m^{-2}})$
$\mathbf{J}_{\mathrm{c}}$	conduction (ohmic) current density phasor $(\mathrm{A \cdot m^{-2}})$
$\mathcal{J}_{\mathrm{c}}$	conduction (ohmic) current density $(\mathrm{A \cdot m^{-2}})$
$\mathcal{J}_{\mathrm{i}}$	time-dependent impressed electric current density (source) $(\mathrm{A \cdot m^{-2}})$
$\mathbf{M}_{\mathrm{i}}$	impressed magnetic current density phasor (source) $(\mathrm{V \cdot m^{-2}})$
$\mathcal{M}_{\mathrm{i}}$	time-dependent impressed magnetic current density (source) $(\mathrm{V \cdot m^{-2}})$
$\boldsymbol{\rho}_{\mathrm{e}}$	electric charge density phasor $(\mathrm{C \cdot m^{-3}})$
ρ_{e}	time-dependent electric charge density $(\mathrm{C \cdot m^{-3}})$
$\boldsymbol{\rho}_{\mathrm{i}}$	impressed electric charge density phasor (source) $(\mathrm{C \cdot m^{-3}})$
ρ_{i}	impressed time-dependent electric charge density (source) $(\mathrm{C \cdot m^{-3}})$
$\boldsymbol{\rho}_{\mathrm{s,e}}$	electric surface charge density phasor $(\mathrm{C \cdot m^{-2}})$
ρ_{s}	time-dependent surface electric charge density $(\mathrm{C \cdot m^{-2}})$
$\mathcal{I}$	time-dependent electric current (A)
$\mathcal{I}_{\mathrm{m}}$	time-dependent (fictitious) magnetic current (V)

$\kappa = \boldsymbol{\mu}_{\mathrm{r,s}}\boldsymbol{\varepsilon}_{\mathrm{r,s}} - \boldsymbol{\mu}_{\mathrm{r,b}}\boldsymbol{\varepsilon}_{\mathrm{r,b}}$	property contrast, dimensionless
$\breve{a}$	adjoint constant
λ	wavelength (m^{-1})
ω	angular frequency ($\mathrm{rad} \cdot \mathrm{s}^{-1}$)
ϵ_0	absolute permittivity of vacuum, $\approx 8.854187817 \cdot 10^{-12}\ \mathrm{F} \cdot \mathrm{m}^{-1}$
μ_0	absolute permeability of vacuum, $4\pi \cdot 10^{-7}\ \mathrm{H} \cdot \mathrm{m}^{-1}$
ϵ	dielectric permittivity ($\mathrm{F} \cdot \mathrm{m}^{-1}$)
ϵ_r	relative dielectric permittivity, dimensionless
ε	complex dielectric permittivity ($\mathrm{F} \cdot \mathrm{m}^{-1}$)
ε_r	complex relative dielectric permittivity, dimensionless
μ	magnetic permeability ($\mathrm{H} \cdot \mathrm{m}^{-1}$)
μ_r	relative magnetic permeability, dimensionless
$\boldsymbol{\mu}$	complex magnetic permeability ($\mathrm{H} \cdot \mathrm{m}^{-1}$)
$\boldsymbol{\mu}_r$	complex relative magnetic permeability, dimensionless
σ_{e}	electric conductivity ($\mathrm{S} \cdot \mathrm{m}^{-1}$)
σ_{m}	magnetic conductivity ($\Omega \cdot \mathrm{m}^{-1}$)
$\tan \delta_{\mathrm{d}}$	dielectric loss tangent, dimensionless
$\tan \delta_{\mathrm{m}}$	magnetic loss tangent, dimensionless
$\eta_0 = \sqrt{\frac{\mu_0}{\epsilon_0}}$	intrinsic impedance of vacuum (Ω)
η	intrinsic impedance of medium (Ω)
$\mathbb{R}^n$	n-dimensional real space
$\mathbb{C}^n$	n-dimensional complex space

Acronyms

AC	alternating current
BIM	Born iterative method
CO	calibration object
CP-MCT	chirp-pulse microwave computed tomography
CT	computed tomography
DAS	delay and sum
DBIM	distorted Born iterative method
DFT	discrete Fourier transform
DT	diffraction tomography
EFIE	electric field integral equation
EM	electromagnetic
FDTD	finite-difference time-domain
FFR	feature-free region
FFT	fast Fourier transform
FT	Fourier transform
GPR	ground-penetrating radar
MFIE	magnetic field integral equation
MoM	method of moments
MRI	magnetic resonance imaging
MW	microwave
OUT	object under test
PEC	perfect electric conductor
PMC	perfect magnetic conductor
PSF	point-spread function
RF	radio frequency
RMS	root mean square
RO	reference object
ROI	region of interest
Rx	receiver or receiving
SAR	synthetic aperture radar
SNR	signal-to-noise ratio
TE	transverse electric
TEM	transverse electromagnetic
TLM	transmission-line matrix

TM	transverse magnetic
TR	time reversal
TRM	time-reversal mirror
Tx	transmitter or transmitting
UWB	ultra-wideband
VNA	vector network analyzer

1 Scalar-Wave Models in Electromagnetic Scattering

This chapter is concerned with the scalar forward models used in microwave (MW) imaging. These are mathematical models of varying degrees of accuracy that predict the field based on a known source of radiation in a known environment. They are called *forward* because they describe the causal (or forward-in-time) relationship in a phenomenon we could express as *cause* → *effect*. In imaging, the *cause* is described by the *model parameters*, i.e., (i) the parameters of the sources generating the field and (ii) the parameters of the environment where this field exists or propagates.[1] The *effect* is described by the *observation data*, or simply, the *data*. These are signals acquired through measurements. Thus, in imaging, the forward model predicts the data, provided the model parameters are known.

The object of imaging, however, is the *inverse problem*, which, in contrast to the forward problem, is expressed as *effect* → *cause*. Finding what caused an effect is not an easy task. The second part of this book is dedicated to the mathematical methods used to accomplish this task. For now, it suffices to say that we first need to have a forward model of a phenomenon before we can start solving inverse problems based on this phenomenon. To illustrate this point, imagine that you are listening to a recording of a symphony; in order to tell which instruments play at any given time, you have first to have heard the sound of each instrument.

The phenomenon of interest in the forward models of MW imaging is the scattering of the high-frequency electromagnetic (EM) field by objects.[2] The scattering objects are often referred to as *targets*, especially in radar, or as *scatterers*. In this chapter, we discuss the mathematical scalar models of scattering.

The EM field is a vectorial field fundamentally described by Maxwell's equations [1, 2, 3, 4, 5]. For a summary of Maxwell's equations, see Appendix A. However, to simplify the analysis, scalar approximations are often made, and here we start with these simpler models. The scalar-wave model is very useful as an intermediate step toward the understanding of the vectorial wave model. It can also serve as a bridge to understanding acoustic and elastic wave phenomena, which are widely used in imaging.

Strictly speaking, the scalar-wave model in electromagnetism is limited to the case of a uniform isotropic medium, which becomes apparent when one attempts to reduce

[1] The term *propagation* refers to the manner in which a field develops in space and time.

[2] The *scattering* of waves refers to the way the original, or *incident*, waves interact with obstacles. This interaction produces secondary, or *scattered*, waves that often spread away from the object in various directions; thus, the term *scatter*.

Maxwell's equations to decoupled scalar second-order partial differential equations [1, 3, 5]. In such a medium, as discussed later, we may work with two types of scalar functions: (i) the *Cartesian* components of the electric and magnetic field vectors and (ii) the so-called *wave functions*.

One may wonder why we are interested in the simple scenario of uniform isotropic medium, bearing in mind that MW imaging is inherently concerned with nonuniform objects. The short answer is that most MW imaging methodologies assume that the object under test (OUT) is immersed in a uniform medium. And in most applications, this uniform medium is predominantly isotropic; for example, air, concrete, sea water, soil, etc. Even if the assumption of uniformity is invalid, which may be the case when we deal with imaging in complex environments (e.g., living tissue or concealed weapon detection), it helps to first understand how imaging is done in a uniform background and then move on to complex environments.

1.1 Partial Differential Equations for Scalar Waves in the Time Domain

Before exploring in depth the mathematics of scalar waves, the reader should be aware of an important physical limitation of the analytical time-domain models of electromagnetism discussed later: they are applicable only if the frequency dependence (or the dispersion) of the medium properties is negligible. This, of course, cannot be true throughout the spectrum; however, it could be approximately true for the bandwidth of the radiation (the bandwidth of the excitation sources). Then, these models can be useful. Time-domain modeling is particularly important in imaging with pulsed radar. More notes on dispersion are given later as appropriate.

The time-domain Maxwell's equations, when applied to the *Cartesian* components of the field vectors in a uniform isotropic medium, lead to the second-order partial differential equation [3, 5] (see Appendix B for its derivation):

$$\left[\nabla^2 - \mu\epsilon\frac{\partial^2}{\partial t^2} - (\mu\sigma_e + \epsilon\sigma_m)\frac{\partial}{\partial t} - \sigma_e\sigma_m\right] u_\xi(\mathbf{r}, t) = \mathcal{S}_\xi(\mathbf{r}, t), \tag{1.1}$$

where $\mathbf{r}$ is position, t is time, $u \equiv \mathcal{E}$ or $\mathcal{H}$ ($\mathcal{E}$ or $\mathcal{H}$ denote the type of field, electric or magnetic, respectively), and $\xi \equiv x, y, z$ is the vector component. The constitutive parameters ϵ, μ, σ_e, and σ_m are permittivity, permeability, electric conductivity, and magnetic conductivity, respectively,[3] all of them being constant in $\mathbf{r}$ as per the assumption of a

[3] In high-frequency problems, the conductivities σ_e and σ_m are often referred to as *equivalent* conductivities, and they represent losses due to the conversion of EM energy into heat (dissipation). The equivalent magnetic conductivity σ_m is zero at DC (direct, or steady, current), reflecting the fact that magnetic charges and, therefore, magnetic conduction currents do not exist. At higher frequencies, however, magnetic materials do exhibit polarization loss analogous to the one observed in polarizable dielectrics. The difference with lossy dielectrics is that, in the latter, loss due to charge transport (electric conduction) is present and this loss mechanism exists all the way down to DC.

uniform medium. The excitation term is

$$\mathcal{S}_\xi(\mathbf{r},t) = \begin{cases} \left(\mu\frac{\partial}{\partial t} + \sigma_\mathrm{m}\right)\mathcal{J}_{\mathrm{i},\xi}(\mathbf{r},t) + \left(\epsilon\frac{\partial}{\partial t} + \sigma_\mathrm{e}\right)^{-1}\frac{\partial \rho_\mathrm{i}}{\partial \xi} & \text{if } u = \mathcal{E} \\ -(\nabla \times \boldsymbol{\mathcal{J}}_\mathrm{i})_\xi & \text{if } u = \mathcal{H}, \end{cases} \tag{1.2}$$

where ρ_i and $\boldsymbol{\mathcal{J}}_\mathrm{i}$ are the *impressed* electric charge and current densities, respectively. As discussed in Appendix B, $(\epsilon\frac{\partial}{\partial t} + \sigma_\mathrm{e})^{-1}$ denotes the inverse of the differential operator $(\epsilon\frac{\partial}{\partial t} + \sigma_\mathrm{e})$.

Eq. (1.1) is a linear second-order partial differential equation describing damped waves, and it appears in analogous forms in various physical fields; see, for example, [6, 7, 8].

It is important to note that Eq. (1.1) follows from Maxwell's curl equations only after imposing the conditions (see Appendix B)

$$\nabla \cdot \boldsymbol{\mathcal{B}}(\mathbf{r},t) = 0 \tag{1.3}$$

$$\nabla \cdot \boldsymbol{\mathcal{D}}(\mathbf{r},t) = \rho_\mathrm{e}(\mathbf{r},t), \tag{1.4}$$

where $\boldsymbol{\mathcal{D}}$ is the electric flux density, $\boldsymbol{\mathcal{B}}$ is the magnetic flux density, and ρ_e is the *total* electric charge density. Eq. (1.3) and Eq. (1.4) are Maxwell's divergence equations. For a summary of Maxwell's equations, see Appendix A.

The solutions to Eq. (1.1) satisfy Maxwell's curl equations if and only if they also satisfy Eq. (1.3–1.4) [1]. In other words, *for a solution to Eq. (1.1) to be admissible, it must be checked against Eq. (1.3–1.4).*

Another limitation of Eq. (1.1) is that it holds for the *Cartesian* components of the field only. This is inconvenient when the field has spherical or cylindrical symmetries. This limitation is overcome by the use of the *wave functions*. The wave functions are in effect the values of two collinear vector potentials, the magnetic vector potential $\boldsymbol{\mathcal{A}}$ and the electric vector potential $\boldsymbol{\mathcal{F}}$, which are so defined as to have a fixed known direction $\hat{\mathbf{u}}$, i.e., $\boldsymbol{\mathcal{A}} = \hat{\mathbf{u}}\mathcal{A}$, $\boldsymbol{\mathcal{F}} = \hat{\mathbf{u}}\mathcal{F}$. The wave functions are $\mathcal{A}$ and $\mathcal{F}$. The construction of EM solutions in a uniform source-free medium using Cartesian, cylindrical, and spherical scalar-wave functions is described in detail in [1, 5]. Also, Appendix C summarizes the methods used to reduce the EM model to two decoupled scalar-wave equations. What matters here is that the two wave functions satisfy the wave equation (see Eq. (C.20) in Appendix C):

$$\left[\nabla^2 - \mu\epsilon\frac{\partial^2}{\partial t^2} - (\mu\sigma_\mathrm{e} + \epsilon\sigma_\mathrm{m})\frac{\partial}{\partial t} - \sigma_\mathrm{e}\sigma_\mathrm{m}\right]u(\mathbf{r},t) = 0, \tag{1.5}$$

where $u(\mathbf{r},t) \equiv \mathcal{A}, \mathcal{F}$ is the wave function.

It must be emphasized that, with the proper choice of the wave functions and their polarization $\hat{\mathbf{u}}$, the above equation Eq. (1.5) can be used in rectangular, cylindrical, or spherical coordinate systems. Moreover, unlike Eq. (1.1), the wave-function model needs to solve at the most two (not three) decoupled scalar equations. Often, one equation suffices, for example for a transverse wave in a uniform source-free medium. Thus the main advantages of the wave-function models are that (i) they can be employed

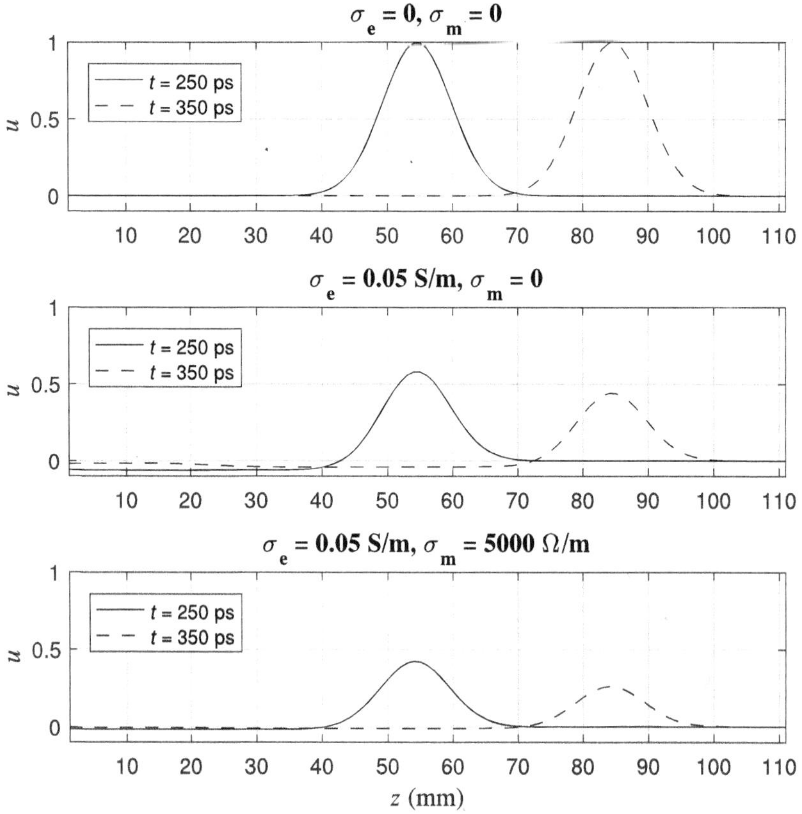

Figure 1.1 Illustration of the impact of the electric and magnetic conductivities on the attenuation of waves: the spatial distribution of a uniform plane wave simulated with MEFiSTo-3D [10].

not only in rectangular but also in curvilinear coordinate systems and (ii) at the most two decoupled scalar equations need to be solved as opposed to three such equations when solving for the vector field in Cartesian coordinates. Notice also that the left side of Eq. (1.5) contains the same partial differential operator (knowns as the *wave* or *d'Alambert* operator) as that in Eq. (1.1).

Let me emphasize that in general the constitutive parameters in Eq. (1.1) and Eq. (1.5) exhibit frequency dependence [9]. This is why Eq. (1.1) and Eq. (1.5) are used only if the excitation signals span a frequency range within which the constitutive parameters are sufficiently constant. The most obvious reminder of the frequency dispersion is the fact that at static and quasi-static (very low frequency) regimes, $\sigma_{\mathrm{m}} = 0$. In such regimes, the time-derivative terms in Eq. (1.1) and Eq. (1.5) are set to zero, reducing these equations to the Laplace form $\nabla^2 u = 0$, *not* to $(\nabla^2 - \sigma_{\mathrm{e}}\sigma_{\mathrm{m}})u = 0$.

To illustrate the impact of the conductivity terms σ_{e} and σ_{m} in the wave equation Eq. (1.5), let us consider a uniform plane wave propagating along z. Fig. 1.1 shows the wave as a function of position in two different instances, $t = 250$ ps and $t = 350$ ps, in three cases: (a) when the medium (vacuum, $\epsilon = \epsilon_0$, $\mu = \mu_0$) has no loss ($\sigma_{\mathrm{e}} = 0$,

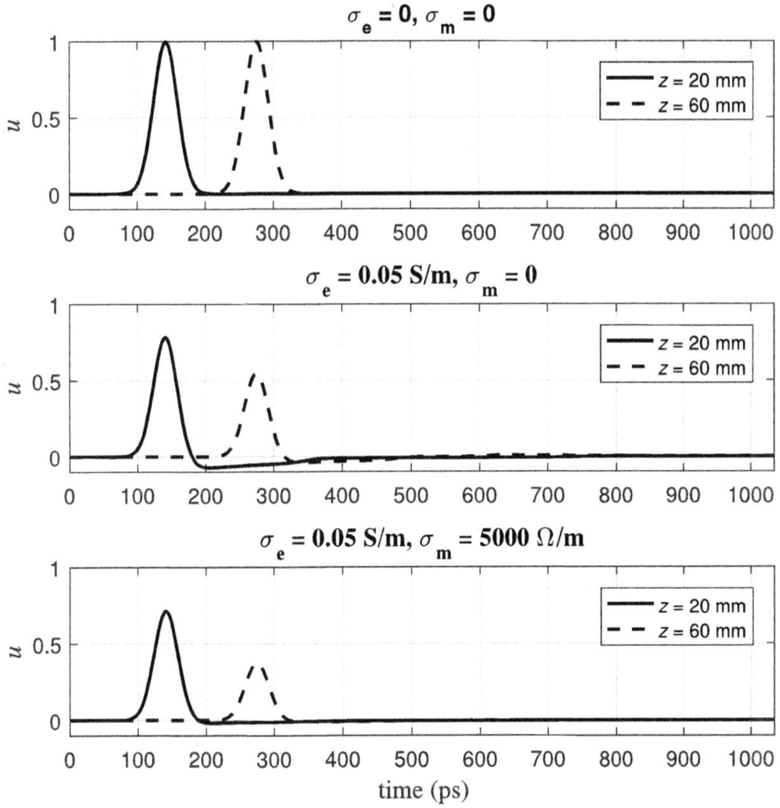

Figure 1.2 Illustration of the impact of the electric and magnetic conductivities on the attenuation of waves: the temporal distribution of a uniform plane wave simulated with MEFiSTo-3D [10].

$\sigma_m = 0$); (b) when the medium has only loss due to electric conductivity ($\sigma_e = 0.05$ S/m, $\sigma_m = 0$); and (c) when the medium has both electric and magnetic loss ($\sigma_e = 0.05$ S/m, $\sigma_m = 5000$ Ω/m). Fig. 1.2 shows the same wave as a function of time measured at two locations: $z = 20$ mm (closer to the source) and $z = 60$ mm. It is clear that losses are responsible for a decrease in strength as the wave propagates away from the source. This decrease is referred to as *dissipation* (the conversion of EM energy into heat). Magnetic and electric specific conductivities are both mathematical means of describing dissipation. Loss also causes changes in the shape of the signal. This is referred to as *dispersion*. Notice the tails behind the main pulse that are clearly visible in the case of losses in both Fig. 1.1 and Fig. 1.2.

1.2 Plane, Spherical, and Cylindrical Waves in the Time Domain

Let us consider the solutions of the wave equations in cases where the wave is independent in two of the three spatial coordinates. They give the mathematical form of the plane, spherical, and cylindrical waves depending on the chosen coordinate system. These solutions are important because they are often used to approximate the field due to distant RF or MW sources. In imaging, these are approximations of what is referred to as the *incident field*, i.e., the field that exists in the background medium when no scattering objects are present. The time-dependent plane and spherical-wave solutions are widely used in pulsed radar imaging.

For simplicity, let us focus on the time-domain symmetric wave solutions in the loss-free case ($\sigma_e = 0$ and $\sigma_m = 0$) when the wave equation is

$$\left(\nabla^2 - \frac{1}{\upsilon^2}\frac{\partial^2}{\partial t^2}\right) u(\mathbf{r}, t) = s(\mathbf{r}, t). \tag{1.6}$$

Here, υ is the speed at which the wavefront advances, and $\upsilon^{-2} = \mu\epsilon$. This case has practical significance. It is true that matter always exhibits loss (or dissipation), however minuscule it may be. Yet, for some forms of matter (e.g., air, most ceramics), the assumption of no loss holds very well at MW frequencies. In addition, the general solutions to Eq. (1.5), where the loss terms are nonzero, are not available in a closed analytical form, unless some assumptions are made; for example, the assumption of low loss [7, 8]. In any case, due to the significant frequency dependence of the damping rates, the analytical modeling of a lossy medium is best done in the frequency domain, which we pursue in the subsequent sections.

A. *Plane-Wave Solution in the Time Domain*

If the wave field is independent of two Cartesian spatial variables, it is referred to as a uniform plane wave, described by what is known as the equation of the vibrating string. For example, if the wave is independent of x and y, then the source-free (or homogeneous) form of Eq. (1.6) is

$$\left(\frac{\partial^2}{\partial z^2} - \frac{1}{\upsilon^2}\frac{\partial^2}{\partial t^2}\right) u(z, t) = 0. \tag{1.7}$$

The general solution of Eq. (1.7) is [11]

$$u(z, t) = f_+(z - \upsilon t) + f_-(z + \upsilon t). \tag{1.8}$$

Here, f_+ and f_- are arbitrary differentiable functions representing waves propagating in the positive and in the negative z directions, respectively. It is common to refer to the first term, the argument of which is $p^+ = z - \upsilon t$, as the *incident* wave, while the second term, of argument $p^- = z + \upsilon t$, is the *reflected* wave.[4]

[4] The reader is reminded that the particular form of f_+ and f_- is determined once the initial or the boundary conditions are given. For example, if nonzero initial conditions $u(z, 0) = u_0(z)$ and $\frac{\partial u(z,0)}{\partial t} = g_0(z)$ are imposed, then [11]

$$u(z, t) = \frac{1}{2}\left[u_0(z - \upsilon t) + u_0(z + \upsilon t) + \frac{1}{\upsilon}\int_{x-\upsilon t}^{x+\upsilon t} g_0(\xi) d\xi\right].$$

Here, $u_0(\xi)$ and $g_0(\xi)$ are known functions. In a different scenario, the initial conditions may be zero, but a solution may be required for $z \geq 0$ such that $u(0, t) = b(t)$ where $b(t)$ is a known time-varying boundary condition at $z = 0$. Then $u(z, t) = b(t - z/\upsilon)$, $0 \leq z < \infty$.

Electrical engineers will recognize this one-dimensional (1D) wave behavior as the solution to the telegrapher's equation describing the propagation of voltage and current signals along a loss-free transmission line [12]. Eq. (1.8) also represents a uniform plane wave.

Another important class of EM waves, the *transverse electromagnetic (TEM)* waves,[5] also exhibit this behavior along z. In contrast to the uniform plane wave, the TEM wave may vary along the transverse coordinates x and y (as a harmonic function). More specifically, a TEM field component F_ξ, where ξ can be either x or y but *not* z, has the form $F_\xi = \ell(x, y) \cdot f(z \mp vt)$ (the sign depends on whether the wave is incident or reflected), where $\ell(x, y)$ must satisfy the 2D Laplace equation in the xy plane [1, 13]. Notice that the uniform plane wave is a particular case of the TEM wave for which $\ell(x, y) = const$.

The solution of the 3D wave equation Eq. (1.6) for a uniform plane wave propagating in any direction given by the unit vector $\hat{\mathbf{u}}$ is known as the general *one-way wave solution* [11]:

$$u(t, \mathbf{r}) = f(\hat{\mathbf{u}} \cdot \mathbf{r} - vt), \tag{1.9}$$

where $\mathbf{r} = (x, y, z)$ denotes the observation location. The first term in Eq. (1.8) is a special case of Eq. (1.9) when $\hat{\mathbf{u}} = \hat{\mathbf{z}}$, while the second term corresponds to the case when $\hat{\mathbf{u}} = -\hat{\mathbf{z}}$.

The plane-wave solution Eq. (1.9) is widely used in radar imaging to approximate the EM field in a region, which is (a) in open space and far from the sources and (b) sufficiently small to ignore the typical far-zone ($\sim 1/r$) behavior of the field intensity (r being the distance to the source).[6]

B. *Spherical-Wave Solution in the Time Domain*

Another symmetric solution of Eq. (1.6) arises when the field propagates as a spherical wave, i.e., it is independent of the two angular coordinates θ and ϕ of a spherical coordinate system. It is a function of the radial distance r and the time t. This solution plays an important role in the modeling of 3D open problems[7] where the spherical coordinate system is convenient to use. The usual definition of a spherical coordinate system is illustrated in Fig. 1.3. In this case, the Laplacian operator in Eq. (1.6) is written out in spherical coordinates, and the derivatives with respect to

5 The term *transverse electromagnetic* (TEM) indicates that both the electric and the magnetic field vectors are perpendicular to the direction of propagation.

6 The far zone of a radiating structure (an antenna) is all space beyond a distance r_f from the antenna such that it satisfies all of the following conditions: $r_f \gg \lambda$, $r_f \gg D_A$, and $r_f \geq 2D_A^2/\lambda$, where λ is the wavelength of the radiation and D_A is the maximum dimension of the antenna. In practice, the $\gg$ inequality above is usually taken as $> 10\times$. We recall that the far-zone EM field in an unbounded medium is a TEM field i.e., both the $\boldsymbol{\mathcal{E}}$ and the $\boldsymbol{\mathcal{H}}$ field vectors are transverse to $\hat{\mathbf{u}}$ [5]. It is these transverse field components that are approximated by Eq. (1.9). The transverse nature of the far-zone EM wave, together with the assumption that its dependence on the transverse coordinates can be neglected, ensures that the uniform plane-wave solution satisfies not only the wave equation but also Eq. (1.3–1.4).

7 An *open* (or *radiation*) problem is that of analyzing the field in an infinite (or unbounded) region, the boundary of which extends to infinity.

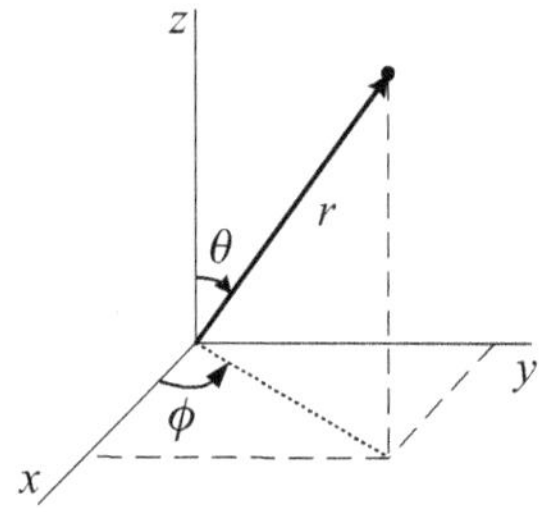

Figure 1.3 The spherical coordinate system.

θ and ϕ are set to zero. For the case of zero sources, the result is

$$\frac{1}{r^2}\frac{\partial}{\partial r}\left(r^2\frac{\partial u}{\partial r}\right) - \frac{1}{\upsilon^2}\frac{\partial^2 u}{\partial t^2} = 0. \tag{1.10}$$

The general solution to this homogeneous equation is easily found since it can be reduced to the equation of the vibrating string Eq. (1.7) by rewriting it as [14]

$$\left(\frac{\partial^2}{\partial r^2} - \frac{1}{\upsilon^2}\frac{\partial^2}{\partial t^2}\right)(ru) = 0. \tag{1.11}$$

The general spherical-wave solution is then

$$u(r,t) = \frac{f_+(r-\upsilon t)}{r} + \frac{f_-(r+\upsilon t)}{r}, \quad \xi = x, y, z. \tag{1.12}$$

The argument $p^+ = r - \upsilon t$ in the first term of Eq. (1.12) implies a wave diverging from the center of the coordinate system, i.e., propagating in the positive radial direction $\hat{\mathbf{r}}$. This case corresponds to the outgoing wave of a point source of waveform $f_+(t)$ located at the origin (0, 0, 0). Thus, the first term in Eq. (1.12) is a physically valid *causal* solution to the open problem with a point source at the origin.[8] In contrast, $p^- = r + \upsilon t$ implies a spherical wave converging toward the origin with time. It could be understood as a spherical wave propagating backward (collapsing) toward the point source. This solution is not causal in an unbounded medium and is normally excluded from the forward model. Note, however, that a region bounded by a spherical reflecting surface (for example, a metallic shell) can support perfectly well the solution of the collapsing wave. In this case, the source of the collapsing wave is the spherical shell boundary, and the collapsing wave is in fact causal.

It is instructive to examine the validity of the scalar spherical-wave approximation of the EM waves in the form of Eq. (1.12) with regard to the conditions in Eq. (1.3–1.4). First, we note that the spherical components of the EM field do not satisfy the wave equation Eq. (1.1): their Cartesian components do. Therefore, Eq. (1.12) is not applicable to the wave Cartesian components in general.

[8] A causal solution describes a response (such as a field value) that never precedes its excitation source. For example, if the source function is identically zero everywhere in space before some initial time t_0, then the causal field solution must be identically zero everywhere in space for all $t \leq t_0$.

The spherical-wave functions $\mathcal{A}$ and $\mathcal{F}$, however, do satisfy the wave equation in spherical coordinates (see Appendix C). As a consequence, it can be shown that all the components of the EM field that observe the $\sim 1/r$ behavior (these are referred to as the *far-zone* or *far-field* components) are transverse to the radial direction, i.e., the direction of propagation [5]. Thus, the far-zone wave is a TEM wave.[9] This result is often used in antenna engineering [15]. In particular, in a spherical coordinate system, the far field has only θ and ϕ components: u_ξ, $u \equiv \mathcal{E}, \mathcal{H}$, $\xi = \theta, \phi$, both of which, in general, are functions of θ and ϕ[10] in addition to their dependence on the distance as $\sim 1/r$. These field components are often approximated in the form of the first (outgoing) term in Eq. (1.12). Such a field, however, does violate Maxwell's divergence equations, Eq. (1.3–1.4).

Take as an example the far-zone electric field due to a current element[11] oriented along z and centered at the origin. Its far-zone $\boldsymbol{\mathcal{E}}$-field has a θ component only, and it behaves as $\mathcal{E}_\theta \sim \sin(\theta)/r$. The divergence of this $\boldsymbol{\mathcal{E}}$-field, found to be $\sim \cos(\theta)/r^2$, is not zero for all θ, thus violating Eq. (1.4). If this result is viewed as an error in the field approximation, we can state that this error decreases with distance as $\sim 1/r^2$ and is zero at $\theta = 90^\circ$. Notice that the plane $\theta = 90^\circ$ (the $z = 0$ plane) is a plane of symmetry in the field pattern where the radiation attains its maximum. This result can be generalized for any antenna by viewing it as a collection of current elements. *The spherical-wave approximation of any of the transverse field components is valid for any observation direction as long as r is sufficiently large. Its accuracy improves for directions close to the direction of maximum radiation* (the maximum of the radiation pattern) where the derivatives of the far field with respect to the angles θ and ϕ are zero.

C. *Cylindrical-Wave Solution in the Time Domain*

Analogously to the plane and spherical waves, the cylindrical wave is described by the solution of the wave equation in cylindrical coordinates, where the wave is set to be independent of two coordinates: the vertical coordinate z and the angular coordinate ϕ. The usual definition of the cylindrical coordinate system is illustrated in Fig. 1.4. The wavefront of the cylindrical wave is an infinitely long (along z) cylindrical surface that advances radially away from the z axis along the positive radial direction $\hat{\rho}$. Clearly, such an infinite wavefront is an approximation that can be only locally valid. This approximation can be useful in the two-dimensional (2D)

[9] Strictly speaking, the far-zone EM field is only *predominantly* TEM because the longitudinal field components (either $\mathcal{E}_r$ or $\mathcal{H}_r$ or both) are only asymptotically zero and therefore negligible in comparison with the transverse components.

[10] The far-field dependence on the observation direction (θ, ϕ) is described by what is known as the *antenna radiation pattern*.

[11] A current element, also referred to as an infinitesimal electric dipole, is a vanishingly short line segment supporting alternating current (AC). The practical implementation of a current element is a very short (compared to the wavelength λ) piece of wire excited by an RF source at its midpoint. The current element, measured in units of (A $\times$ m), is the fundamental point source in radiation theory analogous to the point charge in electrostatics.

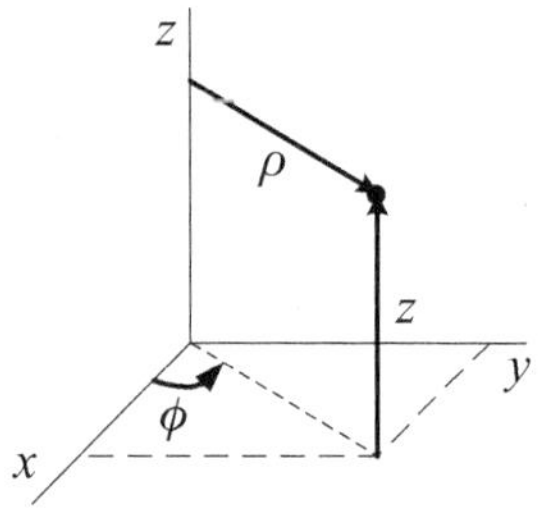

Figure 1.4 The cylindrical coordinate system.

time-domain modeling of open problems when the field and its sources can be assumed to be independent of the vertical variable z.

In this case, the Laplacian operator ∇^2 in Eq. (1.6) is written out in cylindrical coordinates, and the partial derivatives with respect to ϕ and z are set to zero. The result is

$$\frac{1}{\rho}\frac{\partial}{\partial \rho}\left(\rho \frac{\partial u}{\partial \rho}\right) - \frac{1}{v^2}\frac{\partial^2 u}{\partial t^2} = 0. \tag{1.13}$$

As it turns out, the general wave solution in this case is not as simple as in the cases of planar and spherical symmetry; it appears in the form of convolution integrals [16]:

$$u(\rho, t) = \int_{-\infty}^{t-\rho/v} \frac{f_+(\tau)d\tau}{\sqrt{(t-\tau)^2 - (\rho/v)^2}} + \int_{t+\rho/v}^{\infty} \frac{f_-(\tau)d\tau}{\sqrt{(\tau - t)^2 - (\rho/v)^2}}. \tag{1.14}$$

That the first term represents an outgoing wave becomes apparent from the fact that it is the *past* values of f_+ that contribute to $u(\rho, t)$. Note that ρ/v is real-positive and that the upper limit of the integral ensures that the integrand remains real-valued for all τ. In contrast, in the second term, the integration is over the *future* values of f_-. In problems unbounded in ρ ($0 \leq \rho < \infty$), the second term is nonphysical because it is acausal.

To understand better the behavior of the cylindrical wave, let us consider a particular outgoing wave solution when $f_+(t) = \delta(t)$ and $f_-(t) = 0$. Then, as per the sampling property of the δ-function, we have

$$u(\rho, t) = \begin{cases} 0 & \text{if } 0 < t < \rho/v \\ \frac{1}{\sqrt{t^2 - (\rho/v)^2}} & \text{if } t > \rho/v. \end{cases} \tag{1.15}$$

As we will see shortly, this is in fact a solution proportional to Green's function of Eq. (1.13) [17]. This solution can be interpreted as an outward propagating wave behaving as $\sim 1/\sqrt{\rho}$. This becomes apparent if $u(\rho, t)$ is expressed as

$$u(\rho, t) = \begin{cases} 0 & \text{if } 0 < t < \rho/v \\ \frac{1}{\sqrt{t - \rho/v}} \cdot \frac{1}{\sqrt{t + \rho/v}} & \text{if } t > \rho/v, \end{cases} \tag{1.16}$$

where the term with argument $p^+ = t - \rho/v$, $t > \rho/v$, is a forward propagating wave. Assume an observation distance $\rho = \rho_0$. The time variable t can then be expressed as $t = \rho_0/v + \Delta t$. We could call Δt *the time after arrival*. Then, at $\rho = \rho_0$, the observed waveform is

$$u(\rho_0, \Delta t) = \begin{cases} 0 & \text{if } \Delta t < 0 \\ \frac{1}{\sqrt{\Delta t}} \cdot \frac{1}{\sqrt{2\rho_0/v + \Delta t}} & \text{if } \Delta t > 0. \end{cases} \tag{1.17}$$

If $\rho_0/v \gg \Delta t$, i.e., at the initial development of the waveform, its value behaves as $\sim 1/\sqrt{\rho_0}$. When $\Delta t \gg \rho_0/v$, i.e., during the late time response, we see attenuation with time given by $\sim 1/\Delta t$.

1.3 Partial Differential Equations for Scalar Waves in the Frequency Domain

Time-harmonic fields are analyzed conveniently by making use of *phasors*. A scalar phasor F is a complex number, the real and imaginary parts of which are the real-valued Fourier coefficients [11] that represent the periodic (in this case, time-harmonic) function $f(t)$ in the frequency domain. The phasor relates to its time-harmonic counterpart through

$$f(t) = \Re\left\{F e^{i\omega t}\right\}, \tag{1.18}$$

where ω is the angular frequency, measured in *radians per second*. Thus, if $f(t) = f_m \cos(\omega t + \Phi) = a\cos(\omega t) + b\sin(\omega t)$, then $F = f_m e^{i\Phi} = a - ib$, where $a = f_m \cos\Phi$ and $b = -f_m \sin\Phi$.

Phasors are widely used in electrical engineering to analyze linear time-harmonic circuits. They simplify the analysis by reducing the associated linear systems of ordinary differential equations to systems of linear (complex-valued) algebraic equations.

Phasors in EM analysis are complex-valued functions of space that relate to their time-harmonic counterparts through the same mapping into complex space as in Eq. (1.18). The phasors of the field scalar and vector functions are introduced through

$$u(\mathbf{r}, t) = \Re\left\{U(\mathbf{r}) e^{i\omega t}\right\} \tag{1.19}$$

$$\mathcal{F}(\mathbf{r}, t) = \Re\left\{\mathbf{F}(\mathbf{r}) e^{i\omega t}\right\}. \tag{1.20}$$

Notice the upper-case notation for the scalar phasors as opposed to the lower-case notation for their time-dependent counterparts in Eq. (1.19). For vector fields, there is a change of font in the notation of the phasors compared to the time-dependent vectors; see Eq. (1.20).

The time-domain equations for a time-harmonic EM field can be readily cast into a phasor (or frequency-domain) form by making use of the relationship

$$\frac{\partial^n u(\mathbf{r}, t)}{\partial t^n} = \Re\left\{(i\omega)^n U(\mathbf{r}) e^{i\omega t}\right\}. \tag{1.21}$$

See, for example, how Maxwell equations are cast in a time-domain and in a frequency-domain form in Appendix A.

Here, we focus on the Helmholtz equation, which is the frequency-domain counterpart of the wave equation and which is often used to describe the high-frequency EM field in an isotropic uniform medium (see Appendix B and Appendix C).

The wave equation Eq. (1.5) assumes the form of the Helmholtz equation when written for the scalar phasor function $U(\mathbf{r})$:

$$\left(\nabla^2 - \gamma^2\right) U(\mathbf{r}) = \mathcal{S}(\mathbf{r}), \tag{1.22}$$

where γ is the complex propagation constant and $\mathcal{S}(\mathbf{r})$ is the source function. This is exactly the equation governing the Cartesian components of the field phasors $\mathbf{E}$ and $\mathbf{H}$ in a uniform isotropic medium; see Eq. (B.14) in Appendix B.

As per Eq. (1.5) and Eq. (1.21), γ relates to the constitutive parameters as (see also Appendix B, Section B.2)

$$\gamma^2 = -\omega^2 \mu\epsilon + \sigma_{\mathrm{e}}\sigma_{\mathrm{m}} + \mathrm{i}\omega(\mu\sigma_{\mathrm{e}} + \epsilon\sigma_{\mathrm{m}}). \tag{1.23}$$

In the frequency domain, however, it is more convenient to introduce complex constitutive parameters $\varepsilon(\omega) = \varepsilon'(\omega) - \mathrm{i}\varepsilon''(\omega)$ and $\mu(\omega) = \mu'(\omega) - \mathrm{i}\mu''(\omega)$, which readily incorporate the frequency dispersion of the EM properties of matter; see Appendix A. The real parts of the complex permittivity and permeability, ε' and μ', are the same as ϵ and μ in Eq. (1.23), respectively. The effective frequency-dependent conductivities are now incorporated in the respective imaginary parts of the complex permittivity and permeability as

$$\sigma_{\mathrm{e}}(\omega) = \omega\varepsilon''(\omega) \tag{1.24}$$

$$\sigma_{\mathrm{m}}(\omega) = \omega\mu''(\omega). \tag{1.25}$$

For a number of reasons, MW engineers prefer the description of the material losses in terms of the so-called loss tangents:

$$\tan\delta_{\mathrm{d}} = \varepsilon''/\varepsilon' \tag{1.26}$$

$$\tan\delta_{\mathrm{m}} = \mu''/\mu', \tag{1.27}$$

rather than the effective conductivities. In Eq. (1.26–1.27), $\tan\delta_{\mathrm{d}}$ is referred to as the dielectric loss tangent, and $\tan\delta_{\mathrm{m}}$ is the magnetic one. For many materials, the loss tangents exhibit less frequency dependence than the effective conductivities in the MW bands. Moreover, the reference to *conductivity* may appear misleading at high frequencies because (i) the dissipation in dielectrics is largely due to alternating polarization effects, not conduction, and (ii) the dissipation in magnetic materials is entirely due to (alternating) magnetization because there is no magnetic conduction. Occasionally, the distinction between conduction and polarization loss in dielectrics is made explicit by writing the complex permittivity as

$$\varepsilon = \epsilon' - \mathrm{i}\left(\epsilon' \tan\delta_{\mathrm{d}} + \frac{\sigma_{\mathrm{e}}}{\omega}\right). \tag{1.28}$$

However, quantifying this distinction through material measurements is difficult.

By substituting Eq. (1.24–1.25) into Eq. (1.23), we find that the propagation constant γ relates to the complex permittivity and permeability simply as

$$\gamma^2 = -\omega^2 \mu \varepsilon. \tag{1.29}$$

The complex propagation constant γ streamlines the solution of the scalar EM problem in the frequency domain since its mathematical form accommodates both loss-free and lossy media as described in the next section.

Finally, we note that the scalar Helmholtz equation for the Cartesian components of the EM field in the form of Eq. (1.22), where γ is given by Eq. (1.29), can be obtained directly from the frequency-domain Maxwell equations [1, 5]; see Appendix B. Additionally, Appendix C shows that the phasors of the wave functions satisfy the scalar Helmholtz equation too.

1.4 Plane, Spherical, and Cylindrical Waves in the Frequency Domain

This section summarizes the symmetric solutions of the Helmholtz equation in the cases of Cartesian, spherical, and cylindrical coordinate systems. These three cases are treated by the same procedure as the one used with the wave equation, i.e., by postulating that the wave is independent of two of the three spatial variables and setting to zero the respective partial derivatives in the homogeneous Helmholtz equation:

$$\left(\nabla^2 - \gamma^2\right) U(\mathbf{r}) = 0. \tag{1.30}$$

A. Plane-Wave Solution in the Frequency Domain

The phasor $U(z)$ of a uniform plane wave propagating along z is the solution to the 1D Helmholtz equation,

$$\frac{d^2U}{dz^2} - \gamma^2 U = 0, \tag{1.31}$$

and is given by

$$U(z) = Ae^{-\gamma z} + Be^{+\gamma z}, \tag{1.32}$$

where A and B are complex constants. Let $\gamma = \alpha + \mathrm{i}\beta$. Then, in accordance with the phasor definition in Eq. (1.19), the time-harmonic solution corresponding to Eq. (1.32) is

$$u(z,t) = Ae^{-\alpha z}\cos(\omega t - \beta z) + Be^{\alpha z}\cos(\omega t + \beta z). \tag{1.33}$$

By choosing

$$\gamma = \alpha + \mathrm{i}\beta = \mathrm{i}\omega\sqrt{\mu\varepsilon}, \tag{1.34}$$

which is one of the two possible solutions of Eq. (1.29), it is set that the first term in Eq. (1.33) or, equivalently, in Eq. (1.32) represents a plane wave traveling in the positive z direction (the incident wave), while the second term corresponds to a propagation direction of $-z$ (the reflected wave). Note that choosing $\gamma = -i\omega\sqrt{\mu\varepsilon}$ is also possible and will simply result in reversing the physical meaning of the two terms.

Hereafter, Eq. (1.34) is the solution of choice, implying that the condition $\Re\gamma = \alpha \geq 0$ must hold to ensure that the wave is not growing in magnitude as it propagates.

The parameter α (measured in *Nepers per meter*, Np·m^{-1})[12] is referred to as the *attenuation constant*. As evident from Eq. (1.33), it describes the rate of exponential decay of the wave magnitude with distance. If $\alpha = 0$, the wave is propagating without attenuation, and the propagation constant is purely imaginary, namely $\gamma = i\omega\sqrt{\mu'\epsilon'}$, since $\mu'' = \epsilon'' = 0$.

$\Im\gamma = \beta$ is the *phase constant* (measured in *radians per meter*, rad·m^{-1}), and it determines the wavelength λ as per $\beta = 2\pi/\lambda$. The phase constant describes the (linear) rate of change in the wave's phase with distance. When β describes propagation in open space, it is often denoted as k and is called the *wavenumber*. The phase constant (or the wavenumber) is in effect the "spatial frequency" of the wave. The analogy between ω (the temporal frequency) and β (the spatial frequency) is obvious from Eq. (1.33), where ωt and βz have analogous impact on the wave's phase.

The generalization of the plane-wave solution in Eq. (1.32) to a plane wave propagating in any given direction specified by the unit vector $\hat{\mathbf{u}}$ is analogous to that in the time domain, see Eq. (1.9), and is given by

$$U(\mathbf{r}) = Ae^{-\gamma\hat{\mathbf{u}}\cdot\mathbf{r}}, \tag{1.35}$$

where A is a complex constant. Note that this solution can accommodate loss through $\Re\gamma$. In the case of loss-free propagation, the notation $U(\mathbf{r}) = Ae^{-i\mathbf{k}\cdot\mathbf{r}}$ is often used, where $\mathbf{k} = \hat{\mathbf{u}}\beta$ is referred to as the *wave vector*.

B. *Spherical-Wave Solution in the Frequency Domain*

The Helmholtz equation for a scalar wave independent of the angular coordinates θ and ϕ reduces to

$$\frac{1}{r^2}\frac{d}{dr}\left(r^2\frac{dU}{dr}\right) - \gamma^2 U = 0. \tag{1.36}$$

Similarly to the time-domain spherical-wave solution, Eq. (1.36) can be equivalently written as

$$\frac{d^2(rU)}{dr^2} - \gamma^2(rU) = 0, \tag{1.37}$$

which is now in the form of the 1D Helmoltz equation Eq. (1.31). Thus, the phasor form of the spherical-wave solution (for the causal wave propagating along $\hat{\mathbf{r}}$) is

12 The Neper (Np) is a logarithmic unit for ratios similar to the decibel (dB). The ratio $\frac{s_1}{s_2}$ is calculated in Nepers using its natural logarithm, i.e., $R_{\mathrm{Np}} = \ln(\frac{s_1}{s_2})$. In contrast, the ratio in dB uses the base-10 logarithm, $R_{\mathrm{dB}} = 10\log_{10}(\frac{s_1}{s_2})$.

found to be

$$U(\mathbf{r}) = A\frac{e^{-\gamma r}}{r}, \tag{1.38}$$

where A is a complex constant.

C. Cylindrical-Wave Solution in the Frequency Domain

The Helmholtz equation for a scalar wave independent of ϕ and z in cylindrical coordinates is

$$\frac{1}{\rho}\frac{d}{d\rho}\left(\rho\frac{dU}{d\rho}\right) - \gamma^2 U = 0, \tag{1.39}$$

which can be equivalently written as

$$\xi^2\frac{d^2 f}{d\xi^2} + \xi\frac{df}{d\xi} + \xi^2 f = 0, \tag{1.40}$$

provided that $U(\rho) = f(k_\rho \rho)$, $\xi = k_\rho \rho$ and $k_\rho = -\mathrm{i}\gamma$. We recognize Eq. (1.40) as the Bessel ordinary differential equation of order $\nu = 0$. Thus, any Bessel function of order zero of the argument $(k_\rho \rho)$ is a solution to Eq. (1.39). However, only one of these Bessel functions, the Hankel function of the second kind $H_0^{(2)}(k_\rho \rho)$, vanishes at infinity for any complex k_ρ [1, 5]. Its asymptotic behavior for large arguments is

$$H_0^{(2)}(\xi) \underset{\xi\to\infty}{\longrightarrow} \sqrt{\frac{2\mathrm{i}}{\pi\xi}}e^{-\mathrm{i}\xi},\ \xi = k_\rho \rho. \tag{1.41}$$

In a loss-free medium, $\gamma = \mathrm{i}\beta$, and $k_\rho = \beta = \omega\sqrt{\mu'\epsilon'}$ is the (real-positive) wavenumber. In this case, at large distances, $H_0^{(2)}(\beta\rho)$ behaves as an outward propagating wave, which decays with distance as $\sim 1/\sqrt{\rho}$—a result consistent with the time-domain solution for the cylindrical wave.

1.5 Green's Functions: The Basics

1.5.1 Green's Function as an Impulse Response

In very general terms, Green's function of a system is the complete description of this system's impulse response. The impulse response is the response to an excitation at a single point. For example, Green's function of the equation of the string describes the string vibration as a result of being plucked. Pulling on the string at a single point for a very brief period of time is an example of an excitation at a single point in space-time.[13]

Let us put this concept in a mathematical form. Let $\mathbf{x}$ denote an *observation point* in the n-dimensional domain $\mathcal{D}$ of the system. For example, $\mathcal{D}$ may be part of the entire

[13] Green's functions in conjunction with the superposition principle offer a powerful method for solving inhomogeneous *linear partial differential equations*. The method was first developed by the English mathematician George Green in the 1830s and since then has become an essential tool in mathematical physics for solving boundary-value problems [18].

3D Euclidean space; then a point in $\mathcal{D}$ is given by $\mathbf{x} = (x, y, z) \in \mathbb{R}^3$. This is the case of the Helmholtz equation. In the case of the wave equation, $\mathcal{D}$ is the 4D domain of space-time with $\mathbf{x} = (x, y, z; t) \in \mathbb{R}^4$. Further, let $\mathbf{x}'$ denote a *source point*. Green's function depends on both where the excitation occurs ($\mathbf{x}'$) and where the response is observed ($\mathbf{x}$) and is thus denoted as $\mathcal{G}(\mathbf{x}, \mathbf{x}')$. The convention is to use the first element of the pair $(\mathbf{x}, \mathbf{x}')$ for the observation point, while the second is for the source point. Adhering to this convention is strongly advised.

1.5.2 Toward the Construction of Solutions Using Green's Functions

Consider the partial differential equation

$$\mathcal{L}u(\mathbf{x}) = s(\mathbf{x}), \tag{1.42}$$

where $\mathcal{L}$ is a linear differential operator, u is the unknown function, and s is the source function. As an example, in the case of the wave equation in a loss-free medium, $\mathcal{L} = \nabla^2 - \frac{1}{v^2}\frac{\partial^2}{\partial t^2}$. In the case of the Helmholtz equation, $\mathcal{L} = \nabla^2 - \gamma^2$.

According to the definition given earlier, Green's function of this equation satisfies

$$\mathcal{L}\mathcal{G}(\mathbf{x}, \mathbf{x}') = \delta(\mathbf{x} - \mathbf{x}'), \tag{1.43}$$

where, in an n-dimensional domain, $\delta(\mathbf{x} - \mathbf{x}') = \delta(x_1 - x_1') \cdot \delta(x_2 - x_2') \cdot \cdots \cdot \delta(x_n - x_n')$, with δ being Dirac's delta function.

Multiply both sides of Eq. (1.42) by $\mathcal{G}(\mathbf{x}, \mathbf{x}')$, multiply both sides of Eq. (1.43) by $u(\mathbf{x})$, subtract the two, and integrate over $\mathbf{x}$ in the entire domain $\mathcal{D}$. Assuming that both $\mathbf{x}$ and $\mathbf{x}'$ reside in $\mathcal{D}$ and using the sampling property of the δ-function, we arrive at

$$u(\mathbf{x}') = \int_{\mathcal{D}} s(\mathbf{x})\mathcal{G}(\mathbf{x}, \mathbf{x}')d\Omega + \int_{\mathcal{D}} \left[u(\mathbf{x})\mathcal{L}\mathcal{G}(\mathbf{x}, \mathbf{x}') - \mathcal{G}(\mathbf{x}, \mathbf{x}')\mathcal{L}u(\mathbf{x})\right] d\Omega, \ \mathbf{x}' \in \mathcal{D}, \tag{1.44}$$

where $d\Omega$ indicates the differential volume element in the domain $\mathcal{D}$ with the integration variable being $\mathbf{x}$. Finally, we notice that nothing will change in Eq. (1.44) if we replace $\mathbf{x}$ with $\mathbf{x}'$ and $\mathbf{x}'$ with $\mathbf{x}$ to obtain

$$u(\mathbf{x}) = \int_{\mathcal{D}} s(\mathbf{x}')\mathcal{G}(\mathbf{x}', \mathbf{x})d\Omega' + \int_{\mathcal{D}} \left[u(\mathbf{x}')\mathcal{L}\mathcal{G}(\mathbf{x}', \mathbf{x}) - \mathcal{G}(\mathbf{x}', \mathbf{x})\mathcal{L}u(\mathbf{x}')\right] d\Omega', \ \mathbf{x} \in \mathcal{D}. \tag{1.45}$$

This form serves us better because we already adopted the convention of denoting the observation point with $\mathbf{x}$ and the source point with $\mathbf{x}'$.

The first integral in Eq. (1.45) is clearly the contribution of the source. We next show that the second integral is a result of the boundary/initial conditions that must complement both Eq. (1.42) and Eq. (1.43) for each of them to have a unique solution.

1.5.3 Green's Theorem and Integral Solution for Helmholtz's Operator

To transform the second integral in Eq. (1.45), we use the well-known Green's integral theorem [11]:

$$\iiint_V \left(\Phi\nabla^2\Psi - \Psi\nabla^2\Phi\right) d\Omega = \oiint_S (\Phi\nabla\Psi - \Psi\nabla\Phi)\cdot\hat{\mathbf{n}}ds, \tag{1.46}$$

where Φ and Ψ are functions of the position $\mathbf{r}$ in 3D space, V is a region of space, and S is its boundary. The surface element ds of S has its unit normal $\hat{\mathbf{n}}$ directed *out* of the volume V.

Assume now that the linear operator $\mathcal{L}$ in Eq. (1.45) is $\mathcal{L} \equiv \nabla^2$, i.e., both u and $\mathcal{G}$ satisfy Poisson's equation. In this case, $\mathbf{x} \equiv \mathbf{r}$ is simply the position in 3D space. Using Eq. (1.46), we can readily convert the second integral in Eq. (1.45) from a volume to a surface one, obtaining the integral equation

$$u(\mathbf{r}) = \iiint_V s(\mathbf{r}')\mathcal{G}(\mathbf{r}',\mathbf{r})dv' + \oiint_S \left[u(\mathbf{r}')\nabla\mathcal{G}(\mathbf{r}',\mathbf{r}) - \mathcal{G}(\mathbf{r}',\mathbf{r})\nabla u(\mathbf{r}')\right]\cdot\hat{\mathbf{n}}ds',\ \mathbf{r}\in V. \tag{1.47}$$

This integral solution of Poisson's equation allows one to compute the field at any point $\mathbf{r} \in V$ from its source $s(\mathbf{r})$ and its boundary values along with those of $\mathcal{G}$. Of course, we need to know Green's function $\mathcal{G}$, but, at this stage, we will naively assume that this is a given.

How do we achieve the same transformation in the case of Helmholtz's or the wave-equation operators that are relevant to us? As it turns out, Green's theorem can be generalized to include all partial-differential operators of the form [11]

$$\mathcal{L} = \nabla\cdot[p(\mathbf{r})\nabla] + q(\mathbf{r}). \tag{1.48}$$

This generalized form of Eq. (1.46) can be written as

$$\iiint_V (\Phi\mathcal{L}\Psi - \Psi\mathcal{L}\Phi)d\Omega = \oiint_S p(\Phi\nabla\Psi - \Psi\nabla\Phi)\cdot\hat{\mathbf{n}}ds. \tag{1.49}$$

Helmholtz's operator $\mathcal{L} = \nabla^2 - \gamma^2$ belongs to the type defined in Eq. (1.48) where $p(\mathbf{r}) = 1$ and $q(\mathbf{r}) = -\gamma^2(\mathbf{r})$. Therefore, just as in the case of Poisson's equation, Eq. (1.49) is used to reduce the second integral in Eq. (1.45) from a volume to a surface one in Helmholtz's equation, the integral solution of which can now be written in the form

$$U(\mathbf{r}) = \iiint_V S(\mathbf{r}')G(\mathbf{r}',\mathbf{r})dv' + \oiint_S \left[U(\mathbf{r}')\nabla G(\mathbf{r}',\mathbf{r}) - G(\mathbf{r}',\mathbf{r})\nabla U(\mathbf{r}')\right]\cdot\hat{\mathbf{n}}ds',\ \mathbf{r}\in V. \tag{1.50}$$

Here, we have adopted the specific notation G for Green's function of the Helmholtz equation. This notation is used throughout the book.

1.5.4 Green's Theorem and Integral Solution for the Wave Operator

We now turn to the loss-free wave equation, the operator of which is $\mathcal{L} = \nabla^2 - \frac{1}{v^2}\frac{\partial^2}{\partial t^2}$. The 4D domain $\mathcal{D}$ of the wave solution involves the three spatial variables of position and the temporal variable, i.e., a point in $\mathcal{D}$ is specified by $\mathbf{x} \equiv (\mathbf{r}, t)$. The left-hand side of Green's generalized theorem Eq. (1.49) involves the integrand $\Phi\mathcal{L}\Psi - \Psi\mathcal{L}\Phi$, which, in the case of the wave operator, is expanded as

$$\Phi\mathcal{L}\Psi - \Psi\mathcal{L}\Phi = \left(\Phi\nabla^2\Psi - \Psi\nabla^2\Phi\right) - \frac{1}{v^2}\left(\Phi\frac{\partial^2\Psi}{\partial t^2} - \Psi\frac{\partial^2\Phi}{\partial t^2}\right). \tag{1.51}$$

Employing the equalities,

$$\Phi\nabla^2\Psi - \Psi\nabla^2\Phi = \nabla\cdot\left(\Phi\nabla\Psi - \Psi\nabla\Phi\right), \tag{1.52}$$

and

$$\Phi\frac{\partial^2\Psi}{\partial t^2} - \Psi\frac{\partial^2\Phi}{\partial t^2} = \frac{\partial}{\partial t}\left(\Phi\frac{\partial\Psi}{\partial t} - \Psi\frac{\partial\Phi}{\partial t}\right), \tag{1.53}$$

Eq. (1.51) can be integrated over space and time to produce

$$\int_{T_{\min}}^{T_{\max}}\iiint_V \left(\Phi\mathcal{L}\Psi - \Psi\mathcal{L}\Phi\right)dv dt = \int_{T_{\min}}^{T_{\max}}\oint\!\!\!\oint_S \left(\Phi\nabla\Psi - \Psi\nabla\Phi\right)\cdot\hat{\mathbf{n}}ds dt - \iiint_V \frac{1}{v^2}\left[\Phi\frac{\partial\Psi}{\partial t}, -\Psi\frac{\partial\Phi}{\partial t}\right]_{T_{\min}}^{T_{\max}} dv. \tag{1.54}$$

In fact, in the 4D domain $\mathcal{D}$, we can define the 4D gradient [19]

$$\bar{\nabla} = \nabla - \frac{1}{v}\frac{\partial}{\partial t}\hat{\mathbf{t}}, \tag{1.55}$$

where $\hat{\mathbf{t}}$ is the temporal unit vector. This allows us to treat the field initial and terminal values at $T_{\min}$ and $T_{\max}$, respectively, as boundary values. This is illustrated in Fig. 1.5. For example, $T_{\max}$ defines a boundary in the 4D space-time, the unit normal of which is $\hat{\mathbf{n}} = \hat{\mathbf{t}}$. Meanwhile, at the $T_{\min}$ boundary, $\hat{\mathbf{n}} = -\hat{\mathbf{t}}$. Both of these boundaries are part of the domain boundary $\partial\mathcal{D}$.

With this convention, Green's theorem in Eq. (1.54) can be generalized in 4D space as

$$\int_{\mathcal{D}} \left(\Phi\mathcal{L}\Psi - \Psi\mathcal{L}\Phi\right)d\Omega = \oint_{\partial\mathcal{D}} \left(\Phi\bar{\nabla}\Psi - \Psi\bar{\nabla}\Phi\right)\cdot\hat{\mathbf{n}}d\sigma, \tag{1.56}$$

to include the wave operator. Here, $d\Omega$ is the differential volume in $\mathcal{D}$, and $d\sigma$ is a differential element of the 4D surface $\partial\mathcal{D}$.

Making use of Eq. (1.56), we can write Eq. (1.45) in the case of the wave equation as

$$u(\mathbf{x}) = \int_{\mathcal{D}} s(\mathbf{x}')g(\mathbf{x}',\mathbf{x})d\Omega' + \oint_{\partial\mathcal{D}} \left[u(\mathbf{x}')\bar{\nabla}g(\mathbf{x}',\mathbf{x}) - g(\mathbf{x}',\mathbf{x})\bar{\nabla}u(\mathbf{x}')\right]\cdot\hat{\mathbf{n}}d\sigma', \mathbf{x}\in\mathcal{D}. \tag{1.57}$$

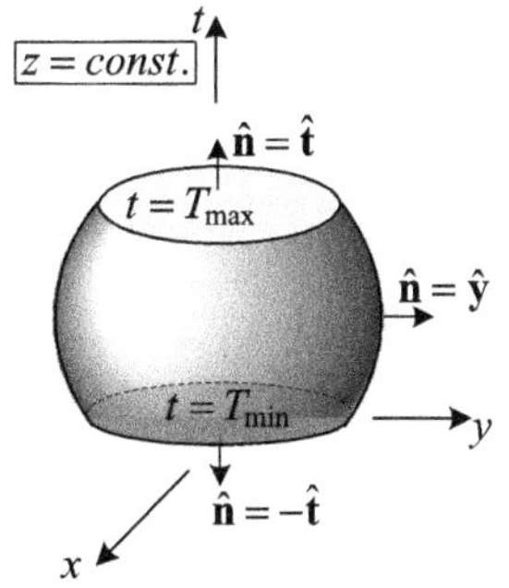

Figure 1.5 Illustration of the domain $\mathcal{D}$ of space and time: a "cut" in a $z = const.$ plane.

Here, Green's function notation for the wave equation is $\mathcal{G} \equiv g$. This notation is used throughout the book. For clarity, let us write Eq. (1.57) in explicit spatial and temporal terms:

$$\begin{aligned} u(\mathbf{r}, t) = &\int_{T_{\min}}^{T_{\max}}\iiint_V s(\mathbf{r}', t')g(\mathbf{r}', t'; \mathbf{r}, t)dv'dt' \\ &+ \int_{T_{\min}}^{T_{\max}}\oiint_S \left[u(\mathbf{r}', t')\nabla g(\mathbf{r}', t'; \mathbf{r}, t) - g(\mathbf{r}', t'; \mathbf{r}, t)\nabla u(\mathbf{r}', t')\right] \cdot \hat{\mathbf{n}} ds' dt' \\ &- \iiint_V \frac{1}{v^2}\left[u(\mathbf{r}', t')\frac{\partial g(\mathbf{r}', t'; \mathbf{r}, t)}{\partial t'} - g(\mathbf{r}', t'; \mathbf{r}, t)\frac{\partial u(\mathbf{r}', t')}{\partial t'}\right]_{t'=T_{\min}}^{T_{\max}} dv', \\ &\mathbf{r} \in V \wedge t \in [T_{\min}, T_{\max}]. \quad (1.58) \end{aligned}$$

1.5.5 Interior and Exterior Integral Solutions of the Wave Equation

The previous discussion of the application of Green's generalized theorem to the Poisson, Helmholtz, and wave equations makes it apparent that they all can be treated in the framework of the generalized n-dimensional linear operator of the form

$$\mathcal{L} = \bar{\nabla} \cdot \left[p(\mathbf{x})\bar{\nabla}\right] + q(\mathbf{x}), \quad (1.59)$$

which is an extension of the operator in Eq. (1.48) to n-dimensional space. Here, $\bar{\nabla}$ denotes an n-dimensional gradient. For all these cases, the integral solution $u(\mathbf{x})$, when $\mathbf{x}$ is inside the domain $\mathcal{D}$, can be written in the generalized form

$$u(\mathbf{x}) = \int_{\mathcal{D}} s(\mathbf{x}')\mathcal{G}(\mathbf{x}', \mathbf{x})d\Omega' + \oint_{\partial\mathcal{D}} \left[u(\mathbf{x}')\bar{\nabla}\mathcal{G}(\mathbf{x}', \mathbf{x}) - \mathcal{G}(\mathbf{x}', \mathbf{x})\bar{\nabla}u(\mathbf{x}')\right] \cdot \hat{\mathbf{n}}d\sigma', \mathbf{x} \in \mathcal{D}. \quad (1.60)$$

Eq. (1.60) is known as the *interior integral solution* of Eq. (1.42). The term *interior* emphasizes the fact that $\mathbf{x}$ resides in $\mathcal{D}$. It allows us to compute the field anywhere in $\mathcal{D}$, provided we know the sources inside $\mathcal{D}$ and the boundary values on $\partial\mathcal{D}$.

Following the same procedure but under the assumption that $\mathbf{x} \notin \mathcal{D}$ while the integration is over $\mathbf{x}' \in \mathcal{D}$, results in the *exterior integral solution*:

$$0 = \int_{\mathcal{D}} s(\mathbf{x}')\mathcal{G}(\mathbf{x}', \mathbf{x})d\Omega' + \oint_{\partial\mathcal{D}} \left[u(\mathbf{x}')\bar{\nabla}\mathcal{G}(\mathbf{x}', \mathbf{x}) - \mathcal{G}(\mathbf{x}', \mathbf{x})\bar{\nabla}u(\mathbf{x}')\right] \cdot \hat{\mathbf{n}}d\sigma', \quad \mathbf{x} \notin \mathcal{D}. \tag{1.61}$$

The vanishing left side in Eq. (1.61) is a result of the fact that the impulse excitation $\delta(\mathbf{x} - \mathbf{x}')$ of Green's function is identically zero for $\forall\mathbf{x}$ since $\mathbf{x} \notin \mathcal{D}$ while $\mathbf{x}' \in \mathcal{D}$.

The exterior solution is also very useful. It is the basis of what is known as the *surface equivalence principle*. The equivalence principle states that the contribution from sources residing *outside* a domain within which we seek a solution (see the first integral in Eq. (1.61)) can be replaced by the contribution of *equivalent* surface sources residing on the boundary enclosing the domain (see the second integral in Eq. (1.61)). The equivalent surface sources are $u(\mathbf{x}' \in \partial\mathcal{D})$ and $\bar{\nabla}u(\mathbf{x}' \in \partial\mathcal{D})$. This concept is discussed in more detail later in the context of Helmholtz's and wave equations.

1.5.6 Uniqueness of Green's Functions

Green's function as a solution to the inhomogeneous equation Eq. (1.43) is not unique. One can add any other function $\mathcal{G}_0(\mathbf{x}, \mathbf{x}')$ that solves the homogeneous equation $\mathcal{L}\mathcal{G}_0 = 0$, to Green's existing function $\mathcal{G}(\mathbf{x}, \mathbf{x}')$ that solves Eq. (1.43), and the result will be another valid Green's function.

This is often used to construct Green's function that would make the solution to a particular problem easier. For example, one can choose Green's function such that it vanishes on the boundary $\partial\mathcal{D}$ along with the normal component of its gradient $\nabla\mathcal{G}(\mathbf{x}', \mathbf{x}) \cdot \hat{\mathbf{n}}$. This will result in a zero surface integral in Eq. (1.60). Another choice, which makes the surface integral vanish, is the one where Green's function satisfies the same boundary conditions as the solution $u(\mathbf{x}')$. This makes the two terms in the integrand of the surface integral the same, resulting in their mutual cancellation. These two approaches are commonly used when constructing integral solutions to EM problems.

1.5.7 Reciprocity of Green's Functions

An important property of Green's functions for the operators of the type in Eq. (1.59) is that they possess symmetry. This symmetry is referred to as *reciprocity*. Reciprocity means that exchanging the location of source and observation points does not change Green's function:

$$\mathcal{G}(\mathbf{x}, \mathbf{x}') = \mathcal{G}(\mathbf{x}', \mathbf{x}). \tag{1.62}$$

Do not make the mistake of assuming that reciprocity holds only in uniform media. If $\mathcal{D}$ is simply 3D space, exchanging the locations of the source and the observer will obviously change nothing. Reciprocity holds in heterogeneous environments as well.

The proof of Eq. (1.62) is fairly straightforward, and it exploits Green's generalized theorem Eq. (1.56). Consider Green's function in two instances: the δ-source is at $\mathbf{x}_1$ and then it is at $\mathbf{x}_2$. The respective equations are

$$\mathcal{L}\mathcal{G}(\mathbf{x}, \mathbf{x}_1) = \delta(\mathbf{x} - \mathbf{x}_1), \tag{1.63}$$

$$\mathcal{L}\mathcal{G}(\mathbf{x}, \mathbf{x}_2) = \delta(\mathbf{x} - \mathbf{x}_2). \tag{1.64}$$

Multiply Eq. (1.63) by $\mathcal{G}(\mathbf{x}, \mathbf{x}_2)$, multiply Eq. (1.64) by $\mathcal{G}(\mathbf{x}, \mathbf{x}_1)$, subtract the obtained equations, integrate the result over $\mathbf{x} \in \mathcal{D}$, make use of the sampling property of the δ-functions, and apply Green's generalized theorem Eq. (1.56). The result is

$$\mathcal{G}(\mathbf{x}_1, \mathbf{x}_2) - \mathcal{G}(\mathbf{x}_2, \mathbf{x}_1) = \oint_{\partial\mathcal{D}} \left[\mathcal{G}(\mathbf{x}, \mathbf{x}_2)\bar{\nabla}\mathcal{G}(\mathbf{x}, \mathbf{x}_1) - \mathcal{G}(\mathbf{x}, \mathbf{x}_1)\bar{\nabla}\mathcal{G}(\mathbf{x}, \mathbf{x}_2)\right] \cdot \hat{\mathbf{n}} d\sigma. \tag{1.65}$$

The right-hand side vanishes because both $\mathcal{G}(\mathbf{x}, \mathbf{x}_1)$ and $\mathcal{G}(\mathbf{x}, \mathbf{x}_2)$ satisfy the same boundary conditions. Therefore,

$$\mathcal{G}(\mathbf{x}_1, \mathbf{x}_2) = \mathcal{G}(\mathbf{x}_2, \mathbf{x}_1), \tag{1.66}$$

which proves Eq. (1.62).

The intuitive understanding of reciprocity in Helmholtz's (or Poisson's) equation is not difficult, especially for electrical engineers who are familiar with the reciprocity of linear circuits. However, the reciprocity of Green's function for the wave equation may be puzzling: it implies that a source in the future produces a result in the past just like the result that the same source in the past would produce in the future. In other words, the loss-free wave equation does not discriminate between past and future. It allows for a wave to diverge from its source in both directions of time in the same way it allows for it to diverge in space. For example, Green's function for an impulse excitation $\delta(\mathbf{r}, t)$ at the origin of space-time,

$$g(\mathbf{r}, t) = \frac{\delta(t - r/v)}{4\pi r}, r = \mid \mathbf{r} \mid, \tag{1.67}$$

solves the loss-free wave equation for both $t \geq 0$ and $t \leq 0$. This peculiarity is a mathematical fact that goes against the principle of causality and needs to be resolved.

Causality requires that the cause precedes the effect. Green's function $g(\mathbf{r}, t; \mathbf{r}', t')$, which is zero at $t < t'$, $\forall \mathbf{r}$, is said to be *causal*. On the contrary, if Green's function $g(\mathbf{r}, t; \mathbf{r}', t') \neq 0$ anywhere at any $t < t'$, then it is *acausal*. Such Green's function violates causality because it allows an effect to appear earlier than its cause. Acausal solutions are nonphysical and must be discarded in the construction of forward models. As we see later, they are useful, however, as mathematical tools in solving inverse problems.

Differentiating between causal and acausal Green's functions presents a dilemma as far as the reciprocity relation Eq. (1.62) is concerned. Clearly, this relation cannot

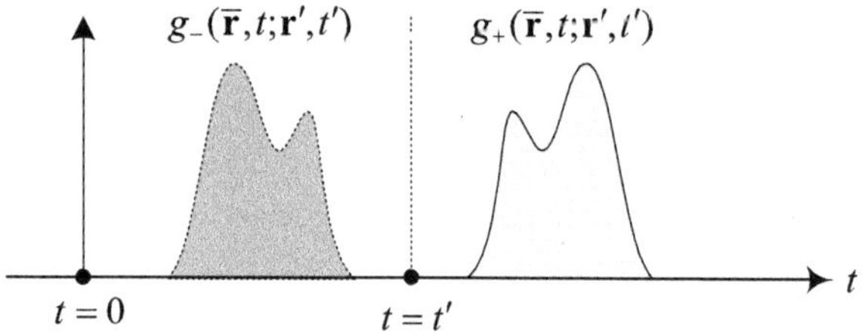

Figure 1.6 Illustration of Green's causal and acausal functions of the wave equation and their mirror symmetry in time. The bar above the position vectors indicates that the waveforms correspond to a fixed position in space. $t = t'$ marks the time of excitation.

hold if $g(\mathbf{r}, t; \mathbf{r}', t')$ is either causal or acausal. It is customary to denote causal solutions with a "+" subscript, while acausal ones are distinguished by a "−" subscript. Assume $g_+(\mathbf{r}, t; \mathbf{r}', t')$ is Green's causal function so that it can attain nonzero values if $t \geq t'$. However, it is identically zero for $t < t'$, $\forall \mathbf{r}$. Then, for $t \geq t'$, its reciprocal, $g_+(\mathbf{r}', t'; \mathbf{r}, t)$, is identically zero. Therefore, $g_+(\mathbf{r}, t; \mathbf{r}', t') = g_+(\mathbf{r}', t'; \mathbf{r}, t)$ cannot hold.

This dilemma is resolved by "splitting" Green's function into a causal wave and an acausal wave, which are mirror images of each other in time, with the point of symmetry being t' (the excitation instance):

$$g(\mathbf{r}, t; \mathbf{r}', t') = \begin{cases} g_+(\mathbf{r}, t; \mathbf{r}', t'), & \text{if } t > t' \\ g_-(\mathbf{r}, t; \mathbf{r}', t'), & \text{if } t < t' \end{cases} \tag{1.68}$$

where

$$g_-(\mathbf{r}, -t; \mathbf{r}', -t') = g_+(\mathbf{r}, t; \mathbf{r}', t'). \tag{1.69}$$

This is illustrated in Fig. 1.6. Green's causal wave g_+ describes the impulse response developing toward the future (positive time flow). In contrast, Green's acausal wave g_- describes the impulse response developing toward the past (negative time flow).

Using Green's theorem, it can be shown that Green's causal function satisfies [20]

$$g_+(\mathbf{r}, t; \mathbf{r}', t') = g_+(\mathbf{r}', -t'; \mathbf{r}, -t). \tag{1.70}$$

This relation is commonly referred to as the *wave-equation reciprocity condition* because the reciprocity in Eq. (1.62) cannot be directly employed. The change of sign in the temporal variables of the source and the observation allows for preserving the causality of the function. The same reciprocity relation holds for Green's acausal function too, i.e.,

$$g_-(\mathbf{r}, t; \mathbf{r}', t') = g_-(\mathbf{r}', -t'; \mathbf{r}, -t). \tag{1.71}$$

The wave-equation reciprocity shows that its Green's functions feature not only symmetry with respect to the spatial positions of source and observation (just like Green's functions of Helmholtz's equation) but also temporal antisymmetry.

Another important relationship between Green's causal and acausal functions arises when the wave-equation reciprocity Eq. (1.70)–(1.71) is applied to Eq. (1.69):

$$g_+(\mathbf{r}, t; \mathbf{r}', t') = g_-(\mathbf{r}', t'; \mathbf{r}, t). \tag{1.72}$$

The important implication here is that for every Green's causal function of the wave equation there exists an acausal counterpart (and vice versa), which is obtained by a simple interchange of excitation and observation points.

1.5.8 Green's Theorem and Integral Solution for the Damped-Wave Operator

So far, the wave equation has been considered in its loss-free form where $\mathcal{L} = \nabla^2 - \frac{1}{\upsilon^2}\frac{\partial^2}{\partial t^2}$. In practice, however, we often encounter cases where we cannot ignore the damping effect of the losses. The general damped-wave operator (known as d'Alembert's operator) is given by

$$\mathcal{L} = \nabla^2 - \frac{1}{\upsilon^2}\frac{\partial^2}{\partial t^2} - a\frac{\partial}{\partial t} - b, \tag{1.73}$$

where a and b are assumed constant in time just like υ. We already encountered this operator in the scalar models of electromagnetism discussed in Section 1.1.

D'Alembert's operator is not in the form Eq. (1.59). This implies that Green's generalized 4D theorem Eq. (1.56) would not hold, and therefore, we have no means of converting the second integral in Eq. (1.45) over the domain $\mathcal{D}$ into a boundary integral over $\partial\mathcal{D}$.

Let us examine what exactly happens with Green's theorem in this case. This is going to help us find out what modifications are necessary to arrive at an integral solution similar to the one for Helmholtz's and the loss-free wave equations. Recall that the left-hand side of Green's theorem involves the integrand $\mathcal{I} = \Phi\mathcal{L}\Psi - \Psi\mathcal{L}\Phi$. An attempt to employ $\mathcal{L}$ defined in Eq. (1.73) with this integrand results in

$$\mathcal{I} = \left(\Phi\nabla^2\Psi - \Psi\nabla^2\Phi\right) - \frac{1}{\upsilon^2}\left(\Phi\frac{\partial^2\Psi}{\partial t^2} - \Psi\frac{\partial^2\Phi}{\partial t^2}\right) - a\left(\Phi\frac{\partial\Psi}{\partial t} - \Psi\frac{\partial\Phi}{\partial t}\right) - b\left(\Phi\Psi - \Psi\Phi\right). \tag{1.74}$$

The last term is clearly zero. We can employ the identities Eq. (1.52) and Eq. (1.53) to the first and the second term, respectively, as we did in the case of the wave equation. The third term, however, cannot be reduced to the form $\frac{\partial}{\partial t}(\cdot)$, which would allow for eliminating the integration over time. Therefore, it is the first-order temporal derivative in d'Alembert's operator that prevents the direct application of Green's theorem.

This problem is easily resolved if we define an auxiliary differential operator of the form

$$\mathcal{L}^\dagger = \nabla^2 - \frac{1}{\upsilon^2}\frac{\partial^2}{\partial t^2} + a\frac{\partial}{\partial t} - b. \tag{1.75}$$

Then, we can modify the integrand $\mathcal{I} = \Phi\mathcal{L}\Psi - \Psi\mathcal{L}\Phi$ as $\mathcal{I}^\dagger = \Phi\mathcal{L}\Psi - \Psi\mathcal{L}^\dagger\Phi$, which is expanded as

$$\mathcal{I}^\dagger = \left(\Phi\nabla^2\Psi - \Psi\nabla^2\Phi\right) - \frac{1}{\upsilon^2}\left(\Phi\frac{\partial^2\Psi}{\partial t^2} - \Psi\frac{\partial^2\Phi}{\partial t^2}\right) - a\left(\Phi\frac{\partial\Psi}{\partial t} + \Psi\frac{\partial\Phi}{\partial t}\right) - b\left(\Phi\Psi - \Psi\Phi\right). \tag{1.76}$$

Now, the third term can be represented as

$$\Phi \frac{\partial \Psi}{\partial t} + \Psi \frac{\partial \Phi}{\partial t} = \frac{\partial}{\partial t}(\Psi \Phi). \tag{1.77}$$

After applying the identities Eq. (1.52), Eq. (1.53), and Eq. (1.77) to the first, second, and third term of Eq. (1.76), respectively, and integrating over $\mathcal{D}$, we obtain yet another generalization of Green's theorem:

$$\int_{T_{\min}}^{T_{\max}}\iiint_V \left(\Phi \mathcal{L}\Psi - \Psi \mathcal{L}^\dagger \Phi\right) dvdt = \int_{T_{\min}}^{T_{\max}}\oiint_S (\Phi \nabla \Psi - \Psi \nabla \Phi)\cdot \hat{\mathbf{n}}ds$$
$$- \iiint_V \frac{1}{\upsilon^2}\left[\Phi \frac{\partial \Psi}{\partial t} - \Psi \frac{\partial \Phi}{\partial t}\right]_{T_{\min}}^{T_{\max}} dv - \iiint_V a\,[\Phi\Psi]_{T_{\min}}^{T_{\max}}\, dv. \tag{1.78}$$

We can use Eq. (1.78) to construct an integral solution to the damped-wave equation in a manner analogous to that for Helmholtz's and the loss-free wave equation. We seek a solution to

$$\mathcal{L}u(\mathbf{r}, t) = s(\mathbf{r}, t), \tag{1.79}$$

where $\mathcal{L}$ is defined in Eq. (1.73). Let Green's function $g^\dagger$ satisfy

$$\mathcal{L}^\dagger g^\dagger(\mathbf{r}, t; \mathbf{r}', t') = \delta(\mathbf{r} - \mathbf{r}', t - t'), \tag{1.80}$$

where $\mathcal{L}^\dagger$ is defined in Eq. (1.75). With the help of Eq. (1.78) and the sampling property of the δ-function, we obtain the interior solution as

$$u(\mathbf{r}, t) = \int_{T_{\min}}^{T_{\max}}\iiint_V s(\mathbf{r}', t')g^\dagger(\mathbf{r}', t'; \mathbf{r}, t)dv'dt' + \mathcal{I}_{\partial\mathcal{D}},\ \mathbf{r} \in V \wedge t \in [T_{\min}, T_{\max}], \tag{1.81}$$

where

$$\mathcal{I}_{\partial\mathcal{D}} = \int_{T_{\min}}^{T_{\max}}\oiint_S \left[u(\mathbf{r}', t')\nabla g^\dagger(\mathbf{r}', t'; \mathbf{r}, t) - g^\dagger(\mathbf{r}', t'; \mathbf{r}, t)\nabla u(\mathbf{r}', t')\right]\cdot \hat{\mathbf{n}}ds'dt'$$
$$- \iiint_V \frac{1}{\upsilon^2}\left[u(\mathbf{r}', t')\frac{\partial g^\dagger(\mathbf{r}', t'; \mathbf{r}, t)}{\partial t'} - g^\dagger(\mathbf{r}', t'; \mathbf{r}, t)\frac{\partial u(\mathbf{r}', t')}{\partial t'}\right]_{t'=T_{\min}}^{T_{\max}} dv'$$
$$- \iiint_V a\left[g^\dagger(\mathbf{r}', t'; \mathbf{r}, t)\cdot u(\mathbf{r}', t')\right]_{t'=T_{\min}}^{T_{\max}} dv'. \tag{1.82}$$

Correspondingly, the exterior solution to the damped-wave equation is

$$0 = \int_{T_{\min}}^{T_{\max}}\iiint_V s(\mathbf{r}', t')g^\dagger(\mathbf{r}', t'; \mathbf{r}, t)dv'dt' + \mathcal{I}_{\partial\mathcal{D}},\ \mathbf{r} \notin V \vee t \notin [T_{\min}, T_{\max}]. \tag{1.83}$$

The utility of these integral solutions may not be obvious because now we have to deal with Green's auxiliary functions $g^{\dagger}$. However, as we will soon see, these auxiliary functions are closely related to Green's original functions.

1.5.9 The Mathematical Perspective: Adjoint and Self-Adjoint Operators and Green's Functions

In a mathematical language, the auxiliary operator $\mathcal{L}^{\dagger}$, which was introduced in Eq. (1.75) to accommodate the case of damped waves, is the operator *adjoint* to d'Alembert's operator in Eq. (1.73). For a rigorous treatment of the subject of adjoint operators and their application to the solution of partial differential equations, the reader is referred to [19]. Here, let me point out some relevant results.

1. Adjoint operators allow for generalizing Green's theorem over the whole class of linear partial differential equations.
2. There are simple rules allowing for the derivation of an adjoint operator once we know the original one. $\mathcal{L}^{\dagger}$ in Eq. (1.75) was obtained from $\mathcal{L}$ in Eq. (1.73) following these rules.
3. Green's function $\mathcal{G}^{\dagger}$ solving $\mathcal{L}^{\dagger}\mathcal{G}^{\dagger} = \delta$ is termed Green's adjoint function.
4. Green's function and its adjoint counterpart satisfy the relation

$$\mathcal{G}(\mathbf{x}, \mathbf{x}') = \mathcal{G}^{\dagger}(\mathbf{x}', \mathbf{x}). \tag{1.84}$$

   This means that an integral solution to any linear partial differential equation can be constructed using either $\mathcal{G}$ or its adjoint counterpart $\mathcal{G}^{\dagger}$. For example, we have already shown through the damped-wave equation that the contribution of a volume source distribution $s(\mathbf{x}')$ is given by

$$u(\mathbf{x}) = \int_{\mathcal{D}} s(\mathbf{x}')\mathcal{G}^{\dagger}(\mathbf{x}', \mathbf{x})d\Omega'. \tag{1.85}$$

   However, it can also be written as

$$u(\mathbf{x}) = \int_{\mathcal{D}} s(\mathbf{x}')\mathcal{G}(\mathbf{x}, \mathbf{x}')d\Omega'. \tag{1.86}$$

   You will recognize this result as the familiar superposition integral describing the contribution of a distributed source.
5. If Green's function for a given operator is symmetric, i.e., it satisfies the reciprocity condition Eq. (1.62), $\mathcal{G}(\mathbf{x}, \mathbf{x}') = \mathcal{G}(\mathbf{x}', \mathbf{x})$, the operator is termed *self-adjoint* because this implies that Green's adjoint function is the same as the original one:

$$\mathcal{G}^{\dagger}(\mathbf{x}', \mathbf{x}) = \mathcal{G}(\mathbf{x}, \mathbf{x}') = \mathcal{G}(\mathbf{x}', \mathbf{x}). \tag{1.87}$$

   Also, the adjoint operator in this case is the same as the original one i.e., $\mathcal{L}^{\dagger} \equiv$. For example, Helmholtz's operator is self-adjoint. This is why, the contribution of a volume source distribution $\mathcal{S}(\mathbf{x}')$, which in Eq. (1.50) is given as

$$U(\mathbf{r}) = \iiint_V \mathcal{S}(\mathbf{r}')G(\mathbf{r}', \mathbf{r})dv', \tag{1.88}$$

can be equivalently stated as

$$U(\mathbf{r}) = \iiint_V \mathcal{S}(\mathbf{r}')G(\mathbf{r}, \mathbf{r}')dv'. \tag{1.89}$$

6. When causality is enforced, the loss-free wave equation is not self-adjoint because its reciprocity—see Eqs. (1.70)–(1.71)—features spatial symmetry but temporal antisymmetry. Its Green's causal function and its acausal counterpart relate through Eq. (1.72), which is identical to that between Green's function and its adjoint; see Eq. (1.84). This is why the acausal counterpart of Green's causal function is often referred to as its adjoint. Of course, one could also state that the causal counterpart of Green's acausal function is its adjoint.
7. The damped-wave equation is not self-adjoint because the operators in Eq. (1.73) Eq. (1.75) are not the same. However, as follows from Eq. (1.84), Green's adjoint function can be easily obtained from Green's function of d'Alembert's original operator by simply exchanging the excitation and observation points in space-time:

$$g^\dagger(\mathbf{r}', t'; \mathbf{r}, t) = g(\mathbf{r}, t; \mathbf{r}', t'). \tag{1.90}$$

In view of causality, which must hold in the case of damped waves as well, the above relation means that if $g(\mathbf{r}, t; \mathbf{r}', t')$ is causal, i.e., it vanishes for $t < t'$, then $g^\dagger(\mathbf{r}', t'; \mathbf{r}, t)$ is acausal and it vanishes for $t > t'$. In Appendix D, an explicit proof is developed showing that indeed, for every causal Green's function $g_+(\mathbf{r}, t; \mathbf{r}', t')$ of the damped-wave operator Eq. (1.73), there exists an acausal counterpart g_- such that

$$g_-(\mathbf{r}', t'; \mathbf{r}, t) = g_+(\mathbf{r}, t; \mathbf{r}', t'). \tag{1.91}$$

Here, g_- is Green's function of d'Alembert's adjoint operator Eq. (1.75).
8. Green's functions of both d'Alembert's original and adjoint operators satisfy the same reciprocity relation featuring temporal antisymmetry as Green's functions of the loss-free equation; see Eqs. (1.70)–(1.71).

This is easy to see. Consider the equation satisfied by Green's causal function:

$$\left(\nabla^2 - \frac{1}{\upsilon^2}\frac{\partial^2}{\partial t^2} - a\frac{\partial}{\partial t} - b\right) g_+(\mathbf{r}, t; \mathbf{r}', t') = \delta(\mathbf{r} - \mathbf{r}')\delta(t - t'). \tag{1.92}$$

Change the temporal variables using $\tau = -t$ and $\tau' = -t'$. The resulting equation,

$$\left(\nabla^2 - \frac{1}{\upsilon^2}\frac{\partial^2}{\partial \tau^2} + a\frac{\partial}{\partial \tau} - b\right) g_+(\mathbf{r}, \tau; \mathbf{r}', \tau') = \delta(\mathbf{r} - \mathbf{r}')\delta(\tau - \tau'), \tag{1.93}$$

is the same as that for Green's acausal (or adjoint) function:

$$\left(\nabla^2 - \frac{1}{\upsilon^2}\frac{\partial^2}{\partial t^2} + a\frac{\partial}{\partial t} - b\right) g_-(\mathbf{r}, t; \mathbf{r}', t') = \delta(\mathbf{r} - \mathbf{r}')\delta(t - t'), \tag{1.94}$$

with the only difference being in the reversed direction of time, i.e., τ is replaced by t. Requiring that both the causal and acausal functions satisfy the same boundary conditions and the same initial/terminal conditions (in their respective temporal spaces) ensures that

$$g_+(\mathbf{r}, \tau; \mathbf{r}', \tau') = g_-(\mathbf{r}, t; \mathbf{r}', t'), \quad \tau = -t\,, \quad \tau' = -t'. \tag{1.95}$$

We now employ Eq. (1.84) and obtain

$$g_+(\mathbf{r}, -t; \mathbf{r}', -t') = g_-(\mathbf{r}, t; \mathbf{r}', t') = g_+(\mathbf{r}', t'; \mathbf{r}, t). \tag{1.96}$$

This is indeed the same wave-equation reciprocity that we already stated in Eq. (1.70).

An analogous derivation can be carried out for Green's acausal function $g_-(\mathbf{r}, t; \mathbf{r}', t')$, starting from the equation it satisfies. i.e., d'Alembert's adjoint equation. Therefore, the wave-equation reciprocity holds for both Green's causal and acausal functions in the case of wave damping as well:

$$g_{+/-}(\mathbf{r}, t; \mathbf{r}', t') = g_{+/-}(\mathbf{r}', -t'; \mathbf{r}, -t). \tag{1.97}$$

Another important result that emerges from this analysis is that the relationship between Green's original (causal) and adjoint (acausal) functions, Eq. (1.90), can be extended to include time reversal:

$$g_+(\mathbf{r}, t; \mathbf{r}', t') = g_-(\mathbf{r}', t'; \mathbf{r}, t) = g_-(\mathbf{r}, -t; \mathbf{r}', -t'). \tag{1.98}$$

This result shows that for every Green's causal function there exists an acausal counterpart, which is its mirror image in time with the point of antisymmetry being the excitation instance. Note that the same relation was already established in the case of the loss-free wave equation; see Eq. (1.69).

1.6 Integral Solutions to the Wave Equation

Since Green's functions of the loss-free and lossy wave equations satisfy the same reciprocity condition and the same relationship between causal and acausal counterparts, from now on we treat both equations in a common framework. The term *wave equation* will refer to both the loss-free and the lossy case unless stated otherwise.

In view of the relationship Eq. (1.90) between Green's function and its adjoint counterpart, the interior solution to the wave equation, previously given in Eq. (1.58) and Eq. (1.81)–(1.82), is now written as

$$u(\mathbf{r}, t) = \iiiint\limits_{T_{\min}V}^{T_{\max}} g(\mathbf{r}, t; \mathbf{r}', t') s(\mathbf{r}', t') dv' dt' + \mathcal{I}_{\partial D}\,, \quad \mathbf{r} \in V \wedge t \in [T_{\min}, T_{\max}], \tag{1.99}$$

where

$$\mathcal{I}_{\partial\mathcal{D}} = \int_{T_{\min}}^{T_{\max}} \oiint_S \left[u(\mathbf{r}',t')\nabla g(\mathbf{r},t;\mathbf{r}',t') - g(\mathbf{r},t;\mathbf{r}',t')\nabla u(\mathbf{r}',t')\right] \cdot \hat{\mathbf{n}} ds' dt'$$
$$- \iiint_V \frac{1}{v^2}\left[u(\mathbf{r}',t')\frac{\partial g(\mathbf{r},t;\mathbf{r}',t')}{\partial t'} - g(\mathbf{r},t;\mathbf{r}',t')\frac{\partial u(\mathbf{r}',t')}{\partial t'}\right]_{t'=T_{\min}}^{T_{\max}} dv'$$
$$- \iiint_V a\left[g(\mathbf{r},t;\mathbf{r}',t') \cdot u(\mathbf{r}',t')\right]_{t'=T_{\min}}^{T_{\max}} dv'. \quad (1.100)$$

Note that $a = 0$ in the loss-free wave equation.

At this stage, we are not committing to specifying Green's function $g(\mathbf{r},t;\mathbf{r}',t')$ or the solution $u(\mathbf{r},t)$ as causal or acausal. As previously discussed, the wave equation allows for both. In practice, however, physically meaningful wave solutions in a lossy medium are those decaying with time, which implies causality. In Eq. (1.99), if $u(\mathbf{r},t)$ is causal, then $g(\mathbf{r},t;\mathbf{r}',t')$ must be causal and vice versa. As per Eq. (1.90), we could employ $g^{\dagger}(\mathbf{r}',t';\mathbf{r},t)$ instead of $g(\mathbf{r},t;\mathbf{r}',t')$. In this case, however, if $u(\mathbf{r},t)$ is causal, then $g^{\dagger}(\mathbf{r}',t';\mathbf{r},t)$ must be acausal and vice versa. We should note that the integral solution in the form of Eq. (1.99)–(1.100), which employs g, is more common than that in Eq. (1.81)–(1.82), which employs $g^{\dagger}$.

Eq. (1.99)–(1.100) is essentially a statement of superposition describing waves due to three possible causes. First, we have the quadruple integral in Eq. (1.99), which describes the portion of the solution attributed to a known source $s(\mathbf{r}',t')$.

Second, in Eq. (1.100), we have the triple integral over S and time, which describes the portion of the solution due to nonzero boundary values of the field, $u(\mathbf{r}' \in S,t')$ and $\frac{\partial u(\mathbf{r}' \in S,t')}{\partial n}$. These boundary values account for all sources that may reside *outside* V and, therefore, cannot be taken into account by the quadruple integral. An explaination of how this works is provided shortly. For now, it suffices to remember that *it is a mistake* to include the boundary values $u(\mathbf{r}' \in S,t')$ and $\frac{\partial u(\mathbf{r}' \in S,t')}{\partial n}$ due to the *interior* sources $s(\mathbf{r}',t')$ in the surface-time integral of Eq. (1.99)–(1.100)! Plugging in the boundary values together with the source generating them simply means that you are computing the same effect twice. In fact, with the surface unit normal $\hat{\mathbf{n}}$ pointing outward, the two contributions will cancel, producing zero field inside V. If there are no sources external to V, this term is zero.

Third, we have the triple integral over V for the initial and terminal field values at $t' = T_{\min}$ and $t' = T_{\max}$, respectively. The initial conditions $u(\mathbf{r} \in V, t' = T_{\min})$ and $\frac{\partial u(\mathbf{r} \in V, t' = T_{\min})}{\partial t'}$ account for the impact of all sources that may have existed before $T_{\min}$. Such sources are different from $s(\mathbf{r}',t')$ and precede it! If such sources are not present, then the initial conditions are zero. Similarly, the terminal conditions $u(\mathbf{r} \in V, t' = T_{\max})$ and $\frac{\partial u(\mathbf{r} \in V, t' = T_{\max})}{\partial t'}$ account for the impact of all sources that may exist after $T_{\max}$. Such sources cannot have any effect on the observed field in the interval $[T_{\min}, T_{\max}]$, if this field is causal because, by definition, the causal field must lag behind the sources. This is taken care of by choosing the causal Green function, which in turn means that we do not need terminal conditions.

The *exterior* integral solution is obtained when $\mathbf{r}$ is outside V and/or t is outside $[T_{\min}, T_{\max}]$. Just as with the interior solution, it is common to formulate it in terms of Green's original function rather than the adjoint one. Our original derivation in Eq. (1.81), which is in terms of Green's adjoint function, can be equivalently written as

$$0 = \int_{T_{\min}}^{T_{\max}}\iiint_V g(\mathbf{r}, t; \mathbf{r}', t')s(\mathbf{r}', t')dv'dt' + \mathcal{I}_{\partial\mathcal{D}}\,, \quad \mathbf{r} \notin V \vee t \notin [T_{\min}, T_{\max}], \tag{1.101}$$

using the relation in Eq. (1.90). Here, $\mathcal{I}_{\partial\mathcal{D}}$ is the same as in Eq. (1.100). The exterior solution is fundamental in understanding the physical nature of the boundary conditions and the initial/terminal conditions used in the integral equation models.

Let V^{e} be the complement of V such that $V \cup V^{\mathrm{e}}$ is all space and let $\mathbf{r} \in V^{\mathrm{e}}$. Also, let the source $s(\mathbf{r}', t')$ be nonzero only in V, i.e., $s(\mathbf{r}', t') = 0$, $\forall \mathbf{r}' \in V^{\mathrm{e}}$. For simplicity, also assume that the initial and terminal field values are zero. Then, when the observation point is inside V^{e}, Eq. (1.101) provides a straightforward relation between the external sources $s(\mathbf{r}', t')$ and the boundary conditions:

$$\int_{T_{\min}}^{T_{\max}}\iiint_V g(\mathbf{r}, t; \mathbf{r}', t')s(\mathbf{r}', t')dv'dt' = -\int_{T_{\min}}^{T_{\max}}\oiint_S \left(u\frac{\partial g}{\partial n} - g\frac{\partial u}{\partial n}\right)ds'dt'\,, \quad \mathbf{r} \in V^{\mathrm{e}}. \tag{1.102}$$

Now, let us apply the *interior* solution Eq. (1.99) to the points $\mathbf{r} \in V^{\mathrm{e}}$:

$$u(\mathbf{r}, t) = \int_{T_{\min}}^{T_{\max}}\oiint_S \left(u\frac{\partial g}{\partial n_{\mathrm{e}}} - g\frac{\partial u}{\partial n_{\mathrm{e}}}\right)ds'dt'\,, \quad \mathbf{r} \in V^{\mathrm{e}}. \tag{1.103}$$

Here, $\hat{\mathbf{n}}_{\mathrm{e}} = -\hat{\mathbf{n}}$ points from V^{e} into V. Comparing Eq. (1.103) to Eq. (1.102), we see that the boundary integral in the right side of Eq. (1.103) is nothing but an equivalent representation of the impact of the sources outside V^{e} on the field inside V^{e}. This explains my earlier statement that the boundary values are representative of the effect that exterior sources have on the interior solution.

Similar arguments apply to the physical meaning of the initial and terminal conditions in Eq. (1.100). The initial conditions act as equivalent "temporal boundary" sources representing the impact on a causal solution due to sources that may exist prior to $T_{\min}$. Similarly, the terminal conditions act as equivalent "temporal boundary" sources representing the impact on an acausal solution due to sources that may exist after $T_{\max}$.

1.6.1 Causal and Acausal Waves and Green's Functions

To understand better the relationship between Green's causal and acausal functions for the wave equation, here we discuss in greater detail the concepts of causal (retarded) and acausal (advanced) waves. As discussed earlier, the wave equation does not discriminate

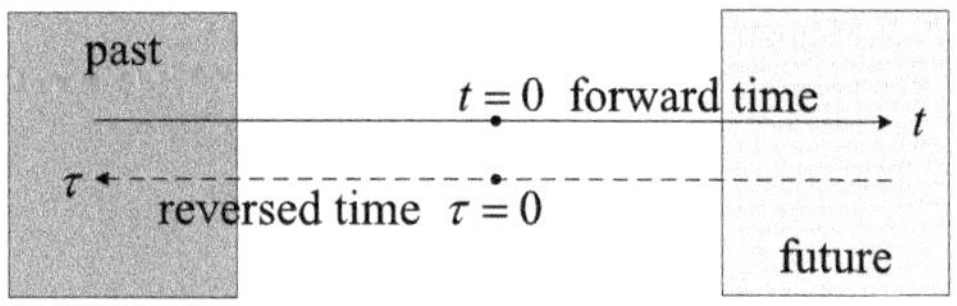

Figure 1.7 Illustration of the two time axes: t is the forward time axis, and τ shows the reversed time axis. Here, $\tau = -t$, and the two axes have coincident origins.

between past and future. Based on our experience, we are compelled to assume that time flow is unidirectional from cause (in the past) toward effect (in the future), which sets the positive direction of the time axis t. Mathematically, however, nothing prevents us from setting the positive direction of the time axis from future toward past. Such a time variable is termed *reversed time* and is denoted as τ as opposed to t, which denotes the usual *forward time*. Fig. 1.7 illustrates the two temporal variables and their axes when $\tau = -t$. In this case, the origins of the two axes, $t = 0$ and $\tau = 0$, coincide. If convenient, choosing the origin $\tau = 0$ so that it does not coincide with $t = 0$ is also possible. For example, if $\tau = 0$ when $t = T_{\max}$, then $\tau = T_{\max} - t$. What really matters here is that for any waveform $u(t)$, we can state that $\frac{du(t)}{dt} = -\frac{du(\tau)}{d\tau}$.

The relation $\frac{du(t)}{dt} = -\frac{du(\tau)}{d\tau}$ in Fig. 1.7 between the first-order temporal derivatives with respect to t and τ means that if a causal function $u_+(\mathbf{r}, t)$ satisfies the wave equation in forward time,

$$\left(\nabla^2 - \frac{1}{\upsilon^2}\frac{\partial^2}{\partial t^2} - a\frac{\partial}{\partial t} - b\right) u_+(\mathbf{r}, t) = s(\mathbf{r}, t), \tag{1.104}$$

then it satisfies the adjoint equation in reversed time:

$$\left(\nabla^2 - \frac{1}{\upsilon^2}\frac{\partial^2}{\partial \tau^2} + a\frac{\partial}{\partial \tau} - b\right) u_+(\mathbf{r}, \tau) = s(\mathbf{r}, \tau). \tag{1.105}$$

Let us now admit the existence of a fictitious auxiliary wave-equation solution $u_-(\mathbf{r}, \tau)$ in reversed time such that it satisfies

$$\left(\nabla^2 - \frac{1}{\upsilon^2}\frac{\partial^2}{\partial \tau^2} - a\frac{\partial}{\partial \tau} - b\right) u_-(\mathbf{r}, \tau) = s^\dagger(\mathbf{r}, \tau). \tag{1.106}$$

Here, $s^\dagger(\mathbf{r}, \tau) = s(\mathbf{r}, t)$ is an auxiliary source, termed the *adjoint source*, which is a "mirror" temporal image of the original source. Its distribution in the reversed-time (or adjoint) domain $\mathcal{D}^\dagger = \left\{(\mathbf{r}, \tau) \in \mathbb{R}^4\right\}$ is identical to the distribution of the original source $s(\mathbf{r}, t)$ in the forward-time domain $\mathcal{D} = \left\{(\mathbf{r}, t) \in \mathbb{R}^4\right\}$. Clearly, u_- is acausal because it develops along τ (toward the past) in the same way u_+ develops along t (toward the future). Changing variables from τ to t results in u_- satisfying the adjoint wave equation in t:

$$\left(\nabla^2 - \frac{1}{\upsilon^2}\frac{\partial^2}{\partial t^2} + a\frac{\partial}{\partial t} - b\right) u_-(\mathbf{r}, t) = s^\dagger(\mathbf{r}, t). \tag{1.107}$$

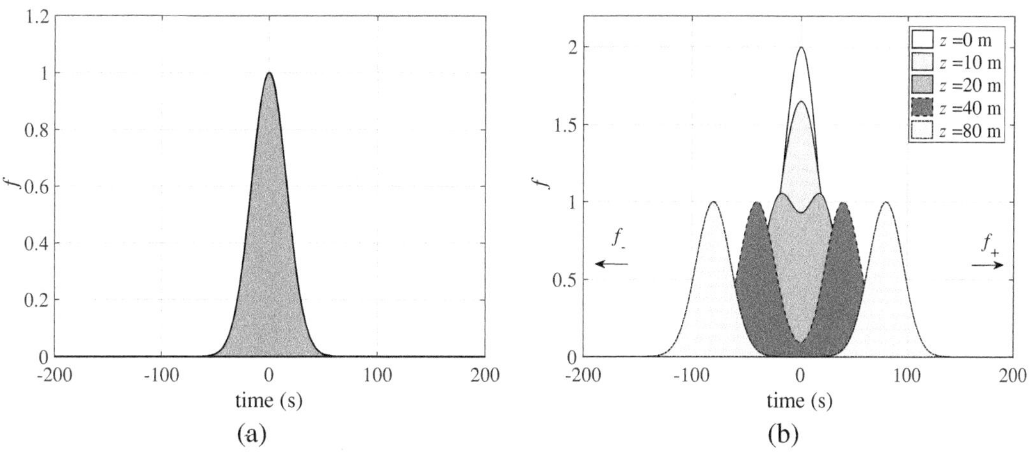

Figure 1.8 Illustration of the plane-wave solutions having Gaussian waveforms of the same magnitude and traveling forward and backward in time. (a) The Gaussian waveform $f(\xi) = \exp(-\alpha\xi^2)$, where α is a constant, centered at $t = 0$, i.e., $\xi = t$. (b) The superposition of the waves $f_+(z - vt)$ and $f_-(z + vt)$, both of which are valid solutions to the wave equation, as functions of time plotted for different positions along $z \geq 0$.

It is now clear that the two wave solutions u_+ and u_- are mutually adjoint: in $\mathcal{D}$, u_+ satisfies the wave equation, while u_- satisfies the adjoint wave equation. Conversely, in $\mathcal{D}^\dagger$, it is u_- that satisfies the wave equation while u_+ satisfies the adjoint one.

Moreover, $u_+(\mathbf{r}, t)$ and $u_-(\mathbf{r}, \tau)$ appear identical in their respective domains, i.e., $u_+(\mathbf{r}, t) = u_-(\mathbf{r}, \tau)$, provided they have the same sources and boundary conditions in $\mathcal{D}$ and $\mathcal{D}^\dagger$, respectively. In Appendix D, a rigorous proof is provided that for every causal solution $u_+(\mathbf{r}, t)$ there exists an acausal solution $u_-(\mathbf{r}, \tau)$ such that

$$u_-(\mathbf{r}, \tau) = u_+(\mathbf{r}, t), \tag{1.108}$$

where, if $t \in [0, T_{\max}]$ and $T_{\max}$ is finite, then $\tau = T_{\max} - t$, $\tau \in [0, T_{\max}]$. Alternatively, if $t \in (-\infty, +\infty)$, then $\tau = -t$, $\tau \in (-\infty, +\infty)$.

Waves running backward in time are rather counter-intuitive. So, let us take the time and consider an example with the plane-wave solution already discussed in Section 1.2, Eq. (1.8). Let the two terms of the general solution, $f_+(z - vt)$ and $f_-(z + vt)$, have the same magnitudes and waveforms. For simplicity, we set $v = 1$ m/s. Let the waveform be described by the truncated Gaussian function shown in Fig. 1.8-a. The truncation is done at a very small value, but it is necessary to define clearly the temporal extent of the pulse from $-T_p$ to $+T_p$ ($T_p = 80$ s).

Let us now assume that the waveform recorded at $z = 0$, $f(t) = 2\exp(-\alpha t^2)$, is the "event" that launches the waves in space-time. This is an excitation in the form of a boundary condition. The location $z = 0$ is the boundary for both the $z \geq 0$ and the $z \leq 0$ regions. Fig. 1.8-b shows what waveforms we would observe at various positions in the region $z \geq 0$ if acausal waves could indeed be launched in addition to the causal ones.

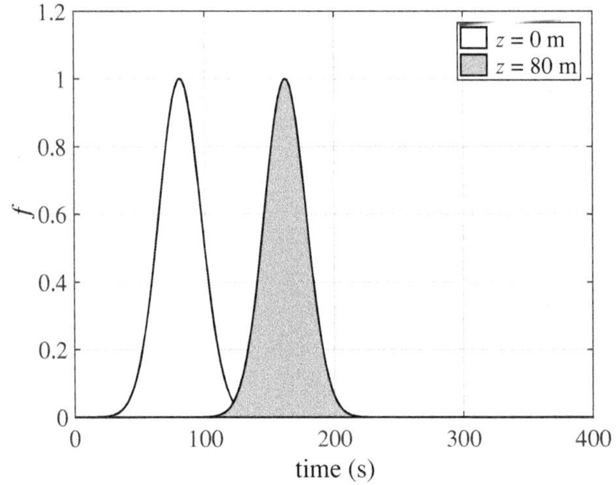

Figure 1.9 Illustration of an excitation pulse (at $z = 0$) set to have nonzero values for $t > 0$ only, together with the resulting causal waveform recorded some distance away.

For example, at $z = 80$ m, our recording device would show a clear pulse that peaks 80 seconds *before* our source peaks, and then it would show another pulse that peaks 80 seconds *after* the source peaks. The former pulse is the effect of the acausal (advanced) wave, while the latter pulse is due to the familiar causal (retarded, delayed) wave. The acausal wave propagates from the present into the past and is therefore nonphysical. Adding an acausal solution to a causal one, as we did here, makes the overall result nonphysical.

When we construct forward models, we use only causal solutions because they are the solutions that relate to the physical reality. Acausal solutions are virtual models, and they are used strictly as mathematical tools aiding in the solution of the inverse problem.

In time-domain wave analysis, especially in numerical simulations, it is customary to associate the origin of the time axis ($t = 0$) with the initialization of all sources and, therefore, all causal fields. Thus, the source functions are defined so that they are nonzero only for $t > 0$. Fig. 1.9 shows an example of a "properly" set excitation Gaussian waveform at $z = 0$, as well as the causal waveform observed at $z = 80$ m.

Causal forward models such as those produced by EM time-domain simulators do not allow for acausal solutions, i.e., all field values and all observed responses are zero for $t < 0$. However, Eq. (1.108) provides an easy way to obtain the acausal (adjoint) counterpart $u_-(\mathbf{r}, t)$ of the causal wave $u_+(\mathbf{r}, t)$. Since in the simulation $t \in [0, T_{\max}]$, the adjoint time variable is set as $\tau = T_{\max} - t$, and it, too, belongs to $[0, T_{\max}]$. If u_+ at $\mathbf{r}$ is a discrete time sequence, all that is needed to obtain u_- at $\mathbf{r}$ is to reverse the time sequence. Fig. 1.10 shows the time-dependent causal pulse (top plot) of a uniform plane wave propagating along z in a lossy medium. The excitation is at $z = 0$, while the shown pulses are recorded at $z = 20$ mm and $z = 60$ mm. We observe that the causal pulse is increasingly delayed and attenuated with distance. The respective acausal pulse

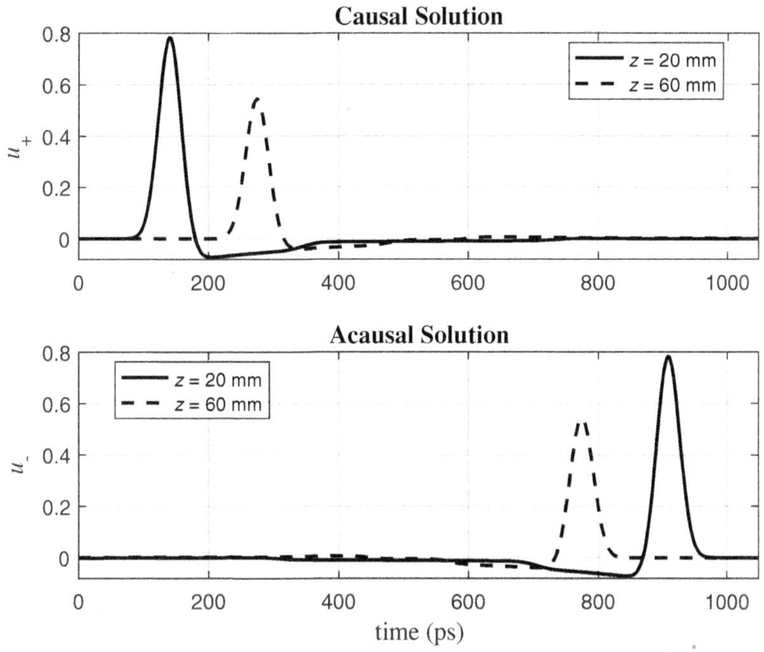

Figure 1.10 Illustration of a causal 1D wave in a lossy medium together with its acausal counterpart. The wave pulses are recorded at two distances from the source: $z = 20$ mm and $z = 60$ mm. The causal wave is simulated with MEFiSTo-3D [10]. Its acausal counterpart is obtained using Eq. (1.108).

is shown in the bottom plot. It has been obtained from the causal one using Eq. (1.108) where $\tau = T_{\max} - t$. We observe that as time advances the acausal pulse moves *toward* the source and grows in strength. Both of these features are nonphysical. Fig. 1.11 shows the same pulses, but as a function of the distance z from the source.

We can now return to the interior and exterior wave-equation solutions in Eq. (1.99) and Eq. (1.101), respectively. These integral solutions allow for both causal and acausal Green's functions. For example, to obtain the contribution of the sources $s(\mathbf{r}', t')$ in the first term of Eq. (1.99), we could employ either $g_+(\mathbf{r}, t; \mathbf{r}', t')$ or $g_-(\mathbf{r}, t; \mathbf{r}', t')$.

What is the impact of this choice on our solution? We observe that $g_+(\mathbf{r}, t; \mathbf{r}', t')$ can be nonzero only if $t > t'$ and is thus able to contribute to a causal field solution only. On the contrary, $g_-(\mathbf{r}, t; \mathbf{r}', t')$ is zero for $\forall t > t'$ and is thus incapable of describing the source contribution to a causal field. It could be used to describe a virtual acausal field that this same source would launch backward in time. Such acausal fields are a fine mathematical tool in solving inverse problems and sensitivity-analysis problems.

From now on, a "plus" or a "minus" subscript will be used in the notation for the solution u to clearly distinguish between the causal wave (u_+) and the acausal one (u_-).

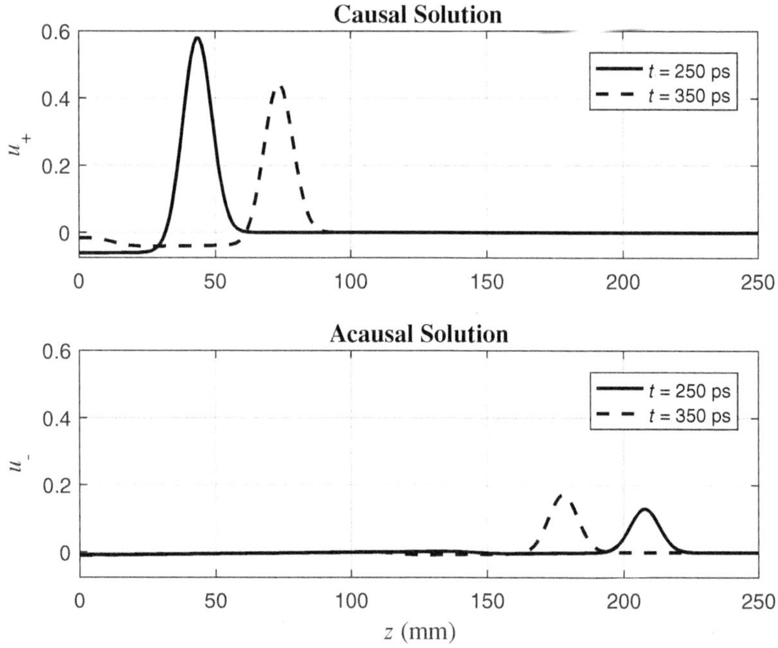

Figure 1.11 The same causal and acausal waves as in Fig. 1.10, this time plotted as functions of distance z from the source at two instances: $t = 250$ ps and $t = 350$ ps.

1.6.2 Causal Interior and Exterior Integral Solutions to the Wave Equation

The causal form of the interior and exterior integral solutions, Eq. (1.99) and Eq. (1.101), is

$$u_+(\mathbf{r}, t) = \int_{T_{\min}}^{T_{\max}}\iiint_V g_+(\mathbf{r}, t; \mathbf{r}', t')s(\mathbf{r}', t')dv'dt' + \mathcal{I}_{\partial D}, \quad \mathbf{r} \in V \ \wedge\ t \in [T_{\min}, T_{\max}], \tag{1.109}$$

and

$$0 = \int_{T_{\min}}^{T_{\max}}\iiint_V g_+(\mathbf{r}, t; \mathbf{r}', t')s(\mathbf{r}', t')dv'dt' + \mathcal{I}_{\partial D}, \quad \mathbf{r} \notin V \ \vee t \notin [T_{\min}, T_{\max}], \tag{1.110}$$

respectively, where

$$\mathcal{I}_{\partial D} = \int_{T_{\min}}^{T_{\max}}\oiint_S \left(u_+\frac{\partial g_+}{\partial n} - g_+\frac{\partial u_+}{\partial n}\right)ds'dt' - \iiint_V \frac{1}{v^2}\left[u_+\frac{\partial g_+}{\partial t'} - g_+\frac{\partial u_+}{\partial t'}\right]_{T_{\min}}^{T_{\max}} dv'. \tag{1.111}$$

The arguments of u_+ and g_+ in Eq. (1.111) are $(\mathbf{r}', t')$ and $(\mathbf{r}, t; \mathbf{r}', t')$, respectively. It should be reiterated that Eq. (1.109)–Eq. (1.111), can equivalently make use of

$g_-(\mathbf{r}', t'; \mathbf{r}, t)$. However, since a causal system is naturally described by Green's causal function, the above integral forms of the causal field solutions are preferred.

1.6.3 Integral-Equation Models of Transient Radiation

In accordance with our agreement that all physical sources are nonzero only for positive time, we set the temporal extent of the source $s(\mathbf{r}', t')$ as $0 \le t' \le T_s$. T_s is referred to as the source *turn-off time* and $s(\mathbf{r}', t') = 0$ for $t' > T_s$, $\forall \mathbf{r}'$. Let us expand the upper limit of the time integration in the integral solutions to infinity, i.e., $T_{\max} \to +\infty$, while $T_{\min} = 0$. We are seeking a causal solution; therefore, the time interval of interest is $t \in [0, \infty)$. Thus, whether the problem is interior or exterior depends only on whether $\mathbf{r}$ belongs to V or not. Here, we are interested in the interior problem where the region of interest V contains all sources, is simply connected, and is open to infinity as shown in Fig. 1.12-a.

As per Eq. (1.109), the radiation field is now obtained as

$$u_+(\mathbf{r}, t) = \int_0^{T_s}\iiint_V g_+(\mathbf{r}, t; \mathbf{r}', t')s(\mathbf{r}', t')dv'dt', \qquad (1.112)$$

where g_+ is Green's causal function for the specific open medium. Here, the contribution of $\mathcal{I}_{\partial\mathcal{D}}$ in Eq. (1.111) is zero. The first term in $\mathcal{I}_{\partial\mathcal{D}}$ is zero because it is associated with the boundary S_∞, which extends to infinity. The distance between the observation point at $\mathbf{r}$ and the boundary is infinite. Since $g_+(\mathbf{r}, t; \mathbf{r}' \to \infty, t')$ must be zero for $t' > t$, the integration over t' is limited within $[0, t]$. However, unless the velocity υ is infinite, the system's impulse response g_+ to an excitation that is infinitely far away is zero for $t' \in [0, t]$. Further, the second term in Eq. (1.111) is associated with the field initial values (u_+ and $\frac{\partial u_+}{\partial t'}$) at $t' = T_{\min} = 0$ and its terminal values at $t' = T_{\max} \to \infty$. The initial values of the causal field are zero because its sources are zero at $t' = 0$. At the same time, at $t' = T_{\max} \to \infty$, it is Green's function and its time derivative that are zero (due to causality). Thus, the second term of $\mathcal{I}_{\partial\mathcal{D}}$ is also zero.

Eq. (1.112) is a mathematical expression of the superposition of the contributions of all sources distributed over all space and positive time.

In another scenario, the region of interest V contains no sources, is multiply-connected,[14] and is again open to infinity as shown in Fig. 1.12-b. The sources are in the exterior simply-connected volume V^e, which is enclosed by the boundary S. However, in the interior solution, the integration with respect to $\mathbf{r}'$ is over V. Thus there is no direct contribution to the field solution from the sources in V^e. Their contribution is now indirect in the form of inhomogeneous boundary conditions on S, which "absorb" the source contributions not only from all $\mathbf{r}' \notin V$ but also for all $t' \le 0$ (if any). We assume

[14] A multiply-connected domain is the one that contains closed paths (or contours) that cannot be continuously deformed to a point. Such a closed path is shown in Fig. 1.12-b with a dotted line. A simply-connected domain is a domain that is not multiply-connected.

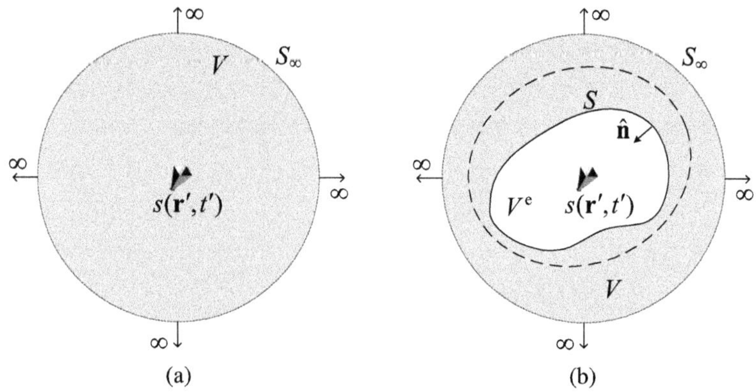

Figure 1.12 Illustration of the regions of interest in two radiation problems. (a) The region of interest is all space represented by the simply-connected region V, the boundary of which S_∞ extends to infinity. (b) The region of interest V is multiply-connected and open to infinity. The sources are outside of V enclosed in the volume V^e bounded by the surface S. The dotted-line contour is an example of a closed path in a multiply-connected region that cannot be shrunk to a point.

that we have no information about the turn-on and turn-off times of the sources exterior to V, and we set the limits of the time integration as $T_{\min} \to -\infty$ and $T_{\max} \to +\infty$. The resulting integral expression is

$$u_+(\mathbf{r},t) = \int_{-\infty}^{+\infty} \oiint_S \left(u_+(\mathbf{r}',t') \frac{\partial g_+(\mathbf{r},t;\mathbf{r}',t')}{\partial n} - g_+(\mathbf{r},t;\mathbf{r}',t') \frac{\partial u_+(\mathbf{r}',t')}{\partial n} \right) ds' dt'. \tag{1.113}$$

Here, it is the first (boundary-value) term of $\mathcal{I}_{\partial\mathcal{D}}$, see Eq. (1.111), that contributes to the field solution. The second term of $\mathcal{I}_{\partial\mathcal{D}}$ is zero. Similarly to the previous case, this is because at $t' = T_{\min} = -\infty$, the initial values of the field (u_+ and $\frac{\partial u_+}{\partial t'}$) are zero, while at $t' = T_{\max} \to \infty$, Green's function and its time derivative are zero.

Eq. (1.113) is the time-domain form of the Kirchhoff–Helmholtz integral solution of the interior problem. It shows that the original problem of finding the radiation field of a given source can be replaced by the equivalent boundary-value problem where these sources are now external to the domain of interest and their impact is represented through the field boundary values at S. This source representation through boundary conditions is the basis of the *surface equivalence principle* as applied to the wave equation. In-depth discussion on the surface equivalence is given later in the framework of Helmholtz's equation.

1.6.4 Integral-Equation Model of Transient Back-Propagation

Let us consider the region V bounded by the finite surface S as shown in Fig. 1.13-a. V contains sources that are of limited extent in space and in time, i.e., $t' \in [0, T_s]$,

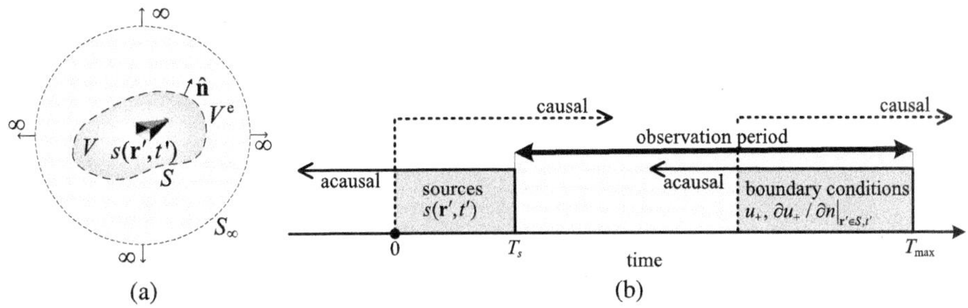

Figure 1.13 Illustration of the region of interest in the back-propagation problem. (a) The region of interest is the simply-connected volume V, the boundary of which S is finite. V contains unknown sources described by $s(\mathbf{r}', t')$, which are nonzero in the interval $[0, T_s]$. (b) The time intervals in which the causal and acausal waves exist as excited by the sources and by the boundary conditions on S.

see Fig. 1.13-b. Here, we assume that the sources are unknown. This is the case in the inverse source problem where the sources are to be reconstructed from field measurements on a surface (the surface S) enclosing the source but well removed from it. These measurements are referred to as the *data*, and the surface S is the *acquisition surface*.

The main idea of the back-propagation method for the solution of the inverse source problem is to find the back-propagated replica of the causal field $u_+(\mathbf{r}, t)$ from the data i.e., its boundary values, recorded during some time period $[0, T_{\text{max}}]$. It relies on the fact that, in principle, it is possible to acquire and store the field values as a function of time at a boundary surface removed from the sources. Back-propagating these values allows for retracing the field history back to its sources. The process is similar to a movie being played backward.

Fig. 1.14 illustrates the forward and backward temporal progressions of a 1D causal field $u_+(z, t) = f(t - z/v)$, $z \geq 0$, excited by a point source at $z = 0$. Forward time flow (increasing t) leads to the waveform translation toward greater z values, i.e., away from the source. Mathematically, this is facilitated by Green's causal function $g_+(z, t; z', t')$.

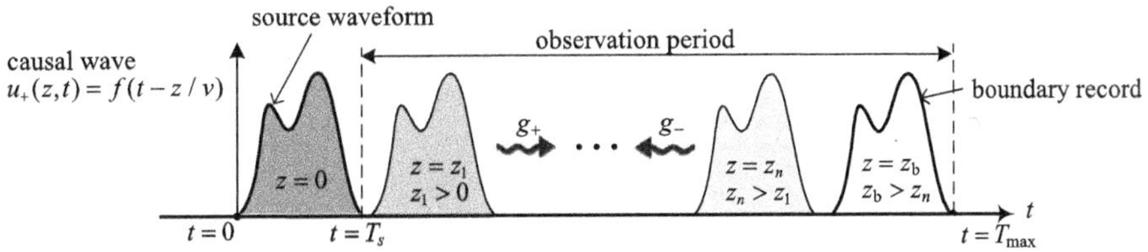

Figure 1.14 Illustration of the temporal dependence of a 1D causal wave propagating along $+z$. The source is at the origin $z = 0$, while the boundary point is at $z_b > 0$. Increasing time results in a waveform traveling forward, i.e., away from the source and toward the boundary, while decreasing time translates the waveform backward, i.e., toward the source. The waveform translation away from the source as time increases is mathematically described by Green's causal function g_+ in the integral solution. Green's acausal function g_- can "translate" the waveform in an equivalent manner, i.e., toward the source as time decreases. However, this acausal, yet equivalent, solution is excited by the boundary conditions.

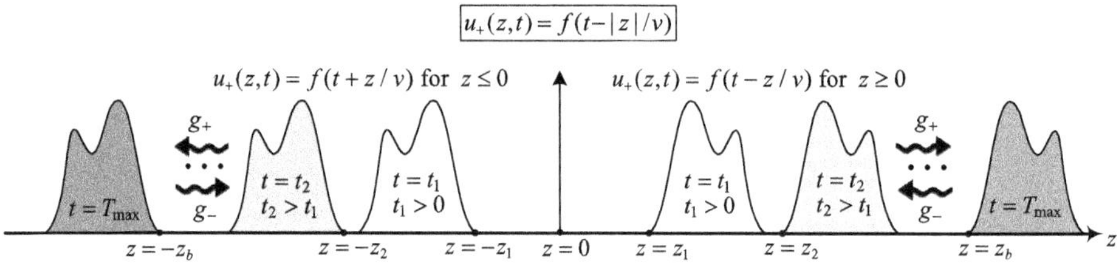

Figure 1.15 Illustration of the spatial dependence of a 1D causal wave spreading away from a source centered at $z = 0$. Notice that the disturbance clears out the region of interest $[-z_b, z_b]$ after T_{max}.

Eventually, the disturbance passes through the boundary point z_b, where its waveform is recorded to produce the boundary data. The same wave field, however, can start on the right of the time axis (at $t = T_{max}$) at the boundary point z_b. Thereafter, by decreasing t (time flow toward past), the waveform $f(t - z/v)$ is being translated toward smaller z values, away from the boundary and toward the source. Mathematically, this is facilitated by Green's acausal function $g_-(z, t; z', t')$. This latter process describes exactly the same wave solution $u_+(z, t)$, only that time "steps" backward. This is because g_- and g_+ relate through Eq. (1.96), i.e., $g_-(z, t; z', t') = g_+(z, -t; z', -t')$.

Note that a point source in 1D open space excites two waves: one along $z \geq 0$ (as described earlier) and one along $-z$. Both waves travel away from the source because both are causal. They are mirror images of each other: $f(t - z/v)$ is the wave traveling along $z \geq 0$, while $f(t + z/v)$ is the wave traveling along $z \leq 0$. This is illustrated in Fig. 1.15, which shows the wave development in space, i.e., along z. The change in the position of the wave along z is shown when time steps forward (arrow with g_+) and when time steps backward (arrow with g_-).

Assume that T_{max} is sufficiently large to allow for a complete recording of the boundary data. This means that after T_{max} the data have decayed sufficiently to be considered negligible and to be set to zero, i.e., $u_+ = 0$ and $\frac{\partial u_+}{\partial n} = 0$ on S for $t > T_{max}$. Since u_+ is causal and propagating away from the source, this also implies that it is zero inside V for $t > T_{max}$. Due to the finite speed of propagation, the data at S are delayed with respect to the sources, as is illustrated in Fig. 1.13-b and in Fig. 1.14. The time separation between the two would be greater for greater distances between the source region and the boundary.

Mathematically, back-propagation rests on the uniqueness theorem of the wave equation. Consider the wave solution u_+ in V in the observation period $T_s < t \leq T_{max}$ (indicated in both Fig. 1.13-b and Fig. 1.14). In this period, there are no active sources, and u_+ satisfies the homogeneous wave equation. On one hand, it can be expressed through the source function $s(\mathbf{r}, t)$ and Green's causal function as

$$u_+(\mathbf{r}, t) = \iiiint_{0\,V}^{T_s} g_+(\mathbf{r}, t; \mathbf{r}', t') s(\mathbf{r}', t') dv' dt' \,, \ \mathbf{r} \in V \wedge t \in (T_s, T_{max}] \,, \qquad (1.114)$$

as per Eq. (1.112). On the other hand, it can also be expressed in terms of the boundary data and Green's acausal function g_-, which relates to g_+ via Eq. (1.91). This is the back-propagated field:

$$u_b(\mathbf{r}, t) = \int_0^{T_{max}} \oiint_S \left(u_+ \frac{\partial g_-}{\partial n} - g_- \frac{\partial u_+}{\partial n} \right) ds' dt', \mathbf{r} \in V \wedge t \in (T_s, T_{max}], \tag{1.115}$$

where the arguments of u_+ and g_- in the integral are $(\mathbf{r}', t')$ and $(\mathbf{r}, t; \mathbf{r}', t')$, respectively. According to the uniqueness theorem, if both solutions, u_+ and u_b, satisfy the same homogeneous wave equations and the same boundary conditions on S, they are identical. Since the boundary values u_+ and $\frac{\partial u_+}{\partial n}$ in Eq. (1.115) are assigned as those of u_b, the equivalence of the two solutions is ensured.

Note that for $t \in (T_s, T_{max}]$, u_+ and u_b are both causal with respect to the source $s(\mathbf{r}', t'), t' \in [0, T_s]$ whereas they precede the boundary data $u_+(\mathbf{r} \in S, t'), t' > t$. Thus, if the boundary data are seen as a surface source, u_+ and u_b represent an acausal solution. As a word of caution, the back-propagated field $u_b(\mathbf{r}, t)$, although it employs Green's acausal function, is *not* the adjoint counterpart $u_-(\mathbf{r}, t)$ of $u_+(\mathbf{r}, t)$. For a detailed discussion of the adjoint acausal solutions, refer to Appendix D. The adjoint solution u_- is *acausal with respect to the source* $s(\mathbf{r}', t')$ and is defined by

$$u_-(\mathbf{r}, t) = \int_0^{T_s} \iiint_V g_-(\mathbf{r}, t; \mathbf{r}', t') s(\mathbf{r}', t') dv' dt' = \int_0^{T_s} \iiint_V g_+(\mathbf{r}', t'; \mathbf{r}, t) s(\mathbf{r}', t') dv' dt'. \tag{1.116}$$

We see that u_- in Eq. (1.116) is not the same as u_+ in Eq. (1.114) because $g_+(\mathbf{r}', t'; \mathbf{r}, t) \neq g_+(\mathbf{r}, t; \mathbf{r}', t')$. Since $u_b \equiv u_+$ for $t \in (T_s, T_{max}]$, u_- is also not the same as u_b.

In fact, as shown in Appendix D, $u_-(\mathbf{r}, t)$ can be expressed as

$$u_-(\mathbf{r}, \tau) = \int_0^{T_s} \iiint_V g_+(\mathbf{r}, \tau; \mathbf{r}', \tau') s^\dagger(\mathbf{r}', \tau') dv' d\tau', \tag{1.117}$$

where $\tau = T_{max} - t$ is the reversed time variable and $s^\dagger(\mathbf{r}', \tau') = s(\mathbf{r},' T_{max} - t')$ is the adjoint (or time-reversed) source. Thus, u_- relates to u_+ and therefore to the back-propagated field u_b as

$$u_-(\mathbf{r}, T_{max} - t) = u_+(\mathbf{r}, t) = u_b(\mathbf{r}, t), \; t \in (T_s, T_{max}]. \tag{1.118}$$

As such, at each observation point $\mathbf{r}$, the u_- temporal dependence appears as a "flipped" (or mirror) copy of u_+ (or u_b). For an illustration, see Appendix D, Fig. D.1.

This discussion about the adjoint field is not without purpose. Just like u_b in Eq. (1.115), the adjoint solution u_- can be expressed in terms of the field boundary values as

$$u_-(\mathbf{r},t)=\int_0^{T_{\max}}\oiint_S\left[u_+(\mathbf{r},T_{\max}-t')\frac{\partial g_-(\mathbf{r},t;\mathbf{r}',t')}{\partial n}\right.$$
$$\left.-g_-(\mathbf{r},t;\mathbf{r}',t')\frac{\partial u_+(\mathbf{r},T_{\max}-t')}{\partial n}\right]ds'dt',\ \mathbf{r}\in V\wedge t\in(T_s,T_{\max}]. \quad (1.119)$$

In view of Eq. (1.118) and the relation between Green's causal and acausal functions Eq. (1.98), the back-propagated field from Eq. (1.119) can be expressed as

$$u_b(\mathbf{r},\tau)=\int_0^{T_{\max}}\oiint_S\left[u_+(\mathbf{r},T_{\max}-t')\frac{\partial g_+(\mathbf{r},-t;\mathbf{r}',-t')}{\partial n}\right.$$
$$\left.-g_+(\mathbf{r},-t;\mathbf{r}',-t')\frac{\partial u_+(\mathbf{r},T_{\max}-t')}{\partial n}\right]ds'dt',\ \tau=T_{\max}-t,\ \mathbf{r}\in V\wedge t\in(T_s,T_{\max}]. \quad (1.120)$$

If the medium is loss-free, Green's function satisfies $g_+(\mathbf{r},-t;\mathbf{r}',-t')=g_+(\mathbf{r},t;\mathbf{r}',t')$, and Eq. (1.119) can be written as

$$u_b(\mathbf{r},\tau)=\int_0^{T_{\max}}\oiint_S\left[u_+(\mathbf{r},T_{\max}-t')\frac{\partial g_+(\mathbf{r},t;\mathbf{r}',t')}{\partial n}\right.$$
$$\left.-g_+(\mathbf{r},t;\mathbf{r}',t')\frac{\partial u_+(\mathbf{r},T_{\max}-t')}{\partial n}\right]ds'dt',\ \tau=T_{\max}-t,\ \mathbf{r}\in V\wedge t\in(T_s,T_{\max}]. \quad (1.121)$$

Eq. (1.121) is a remarkable result. It gives us a model that employs Green's causal function while producing the desired back-propagated field. What we need in Eq. (1.121) is the boundary data in *reversed order in time*. Reversing the time sequence is implemented through the last-in-first-out sequencing of the boundary excitation represented by $u_+(\mathbf{r}\in S,\tau')$ or $\frac{\partial u_+(\mathbf{r}\in S,\tau')}{\partial n}$, $\tau'=T_{\max}-t'$. Remember that, as per the uniqueness theorem of the wave equation, we need *either* u_+ at S *or* its normal derivative $\frac{\partial u_+}{\partial n}$, not both, in order to solve for u_b. The model runs in forward time, but the solution observed is the wave propagating back toward the source so that in early time (τ is large) the wave is observed close to the boundary and in late time (τ is small) the wave enters the source region. This is just like a movie being played backward.

This model is useful in two ways. First, we can generate the back-propagated field experimentally by "injecting" the boundary data in *reversed time sequence* into the region of interest V. The so-generated field converges toward the sources that generated it in the first place. For instance, in a scattering experiment, an unknown object is illuminated and the field it scatters is recorded by receivers at an acquisition surface S, which defines the boundary of V. In the subsequent back-propagation experiment, the receivers act as transmitters where the time-reversed recorded waveforms are injected back into V. The so-generated field is a very special kind of illumination because it naturally "focuses" onto the scatterers, "ignoring" heterogeneities in the background medium. As a result, the scattering from the unknown object is enhanced in this second experiment, and it can be further processed to produce an image. Such a strategy for the solution of the inverse scattering problem is known as *experimental time reversal*.

Second, we can use a time-domain field simulator to model numerically the back-propagated field. Just like in the back-propagation experiment, the observation points at S are used as excitation points, the waveforms of which are the time-reversed responses. The simulation shows explicitly the locations toward which the back-propagated field converges, i.e., the locations of the scatterers. At the end of this section, a simple 1D example illustrates the concept of the back-propagated field through time-domain simulations that employ the excitation with reversed-in-time boundary data. This strategy for the solution of the inverse scattering problem is known as *computational time-reversal imaging*.

It should be reiterated that the time-reversal strategies rely on the assumption that the losses in the propagation medium are negligible. This limits their applicability.

We now return to Eq. (1.114) and Eq. (1.115) to match their right-hand sides and to obtain an integral equation relating the unknown source $s(\mathbf{r}, t)$ to the data:

$$\int_0^{T_s}\iiint_V g_+ s\, dv' dt' = \int_0^{T_{\max}}\oiint_S \left(u_+\frac{\partial g_-}{\partial n} - g_-\frac{\partial u_+}{\partial n}\right) ds' dt', \mathbf{r} \in V \wedge t \in (T_s, T_{\max}]. \tag{1.122}$$

This integral equation expresses the underlying principle of the back-propagation (or time-reversal) methods used to solve inverse source problems. It shows that there is a linear relationship between the unknown source $s(\mathbf{r}, t)$ and the field boundary values at S (the data). Provided that the medium's Green function g_+ and its acausal counterpart g_- are known, in principle it is possible to reconstruct the source from the measured data. Here, brief comments are given on some misunderstandings and principal difficulties associated with this solution.

First, it is important to remember that Eq. (1.122) holds only for $t \in (T_s, T_{\max}]$. For $t \leq T_s$, the sources are not zero, and the back-propagated solution obtained from the data no longer matches the causal solution u_+. This indicates the first difficulty with practical implementation—the start and end times of the excitation ($t = 0$ and $t = T_s$) are known only approximately at best. As long as $t \in (T_s, T_{\max}]$, the back-propagated wave u_b is an exact copy of the causal solution u_+ and it converges toward the sources as time "steps" backward. Once t gets smaller than T_s, however, the differences between u_+ and u_b start to increase. Finally, when $t < 0$, the back-propagated field starts to diverge from the source region, and it has nothing in common with the real solution, which is zero.

These points are illustrated through an example with a uniform plane wave (1D wave) generated by a point source at $x = 0$ and propagating along $+x$ and $-x$. The simulation

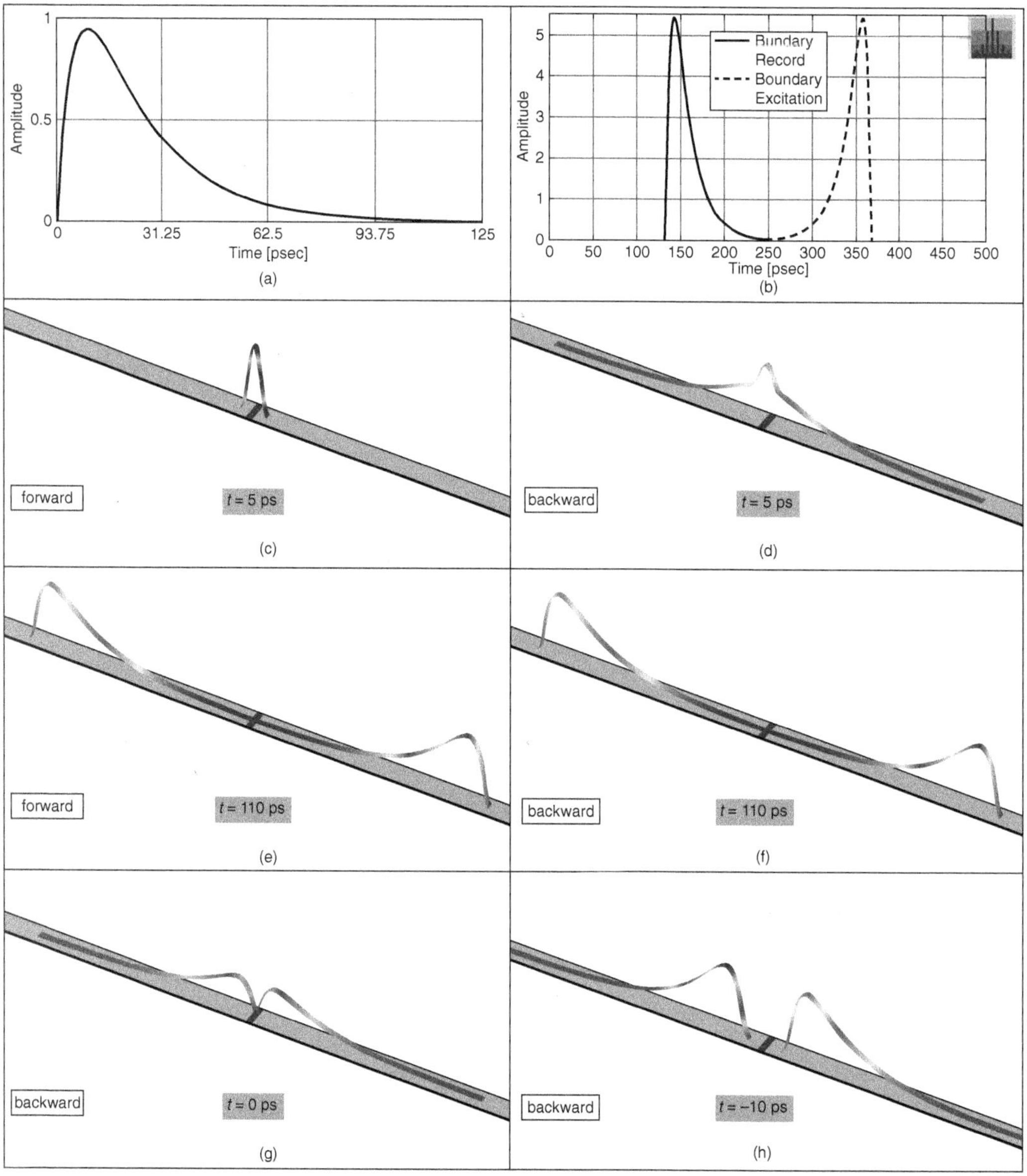

Figure 1.16 1D example illustrating the differences between the original solution $u_+(x, t)$ due to an excitation at $x = 0$ (shown as a red strip in the middle of the line) and the back-propagated solution $u_b(x, t)$. (a) The excitation waveform. (b) The boundary record (the data) together with the boundary excitation, which is a time-reversed version of the data. (c) $u_+(x, t)$ at an early stage of development, $t = 5$ ps. (d) $u_b(x, t)$ at $t = 5$ ps exhibiting significant differences from $u_+(x, t)$ at $t = 5$ ps (see (c) on the left). (e) $u_+(x, t)$ at a later time, $t = 110$ ps, showing the two waves (along $+x$ and along $-x$) diverging from the source. (f) $u_b(x, t)$ at $t = 110$ ps showing no difference from $u_+(x, t)$ (see (e) on the left). (g) $u_b(x, t)$ at $t = 0$ clearly showing nonzero values. Note that at $t = 0$, $u_+(x, 0) = 0$. (h) $u_b(x, t)$ at $t = -10$ ps, which, unlike u_+, is nonzero for negative time and diverges from the source.

is performed with the time-domain EM simulator MEFiSTo-3D [10, 21].[15] Fig. 1.16-a shows the excitation waveform applied at $x = 0$. The source turn-off time T_s is at about 125 ps. The source (indicated by a narrow red strip) is located at the center of the transmission-line structure. Fig. 1.16-b shows the boundary data acquired at $x = 40$ mm. The data at $x = -40$ mm are the same due to the problem symmetry. These data are used to launch the back-propagated wave from both sides (the left and right boundaries) in a forward-in-time simulation as predicted by Eq. (1.121). The reader is reminded that in the back-propagation simulation, the temporal coordinate is assumed to be $\tau = T_{\max} - t$ (reversed time).

Figs. 1.16-c,e show snapshots of the 1D forward wave launched by the source at the origin. This is the original wave that the back-propagated wave must match for $t \geq T_s$. We observe in Fig. 1.16-e that at $t = 110$ ps the wave portions along $+x$ and $-x$ are already well formed and they move away from the source.

Figs. 1.16-d,f,g,h show snapshots of the 1D back-propagated wave as time t decreases (τ increases). In the back-propagation simulation, the source is off at all times—remember that it is unknown in the inverse source problem. At $t = 110$ ps, which is already inside the temporal interval of the excitation ($t < T_s \approx 125$ ps), the forward and the back-propagated fields are still practically indistinguishable; see Fig. 1.16-e, f. However, clear differences are observed between the two waves at $t = 5$ ps, which is well within the time period $[0, T_s]$; compare Fig. 1.16-c with Fig. 1.16-d. Moreover, the back-propagated wave is not extinguished for $t \leq 0$; see Fig. 1.16-g. In fact, it goes through the source region unabated and moves on to diverge from the source as the simulation steps back in time t (forward in τ); see Fig. 1.16-h.

In summary, to solve the inverse source problem in the time domain, it is imperative to have at least approximate knowledge of when the sources to be reconstructed are being turned on and off. In addition, it is important to have sufficient distance between the sources and the surface S, which would provide a long enough observation period $(T_s, T_{\max}]$ during which the match between the original forward solution and its back-propagated version is exact.

The second practical difficulty arises because the boundary data are inevitably incomplete due to limitations in the sampling in space and time. In addition, noise and measurement uncertainties corrupt the data quality. These factors contribute to the ill-posed nature of the inverse problem.[16]

Third, there are certain types of sources, called *nonradiating sources* [26], whose field is zero outside the source support, i.e., the source extent in space and in time. Such sources exist for all physical wave fields, the EM field included. Since these sources do not radiate, the data collected at the acquisition surface are unaffected by their presence. The reader is reminded that the acquisition surface must be spatially well removed from

[15] MEFiSTo-3D is based on the transmission-line matrix method (TLM) [22, 23, 24].

[16] An operator equation $K(f) = g$, $K : \mathcal{H}_1 \to \mathcal{H}_2$, is said to be well posed if [25] (i) for each $g \in \mathcal{H}_2$ there exists a solution $f \in \mathcal{H}_1$; (ii) the solution f is unique; and (iii) the solution is stable with respect to perturbations in g. Condition (iii) means that small changes in g must lead to small changes in f. Mathematically, the operator equation is stable if in the vicinity of a solution f^* satisfying $Kf^* = g^*$, the solution obeys $f \to f^*$ whenever $g \to g^*$. A problem that is not well posed is said to be ill posed.

the source. Consequently, the inverse problem is not unique. One can always add a non-radiating component to the source solution, and the resulting new solution will be valid as well. In MW radiation, even if the source is radiating, its field always has components that attenuate quickly with distance. These are the near-zone and the evanescent field components. The values of these field components are often well below the noise floor of the measurement setup and are thus lost. Due to the loss of near-field data, critical information about the source is often missing, making the solution of the inverse source problem challenging.

As a final note, the back-propagation modeling, just like the forward one, may use any valid numerical method. The integral-equation models, Eq. (1.121) and Eq. (1.122), discussed earlier are used to simply provide a mathematical justification of the solution to the inverse source problem through back-propagation of the boundary data. However, in time-domain analysis, numerical methods based on the volume discretization of partial differential equations, i.e., the wave equation, are common. Examples include the finite-difference time-domain (FDTD) method [27], the finite-element time-domain (FETD) method [28], and the transmission-line matrix (TLM) method [22] (used to generate the example in Fig. 1.16).

1.6.5 The Initial-Value Problem

The initial-value problem arises in the special case of Eq. (1.109) where there are no sources present, $s(\mathbf{r}', t') = 0$, throughout all space (the volume of interest V extends to infinity) and for all time beyond some initial instance T_0, i.e., $t, t' \geq T_0$. In this case, Eq. (1.109) reduces to

$$u_+(\mathbf{r}, t) = -\iiint\limits_V \frac{1}{v^2}\left[u_+(\mathbf{r}', t')\frac{\partial g_+(\mathbf{r}, t; \mathbf{r}', t')}{\partial t'} - g_+(\mathbf{r}, t; \mathbf{r}', t')\frac{\partial u_+(\mathbf{r}', t')}{\partial t'}\right]_{t'=T_0}^{\infty} dv'. \tag{1.123}$$

The contribution of the volume integral above vanishes for the upper limit $t' \to \infty$ due to the causality of Green's function. Thus, the only factors determining the causal field solution are its initial values at T_0:

$$u_+(\mathbf{r}, t) = \iiint\limits_V \frac{1}{v^2}\left[u_+(\mathbf{r}', T_0)\frac{\partial g_+(\mathbf{r}, t; \mathbf{r}', T_0)}{\partial t'} - g_+(\mathbf{r}, t; \mathbf{r}', T_0)\frac{\partial u_+(\mathbf{r}', T_0)}{\partial t'}\right] dv'. \tag{1.124}$$

Eq. (1.124) is referred to as the integral solution to the *Cauchy initial-value problem*. It tells us that if the values of a causal field are known throughout space at a given initial moment of time T_0, then this field can be determined uniquely for all future time instances, $t > T_0$.

1.6.6 The Initial-Value Problem—Back-Propagation

A back-propagated version of the Cauchy initial-value solution can also be obtained. Assume that the field is known in all space at $t = T_{\max}$, and we would like to recover its values for $\forall t < T_{\max}$. Since this solution exists only for time instants preceding $T_{\max}$, it would seem more appropriate to call it a solution to the Cauchy terminal-value problem. Just like the case of back-propagation of the field from its boundary values (see Section 1.6.4), this is facilitated by Green's acausal function:

$$u_{\mathrm{b}}(\mathbf{r},t) = \iiint\limits_V \frac{1}{\upsilon^2}\left[u_+(\mathbf{r}', T_{\max})\frac{\partial g_-(\mathbf{r},t;\mathbf{r}',T_{\max})}{\partial t'} - g_-(\mathbf{r},t;\mathbf{r}',T_{\max})\frac{\partial u_+(\mathbf{r}',T_{\max})}{\partial t'}\right]dv'. \tag{1.125}$$

Here, $u_{\mathrm{b}}(\mathbf{r},t)$ is the back-propagated field while $u_+(\mathbf{r}',T_{\max})$ and $\frac{\partial u_+(\mathbf{r}',T_{\max})}{\partial t'}$ are the terminal values of a causal field that we assume are available through measurement.

Let us now represent the causal field u_+, the terminal values of which are used in Eq. (1.125), as a Cauchy initial-value integral with initial values at $t = 0$. As per Eq. (1.124) where $T_0 = 0$, we have

$$u_+(\mathbf{r},t) = \iiint\limits_V \frac{1}{\upsilon^2}\left[u_+(\mathbf{r}',0)\frac{\partial g_+(\mathbf{r},t;\mathbf{r}',0)}{\partial t'} - g_+(\mathbf{r},t;\mathbf{r}',0)\frac{\partial u_+(\mathbf{r}',0)}{\partial t'}\right]dv',\, t\in[0,T_{\max}]. \tag{1.126}$$

We now prove that $u_{\mathrm{b}}(\mathbf{r},t)$ obtained from Eq. (1.125) using the field *terminal* values at $T_{\max}$ is identical to $u_+(\mathbf{r},t)$ in Eq. (1.126). Using Eq. (1.96), the back-propagated solution is written as

$$u_{\mathrm{b}}(\mathbf{r},-t) = \iiint\limits_V \frac{1}{\upsilon^2}\left[u_+(\mathbf{r}',-T_{\max})\frac{\partial g_+(\mathbf{r},-t;\mathbf{r}',-T_{\max})}{\partial(-t')} - \right.$$
$$\left. g_+(\mathbf{r},-t;\mathbf{r}',-T_{\max})\frac{\partial u_+(\mathbf{r}',-T_{\max})}{\partial(-t')}\right]dv'. \tag{1.127}$$

We now translate all time dependencies by $T_{\max}$ (or, equivalently, shift the origin of the time axis to the left by $T_{\max}$) to obtain

$$u_{\mathrm{b}}(\mathbf{r},\tau) = \iiint\limits_V \frac{1}{\upsilon^2}\left[u_+(\mathbf{r}',0)\frac{\partial g_+(\mathbf{r},\tau;\mathbf{r}',0)}{\partial \tau} - g_+(\mathbf{r},\tau;\mathbf{r}',0)\frac{\partial u_+(\mathbf{r}',0)}{\partial \tau}\right]dv', \tag{1.128}$$

where we have introduced the time variable $\tau = T_{\max} - t$, $\tau \in [0, T_{\max}]$. Comparing Eq. (1.128) and Eq. (1.126), we see that the back-propagated field u_{b} would indeed reproduce the original causal field u_+.

To summarize, in principle, knowing the field "measured" values in all space at a given instant $T_{\max}$ allows for recovering uniquely its *past* values in all space back to a time instant at which all sources are off. In practice, it is difficult to measure and record the field values in all space at a given instant. In simulations, however, this is not a problem. This is true in particular for time-domain simulators that employ volume discretization; for example, the FDTD, FETD, and TLM methods [29]. Many examples in this book use the commercial simulator MEFiSTo-3D [10, 21], and one reason is that this simulator can run both forward and backward in time starting from the field distribution in the analyzed volume at the current iteration (or time step).

1.7 Integral Solutions to the Helmholtz Equation

For the Helmholtz operator $\mathcal{L} = \nabla^2 - \gamma^2$, the domain is 3D space, $\mathcal{D} \subset \mathbb{R}^3$, and $\partial\mathcal{D}$ is the surface enclosing $\mathcal{D}$. To make the notations consistent with those of the wave equation, we replace the symbols $\mathcal{D}$ and $\partial\mathcal{D}$ with V and S, respectively.

1.7.1 Interior and Exterior Kirchhoff–Helmholtz Equations

The *interior* integral solution $U(\mathbf{r})$ to the inhomogeneous Helmholtz equation with a source function $\mathcal{S}(\mathbf{r}')$ is obtained as a particular case of Eq. (1.60):

$$U(\mathbf{r}) = \iiint_V \mathcal{S}(\mathbf{r}')G(\mathbf{r}', \mathbf{r})dv' + \oiint_S \left[U(\mathbf{r}')\frac{\partial G(\mathbf{r}', \mathbf{r})}{\partial n} - G(\mathbf{r}', \mathbf{r})\frac{\partial U(\mathbf{r}')}{\partial n}\right] ds', \ \mathbf{r} \in V. \quad (1.129)$$

In the above, we can replace $G(\mathbf{r}', \mathbf{r})$ with $G(\mathbf{r}, \mathbf{r}')$ and still have a valid solution due to the reciprocity principle; see Eq. (1.62). The reader is reminded that the directional derivatives $\frac{\partial f(\mathbf{r}' \in S)}{\partial n}$, $f \equiv U, G$, are taken with respect to the *outward* unit normal $\hat{\mathbf{n}}$ on S bounding V.

Note that even if $\mathcal{S}(\mathbf{r}') = 0$ (no sources in V), the solution is not trivial if $U(\mathbf{r})$ along with Green's function satisfies nonzero boundary conditions such that $f(\mathbf{r}' \in S) \neq 0$ or $\frac{\partial f(\mathbf{r}' \in S)}{\partial n} \neq 0$, $f \equiv U, G$. This is the *boundary-value problem*.

We can interpret Eq. (1.129) as the superposition of two solutions: (i) the volume integral is the particular solution to the inhomogeneous equation $(\nabla^2 - \gamma^2)U = S$ representing the contribution of all sources *inside* V, and (ii) the surface integral is the solution to the boundary-value problem $(\nabla^2 - \gamma^2)U = 0$, the inhomogeneous boundary conditions of which represent the contribution of all sources *outside* V.

We next consider the *exterior* integral solution to the Helmholtz equation, which is a particular case of Eq. (1.61) and can be stated as

$$0 = \iiint_V \mathcal{S}(\mathbf{r}')G(\mathbf{r}, \mathbf{r}')dv' + \oiint_S \left[U(\mathbf{r}')\frac{\partial G(\mathbf{r}, \mathbf{r}')}{\partial n} - G(\mathbf{r}, \mathbf{r}')\frac{\partial U(\mathbf{r}')}{\partial n}\right] ds', \ \mathbf{r} \notin V. \quad (1.130)$$

Eq. (1.129) and Eq. (1.130) are known as the interior and exterior Kirchhoff–Helmholtz integral equations, respectively.

Note that the condition $\mathbf{r} \notin V$ in Eq. (1.130) implies that V does not occupy all space and a complement exists V^e such that $V \cup V^e$ comprise all space. If V extends to infinity, then the exterior solution becomes irrelevant.

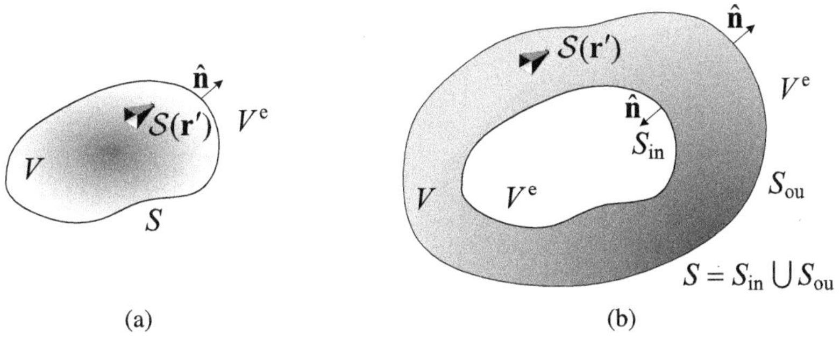

Figure 1.17 Illustrations of the interior V and exterior V^e regions associated with the closed surface S in 3D space. (a) The interior domain V (shaded in gray) is simply-connected while the exterior domain V^e is multiply-connected and infinite. (b) The interior domain V (shaded in gray) is multiply-connected and so is the exterior domain V^e, which has a finite part (enclosed by S_{in}) and an infinite part (outside S_{ou}).

Fig. 1.17 illustrates two possible scenarios for the region of interest V. In Fig. 1.17-a, V is finite and simply-connected. It is enclosed by the surface S, which separates it from V^e, which is multiply-connected. If S expands to infinity, V occupies all space, and the solution in V^e becomes irrelevant. Fig. 1.17-b illustrates a different scenario where V is multiply-connected and "sandwiched" between two non-intersecting closed surfaces, S_{in} and S_{ou}. Now both V and V^e are multiply-connected. Various other arrangements for V and its complement V^e are possible, including, for example, interchanging V and V^e in Fig. 1.17-a and Fig. 1.17-b. What is important to remember is that a point belonging to V cannot belong to V^e and vice versa as well as that the surface unit normal $\hat{\mathbf{n}}$ points from V into V^e. Since the interior integral equation Eq. (1.129) produces a field solution for $\mathbf{r} \in V$, V is often referred to as the *interior region* whereas V^e is the *exterior region*.

It is emphasized that $\mathcal{S}(\mathbf{r}')$ in both Eq. (1.129) and Eq. (1.130) describes only the sources in V, as implied by the fact that the integration over $\mathbf{r}'$ is limited to the region V. This does not mean that there could be no sources in V^e. However, when solving for the field in V, the contribution of such external sources is represented by the inhomogeneous boundary conditions on S. In this regard, notice that the exterior Kirchhoff–Helmholtz equation Eq. (1.130) tells us that the contribution of the boundary conditions $U(\mathbf{r}' \in S) \neq 0$ and $\frac{\partial U(\mathbf{r}' \in S)}{\partial n} \neq 0$ to the field in V^e cannot be zero if sources are present in V. In fact, we can write the *interior* solution for an observation point $\mathbf{r} \in V^e$ as

$$U(\mathbf{r}) = -\oiint_S \left[U(\mathbf{r}') \frac{\partial G(\mathbf{r}, \mathbf{r}')}{\partial n} - G(\mathbf{r}, \mathbf{r}') \frac{\partial U(\mathbf{r}')}{\partial n} \right] ds', \ \mathbf{r} \in V^e. \tag{1.131}$$

In Eq. (1.131), the minus sign takes into account that the unit normal is still set to point from V into V^e. The comparison of Eq. (1.131) and Eq. (1.130) clearly shows that the impact the source in V has on the solution in V^e can be equivalently described through the boundary values of the field it generates.

This is in accord with our interpretation of Eq. (1.129) where we already observed that the interior solution to any problem is a superposition of the contributions of the

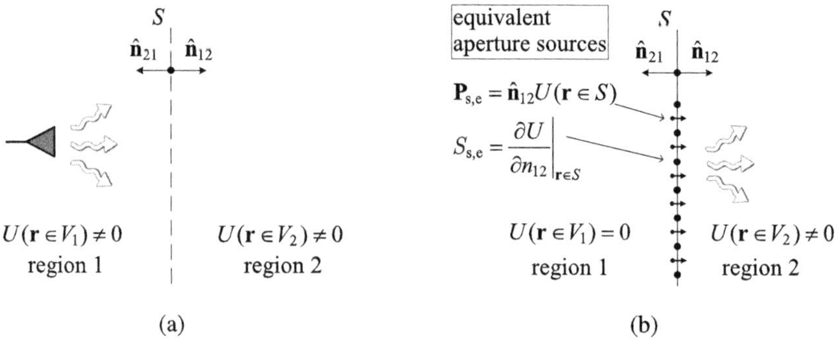

Figure 1.18 Illustrations of the equivalence principle in radiation problems where the emitter is replaced by surface sources on a fictitious surface S (the aperture) separating the source region V_1 from the observation region V_2. (a) The actual problem where the emitter radiates in open space and the field is nonzero in both V_1 and V_2. (b) The equivalent problem where the equivalent aperture sources radiate and where the field in V_2 is exactly the same as that in the actual problem. However, the field in V_1 is zero.

interior sources (the volume-integral term) and the sources lying outside of the considered domain through the boundary-integral term. This is the basis of the *surface equivalence principle* as applied to Helmholtz's equation.

1.7.2 Application of the Surface Equivalence Principle to Radiation

One common application of the equivalence principle is in radiation. Fig. 1.18-a shows the actual problem of an emitter radiating in open space. The boundary S is a fictitious surface at which the field of the emitter can be acquired either through measurements or through simulations. For simplicity, in this example the surface is shown as an infinite plane. Region 1 hosts the actual emitter. Region 2 contains no sources, but it is the region where we want to know certain characteristics of the radiated field. For example, we may want to know what is the far-field radiation pattern of the emitter.[17] Regions 1 and 2 comprise all space where the actual problem exists.

Fig. 1.18-b shows the equivalent problem where the equivalent aperture sources radiate. These sources are defined by $U(\mathbf{r} \in S)$ and $\frac{\partial U(\mathbf{r}\in S)}{\partial n}$ as explained shortly. Similarly to the actual emitter, they radiate in open space. However, they generate zero field in V_1, whereas in V_2 they generate exactly the same field as that of the actual emitter. This equivalent problem is nothing but the physical interpretation of the interior and exterior integral solutions written for region 2.

Table 1.1 summarizes the interior and exterior solutions written for region 1 (integration is over region 1, but the observation point can be either in region 1 or in region 2) and for region 2 (integration is over region 2, but the observation point can be either in region 2 or in region 1). Note that the unit normals $\hat{\mathbf{n}}_{12}$ and $\hat{\mathbf{n}}_{21}$ have opposite directions.

[17] The radiation pattern of an emitter is a plot of the normalized field intensity (or power density) as a function of the angular position of the observation point relative to the emitter.

Table 1.1 Interior and Exterior Solutions for Regions 1 and 2 in Fig. 1.18

	region 1
interior	$U(\mathbf{r}) = \iiint_{V_1} \mathcal{S}(\mathbf{r}')G(\mathbf{r},\mathbf{r}')dv' + \iint_S \left[U(\mathbf{r}')\frac{\partial G(\mathbf{r},\mathbf{r}')}{\partial n_{12}} - \frac{\partial U(\mathbf{r}')}{\partial n_{12}}G(\mathbf{r},\mathbf{r}')\right]ds', \mathbf{r} \in V_1$
exterior	$0 = \iiint_{V_1} \mathcal{S}(\mathbf{r}')G(\mathbf{r},\mathbf{r}')dv' + \iint_S \left[U(\mathbf{r}')\frac{\partial G(\mathbf{r},\mathbf{r}')}{\partial n_{12}} - \frac{\partial U(\mathbf{r}')}{\partial n_{12}}G(\mathbf{r},\mathbf{r}')\right]ds', \mathbf{r} \in V_2$
	region 2
interior	$U(\mathbf{r}) = \iint_S \left[U(\mathbf{r}')\frac{\partial G(\mathbf{r},\mathbf{r}')}{\partial n_{21}} - \frac{\partial U(\mathbf{r}')}{\partial n_{21}}G(\mathbf{r},\mathbf{r}')\right]ds', \mathbf{r} \in V_2$
exterior	$0 = \iint_S \left[U(\mathbf{r}')\frac{\partial G(\mathbf{r},\mathbf{r}')}{\partial n_{21}} - \frac{\partial U(\mathbf{r}')}{\partial n_{21}}G(\mathbf{r},\mathbf{r}')\right]ds', \mathbf{r} \in V_1$

They point from the respective interior region into the respective exterior region; see Fig. 1.18. Thus, the respective directional derivatives have opposite signs.

Notice that the interior solution for region 2 consists of the same surface integral (with a minus sign) as the second term in the exterior solution for region 1. Moreover, this second term is equal to the contribution of the source in region 1 with a minus sign. This is the mathematical expression of the surface equivalence principle when applied to radiation problems, which states that *the source in an actual problem can be equivalently replaced by surface sources derived from the field boundary values at the aperture.*

In order to understand the meaning of the field boundary values on S as equivalent surface sources, we inspect one more time the interior solution for region 2:

$$U(\mathbf{r}) = \iint_S \left[U(\mathbf{r}')\frac{\partial G(\mathbf{r},\mathbf{r}')}{\partial n_{21}} - \frac{\partial U(\mathbf{r}')}{\partial n_{21}}G(\mathbf{r},\mathbf{r}')\right]ds', \mathbf{r} \in V_2. \qquad (1.132)$$

As Fig. 1.18-b indicates, there is a discontinuity of the field at the surface S in the equivalent problem; namely, the field U and its normal derivative $\frac{\partial U}{\partial n}$ change abruptly from zero values in region 1 to nonzero values in region 2. The interface boundary conditions derived from Helmholtz's equations show that such discontinuity arises in the presence of surface distributions of sources; see, for example, [30].

With reference to Fig. 1.19-a, if a surface source distribution $\mathcal{S}_s$ is present at S, then the field normal derivative $\frac{\partial U}{\partial n}$ is discontinuous at S so that

$$\frac{\partial U^{(2)}}{\partial n} - \frac{\partial U^{(1)}}{\partial n} = \mathcal{S}_s, \ \mathbf{r} \in S, \qquad (1.133)$$

where $U^{(1)}$ is on the side of region 1, $U^{(2)}$ is on the side of region 2, and $\hat{\mathbf{n}}$ points from region 1 into region 2. Thus, the equivalent problem depicted in Fig. 1.18-b involves equivalent surface source such that

$$\mathcal{S}_{s,e} = \frac{\partial U}{\partial n_{12}} = -\frac{\partial U}{\partial n_{21}}. \qquad (1.134)$$

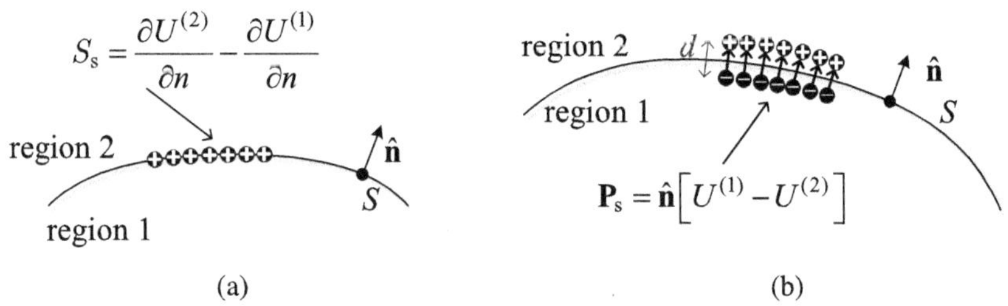

Figure 1.19 An illustration of the boundary conditions associated with sheet sources for Helmholtz's equation. (a) A single-layer surface distribution $\mathcal{S}_s$ causes discontinuity of the field normal derivative. The illustration shows a distribution of surface sources of positive values, but these can assume any complex value. (b) A double-layer (dipole) surface distribution $\mathbf{P}_s$ causes discontinuity of the field itself. By convention, the surface unit normal $\hat{\mathbf{n}}$ defines a boundary double-layer source of positive dipole moment. The distribution, however, can assume any complex value.

This surface source is responsible for the second term in the integral in Eq. (1.132).

Electrical engineers have a good grasp of the concept of surface source distributions (also referred to as source sheets) such as, for example, surface charge density ρ_S measured in *coulombs per square meter*. In general, for a given volume source distribution $\mathcal{S}(\mathbf{r})$ of negligible thickness h, a source sheet is mathematically defined as

$$\mathcal{S}_s(\mathbf{r}) = \lim_{h\to 0} \int_{-h/2}^{h/2} \mathcal{S}(\mathbf{r} + \xi\hat{\mathbf{n}})d\xi \approx \lim_{h\to 0} \left[h \cdot \mathcal{S}(\mathbf{r})\right]. \tag{1.135}$$

Conversely, the volume distribution can be expressed through the surface one using Dirac's δ-function:

$$\mathcal{S}(\mathbf{r}) = \mathcal{S}_s \cdot \delta(\xi - \xi_S), \tag{1.136}$$

where ξ is the coordinate along the surface unit normal and ξ_S is this coordinate at the surface. In the various fields of physics and mathematics, such a sheet source also bears the names *single layer*, *singlet*, or *monopole layer*.

Note that the substitution of Eq. (1.136) as a source term in the Helmholtz equation, $(\nabla^2 - \gamma^2)U = \mathcal{S}_s\delta(\xi - \xi_S)$, leads to the boundary condition Eq. (1.133) in a straightforward manner: the integration of the left side of this equation with respect to ξ over a vanishingly small interval centered on the source sheet yields $(\frac{\partial U^{(2)}}{\partial n} - \frac{\partial U^{(1)}}{\partial n})$ whereas the integration of the right side results in $\mathcal{S}_s$.

The physical meaning of the boundary condition Eq. (1.133) is illustrated in Fig. 1.20-a. There, the positive value of the source sheet $\mathcal{S}_s > 0$ forces the field negative normal derivative in region 1 to abruptly change to a positive one in region 2. To use the electric potential φ as an example, we recall that under Lorenz's gauge, φ satisfies the Helmholtz equation where the source function is $\mathcal{S} \equiv -\frac{\rho_V}{\varepsilon}$, ρ_V being the volume charge density (in C·m^{-3}). Correspondingly, $\mathcal{S}_s \equiv -\frac{\rho_s}{\varepsilon}$, i.e., a positive source sheet $\mathcal{S}_s$ represents a sheet of negative surface charge distribution. The dotted line in Fig. 1.20-a illustrates the

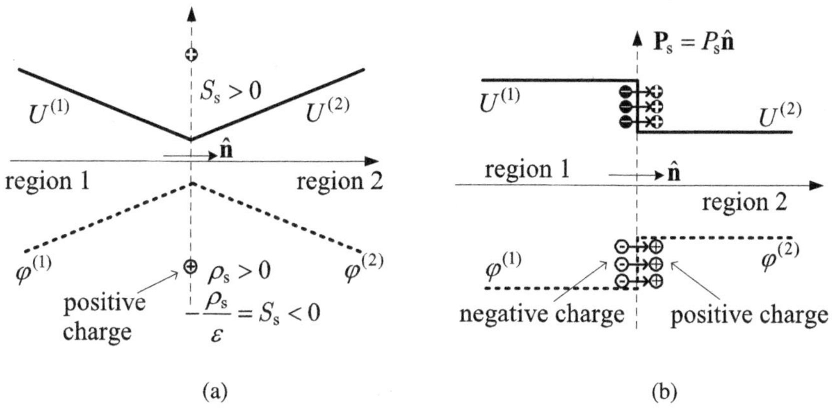

Figure 1.20 Illustration of the field discontinuities associated with the presence of sheet sources for the Helmholtz equation. (a) Single-layer source and the discontinuity in the field normal derivative: the solid line shows the field behavior associated with a positive single-layer source; the dotted line shows the behavior of the electric scalar potential in the vicinity of a positively charged sheet of density ρ_s, which corresponds to a negative surface source. (b) Double-layer source and the field discontinuity: the solid line shows the field discontinuity in the vicinity of a double layer of moment directed along $\hat{\mathbf{n}}$; the dotted line shows the behavior of the electric scalar potential in the vicinity of a layer of electric dipoles, the moment of which is also along $\hat{\mathbf{n}}$. However, this dipole moment corresponds to a double layer along $-\hat{\mathbf{n}}$.

behavior of φ, i.e., the jump in its normal derivative, in the vicinity of a sheet of *positive* surface charge.

Further, the field U itself is discontinuous on S if there is a sheet of dipole sources of moment $\mathbf{P}_s = P_s\hat{\mathbf{n}}$. Such a distribution consists of two single-layer sheets separated by an infinitesimal distance d (see Fig. 1.19-b). The mathematical definition of P_s is

$$P_s(\mathbf{r}) = \lim_{d\to 0} \left[d \cdot \mathcal{S}_s(\mathbf{r})\right], \tag{1.137}$$

and it is measured in the units of the field U itself. The polarity of the associated single-layer sheets and the direction of the moment are indicated in Fig. 1.19-b. Note that the convention is that the dipole moment points from the negative monopole toward the positive one. This sheet source is also referred to as *double layer*, *doublet*, or *dipole layer*.

The boundary condition in this case is [30][18]

$$\left(U^{(1)} - U^{(2)}\right)\hat{\mathbf{n}} = \mathbf{P}_s. \tag{1.138}$$

[18] The derivation of this boundary condition is straightforward. With reference to Fig. 1.20-b, let the position at the immediate left of the negative source sheet, $-\mathcal{S}_s$, be denoted as L (for left), the vanishingly thin region between the negative and positive source sheets be denoted as T (for transition), and the region on the immediate right of the positive source sheet, $+\mathcal{S}_s$, be denoted as R (for right). As per Eq. (1.133), we have that $\frac{\partial U_R}{\partial n} - \frac{\partial U_T}{\partial n} = \mathcal{S}_s$ and $\frac{\partial U_T}{\partial n} - \frac{\partial U_L}{\partial n} = -\mathcal{S}_s$. Adding these two equations leads to the conclusion that $\frac{\partial U_R}{\partial n} = \frac{\partial U_L}{\partial n}$. Further, subtracting them results in $\frac{\partial U_T}{\partial n} = \frac{\partial U_R}{\partial n} - \mathcal{S}_s$. Finally, the latter equation is multiplied by the distance d between the sheet sources and a limit is taken as $d \to 0$. The left-hand side results in $U^{(2)} - U^{(1)}$; the first term in the right-hand side vanishes, whereas the second term produces $-P_s$.

It is illustrated in Fig. 1.20-b. Thus, the equivalent problem in Fig. 1.18-b involves also an equivalent double-layer source such that

$$\mathbf{P}_{s,e} = -U\mathbf{n}_{12}. \tag{1.139}$$

In order to account for double-layer sources, we need the respective integral solution in terms of Green's function. This solution is well known [26, 30]:[19]

$$\iint_S \mathbf{P}_s(\mathbf{r}') \cdot \nabla_{\mathbf{r}'} G(\mathbf{r}, \mathbf{r}') ds'. \tag{1.140}$$

It is important to note that the gradient of Green's function above operates on $\mathbf{r}'$. In the case of the equivalent problem depicted in Fig. 1.18-b, the dipole surface source is given by Eq. (1.139), which leads to a field contribution in region 2 in the form

$$-\iint_S U(\mathbf{r}') \frac{\partial G(\mathbf{r}, \mathbf{r}')}{\partial n_{12}} ds' = \iint_S U(\mathbf{r}') \frac{\partial G(\mathbf{r}, \mathbf{r}')}{\partial n_{21}} ds'. \tag{1.141}$$

Comparing Eq. (1.141) with the first term in the integral solution Eq. (1.132), we see that this term can be interpreted as the contribution of an equivalent double-layer source on S.

A brief comment on the nature of the double-layer source is necessary here because this is a topic that is rarely discussed in electrical engineering courses. One more time, we recall the application with the electric field and its scalar potential φ. The double-layer source in electricity consists of two sheets of surface charge separated by an infinitesimal distance d. A charged battery is an example of a double-layer source in electrostatics. Bearing in mind that a positive single-layer sheet in the Helmholtz equation for φ corresponds to a negative surface charge density, $\mathcal{S}_s = -\frac{\rho_s}{\varepsilon}$, the double-layer source of Eq. (1.137) becomes

$$P_s(\mathbf{r}) = \lim_{d \to 0} \left[-d \cdot \frac{\rho_s}{\varepsilon}(\mathbf{r}) \right]. \tag{1.142}$$

The behavior of φ in the vicinity of a double layer of charge is illustrated in Fig. 1.20-b. We observe that the effect of the double-layer sheet is to create potential difference.

In summary, the radiation from any emitter into a region well removed from it can be replaced equivalently by the radiation from single-layer and double-layer sources on any fictitious surface (the aperture) enclosing the emitter. These sources are related to the field boundary values through Eq. (1.134) and Eq. (1.139).

Before concluding this topic, there is one more important and curious observation regarding the solutions summarized in Table 1.1. The exterior solution for region 2 (last row) consists of a surface integral found also in the interior solution for region 1 (first row). This surface integral apparently produces zero field in region 1. This result

[19] The integral solution for the field generated by a double-layer source $\mathbf{P}_s$ is obtained from the integral solution for a field due to the conventional single-layer source, $U(\mathbf{r}) = \iiint \mathcal{S}(\mathbf{r}')G(\mathbf{r}, \mathbf{r}')dv'$, and the superposition principle.

makes sense because in the original (unbounded) problem, the solution is completely described by the volume integral $\iiint_{V_1} \mathcal{S}(\mathbf{r}')G(\mathbf{r},\mathbf{r}')dv'$. Yet, the fact that the equivalent surface sources radiate into region 2 while producing a zero net field in region 1 may be counterintuitive and confusing. This curious feature of the equivalence principle is known as the *extinction theorem* [14, 31].

1.7.3 Uniqueness Theorem and the Relationship between Single-Layer and Double-Layer Equivalent Surface Sources

The equivalence principle discussed in the previous section employs two types of aperture sources, single-layer and double-layer, which relate to the surface values of the normal derivative and the field itself, respectively. But these two sets of field values are not independent. This is immediately evident from the exterior solution for region 1; see Table 1.1. The equation is restated here in a slightly rearranged form for convenience:

$$\iiint_{V_1} \mathcal{S}(\mathbf{r}')G(\mathbf{r},\mathbf{r}')dv' = \iint_S \left[\frac{\partial U(\mathbf{r}')}{\partial n_{12}}G(\mathbf{r},\mathbf{r}') - U(\mathbf{r}')\frac{\partial G(\mathbf{r},\mathbf{r}')}{\partial n_{12}}\right]ds', \mathbf{r} \in V_2. \tag{1.143}$$

Let us denote the field radiated by the actual source $\mathcal{S}(\mathbf{r}')$ as U^{inc}. This field is expressed as

$$U^{\text{inc}}(\mathbf{r}) = \iiint_{V_1} \mathcal{S}(\mathbf{r}')G(\mathbf{r},\mathbf{r}')dv'. \tag{1.144}$$

Now, assume that we have measured the field values at the surface S (the aperture in the equivalent problem). Since S is fictitious, these are the incident-field values at S, $U^{\text{inc}}(\mathbf{r} \in S)$. Then, when Eq. (1.143) is written for observation points at S, it leads to

$$U^{\text{inc}}(\mathbf{r}) = \iint_S \left[\frac{\partial U^{\text{inc}}(\mathbf{r}')}{\partial n_{12}}G(\mathbf{r},\mathbf{r}') - U^{\text{inc}}(\mathbf{r}')\frac{\partial G(\mathbf{r},\mathbf{r}')}{\partial n_{12}}\right]ds', \mathbf{r} \in V_2 \wedge \mathbf{r} \to S. \tag{1.145}$$

With Green's function known, it is clear that the field normal derivative $\frac{\partial U^{\text{inc}}}{\partial n_{12}}$ at S can be found from the field boundary values U^{inc} using Eq. (1.145). The reverse is also true: if the field normal derivative is known on S, the field itself can be determined from Eq. (1.145).

Equations such as Eq. (1.145) can be solved using various numerical methods such as the boundary-element method (BEM) [32], the finite-element method (FEM) [28, 31], and the method of moments (MoM) [31, 33, 34]. In these solutions, the main problem is how to treat Eq. (1.145) properly using Cauchy's principal value in order to address the singularity of Green's function when $\mathbf{r}' \to \mathbf{r}$. A rigorous treatment of the various boundary-value problems arising in physics can be found in [35].

In summary, it suffices to know either the incident field or its normal derivative at S in order to solve the equivalent problem in region 2 because these two are not independent.

This result is in agreement with the uniqueness theorem for Helmholtz's equation [36], which states that if at any point on the boundary surface S *either* the function value (known as *Dirichlet boundary condition*) *or* its normal derivative (known as *Neumann boundary condition*) is specified, there exists one and only one solution to the Helmholtz equation. Therefore, providing both the Dirichlet and the Neumann conditions at any given point on S overdetermines an interior boundary-value problem; see Eq. (1.132). Nonetheless, Eq. (1.132) is a convenient formulation of the equivalent problem. The only piece of information added by the uniqueness theorem here is that we need not measure both the field *and* its normal derivate on the chosen aperture. Measuring only one of those is sufficient because the other one can be computed.

Example: An emitter positioned in the half-space $z < 0$ (region 1) generates a plane wave, the boundary value of which at $z = 0$ is measured as $U_{\mathrm{B}}^{\mathrm{inc}}$. Find this field's normal derivative at $z = 0$, $\frac{\partial U_{\mathrm{B}}^{\mathrm{inc}}}{\partial n}$, if the medium propagation constant is γ. Use Green's function of the 1D Helmholtz equation $G(z, z') = -\frac{\exp(-\gamma|z-z'|)}{2\gamma}$. Express the field in region 2 ($z \geq 0$) as a function of the boundary values $U_{\mathrm{B}}^{\mathrm{inc}}$ and $\frac{\partial U_{\mathrm{B}}^{\mathrm{inc}}}{\partial n}$.

Solution: This is a 1D problem where the surface unit normal is along z, $\hat{\mathbf{n}} \equiv \hat{\mathbf{z}}$, and the surface integral in Eq. (1.145) reduces to a point so that

$$U^{\mathrm{inc}}(z) = \frac{dU^{\mathrm{inc}}(z')}{dz'} G(z, z') - U^{\mathrm{inc}}(z') \frac{dG(z, z')}{dz'}, z \geq 0 \wedge z \to 0, z' = 0. \tag{1.146}$$

For an observation point at the boundary ($z = 0$), Eq. (1.146) is rearranged to obtain

$$U_{\mathrm{B}}^{\mathrm{inc}} \left[1 + \frac{dG(0, 0)}{dz'} \right] = \frac{dU_{\mathrm{B}}^{\mathrm{inc}}}{dz'} G(0, 0). \tag{1.147}$$

Since $z \geq z'$, Green's function is written as $G(z, z') = -\frac{\exp[-\gamma(z-z')]}{2\gamma}$. Then, $G(0, 0) = -\frac{1}{2\gamma}$ whereas $\frac{dG(0,0)}{dz'} = -\frac{1}{2}$. Upon substitution in Eq. (1.147), the field normal derivative at the boundary is found as

$$\left. \frac{dU_{\mathrm{B}}^{\mathrm{inc}}}{dz'} \right|_{z'=0} = -\gamma U_{\mathrm{B}}^{\mathrm{inc}}. \tag{1.148}$$

This is, of course, an expected result because the incident field is a plane wave traveling along z. However, the point of this simple example is that the field normal derivative can be found from the field boundary value without knowing it as a function of position.

In order to find the field as a function of position in region 2 ($z \geq 0$), we make use of Eq. (1.132), which appears here in the simple 1D form:

$$U(z) = U_{\mathrm{B}}^{\mathrm{inc}} \frac{dG(z, z' = 0)}{d(-z')} - \frac{dU_{\mathrm{B}}^{\mathrm{inc}}}{d(-z')} G(z, z' = 0), z \geq 0. \tag{1.149}$$

Notice that here the normal derivatives are along $-\hat{\mathbf{z}}$, which is the outward unit normal for region 2. After the substitution of Eq. (1.148) and Green's function, we find that

$$U(z \geq 0) = U_{\mathrm{B}}^{\mathrm{inc}} e^{-\gamma z}. \tag{1.150}$$

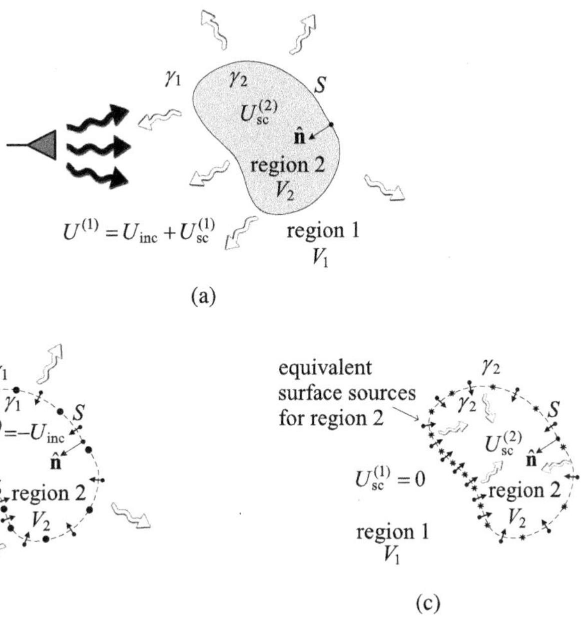

Figure 1.21 Illustration of the application of the equivalence principle in scattering. (a) The actual problem where an emitter in region 1 generates the incident field U_{inc}, which is the field that would have existed if the scatterer were not present. The scattering object (region 2) is the reason for the scattered field $U_{\text{sc}}^{(1)}$ in region 1. The total field in region 1, $U^{(1)}$, is a superposition of the incident and scattered field terms. The field in region 2, $U_{\text{sc}}^{(2)}$, has no explicit incident term because region 2 does not contain sources. (b) The equivalent problem for determining the scattered field in region 1; see Eq. (1.151). The equivalent surface sources reside on S, and they radiate in open space where the propagation constant is that of region 1. They produce the same scattered field $U_{\text{sc}}^{(1)}$ in region 1 as that in the actual problem. In region 2, they produce a scattered field, which cancels the incident field because the total field must be zero. (c) The analogous equivalent problem for determining the scattered field in region 2; see Eq. (1.152).

1.7.4 Application of the Surface Equivalence Principle to Scattering

The surface equivalence principle is also employed to efficiently compute the scattering from an object of a homogeneous interior immersed in a homogeneous background. Its efficiency stems from the fact that the equivalence principle allows for constructing an integral equation for the field values only at the surface of the object. Thus, an intrinsically 3D problem is reduced to a boundary integral equation on a surface where the number of unknowns is orders of magnitude smaller than the number of unknowns in the overall 3D problem. Once the field and its normal derivative are computed on the surface of the scatterer, with the equivalence principle, the field can be computed at any desired point.

Fig. 1.21-a illustrates such a problem. The interior of the scattering object (region 2) is described by the propagation constant γ_2. The propagation constant of the background (region 1) is γ_1. Green's functions for both region 1 and 2 are often chosen, for simplicity, to describe radiation in open space in the respective medium. Since

$\gamma_1 \neq \gamma_2$, Green's functions used in the solutions for the two regions are not the same, $G_1(\mathbf{r}, \mathbf{r}') \neq G_2(\mathbf{r}, \mathbf{r}')$. These solutions are stated next:

- region 1 interior and exterior solutions

$$\underbrace{U_{\text{inc}}(\mathbf{r}) + \oiint\limits_S \left[U^{(1)}(\mathbf{r}') \frac{\partial G_1(\mathbf{r}, \mathbf{r}')}{\partial n} - \frac{\partial U^{(1)}(\mathbf{r}')}{\partial n} G_1(\mathbf{r}, \mathbf{r}') \right] ds'}_{U_{\text{sc}}^{(1)}(\mathbf{r})} = \begin{cases} U^{(1)}(\mathbf{r}) & \mathbf{r} \in V_1 \\ 0 & \mathbf{r} \in V_2. \end{cases} \tag{1.151}$$

- region 2 interior and exterior solutions

$$\underbrace{- \oiint\limits_S \left[U^{(2)}(\mathbf{r}') \frac{\partial G_2(\mathbf{r}, \mathbf{r}')}{\partial n} - \frac{\partial U^{(2)}(\mathbf{r}')}{\partial n} G_2(\mathbf{r}, \mathbf{r}') \right] ds'}_{U_{\text{sc}}^{(2)}(\mathbf{r})} = \begin{cases} U^{(2)}(\mathbf{r}) & \mathbf{r} \in V_2 \\ 0 & \mathbf{r} \in V_1. \end{cases} \tag{1.152}$$

The normal derivatives in the surface integrals above are all taken along the surface unit normal $\hat{\mathbf{n}}$, which points from region 1 into region 2 as shown in Fig. 1.21. Also, U_{inc} in Eq. (1.151) is given by

$$U_{\text{inc}}(\mathbf{r}) = \iiint_{V_1} \mathcal{S}(\mathbf{r}') G_1(\mathbf{r}, \mathbf{r}') dv', \tag{1.153}$$

and it is the field that would have existed at any position (in region 1 or 2) in the absence of the scatterer.

Eq. (1.151) and Eq. (1.152) allow for formulating a system of equations for the field and its normal derivative on S. Note that S here is a physical interface, not just a fictitious surface, as was the case with radiation (see Section 1.7.2).

First, we require that the observation point $\mathbf{r}$ approaches S from the side of V_1 in the case of Eq. (1.151) and from the side of V_2 in Eq. (1.152). Second, we require that a certain boundary condition is fulfilled by $U^{(1)}(\mathbf{r} \in S)$ and $U^{(2)}(\mathbf{r} \in S)$. A common boundary condition imposes the field continuity across S, i.e.,

$$U^{(1)}(\mathbf{r}) = U^{(2)}(\mathbf{r}) = U_{\text{B}}\, , \ \mathbf{r} \in S. \tag{1.154}$$

The subscript in U_{B} emphasizes that this is the field boundary value. Hereafter, the subscript $(\cdot)_{\text{B}}$ will inherently indicate that the point is at the interface, i.e., $\mathbf{r} \in S$. This is, for example, a boundary condition satisfied by the tangential $\mathbf{E}$ and $\mathbf{H}$ field components across interfaces free of sources.

A more general form of the field boundary condition is

$$a_1 U_{\text{B}}^{(1)} = a_2 U_{\text{B}}^{(2)}. \tag{1.155}$$

The $\mathbf{E}$ and $\mathbf{H}$ field normal components satisfy this type of boundary condition where a_i, $i = 1, 2$, are the permittivities or permeabilities of the two regions, respectively.

Eq. (1.151) and Eq. (1.152), for $\mathbf{r} \in V_1$ and $\mathbf{r} \in V_2$, respectively, can now be used in conjunction with Eq. (1.155) to obtain a surface integral equation relating $U_{\text{B}}^{(1)}$ to the field normal derivatives at S:

$$\oiint_S \left\{ U_B^{(1)}(\mathbf{r}') \left[\frac{\partial G_1(\mathbf{r},\mathbf{r}')}{\partial n} + \frac{\partial G_2(\mathbf{r},\mathbf{r}')}{\partial n} \right] - \left[\frac{\partial U_B^{(1)}(\mathbf{r}')}{\partial n} G_1(\mathbf{r},\mathbf{r}') + \frac{a_2}{a_1} \frac{\partial U_B^{(2)}(\mathbf{r}')}{\partial n} G_2(\mathbf{r},\mathbf{r}') \right] \right\} ds' = -U_{\text{inc}}(\mathbf{r}), \mathbf{r} \in S. \quad (1.156)$$

An additional boundary condition relates the field normal derivatives at the interface:

$$b_1 \frac{\partial U_B^{(1)}}{\partial n} = b_2 \frac{\partial U_B^{(2)}}{\partial n}. \quad (1.157)$$

Now, Eq. (1.156) can be written in terms of two unknowns only, $U_B^{(1)}$ and $\frac{\partial U_B^{(1)}}{\partial n}$:

$$\oiint_S \left\{ U_B^{(1)}(\mathbf{r}') \left[\frac{\partial G_1(\mathbf{r},\mathbf{r}')}{\partial n} + \frac{\partial G_2(\mathbf{r},\mathbf{r}')}{\partial n} \right] - \frac{\partial U_B^{(1)}(\mathbf{r}')}{\partial n} \left[G_1(\mathbf{r},\mathbf{r}') + \frac{a_2 b_1}{a_1 b_2} G_2(\mathbf{r},\mathbf{r}') \right] \right\} ds' = -U_{\text{inc}}(\mathbf{r}), \mathbf{r} \in S. \quad (1.158)$$

One more equation is needed at S to solve for the two unknown field quantities. It is available from the extinction theorem of region 2 and is Eq. (1.152) for $\mathbf{r} \in V_1$, where we substitute the boundary conditions Eq. (1.155) and Eq. (1.157). This leads to the relation

$$\oiint_S \left[U_B^{(1)}(\mathbf{r}') \frac{\partial G_2(\mathbf{r},\mathbf{r}')}{\partial n} - \left(\frac{a_2 b_1}{a_1 b_2} \right) \frac{\partial U_B^{(1)}(\mathbf{r}')}{\partial n} G_2(\mathbf{r},\mathbf{r}') \right] ds' = 0, \mathbf{r} \in S \wedge \mathbf{r} \in V_1. \quad (1.159)$$

Note that while $\mathbf{r}$ belongs to S, it resides on the side of region 1, i.e., $\mathbf{r} \in V_1$.

We see that the terms in Eq. (1.159) are contained in Eq. (1.158). Setting those to zero reduces Eq. (1.158) to

$$\oiint_S \left[U_B^{(1)}(\mathbf{r}') \frac{\partial G_1(\mathbf{r},\mathbf{r}')}{\partial n} - \frac{\partial U_B^{(1)}(\mathbf{r}')}{\partial n} G_1(\mathbf{r},\mathbf{r}') \right] ds' = -U_{\text{inc}}(\mathbf{r}), \mathbf{r} \in S \wedge \mathbf{r} \in V_2, \quad (1.160)$$

which is in essence an application of the extinction theorem for region 1 at $\mathbf{r} \in S$; see Eq. (1.151) for $\mathbf{r} \in V_2$.

Eq. (1.159) and Eq. (1.160) provide the two independent equations from which the field $U_B^{(1)}$ and its normal derivative $\frac{\partial U_B^{(1)}}{\partial n}$ are determined. $U_B^{(2)}$ and $\frac{\partial U_B^{(2)}}{\partial n}$ can be subsequently found from Eq. (1.155) and Eq. (1.157). Once the field boundary values and derivatives are known, the interior solutions for regions 1 and 2 in Eq. (1.151) and Eq. (1.152) can provide the field value at any point.

The role of the field boundary values at S as equivalent surface sources here is analogous to that in radiation. For example, $U_{\rm B}^{(1)}$ can be viewed as an equivalent surface double-layer source, $\mathbf{P}_{\rm s,e}^{(1)} = U_{\rm B}^{(1)}\hat{\mathbf{n}}$ in Eq. (1.151), which arises when the field in region 2 is set to zero. On the other hand, $S_{\rm s,e}^{(1)} = -\frac{\partial U_{\rm B}^{(1)}}{\partial n}$ is the equivalent single-layer source in Eq. (1.151). The scattering problem differs from the radiation one in that S traverses an actual interface. As a result, the field boundary values (the equivalent sources) are found by solving a system of two equations, Eq. (1.159) and Eq. (1.160), whereas in the radiation problem these are known since they come from the known field generated by the source in region 1.

Fig. 1.21-b illustrates the equivalence principle in relation to Eq. (1.151): the total field in region 2 is set to zero, which requires respective surface sources, namely, the single-layer source, $-\frac{\partial U_{\rm B}^{(1)}}{\partial n}$, and the double-layer one, $U_{\rm B}^{(1)}\hat{\mathbf{n}}$. These sources radiate in a homogeneous open space with the propagation constant being γ_1. They are responsible for generating the scattered field in region 1. Meanwhile, these sources produce a scattered field in region 2 that cancels the incident one to produce a zero total field—this is the extinction theorem. An analogous interpretation holds with regard to the equivalent surface sources generating the scattered field in region 2 as described by Eq. (1.152). This is illustrated in Fig. 1.21-c.

Example An emitter is positioned in the half-space $z < 0$ (region 1) of propagation constant γ_1. In the absence of obstacles, it generates a uniform plane *incident* wave, the value of which at $z = 0$ is measured as $U_{\rm B}^{\rm inc}$. In a scattering scenario, the medium in the half-space $z \geq 0$ (region 2) has a propagation constant γ_2. At $z = 0$, the total field must satisfy the boundary conditions

$$U_{\rm B}^{(1)} = U_{\rm B}^{(2)} = U_{\rm B} \text{ and } b_1 \frac{U_{\rm B}^{(1)}}{\partial z} = b_2 \frac{U_{\rm B}^{(2)}}{\partial z}.$$

(a) Find $U_{\rm B}$ as well as the normal derivative $\frac{\partial U_{\rm B}^{(1)}}{\partial z}$. Use Green's function of the 1D Helmholtz equation $G(z, z') = -\frac{\exp(-\gamma_i|z-z'|)}{2\gamma_i}$. (b) Express the scattered field in regions 1 and 2 in terms of $U_{\rm B}^{\rm inc}$, γ_1 and γ_2.

Solution: This is a 1D problem where the surface unit normal is along z, $\hat{\mathbf{n}} \equiv \hat{\mathbf{z}}$, and the surface integrals in Eq. (1.159) and Eq. (1.160) reduce to an equation at a point. From Eq. (1.159), we obtain

$$\left[U_{\rm B} \frac{\partial G_2(z, z')}{\partial z'} - \frac{\partial U_{\rm B}^{(2)}}{\partial z'} G_2(z, z') \right]_{z'=0} = 0\,, z = 0_-, \tag{1.161}$$

whereas from Eq. (1.160), we obtain

$$\left[U_{\rm B} \frac{\partial G_1(z, z')}{\partial z'} - \frac{\partial U_{\rm B}^{(1)}}{\partial z'} G_1(z, z') \right]_{z'=0} = -U_{\rm B}^{\rm inc}\,, z = 0_+. \tag{1.162}$$

Here, G_1 and G_2 are Green's functions for regions 1 and 2, respectively. The notation $z = 0_-$ (or $z = 0_+$) means that the observation point is at the boundary, but on the side of region 1 where z is negative (or on the side of region 2 where z is positive). Green's 1D functions for regions 1 and 2 are

$$G_1(z, z') = -\frac{e^{-\gamma_1|z-z'|}}{2\gamma_1}, \quad G_2(z, z') = -\frac{e^{-\gamma_2|z-z'|}}{2\gamma_2}. \tag{1.163}$$

Therefore,

$$G_1(0, 0) = -\frac{1}{2\gamma_1}, \quad G_2(0, 0) = -\frac{1}{2\gamma_2}. \tag{1.164}$$

Next, we find Green's functions' normal derivatives. For $z = 0_+$ and $z' = 0$,

$$\left.\frac{\partial G_1}{\partial z'}\right|_{z=0_+,z'=0} = \frac{\partial}{\partial z'}\left[-\frac{e^{-\gamma_1(z-z')}}{2\gamma_1}\right]_{z=0_+,z'=0} = -\frac{1}{2}. \tag{1.165}$$

Similarly, for $z = 0_-$ and $z' = 0$,

$$\left[\frac{\partial G_2}{\partial z'}\right]_{z=0_-,z'=0} = \frac{\partial}{\partial z'}\left[-\frac{e^{-\gamma_2(z'-z)}}{2\gamma_2}\right]_{z=0_-,z'=0} = \frac{1}{2}. \tag{1.166}$$

Eq. (1.161) and Eq. (1.162) can now be written as a system of equations where U_{B} and $\frac{\partial U_{\mathrm{B}}^{(1)}}{\partial z'}$ are the unknowns:

$$\begin{aligned} \frac{U_{\mathrm{B}}}{2} + \frac{b_1}{b_2}\frac{\partial U_{\mathrm{B}}^{(1)}}{\partial z'}\frac{1}{2\gamma_2} &= 0 \\ \frac{U_{\mathrm{B}}}{2} - \frac{\partial U_{\mathrm{B}}^{(1)}}{\partial z'}\frac{1}{2\gamma_1} &= U_{\mathrm{B}}^{\mathrm{inc}}. \end{aligned} \tag{1.167}$$

Solving Eq. (1.167) provides the answer to part (a) as

$$\begin{aligned} U_{\mathrm{B}} &= \frac{2U_{\mathrm{B}}^{\mathrm{inc}}}{\left(1 + \frac{\gamma_2 b_2}{\gamma_1 b_1}\right)} \\ \frac{\partial U_{\mathrm{B}}^{(1)}}{\partial z'} &= \frac{-2\gamma_1 U_{\mathrm{B}}^{\mathrm{inc}}}{\left(1 + \frac{\gamma_1 b_1}{\gamma_2 b_2}\right)}. \end{aligned} \tag{1.168}$$

When solving part (b), we use the interior solutions for regions 1 and 2, Eq. (1.151) and Eq. (1.152), in order to derive the respective scattered fields. In region 1, Eq. (1.151) yields the scattered-field expression

$$U_{\mathrm{sc}}^{(1)}(z) = U_{\mathrm{B}}\frac{\partial G_1(z, 0)}{\partial z'} - \frac{\partial U_{\mathrm{B}}^{(1)}}{\partial z'}G_1(z, 0), z < 0. \tag{1.169}$$

For $z < z' = 0$,

$$G_1(z, z') = -\frac{e^{-\gamma_1(z'-z)}}{2\gamma_1}. \tag{1.170}$$

Therefore,

$$G_1(z, 0) = -\frac{e^{\gamma_1 z}}{2\gamma_1} \text{ and } \frac{\partial G_1(z, 0)}{\partial z'} = \frac{e^{\gamma_1 z}}{2}. \tag{1.171}$$

Substituting in Eq. (1.169) and rearranging yields

$$U_{\text{sc}}^{(1)}(z) = U_{\text{B}}^{\text{inc}} \left(\frac{b_1\gamma_1 - b_2\gamma_2}{b_1\gamma_1 + b_2\gamma_2} \right) e^{\gamma_1 z}, z < 0. \tag{1.172}$$

Analogously, in region 2, Eq. (1.152) yields the scattered-field expression

$$U_{\text{sc}}^{(2)}(z) = -U_{\text{B}} \frac{\partial G_2(z, 0)}{\partial z'} + \frac{\partial U_{\text{B}}^{(2)}}{\partial z'} G_2(z, 0), z > 0. \tag{1.173}$$

For $z > z' = 0$,

$$G_2(z, z') = -\frac{e^{-\gamma_2(z-z')}}{2\gamma_2}. \tag{1.174}$$

Therefore,

$$G_2(z, 0) = -\frac{e^{-\gamma_2 z}}{2\gamma_2} \text{ and } \frac{\partial G_2(z, 0)}{\partial z'} = -\frac{e^{-\gamma_2 z}}{2}. \tag{1.175}$$

Also, from the total-field boundary conditions,

$$\frac{U_{\text{B}}^{(2)}}{\partial z'} = \frac{b_1}{b_2} \frac{U_{\text{B}}^{(1)}}{\partial z'} = \frac{b_1}{b_2} \frac{\left(-2\gamma_1 U_{\text{B}}^{\text{inc}}\right)}{\left(1 + \frac{\gamma_1 b_1}{\gamma_2 b_2}\right)}. \tag{1.176}$$

In Eq. (1.176), we have used the second equation in Eq. (1.168). The substitution of Eq. (1.175) and Eq. (1.176) in Eq. (1.173) yields

$$U_{\text{sc}}^{(2)}(z) = U_{\text{B}}^{\text{inc}} \left(\frac{2b_1\gamma_1}{b_1\gamma_1 + b_2\gamma_2} \right) e^{-\gamma_2 z}, z > 0. \tag{1.177}$$

It is instructive to discuss the results of our example in the context of an EM wave. Let us assume that the scalar field represents the electric field of a uniform plane wave polarized along x, i.e., $U(z) = E_x(z)$. Let the constitutive parameters of regions 1 and 2 be ε_1, μ_1, and ε_2, μ_2, respectively. The wave is normally incident upon the interface at $z = 0$ between the two regions. This problem can be analyzed entirely in terms of E_x, which satisfies the 1D Helmholtz equation. E_x fulfills the boundary conditions

$$E_x^{(1)} = E_x^{(2)}, \quad \frac{1}{\mu_1}\frac{\partial E_x^{(1)}}{\partial z} = \frac{1}{\mu_2}\frac{\partial E_x^{(2)}}{\partial z}, \tag{1.178}$$

at $z = 0$. The first of these conditions ensures the continuity of the tangential **E** field ($\mathbf{E} = E_x\hat{\mathbf{x}}$), whereas the second one ensures the continuity of the tangential **H** field ($\mathbf{H} = \hat{\mathbf{y}}H_y$). Thus, $b_i = \mu_i^{-1}$ $(i = 1, 2)$, and the expressions for the scattered field in regions 1 and 2 are obtained from Eq. (1.172) and Eq. (1.177) as

$$E_{x,\mathrm{sc}}^{(1)}(z) = E_{x,\mathrm{B}}^{\mathrm{inc}}\left(\frac{\eta_2 - \eta_1}{\eta_1 + \eta_2}\right)e^{\gamma_1 z}, z < 0, \tag{1.179}$$

$$E_{x,\mathrm{sc}}^{(2)}(z) = E_{x,\mathrm{B}}^{\mathrm{inc}}\left(\frac{2\eta_2}{\eta_1 + \eta_2}\right)e^{-\gamma_2 z}, z > 0, \tag{1.180}$$

where $\eta_i = \sqrt{\frac{\mu_i}{\varepsilon_i}}$ is the intrinsic impedance of the respective medium. The electrical engineer will recognize Eqs. (1.179) and (1.180) as the well-known formulas for the reflection and transmission of a uniform plane wave at a planar interface, with the reflection coefficient being $\Gamma = \frac{\eta_2 - \eta_1}{\eta_1 + \eta_2}$ and the transmission coefficient being $T = \frac{2\eta_2}{\eta_1 + \eta_2}$.

The problem of a uniform plane wave normally incident on a planar interface can be analyzed for a wave of *arbitrary polarization* using *one* scalar-wave function; see Appendix C.[20] For example, one could work with the electric vector potential $\boldsymbol{\mathcal{F}}_\epsilon = \mathcal{F}\hat{\mathbf{z}}$ i.e., the wave function is $\mathcal{F}$ and the preferred direction is $\hat{\mathbf{z}}$. $\mathcal{F}$ must satisfy the same boundary conditions as those for E_x in Eq. (1.178). This ensures the fulfillment of the boundary conditions for all **E** and **H** field components. To appreciate that, the reader is referred to the explicit formulas relating the rectangular **E** and **H** field components to $\mathcal{F}$ in Eq. (C.25) in Appendix C. The expressions for the scattered $\mathcal{F}$ waves in regions 1 and 2 are the same as those in Eqs. (1.179) and (1.180).

1.7.5 Fourier Transform and Causal Solutions in the Frequency Domain

As discussed in Section 1.3, Helmholtz's equation solves a field problem one frequency at a time. Thus, the solution represents the system time-harmonic response to a time-harmonic excitation at a given frequency of interest $\omega = \omega_0$. It appears in the form of a complex-valued function of space, a phasor $U(\mathbf{r}, \omega_0)$; see Eq. (1.18).

The phasor itself is nothing but twice the value of the complex-valued Fourier coefficient of a signal. This is easy to see if we recall that a time-harmonic signal component of the form $f(t) = a\cos(\omega_0 t) + b\sin(\omega_0 t)$ is represented by a phasor $F = a - ib$. At the same time, its complex Fourier coefficient $\tilde{F}$ defined by [11],

$$\tilde{F} = \frac{1}{T}\int_{t_0}^{t_0+T} f(t)e^{-i\omega_0 t}\,dt, \tag{1.181}$$

[20] In fact, the reflection/transmission at an infinite planar interface of *any* wave and at *any* angle of incidence can be described with *one* scalar-wave function.

is $\tilde{F} = \frac{1}{2}(a - ib)$. Here, T is the period of $f(t)$ and t_0 is a suitable initial time. As a result, the relationship between the ω_0 harmonic of the wave solution and its complex Fourier coefficient is simply

$$u_{\omega_0}(\mathbf{r}, t) = m_{\omega_0}(\mathbf{r}) \cos\left[\omega_0 t + \Phi_{\omega_0}(\mathbf{r})\right] = \Re\left\{U(\mathbf{r}, \omega_0)e^{i\omega_0 t}\right\} = \Re\left\{2\tilde{U}(\mathbf{r}, \omega_0)e^{i\omega_0 t}\right\}, \tag{1.182}$$

where

$$\tilde{U}(\mathbf{r}, \omega_0) = \frac{1}{T_{\max}} \int_0^{T_{\max}} u(\mathbf{r}, t)e^{-i\omega_0 t} dt. \tag{1.183}$$

In Eq. (1.182), m_{ω_0} and Φ_{ω_0} are the magnitude and the phase of $u_{\omega_0}(\mathbf{r}, t)$, respectively. The limits 0 and $T_{\max}$ in Eq. (1.183) set the temporal observation interval, outside of which $u(\mathbf{r}, t)$ is zero or negligibly small. In view of these constraints on the signal, the limits of the integration can also be extended to $-\infty$ and $+\infty$, respectively. Subject to a certain choice of the normalization factor, i.e., $\frac{1}{T_{\max}}$ in Eq. (1.183), this equation is also the complex Fourier transform (FT) of $u(\mathbf{r}, t)$; see, for example, [11].

It is worth mentioning that Eq. (1.183) is the integral that, when discretized, produces what is known as the *discrete Fourier transform* (DFT) [37]. The DFT is a commonly used numerical tool to extract desired harmonic components of a signal.

In view of the relationship, $U(\mathbf{r}, \omega_0) = 2\tilde{U}(\mathbf{r}, \omega_0)$, between the phasor and the FT, we can derive the phasors of the causal and acausal wave solutions and Green's functions by performing a Fourier transformation of the respective quantities. Let the causal time-domain solution be $u_+(\mathbf{r}, t)$. Dropping the normalization factor, its FT is

$$\tilde{U}_+(\mathbf{r}, \omega) = \int_0^{T_{\max}} u_+(\mathbf{r}, t)e^{-i\omega t} dt. \tag{1.184}$$

Nothing changes if the integration limits are extended to $-\infty$ and $+\infty$. Bearing in mind that the Helmholtz equation Eq. (1.22) can be derived from the wave equation through the Fourier transformation, one can conclude that the causal-signal FT $\tilde{U}_+(\mathbf{r}, \omega_0)$ and its respective phasor $U_+(\mathbf{r})$ satisfy the Helmholtz equation, i.e.,

$$(\nabla^2 - \gamma^2)U_+(\mathbf{r}) = \mathcal{S}(\mathbf{r}), \tag{1.185}$$

where $\mathcal{S}(\mathbf{r})$ is the phasor of the source function.

1.7.6 Acausal Solutions in the Frequency Domain

As discussed before, see Eq. (1.108) and Appendix D, for every causal solution $u_+(\mathbf{r}, t)$, $t \in [0, T_{\max}]$, there exists an acausal adjoint counterpart $u_-(\mathbf{r}, \tau)$ such that $u_-(\mathbf{r}, \tau) = u_+(\mathbf{r}, t)$, where $\tau = T_{\max} - t$, $\tau \in [0, T_{\max}]$ is the reversed time variable. The reader is reminded that if $T_{\max} \to \infty$, i.e., $t \in [0, \infty)$, then $\tau = -t$ and $\tau \in (-\infty, 0]$. The acausal solution $u_-(\mathbf{r}, \tau)$ solves the wave equation, too, but in the adjoint domain $\mathcal{D}^\dagger : \left\{(\mathbf{r}, \tau) \in \mathbb{R}^4\right\}$.

The FT of the acausal solution is

$$\tilde{U}_-(\mathbf{r},\omega)=\int_0^{T_{\max}} u_-(\mathbf{r},t)e^{-\mathrm{i}\omega t}\,dt=\int_0^{T_{\max}} u_+(\mathbf{r},T_{\max}-t)e^{-\mathrm{i}\omega t}\,dt=e^{-\mathrm{i}\omega T_{\max}}\int_0^{T_{\max}} u_+(\mathbf{r},\tau)e^{\mathrm{i}\omega\tau}\,d\tau. \tag{1.186}$$

Comparing Eq. (1.186) to Eq. (1.184), we see that Eq. (1.108), $u_-(\mathbf{r},\tau)=u_+(\mathbf{r},t)$, appears in Fourier space as

$$\tilde{U}_-(\mathbf{r},\omega)=\tilde{U}_+^*(\mathbf{r},\omega)e^{-\mathrm{i}\omega T_{\max}}. \tag{1.187}$$

The asterisk superscript $(\cdot)^*$ denotes conjugation. Eq. (1.187) tells us that *reversing the direction of time flow results in the conjugation of the wave phasor*, i.e.,

$$\mathcal{F}\left\{f(T_{\max}-t)\right\}=\mathcal{F}^*\left\{f(t)\right\}e^{-\mathrm{i}\omega T_{\max}}, \tag{1.188}$$

where $\mathcal{F}$ denotes the FT. Note that if the FT is defined with the lower and upper temporal limits as $-\infty$ and $+\infty$, instead of as 0 and $T_{\max}$, respectively, the reverse-time variable is defined as $\tau=-t$ and the relationship between the FTs of the causal and acausal solutions in Eq. (1.108) is obtained as

$$\tilde{U}_-(\mathbf{r},\omega)=\tilde{U}_+^*(\mathbf{r},\omega). \tag{1.189}$$

Eq. (1.189) follows directly from the identity

$$\tilde{U}(\mathbf{r},\omega)=\tilde{U}^*(\mathbf{r},-\omega), \tag{1.190}$$

which is valid for a real-valued $u(\mathbf{r},t)$.

It is worthwhile to consider the FT of $u_-(\mathbf{r},\tau)$ when the transformed variable is τ ($\tau=T_{\max}-t$ or $\tau=-t$). As shown in Appendix D and discussed in Section 1.6.4, $u_-(\mathbf{r},t)$ satisfies the wave equation *only if time is reversed*, i.e., t is formally replaced by τ in all terms, *the source included*. This means that $u_-(\mathbf{r},\tau)$ satisfies the wave equation in the adjoint domain $\mathcal{D}^\dagger:\{(\mathbf{r},\tau)\}$, *not* in the original one $\mathcal{D}:\{(\mathbf{r},t)\}$.[21] Since $u_-(\mathbf{r},\tau)=u_+(\mathbf{r},t)$, the FT of $u_-(\mathbf{r},\tau)$ with respect to τ,

$$\underset{\sim}{U}_-(\mathbf{r},\omega_0)=\int_0^{T_{\max}} u_-(\mathbf{r},\tau)e^{-\mathrm{i}\omega_0\tau}\,d\tau, \tag{1.191}$$

is exactly the same as the FT of $u_+(\mathbf{r},t)$, $\tilde{U}_+(\mathbf{r},\omega_0)$. Therefore, the respective phasors are identical, i.e., $U_-^\dagger(\mathbf{r})=U_+(\mathbf{r})$. Here, $U_-^\dagger(\mathbf{r})$ is the phasor of $\underset{\sim}{U}_-(\mathbf{r},\omega_0)$.

[21] In the particular case of a loss-free and source-free medium, $u_-(\mathbf{r},t)$ satisfies the wave equation, $(\nabla^2-\frac{1}{v^2}\frac{\partial^2}{dt^2})u=0$, even if time is not reversed.

Similarly, the FT with respect to τ of the time-reversed excitation function $s^\dagger(\mathbf{r}, \tau) = s(\mathbf{r}, t)$,

$$\mathcal{S}^\dagger(\mathbf{r}, \omega_0) = \int_0^{T_{\max}} s^\dagger(\mathbf{r}, \tau) e^{-\mathrm{i}\omega_0 \tau} d\tau, \tag{1.192}$$

is identical to the FT of $s(\mathbf{r}, t)$ with respect to t, i.e., $\mathcal{S}^\dagger(\mathbf{r}, \omega_0) = \tilde{\mathcal{S}}(\mathbf{r}, \omega_0)$. The equality then holds for the respective phasors as well, $\mathcal{S}^\dagger(\mathbf{r}) = \mathcal{S}(\mathbf{r})$.

Consequently, $U_-^\dagger(\mathbf{r})$ solves exactly the same Helmholtz equation Eq. (1.185) as $U_+(\mathbf{r})$:

$$(\nabla^2 - \gamma^2) U_-^\dagger(\mathbf{r}) = \mathcal{S}^\dagger(\mathbf{r}), \tag{1.193}$$

and the frequency-domain result produced by Eq. (1.193) is identical to that of Eq. (1.185).

In summary, *the phasors of the acausal adjoint wave and source are identical to their forward-time counterparts but they correspond to two different time variables, τ and t, respectively*. Explicitly, the two phasors $U_+(\mathbf{r}) = U_-^\dagger(\mathbf{r}) = u_m(\mathbf{r}) e^{\mathrm{i}\Phi(\mathbf{r})}$, correspond to two different time-harmonic waves:

$$U_+(\mathbf{r}) \Leftrightarrow u_+(\mathbf{r}, t) = u_m(\mathbf{r}) \cos\left[\omega_0 t + \Phi(\mathbf{r})\right], \tag{1.194}$$

$$U_-^\dagger(\mathbf{r}) \Leftrightarrow u_-(\mathbf{r}, \tau) = u_m(\mathbf{r}) \cos\left[\omega_0 (T_{\max} - t) + \Phi(\mathbf{r})\right]. \tag{1.195}$$

It is instructive to derive the Helmholtz equation for the acausal phasor $U_-(\mathbf{r})$; see Eq. (1.186). Note that $U_-(\mathbf{r})$ is not the same as $U_-^\dagger(\mathbf{r})$, and, therefore, it does not satisfy the same Helmholtz equation as $U_+(\mathbf{r})$ and $U_-^\dagger(\mathbf{r})$; see Eq. (1.193).

We first express $U_-(\mathbf{r})$ through $U_-^\dagger(\mathbf{r})$ using Eq. (1.187) where we substitute $U_+(\mathbf{r}) = U_-^\dagger(\mathbf{r})$. The result is $U_-(\mathbf{r}) = U_-^{\dagger *}(\mathbf{r}) e^{-\mathrm{i}\omega_0 T_{\max}}$. This, in conjunction with Eq. (1.185), leads to

$$\left[\nabla^2 - (\gamma^*)^2\right] U_-(\mathbf{r}) = S^*(\mathbf{r}) e^{-\mathrm{i}\omega T_{\max}}. \tag{1.196}$$

Note that one should drop the factor $e^{-\mathrm{i}\omega T_{\max}}$ in the source term in Eq. (1.196) if $\tau = -t$.

There are two features of Eq. (1.196) that should be pointed out. First, the propagation constant is now conjugated. This ensures the proper direction of wave propagation (i.e., toward the source)—as opposed to the causal wave propagation away from the source. For example, the phasor of the causal plane wave features the propagation factor $\sim e^{-\mathrm{i}\beta z}$ while that of the acausal wave is $\sim e^{+\mathrm{i}\beta z}$. Second, the source function is conjugated as well. This reflects the requirement that if the field develops in reversed time so does the source.

1.7.7 Green's Causal and Acausal Functions in the Frequency Domain

In the case of $\tau = T_{\max} - t$, Eq. (1.187) allows for the derivation of the relationship between the FTs of Green's causal and (adjoint) acausal functions for the wave equation as[22]

$$\tilde{G}_{-}(\mathbf{r}, \mathbf{r}', \omega) = \tilde{G}_{+}^{*}(\mathbf{r}, \mathbf{r}', \omega). \tag{1.197}$$

The result is the same if $\tau = -t$, in which case we start from Eq. (1.189). Thus, there is a "conjugate" relationship between Green's function causal and acausal counterparts.

1.7.8 Integral-Equation Models of Radiation in the Frequency Domain

There are two cases of interest when describing radiation in open space—the primary radiation from a known source in the infinite region V (see Fig. 1.12-a) and the boundary-value radiation problem where V is an infinite multiply-connected region (see Fig. 1.12-b). These cases have already been considered in the context of the wave equation in Section 1.6.3. The respective frequency-domain integral equations are particular cases of the *interior* Kirchhoff–Helmholtz equation, Eq. (1.129).

The primary radiation field of a known source (the case of Fig. 1.12-a) is obtained as the causal version of the first term (the volume integral) in Eq. (1.129):

$$U_{+}(\mathbf{r}) = \iiint_{V} G_{+}(\mathbf{r}, \mathbf{r}') S(\mathbf{r}') dv', \tag{1.198}$$

where $G_{+}(\mathbf{r}, \mathbf{r}')$ is Green's causal function in open space. The surface integral in Eq. (1.129) is zero because both U_{+} and G_{+} satisfy the same radiation boundary condition [4],[23] leading to mutual cancellation.

Eq. (1.198) can also be obtained through the Fourier transformation of Eq. (1.112).[24]

22 Let the phasor of the causal impulse response at the observation point $\mathbf{r}_o$ be $U_{+}^{\mathrm{IR}}(\mathbf{r}_o)$ while that of the acausal wave is $U_{-}^{\mathrm{IR}}(\mathbf{r}_o)$. The integral solution for the former is $U_{+}^{\mathrm{IR}}(\mathbf{r}_o) = \iiint G_{+}(\mathbf{r}_o, \mathbf{r}_s)\delta(\mathbf{r}_o - \mathbf{r}_s)dv'$. For the latter, it is $U_{-}^{\mathrm{IR}}(\mathbf{r}_o) = \iiint G_{-}(\mathbf{r}_o, \mathbf{r}_s)\delta(\mathbf{r}_o - \mathbf{r}_s)e^{-\mathrm{i}\omega T_{\max}}dv'$. The exponential term here is due to the way the source appears in the respective Helmholtz equation, Eq. (1.196). At the same time, as per Eq. (1.187), the equality $\tilde{U}_{-} = \tilde{U}_{+}^{*}e^{-\mathrm{i}\omega T_{\max}}$ must hold. It follows that $G_{-}(\mathbf{r}_o, \mathbf{r}_s) = G_{+}^{*}(\mathbf{r}_o, \mathbf{r}_s)$.

23 The scalar radiation boundary condition, also known as Sommerfeld's boundary condition, requires that

$$\lim_{r\to\infty} r\left(\frac{\partial U}{\partial r} + \mathrm{i}kU\right) = 0,$$

where k is the wavenumber and r is the distance from the source. In an equivalent form, the above is stated as

$$\frac{\partial U}{\partial r} + \mathrm{i}kU = o\left(\frac{1}{r}\right).$$

The radiation condition states that, far from its source, the wave behaves as an outgoing locally plane wave. For derivation and a milder form of the scalar radiation condition, see [4].

24 The proof exploits the convolution theorem and the fact that the dependence on t and t' in $g_{+}(\mathbf{r}', t'; \mathbf{r}, t)$ appears in the form of the argument $(t - t')$.

The boundary-value radiation field (the case of Fig. 1.12-b) is the causal version of the second (surface) integral in Eq. (1.129):

$$U_+(\mathbf{r}) = \oiint_S \left[U_+(\mathbf{r}') \frac{\partial G_+(\mathbf{r}, \mathbf{r}')}{\partial n} - G_+(\mathbf{r}, \mathbf{r}') \frac{\partial U_+(\mathbf{r}')}{\partial n} \right] ds', \ \mathbf{r} \in V. \tag{1.199}$$

It describes the radiation of the equivalent surface sources in open space, which replaces that of the actual sources enclosed by S.

1.7.9 Integral-Equation Models of Back-Propagation in the Frequency Domain

Similarly to the radiation models, the frequency-domain back-propagation model can be obtained from its transient counterpart Eq. (1.115) through the FT. The result is

$$U_b(\mathbf{r}) = \oiint_S \left(U_+(\mathbf{r}') \frac{\partial G_-(\mathbf{r}, \mathbf{r}')}{\partial n} - G_-(\mathbf{r}, \mathbf{r}') \frac{\partial U_+(\mathbf{r}')}{\partial n} \right) ds', \ \mathbf{r} \in V, \tag{1.200}$$

where $U_+(\mathbf{r}' \in S)$ and $\frac{\partial U_+(\mathbf{r}' \in S)}{\partial n}$ are the phasors of the boundary data. In view of the relation between Green's causal and acausal functions, Eq. (1.197), Eq. (1.200) can also be stated as

$$U_b(\mathbf{r}) = \oiint_S \left(U_+(\mathbf{r}') \frac{\partial G_+^*(\mathbf{r}, \mathbf{r}')}{\partial n} - G_+^*(\mathbf{r}, \mathbf{r}') \frac{\partial U_+(\mathbf{r}')}{\partial n} \right) ds', \ \mathbf{r} \in V. \tag{1.201}$$

As long as the observation point is outside the source region V_s, the phasor of the back-propagated field U_b is equal to that of the causal solution U_+. Thus, we can state the frequency-domain equation of the inverse-source problem as

$$\iiint_{V_s} G_+(\mathbf{r}, \mathbf{r}') S(\mathbf{r}') \, dv' = \oiint_S \left(U_+(\mathbf{r}') \frac{\partial G_-(\mathbf{r}, \mathbf{r}')}{\partial n} - G_-(\mathbf{r}, \mathbf{r}') \frac{\partial U_+(\mathbf{r}')}{\partial n} \right) ds', \mathbf{r} \notin V_s. \tag{1.202}$$

Eq. (1.202) is the frequency-domain expression of the time-domain integral equation of back-propagation Eq. (1.122).

1.8 Incident Wave, Scattered Wave, and Induced Sources of Scattering

So far, we have considered the scalar-wave and Helmholtz equations in a uniform medium along with their Green's functions. In the context of scattering, that medium serves as the background medium (or simply the *background*). By definition, the background is free of scattering objects. The scattering object, or, for short, the *scatterer*, by definition has properties different from those of the background. The scatterer can be viewed as a collection of *induced sources* (or *secondary sources*), i.e., sources that are due to the illumination by the known *external sources* (also referred to as *impressed* or *primary sources*). Understanding the concept of the induced source is critical because it enables the construction of the forward model of scattering. The induced MW sources can be surface currents induced on the surface of metallic scatterers or polarization currents induced in dielectric or magnetic scatterers. Since these sources arise due to the difference (the contrast) between the constitutive parameters of the background and those of the scatterer, these induced sources are often referred to as the *contrast sources*. The vector form of the induced EM sources is described later in Chapter 2. Here, we consider the construction of an induced source in the context of the scalar-wave equation.

Assume that a scatterer exists in a uniform source-free background, the velocity constant of which is given by υ_{b}. The velocity within the scatterer varies with position and is given by $\upsilon_{\mathrm{s}}(\mathbf{r})$. The scalar wave in the presence of the scatterer satisfies

$$\nabla^2 u(\mathbf{r},t) - \frac{1}{\upsilon_{\mathrm{s}}^2(\mathbf{r})}\frac{\partial^2 u(\mathbf{r},t)}{\partial t^2} = 0. \tag{1.203}$$

Introducing the *contrast function*

$$\kappa(\mathbf{r}) = \frac{1}{\upsilon_{\mathrm{s}}^2(\mathbf{r})} - \frac{1}{\upsilon_{\mathrm{b}}^2} \tag{1.204}$$

allows for writing Eq. (1.203) as

$$\nabla^2 u(\mathbf{r},t) - \frac{1}{\upsilon_{\mathrm{b}}^2}\frac{\partial^2 u(\mathbf{r},t)}{\partial t^2} = \kappa(\mathbf{r})\frac{\partial^2 u(\mathbf{r},t)}{\partial t^2}. \tag{1.205}$$

We now introduce the concept of *incident field* $u^{\mathrm{inc}}(\mathbf{r},t)$ as the field that would exist in the absence of the scatterer. This is the field in the background. It satisfies

$$\nabla^2 u^{\mathrm{inc}}(\mathbf{r},t) - \frac{1}{\upsilon_{\mathrm{b}}^2}\frac{\partial^2 u^{\mathrm{inc}}(\mathbf{r},t)}{\partial t^2} = 0. \tag{1.206}$$

Eq. (1.206) is source-free because here the volume of interest is that of the object under test (OUT). This volume does not contain impressed sources. Subtracting Eq. (1.206) from Eq. (1.205) results in

$$\nabla^2 u^{\mathrm{sc}}(\mathbf{r},t) - \frac{1}{\upsilon_{\mathrm{b}}^2}\frac{\partial^2 u^{\mathrm{sc}}(\mathbf{r},t)}{\partial t^2} = \kappa(\mathbf{r})\frac{\partial^2 u(\mathbf{r},t)}{\partial t^2}, \tag{1.207}$$

where $u^{\mathrm{sc}} = u - u^{\mathrm{inc}}$ is the *scattered field*. The name is descriptive because indeed this is a field entirely due to the presence of the scatterer, i.e., due to the nonzero contrast κ. If κ were zero, the particular solution of Eq. (1.207) would be trivial, $u^{\mathrm{sc}}(\mathbf{r},t) = 0$. Finally, $u = u^{\mathrm{inc}} + u^{\mathrm{sc}}$, which satisfies Eq. (1.203), is the *total field*.

The right side of Eq. (1.207) is the induced source. It can be written either in terms of the total-field time derivative (as in Eq. (1.207)) or its Laplacian (as per Eq. (1.203)):

$$s^{\mathrm{ind}}(\mathbf{r},t) = \kappa(\mathbf{r})\frac{\partial^2 u(\mathbf{r},t)}{\partial t^2} = \chi(\mathbf{r})\nabla^2 u(\mathbf{r},t), \tag{1.208}$$

where

$$\chi(\mathbf{r}) = \kappa(\mathbf{r})v_s^2(\mathbf{r}) = \frac{v_b^2 - v_s^2(\mathbf{r})}{v_b^2}. \tag{1.209}$$

We refer to $\chi(\mathbf{r})$ as the *normalized contrast function*.

The induced source described by Eq. (1.208) is different from the impressed sources that we are familiar with in that it is unknown because it depends on the unknown total field $u(\mathbf{r}, t)$. It arises from the interaction of the known incident field u^{inc} (due to the impressed source) with the scatterer. Notice that, in general, this interaction is *not* linear with respect to κ (or χ). For example, doubling the contrast κ does not double the induced source. This is because, in addition to its explicit dependence on κ, s^{ind} depends on it implicitly through the total field u.

The formulation of contrast-dependent induced sources in the frequency domain (the Helmholtz equation) is analogous. The reader is encouraged to show that for a background medium and a scatterer of propagation constants γ_b and γ_s, respectively, the scattered field phasor satisfies the Helmholtz equation

$$\nabla^2 U^{\text{sc}}(\mathbf{r}) - \gamma_b^2 U^{\text{sc}}(\mathbf{r}) = K(\mathbf{r})U(\mathbf{r}), \tag{1.210}$$

where $K = \gamma_s^2 - \gamma_b^2$. Remembering that $\gamma = \mathrm{i}\omega\sqrt{\mu\varepsilon}$, see Eq. (1.34), we can write K as

$$K(\mathbf{r}) = -k_0^2 \left[\mu_{r,s}(\mathbf{r})\varepsilon_{r,s}(\mathbf{r}) - \mu_{r,b}\varepsilon_{r,b}\right], \tag{1.211}$$

where $k_0 = \omega\sqrt{\mu_0\epsilon_0}$ is the wavenumber in vacuum, $\mu_{r,\xi} = \mu_\xi/\mu_0$ is the relative complex permeability, and $\varepsilon_{r,\xi} = \varepsilon_\xi/\epsilon_0$ is the relative complex permittivity, $\xi = \text{b, s}$. $K(\mathbf{r})$ is referred to as the *scattering potential* [14, 26]. Also,

$$\kappa(\mathbf{r}) = \mu_{r,s}(\mathbf{r})\varepsilon_{r,s}(\mathbf{r}) - \mu_{r,b}\varepsilon_{r,b}, \tag{1.212}$$

is the usual definition of the *complex contrast function* in MW scattering. Thus, the induced source representing a scatterer, in the case of Helmholtz's equation, is

$$\mathcal{S}^{\text{ind}}(\mathbf{r}) = K(\mathbf{r}) \cdot U(\mathbf{r}) = -k_0^2\kappa(\mathbf{r})U(\mathbf{r}), \tag{1.213}$$

where $U(\mathbf{r})$ is the total field. Sometimes, it is more convenient to work with the *normalized complex contrast function* $\mathcal{X}(\mathbf{r})$:

$$\mathcal{X} = \frac{K}{\gamma_b^2} = \frac{\gamma_s^2 - \gamma_b^2}{\gamma_b^2} = \frac{\mu_{r,s}\varepsilon_{r,s} - \mu_{r,b}\varepsilon_{r,b}}{\mu_{r,b}\varepsilon_{r,b}}. \tag{1.214}$$

In terms of $\mathcal{X}(\mathbf{r})$, the induced source function becomes

$$\mathcal{S}^{\text{ind}}(\mathbf{r}) = \gamma_b^2\mathcal{X}(\mathbf{r})U(\mathbf{r}). \tag{1.215}$$

Eq. (1.207) and Eq. (1.210) may not appear very useful at first glance since their source functions are unknown—they depend not only on the unknown contrast function but also on the unknown total field. However, their left-hand sides use the same background differential operator that governs the incident field; compare, for example Eq. (1.206) and Eq. (1.207). This allows for the construction of an integral solution to the inhomogeneous problem of Eq. (1.207) or Eq. (1.210), provided we know Green's function of the background.

1.9 Integral-Equation Models of Scattering in Terms of Induced Sources

The integral solutions to the wave and Helmholtz equations discussed in the previous sections form the basis of the scalar forward models of scattering. Next, we show how these models are built from two main components: Green's function of the background and the induced sources in the scatterer. An underlying principle is that the scattered field is the field generated by the unknown induced sources in the known background medium. Here, the background is assumed to be uniform and infinite in extent.

In the frequency-domain analysis, the scattered field is expressed as

$$U^{sc}(\mathbf{r}) = \iiint\limits_V G_+(\mathbf{r}, \mathbf{r}') \underbrace{\left[K(\mathbf{r}')U(\mathbf{r}')\right]}_{\mathcal{S}^{ind}(\mathbf{r}')} dv'. \tag{1.216}$$

This follows from the open-space radiation model of Eq. (1.198) where we have replaced the source $\mathcal{S}(\mathbf{r}')$ with the induced source given by Eq. (1.213). Eq. (1.216) is known as the Lippmann–Schwinger equation of scattering. Occasionally, we also make use of the equivalent form

$$U^{sc}(\mathbf{r}) = \iiint\limits_V G_+(\mathbf{r}', \mathbf{r}) K(\mathbf{r}') U(\mathbf{r}') dv', \tag{1.217}$$

where Green's function reciprocity, $G_+(\mathbf{r}, \mathbf{r}') = G_+(\mathbf{r}', \mathbf{r})$, has been employed. Yet another equivalent form utilizes Green's acausal function, which can replace the causal one using $G_+(\mathbf{r}', \mathbf{r}) = G_-^*(\mathbf{r}', \mathbf{r})$ as per Eq. (1.197).

In the time domain, the application of Eq. (1.112) leads to

$$u^{sc}(\mathbf{r}, t) = \int\limits_0^{T_{max}} \iiint\limits_V g_+(\mathbf{r}, t; \mathbf{r}', t') \underbrace{\left[\chi(\mathbf{r}')\nabla^2 u(\mathbf{r}', t')\right]}_{s^{ind}(\mathbf{r}', t')} dv' dt', \tag{1.218}$$

where the induced source is introduced in accordance with Eq. (1.208). Eq. (1.218) can also be stated in terms of Green's acausal function g_-:

$$u^{sc}(\mathbf{r}, t) = \int\limits_0^{T_{max}} \iiint\limits_V g_-(\mathbf{r}', t'; \mathbf{r}, t) \left[\chi(\mathbf{r}')\nabla^2 u(\mathbf{r}', t')\right] dv' dt', \tag{1.219}$$

where g_- replaces g_+ on the basis of Eq. (1.91).

The induced source functions in the square brackets in Eq. (1.216) and Eq. (1.218) are the unknowns in the inverse scattering problem, whereas $U^{sc}(\mathbf{r})$ or $u^{sc}(\mathbf{r}, t)$ are given by the measured data. Since the induced-source terms are nonlinear functions of the contrast, finding the contrast is intrinsically a nonlinear problem. Mathematically, this is the most important difference between the inverse problem and the forward problem.

In comparison, in the forward problem, the contrast function is known, whereas $U^{sc}(\mathbf{r})$ or $u^{sc}(\mathbf{r}, t)$ are unknown. Eq. (1.216) and Eq. (1.218) are linear with respect to the scattered field since the total field (in the integrand) is a superposition of the scattered field and the known incident field. In fact, they can be explicitly stated as linear integral equations for the total field, e.g.,

$$U(\mathbf{r}) = U^{\text{inc}}(\mathbf{r}) + \iiint\limits_{V} G_{+}(\mathbf{r}, \mathbf{r}')K(\mathbf{r}')U(\mathbf{r}')dv', \tag{1.220}$$

where U^{inc}, G_{+} and K are known. With the proper boundary conditions, the solution of this linear problem is unique. Moreover, with the proper numerical technique, the solution is well posed.

This is not the case with the inverse scattering problem. The nonlinearity of the problem is often associated with nonuniqueness. In addition, there are fundamental physical reasons for the nonuniqueness of the inverse problem, which are associated with the existence of the so-called nonradiating sources [26, 38].[25] See also Section C.3 in Appendix C. Induced nonradiating sources may arise in scattering, which implies nonuniqueness in the reconstruction of the contrast function.

Furthermore, observations of the scattered field (the data) are always incomplete due to the inability to sample the response in all space and at all times (or conversely, at all frequencies). And, unfortunately, when we measure the responses, we inadvertently add noise and uncertainties. While we may be able to deal with nonuniqueness associated with nonradiating sources by imposing constraints, the insufficient sampling and noise remain the root of trouble.

One important question is yet to be answered here with regard to the application of Eq. (1.216) or Eq. (1.218). How do we obtain Green's function for a particular problem? This is not a trivial task, and for the most part we will use ready expressions for Green's function whenever it exists in a closed form. A succinct yet thorough description of methods to determine Green's function analytically can be found in [5]. Here, only a summary is given of the requirements that such a function must meet.

- It must satisfy the partial differential equation of the problem with a point source, i.e., a source distribution given by the Dirac δ-function; see Eq. (1.43).
- It must satisfy boundary conditions such that the integral solution is as simple as possible; usually, the boundary conditions are set to match those of the solution to be found, which eliminates the need to integrate over the domain boundary $\partial\mathcal{D}$.

[25] The nonradiating EM sources are particular configurations of charges and currents, the field of which is zero outside of the source support [39, 40, 41]. Such sources are "invisible" to an observer outside the source volume. Since such sources do not contribute to the observed field, they cannot be identified or excluded from the inverse-source solution.

1.10 Green's Function as an Incident Field

Here, a practical approach is presented that allows for obtaining Green's function associated with the data. It gives the theoretical basis for obtaining Green's function of any acquisition system through measurements or simulations. This approach is particularly important when we deal with problems that do not have Green's function in a closed analytical form, e.g., near-field imaging. It is also instructive because it gives the physical meaning of Green's function in the context of scattering.

Start by writing a general integral expression for the scattered field, which unifies Eq. (1.217) and Eq. (1.219):

$$u^{\text{sc}}(\mathbf{x}) = \int_{\mathcal{D}} \mathcal{G}^{\dagger}(\mathbf{x}', \mathbf{x}) s^{\text{ind}}(\mathbf{x}') d\Omega', \tag{1.221}$$

where $\mathcal{G}^{\dagger}$ is Green's adjoint function of the background and s^{ind} is the induced source representing the scatterer. Our first goal is to determine $\mathcal{G}^{\dagger}$ as a field distribution observed at $\mathbf{x}'$ within the scatterer's volume due to a source at $\mathbf{x}$ on the acquisition surface.

1.10.1 Green's Function in the Case of Point-Wise Sampling

Let us first consider the case where the data consist of point-wise sampling of $u^{\text{sc}}(\mathbf{x})$. From a practical standpoint, point-wise sampling is difficult because it requires calibrated field probes, which are expensive and not amenable to array arrangements. Moreover, they disturb the measured field due to their relatively large size that includes not only a small antenna but also the connector and the RF cable. Nevertheless, this case is worth examining because it is simple theoretically and provides a starting point toward understanding the more complicated practical cases.

In a uniform unbounded medium, the field solutions due to a δ-source for the wave and Helmholtz equations can be found in a closed analytical form. When the δ-source is at the origin of the coordinate system, the respective solutions are referred to as the *fundamental solutions*. These are well known and are briefly reviewed in Section 1.11 and Section 1.12. Thus, a fundamental solution $F(\mathbf{x})$ to a partial differential equation is its Green's function in the case of a uniform unbounded medium with the δ-source at the origin.

Due to the uniformity of the medium and the infinite boundaries, Green's function is invariant to translations in its domain $\mathcal{D}$ and can be readily expressed in terms of the fundamental solution as[26]

$$\mathcal{G}(\mathbf{x}', \mathbf{x}) = -F(\mathbf{x}' - \mathbf{x}). \tag{1.222}$$

[26] Note the minus sign in this relation, which is used throughout the book. As explained in Section 1.11 and Section 1.12, the minus sign is due to the fact that the canonical forms of the wave equation and the Helmholtz operators used to obtain the fundamental solutions are related to the common forms of these operators in engineering through a factor of -1.

Then, Eq. (1.221) can be written as

$$u^{sc}(\mathbf{x}) = -\int_{\mathcal{D}} F^{\dagger}(\mathbf{x}' - \mathbf{x}) s^{ind}(\mathbf{x}') d\Omega'. \tag{1.223}$$

Here, $F^{\dagger}$ is the fundamental solution to the adjoint operator. In the case of Helmholtz's operator, which is self-adjoint, the adjoint and physical fundamental solutions coincide, i.e., $F^{\dagger}(\mathbf{r}' - \mathbf{r}) = F(\mathbf{r}' - \mathbf{r}) = F(\mathbf{r} - \mathbf{r}')$. Mathematically, exchanging $\mathbf{r}$ and $\mathbf{r}'$ in the fundamental solution does not change its value because the spatial dependence is always in the form of $|\mathbf{r}' - \mathbf{r}|$. In the case of the wave equation, where reciprocity involves temporal antisymmetry, $F^{\dagger}(\mathbf{r}' - \mathbf{r}, t' - t) = F(\mathbf{r} - \mathbf{r}', -(t' - t))$. Again, the spatial symmetry is ensured by a spatial dependence of the form $|\mathbf{r}' - \mathbf{r}|$.

In the context of scattering, the uniform unbounded medium is the background. When scatterers are not present, the field solution is by definition the incident field. Thus, the fundamental solution $F(\mathbf{x})$ is the incident field in a uniform unbounded background due to a point source at the origin. By the same token, Green's function $\mathcal{G}(\mathbf{x}, \mathbf{x}') = -F(\mathbf{x} - \mathbf{x}')$ is the incident field at $\mathbf{x}$ when the δ-source is at $\mathbf{x}'$.

Specifically, in the frequency-domain model of scattering, we can state that

$$F^{\dagger}(r) = F(r) = -U^{inc}_{\mathbf{r}}(\mathbf{r}'), \tag{1.224}$$

where $r = |\mathbf{r}' - \mathbf{r}|$. Note that the subscript of $U^{inc}_{\mathbf{r}}$ denotes the location of the δ-source. Thus, Eq. (1.223) can be written as

$$U^{sc}(\mathbf{r}) = \iiint_V U^{inc}_{\mathbf{r}}(\mathbf{r}') \cdot \left[K(\mathbf{r}')U(\mathbf{r}')\right] dv', \tag{1.225}$$

where the induced source is in the square brackets. The fundamental solution to the 3D Helmholtz equation for outgoing waves is $F(r) = \frac{\exp(-\gamma r)}{4\pi r}$, where γ is the propagation constant. Thus, Eq. (1.225) appears in the specific form

$$U^{sc}(\mathbf{r}) = \iiint_V \underbrace{\left[-\frac{\exp(-\gamma r)}{4\pi r}\right]}_{U^{inc}_{\mathbf{r}}(\mathbf{r}')} \cdot \left[K(\mathbf{r}')U(\mathbf{r}')\right] dv'. \tag{1.226}$$

In the time domain, as per Eq. (1.219), $\mathcal{G}^{\dagger}$ is Green's adjoint acausal function, $\mathcal{G}^{\dagger}(\mathbf{x}', \mathbf{x}) \equiv g_-(\mathbf{r}', t'; \mathbf{r}, t)$. It is found from its causal counterpart using Eq. (1.91). In 3D, the causal fundamental solution is well known (see Section 1.11):

$$F_+(r_o, t) = \frac{\delta(r_o/v - t)}{4\pi r_o}, t \geq 0, \tag{1.227}$$

where v is the wave velocity and r_o is the distance from the origin. It follows that Green's causal function is

$$g_+(\mathbf{r}, t; \mathbf{r}', t') = -\frac{\delta\left[r/v - (t - t')\right]}{4\pi r}, t \geq t'. \tag{1.228}$$

As per Eq. (1.91), $F^{\dagger}$ is found as

$$F^{\dagger}(\mathbf{x}'-\mathbf{x}) \equiv -g_{-}(\mathbf{r}',t';\mathbf{r},t) = -g_{+}(\mathbf{r},t;\mathbf{r}',t') = \frac{\delta\left[r/v+(t'-t)\right]}{4\pi r}, t' \le t, \quad (1.229)$$

and the scattered wave model in the time domain is obtained as

$$u^{\text{sc}}(\mathbf{r},t) = \int_{0}^{T_{\max}}\iiint_{V} \underbrace{\left[-\frac{\delta\left[r/v+(t'-t)\right]}{4\pi r}\right]}_{\check{u}^{\text{inc}}_{(\mathbf{r},t)}(\mathbf{r}',t')} \cdot \left[\chi(\mathbf{r}')\nabla^2 u(\mathbf{r}',t')\right] dv'dt', t \ge t'. \quad (1.230)$$

The δ-function term represents the acausal incident field observed at $(\mathbf{r}',t')$ when the δ-source is at $(\mathbf{r},t)$:

$$\check{u}^{\text{inc}}_{(\mathbf{r},t)}(\mathbf{r}',t') = -\frac{\delta\left[r/v+(t'-t)\right]}{4\pi r}, t' \le t. \quad (1.231)$$

Notice that here the adjoint (acausal) field is denoted by a ˇ mark.

In summary, if a response is obtained as a sample of the field at a point, the respective Green's function is nothing but the adjoint incident field generated by a δ-source residing at this sampling point. In the frequency domain, the adjoint incident field phasor is the same as that of the physical incident field. In the time domain, the adjoint incident field can be found from the physical one using time reversal, i.e., Eq. (1.108).

Finally, we return to Eq. (1.230) to simplify it and explicitly state a fundamental time-domain result. Using the sampling property of the δ-function, Eq. (1.230) is written as

$$u^{\text{sc}}(\mathbf{r},t) = -\iiint_{V} \frac{1}{4\pi r} \cdot \underbrace{\chi(\mathbf{r}') \cdot \nabla^2 u(\mathbf{r}',t')}_{s^{\text{ind}}(\mathbf{r}',t')} dv', \quad (1.232)$$

where $t' = t - r/v$ is the *retarded time variable.* Also, we should keep in mind that the induced source is often expressed in the equivalent form $s^{\text{ind}}(\mathbf{r}',t') = \kappa(\mathbf{r}')\frac{\partial^2 u(\mathbf{r}',t')}{\partial t^2}$, as per Eq. (1.208). We see that the scattered field is the result of the induced source, which arises at an earlier time t' as dictated by causality. Moreover, t' is in turn delayed with respect to the time of excitation, i.e., the time at which the illuminating wave was launched. Thus, a scattered response is delayed with respect to the excitation instant by the travel time from the emitter to the scatterer plus the travel time from the scatterer to the sensor.

1.10.2 Green's Function in Complex Background Medium

Let us turn to a more practical scenario. In MW measurements, we rarely sample the field at a point—this would require very small antennas, which are known to be narrow-band, inefficient, and unfocused. One may argue that while the antennas are large, they do eventually "guide" the wave energy to a point, e.g., a pair of terminals or a waveguide connecting it to the electronic circuitry. This argument is correct. However, it also implies that the antenna structure is now part of the background, which can no longer be considered uniform. Moreover, an assumption that the response at the terminals of the sensor is equal or proportional to the field value at, say, the center of the sensor surface, is crude at best. Finally, in many applications, especially short-range imaging, the background medium is assumed to be layered, not uniform. In all these cases, the fundamental solutions are not very useful.

Here, a brief note on terminology should be offered. In the context of scalar waves, it is not appropriate to refer to our receiving devices as antennas since antennas respond not only to the field intensity but also to its polarization. Hereafter, the more general term *sensor* is used for a device that transforms an impinging wave into an electrical signal. Similarly, the term *emitters* is used instead of transmitting antennas.

We consider the structure of the sensor to be part of the background medium. The role of the sensor is to focus the incident wave onto its terminals where it is converted into an electrical signal—voltage, current, or a *power wave*.[27] The sensors are usually linear devices. Thus, the measured response is some linear functional $\mathcal{R}$ (usually involving integration) of the field $u(\mathbf{x})$. In MW engineering, the terminals of each sensor and emitter are commonly referred to as *ports*.[28] An imaging setup usually consists of multiple sensors, the ports of which we distinguish by their index. The j-th port is characterized by a limited observation domain $\mathcal{D}_j$ centered at $\mathbf{x}_j$ (the j-th port center).

In frequency-sweep measurements, where the data are acquired one frequency at a time, the observation domain $\mathcal{D}_j$ is simply the spatial domain of the sensor, e.g., the cross-section S_j of the port. In time-domain measurements, $\mathcal{D}_j$ includes also time. The ports rarely have an instantaneous response, although such an assumption is often acceptable. If the port exhibits significant dispersion (i.e., it responds differently to the different frequency components of the field), its output waveform does not match the shape of the field waveform. Typically, the sensors modify the waveform of the incident field through convolution integrals [42].

Let $\mathcal{R}_j$ describe how the j-th response R_j relates to the field solution:

$$R_j = \mathcal{R}_j \{u(\mathbf{x})\} . \tag{1.233}$$

[27] The root-power waves or simply power waves are phasors describing traveling EM waves in waveguides as a function of position along the waveguide axis [13]. The magnitude of the power wave is the square root of the power carried by the traveling wave (thereby the name). The phase of the power wave corresponds to the phase of the wave, i.e., the phase of the **E** field. Note that in a loss-free waveguide, the phases of the **E** and **H** vectors at any given observation point are the same. In general, for any waveguide, even if lossy, if the phase of **E** is known, that of **H** can be found.

[28] The term *port* originates in network analysis. The port is a point of signal input or output to/from the network. Engineers view an MW imaging setup as a network formed by the sensors, the emitters, and the environment in which they operate. The input to this network comes from the emitters' ports while the outputs are the sensors' ports.

In frequency-sweep measurements, R_j is a complex-valued data point at each frequency. In pulsed measurements, each experiment yields a real-valued waveform; that is, a function of time.

Since $u = u^{\text{inc}} + u^{\text{sc}}$, the response consists of two contributions, R_j^{inc} and R_j^{sc}. We are interested in the latter because it carries the information about the scatterer. In a baseline (or calibration) measurement in the scatterer-free background, we can always acquire R_j^{inc} and subtract it from the measurement with the OUT (R_j) to obtain R_j^{sc} as

$$R_j^{\text{sc}} = R_j - R_j^{\text{inc}}. \tag{1.234}$$

$\mathcal{R}_j$ is next applied to both sides of Eq. (1.221) to obtain an explicit integral solution in terms of R_j^{sc}. Since $\mathcal{R}_j$ operates on $\mathbf{x}$ (not on $\mathbf{x}'$), it can be inserted in the integral, leading to

$$R_j^{\text{sc}} = \int_{\mathcal{D}} \mathcal{R}_j \left\{ \mathcal{G}^{\dagger}(\mathbf{x}', \mathbf{x}) \right\} \cdot s^{\text{ind}}(\mathbf{x}')\, d\Omega'. \tag{1.235}$$

Similarly to the case of point-wise sampling, we view the term $\mathcal{R}_j \left\{ \mathcal{G}^{\dagger}(\mathbf{x}', \mathbf{x}) \right\}$ as an (adjoint) incident field $\check{u}_j^{\text{inc}}(\mathbf{x}')$,

$$\check{u}_j^{\text{inc}}(\mathbf{x}') = \mathcal{R}_j \left\{ \mathcal{G}(\mathbf{x}', \mathbf{x}) \right\}, \tag{1.236}$$

due to some source distribution in $\mathcal{D}_j$. To find out what is that source, $\mathcal{R}_j$ is applied to both sides of the adjoint Green-function equation $\mathcal{L}^{\dagger}\mathcal{G}^{\dagger} = \delta$, where now $\mathbf{x}'$ is the observation point while $\mathbf{x}$ is the source point. This results in

$$\mathcal{L}_{\mathbf{x}'}^{\dagger} \mathcal{R}_j \left\{ \mathcal{G}(\mathbf{x}', \mathbf{x}) \right\} = \mathcal{L}_{\mathbf{x}'}^{\dagger} \check{u}_j^{\text{inc}}(\mathbf{x}') = \mathcal{R}_j \left\{ \delta(\mathbf{x}' - \mathbf{x}) \right\}, \tag{1.237}$$

where $\mathcal{L}_{\mathbf{x}'}^{\dagger}$ operates on $\mathbf{x}'$ and this is emphasized by the added subscript. Since $\mathcal{R}_j$ operates on $\mathbf{x}$, the order in which the operators $\mathcal{L}_{\mathbf{x}'}^{\dagger}$ and $\mathcal{R}_j$ are applied to $\mathcal{G}(\mathbf{x}', \mathbf{x})$ has been exchanged. The rightmost side of Eq. (1.237) identifies the source that would produce the desired field $\check{u}_j^{\text{inc}}(\mathbf{x}')$ defined through Eq. (1.236). Note that this auxiliary source, defined through

$$\check{s}_j(\mathbf{x}') = \mathcal{R}_j \left\{ \delta(\mathbf{x}' - \mathbf{x}) \right\}, \tag{1.238}$$

resides in the domain of observation $\mathcal{D}_j$ in the actual problem setup.

In summary, a response $R_j^{\text{sc}} = \mathcal{R}_j \{u^{\text{sc}}(\mathbf{x})\}$ can be expressed as

$$R_j^{\text{sc}} = \int_{\mathcal{D}} \check{u}_j^{\text{inc}}(\mathbf{x}') \cdot s^{\text{ind}}(\mathbf{x}')\, d\Omega', \tag{1.239}$$

where $\check{u}_j^{\text{inc}}(\mathbf{x}')$ is the adjoint field in the background medium generated by the auxiliary source distribution $\check{s}_j(\mathbf{x}) = \mathcal{R}_j \left\{ \delta(\mathbf{x}' - \mathbf{x}) \right\}$.

The reader is reminded that the adjoint incident phasor coincides with that of the physical solution, i.e., $\check{U}_j^{\text{inc}}(\mathbf{r}') \equiv U_j^{\text{inc}}(\mathbf{r}')$. In the time domain, the adjoint background field is the acausal counterpart of the physical one, i.e., $\check{u}_j^{\text{inc}} \equiv (u_j^{\text{inc}})_-$.

1.10.3 Auxiliary Sources Generating Green's Function (Adjoint Sources)

In Section 1.10.1, we showed that the auxiliary source that generates Green's function in the case of field point sampling is nothing but a point source at the sampling location. Here the auxiliary source is derived in the case of a linear sensor, which responds to the field distribution in its vicinity.

In frequency-domain measurements, the response can be defined as a generic linear functional of the scattered field:

$$R_j^{\text{sc}} = \mathcal{R}_j\left\{U^{\text{sc}}(\mathbf{r})\right\} = \iiint_V M_j(\mathbf{r}) \cdot U^{\text{sc}}(\mathbf{r})\,dv. \tag{1.240}$$

Here, $M_j(\mathbf{r})$ is referred to as j-th port's *modal distribution* (or simply, its *mode*). Note that the integration is over V instead of the port cross-section $S_j \subset V$ because the modal distribution is zero outside S_j, $M_j(\mathbf{r} \notin S_j) = 0$. In waveguides, $M_j(\mathbf{r})$ is usually real-valued but in principle there are no restrictions in using complex-valued modes.

Applying the so-defined linear operator $\mathcal{R}_j$ to the δ-function, see Eq. (1.238), we see that the auxiliary source is the convolution of the modal distribution with the δ-function; therefore,

$$\check{S}_j(\mathbf{r}') = M_j(\mathbf{r}'),\ \mathbf{r}' \in S_j. \tag{1.241}$$

Thus, *the auxiliary source associated with a sensor is its modal distribution.*

Note that if $R_j^{\text{sc}} = u^{\text{sc}}(\mathbf{r}_j)$ (point-wise sampling), then $M_j(\mathbf{r}) = \delta(\mathbf{r} - \mathbf{r}_j)$, and the auxiliary source is the point source $\check{S}_j(\mathbf{r}) = \delta(\mathbf{r} - \mathbf{r}_j)$. Thus, the case considered in Section 1.10.1 is a particular case of the current development. However, in the current framework, the result for the point-wise sampling is more general in the sense that the incident field generated by the auxiliary point source does not have to be the fundamental solution. It is a point source at the sensor's input terminals, and it operates in a nonuniform medium that includes the sensor's structure.

The auxiliary-source derivation in the time domain for a linear nondispersive sensor is similar. The response at the observation time t_o is expressed as

$$r_j^{\text{sc}}(t_o) = \mathcal{R}_j\left\{u^{\text{sc}}(\mathbf{r},t)\right\} = \int_{-\infty}^{\infty}\iiint_V m_j(\mathbf{r}) \cdot \delta(t - t_o) \cdot u^{\text{sc}}(\mathbf{r},t)\,dv\,dt, \tag{1.242}$$

where $m_j(\mathbf{r})$ is the (real-valued) modal distribution in the port cross-section. The nondispersive sensor performs point-wise temporal sampling of the field waveform, and this is represented by $\delta(t - t_o)$. Thus, the sensor's modal distribution in time is a δ-function. Then the adjoint source $\check{s}_j(\mathbf{r}',t') = \mathcal{R}_j\left\{\delta(\mathbf{r}' - \mathbf{r})\delta(t' - t)\right\}$ is

$$\check{s}_j(\mathbf{r}',t') = \int_{-\infty}^{\infty} \delta(t' - t)\delta(t - t_o)dt \times \iiint_V m_j(\mathbf{r}) \cdot \delta(\mathbf{r}' - \mathbf{r})\,dv, \tag{1.243}$$

which results in

$$\check{s}_j(\mathbf{r}',t') = m_j(\mathbf{r}')\delta(t' - t_o). \tag{1.244}$$

In the above, the identity $\int_{-\infty}^{\infty} \delta(a-x)\delta(x-b)dx = \delta(a-b)$ [14] was used. We can now formally replace the observation time variable t_o with t:

$$\check{s}_j(\mathbf{r}', t') = m_j(\mathbf{r}')\delta(t' - t). \quad (1.245)$$

Thus, *the spatial distribution of the time-domain auxiliary source is the j-th port mode whereas in time it is a δ-function*, which is in line with the fact that the sensor performs point-wise temporal sampling.

The auxiliary sources in Eq. (1.241) and Eq. (1.245) are referred to as *adjoint sources*. The fields they generate in the background medium ($\check{U}_j^{\text{inc}}$ and $\check{u}_j^{\text{inc}}$) are the (j-th) *adjoint fields* governed by the adjoint equation $\mathcal{L}^\dagger \check{u} = \check{s}$.

1.10.4 Relationship between Actual Sources and Adjoint Sources

We next show that the adjoint excitation defined by the mode of the j-th port is not just a mathematical abstraction but is also closely related to the actual excitation of the j-th sensor if it were to operate as an emitter.

Let us start with the frequency-domain case. The port's mode $M_j(\mathbf{r})$, describes its operation not only in receiving but also in transmitting mode, i.e., the equivalent source distribution at the transmitting sensor terminals is such that

$$\mathcal{S}_j(\mathbf{r}) \propto M_j(\mathbf{r}). \quad (1.246)$$

This is due to the fact that the feed line (or the waveguide) connecting the sensor to the circuitry supports the same unique mode of field distribution in its cross-section regardless of whether the wave is incoming or outgoing.[29] Comparing Eq. (1.246) to Eq. (1.241) shows a simple proportionality between the adjoint excitation $\check{\mathcal{S}}_j(\mathbf{r})$ and the actual excitation $\mathcal{S}_j(\mathbf{r})$ of the j-th sensor (when it transmits):

$$\check{\mathcal{S}}_j(\mathbf{r}) = \check{A}_j \mathcal{S}_j(\mathbf{r}). \quad (1.247)$$

The proportionality factor $\check{A}_j$, named the *adjoint constant*, depends solely on the intensity and the initial phase of the actual source. The adjoint constant is known. Some examples are given in the next section.

Since Helmholtz's operator is linear and self-adjoint, Eq. (1.247) implies the same linear relationship between the adjoint field $\check{U}_j^{\text{inc}}(\mathbf{r}')$ and the actual field $U_j^{\text{inc}}(\mathbf{r}')$:

$$\check{U}_j^{\text{inc}}(\mathbf{r}') = \check{A}_j U_j^{\text{inc}}(\mathbf{r}'). \quad (1.248)$$

Substituting Eq. (1.248) in the frequency-domain version of the scattering model Eq. (1.235) leads to

$$R_j^{\text{sc}} = \check{A}_j \iiint_V U_j^{\text{inc}}(\mathbf{r}') \cdot \mathcal{S}^{\text{ind}}(\mathbf{r}')dv' = \check{A}_j \iiint_V K(\mathbf{r}') \cdot U_j^{\text{inc}}(\mathbf{r}') \cdot U(\mathbf{r}')dv'. \quad (1.249)$$

[29] Multi-mode waveguides may also be employed although this is rare. This case is far more complicated, in addition to being case-specific, and is not considered here.

The importance of this result cannot be overemphasized. We have just replaced the unrealistic point-wise response $U(\mathbf{r}_j)$ with a quantity (R_j^{sc}) that we can actually measure—voltage, current, power wave. We have also replaced Green's function—which is never known exactly except for an unbounded uniform medium—with something that we can actually simulate or measure, namely, the incident field U_j^{inc} of the j-th sensor when this sensor operates as an emitter in the scatterer-free environment.

We next turn to the case of pulsed (time-domain) systems. Unlike the adjoint source, the actual excitation of a sensor cannot be a δ-function in time due to the technical difficulties associated with the extremely wide bandwidth. Therefore, unlike the frequency-domain case, there is no simple proportionality between the actual source and the adjoint one, the latter being defined in Eq. (1.245). On the other hand, the spatial distributions of the actual and the adjoint sources are the same. Thus, while the j-th adjoint source is $\breve{s}_j(\mathbf{r}, t) = m_j(\mathbf{r})\delta(t' - t)$ as per Eq. (1.245), the actual source is

$$s_j(\mathbf{r}, t) = m_j(\mathbf{r}) w_j(t), \tag{1.250}$$

where $w_j(t)$ is the actual band-limited waveform. Note that we use nonprimed coordinates for the source in accordance with the fact that the adjoint source exists at the sensor's port, i.e., at an observation location. Meanwhile, the primed coordinates are those at which the field is observed because the adjoint field serves as Green's function in the integral solution. Since the adjoint excitation is a δ-function in time $\delta(t' - t)$, the adjoint field $\breve{u}_j(\mathbf{r}', t')$ is the acausal impulse response in the background when the j-th port is excited:

$$\breve{u}_j^{\text{inc}}(\mathbf{r}', t') = \breve{h}_j^{\text{inc}}(\mathbf{r}', t' - t)\,,\ t' \le t. \tag{1.251}$$

On the other hand, the field $u_j^{\text{inc}}(\mathbf{r}', t')$ excited by the actual source $s_j(\mathbf{r}, t)$ (the j-th sensor in transmitting mode) can be expressed through the system's causal impulse response as

$$u_j^{\text{inc}}(\mathbf{r}', t') = h_j^{\text{inc}}(\mathbf{r}', t') * w_j(t'), \tag{1.252}$$

where $*$ denotes convolution. The j-th transfer function $\tilde{H}_j^{\text{inc}}(\mathbf{r})$ can now be determined from the FTs $\tilde{U}_j^{\text{inc}}(\mathbf{r}', \omega)$ and $\tilde{W}_j(\omega)$ of $u_j^{\text{inc}}(\mathbf{r}', t')$ and $w_j(t)$, respectively:

$$\tilde{H}_j^{\text{inc}}(\mathbf{r}', \omega) = \frac{\tilde{U}_j^{\text{inc}}(\mathbf{r}', \omega)}{\tilde{W}_j(\omega)}. \tag{1.253}$$

Therefore, the causal impulse response can be determined from

$$h_j^{\text{inc}}(\mathbf{r}', t') = \mathcal{F}^{-1}\left\{\frac{\tilde{U}_j^{\text{inc}}(\mathbf{r}', \omega)}{\tilde{W}_j(\omega)}\right\}, \tag{1.254}$$

where $\mathcal{F}^{-1}$ denotes the inverse FT. Note that Eq. (1.254) is based on the assumption that within the frequency band of interest $\tilde{W}_j(\omega) \neq 0$.

Further, the adjoint impulse response $\check{h}_j^{\text{inc}}(\mathbf{r}', t')$ in Eq. (1.251) is expressed as

$$\check{h}_j^{\text{inc}}(\mathbf{r}', t') = h_j^{\text{inc}}(\mathbf{r}', \tau'), \tag{1.255}$$

where $\tau' = T_{\max} - t'$ is the reversed time variable. Substituting Eq. (1.255) into Eq. (1.251), we obtain the adjoint incident field in terms of the causal impulse response:

$$\check{u}_j^{\text{inc}}(\mathbf{r}', t') = h_j^{\text{inc}}(\mathbf{r}', \tau' - \tau) = h_j^{\text{inc}}(\mathbf{r}', t - t'). \tag{1.256}$$

Finally, the result in Eq. (1.256) is substituted in the time-domain version of the scattering model Eq. (1.235) to obtain the respective integral equation:

$$r_j^{\text{sc}}(t) = \iiiint\limits_{0\ V}^{T_{\max}} h_j^{\text{inc}}(\mathbf{r}', t - t') \cdot s^{\text{ind}}(\mathbf{r}', t')\, dv'\, dt', \tag{1.257}$$

where $[0, T_{\max}]$ is the observation interval.

Notice that the time integration above appears in the form of convolution with the impulse response. Taking into account that $s^{\text{ind}}(\mathbf{r}', t') = \kappa(\mathbf{r}')\frac{\partial^2 u(\mathbf{r}', t')}{\partial t'^2}$, see Eq. (1.208), Eq. (1.257) can be restated as

$$r_j^{\text{sc}}(t) = \iiint\limits_V \left[h_j^{\text{inc}} * s^{\text{ind}}\right]_{(\mathbf{r}', t)} dv' = \iiint\limits_V \kappa(\mathbf{r}') \left[h_j^{\text{inc}} * u''\right]_{(\mathbf{r}', t)} dv', \tag{1.258}$$

where $u'' \equiv \frac{\partial^2 u}{dt^2}$.

It is worth comparing this result to the frequency-domain one in Eq. (1.249). In phasor form, Eq. (1.258) is written as

$$R_j^{\text{sc}} = \iiint\limits_V H_j^{\text{inc}}(\mathbf{r}') \cdot K(\mathbf{r}') \cdot U(\mathbf{r}')\, dv' = \iiint\limits_V H_j^{\text{inc}}(\mathbf{r}') \cdot S^{\text{ind}}(\mathbf{r}'), \tag{1.259}$$

where $H_j^{\text{inc}}(\mathbf{r}')$ is the phasor of the "transfer function" describing the field distribution under the excitation of the j-th sensor with a source of unit strength. This follows from the fact that, at any given frequency ω, the phasor $H_j^{\text{inc}}(\mathbf{r}', \omega)$ corresponds to the FT of $h_j^{\text{inc}}(\mathbf{r}', t)$ (see Eq. (1.254)), i.e.,

$$H_j^{\text{inc}}(\mathbf{r}', \omega) = \frac{U_j^{\text{inc}}(\mathbf{r}', \omega)}{W_j(\omega)}. \tag{1.260}$$

Thus, $H_j^{\text{inc}}(\mathbf{r}')$ is exactly $\check{A}_j U_j^{\text{inc}}(\mathbf{r}')$ in Eq. (1.249), leading to the expected agreement between the time-domain and frequency-domain models. Eq. (1.260) also elucidates the meaning of the adjoint constant $\check{A}_j$ as a constant normalizing the field strength and phase to those of the excitation waveform w_j at each frequency.

1.10.5 Examples of Adjoint Sources

Eq. (1.249) and Eq. (1.258) allow for dispensing with most of the approximations regarding the acquisition setup, i.e., the sensors, the transmitters, and the environment in which they operate. Some of the common approximations are: the sensors and the emitters are infinitesimal, the field propagates in a uniform unbounded medium, Green's function is a spherical or a cylindrical wave, and evanescent field components are negligible. Often these approximations corrupt the forward model and lead to inaccurate inversion results. Moreover, they are not applicable in near-field imaging or imaging in bounded regions such as within metallic enclosures.

Let us consider the scenario of multiple emitters and sensors. A response in this case depends not only on which sensor receives the scattered signal but also on which emitter illuminates the imaged object. It is customary to denote a response as R_{jk} where the first subscript (j) indexes the sensor while the second subscript (k) indexes the emitter. We continue to employ upper-case letters for frequency-domain (phasor) responses while the lower-case symbol $r_{jk}(t)$ denotes a time-domain response. Also, to simplify notations, we drop the superscript ^{sc} in the notation of a response. Unless stated otherwise, a response will be associated with the scattered field component only.

For phasors, Eq. (1.249) is expressed in terms of the induced source as

$$R_{jk} = \breve{A}_j \iiint\limits_V U_j^{\text{inc}}(\mathbf{r}') \cdot K(\mathbf{r}')U_k(\mathbf{r}')dv'. \tag{1.261}$$

Note that we have indexed the field distributions as these are generated by different sources: U_j^{inc} is the field due to the j-th sensor when it operates as an emitter in the background; U_k is the total field, which is generated by the k-th emitter.

In time-domain measurements of pulsed fields, Eq. (1.257) takes the form

$$r_{jk}(t) = \int\limits_0^{T_{\max}} \iiint\limits_V h_j^{\text{inc}}(\mathbf{r}', t-t') \cdot \chi(\mathbf{r}')\nabla^2 u_k(\mathbf{r}', t')dv'dt', \tag{1.262}$$

where the normalized contrast function $\chi(\mathbf{r}')$ is assumed to be frequency-independent.

We next consider two frequency-domain examples where we derive the adjoint source and construct the integral equation for the respective responses R_{jk}.

Example 1: The j-th sensor, located at $\mathbf{r}_j$, has terminals that are sufficiently small to provide point-like field sampling. The response of interest is the field sampled at $\mathbf{r}_j$ when the k-th emitter is active, i.e., $R_{jk} = \iiint_V \delta(\mathbf{r}-\mathbf{r}_j)U_k^{\text{sc}}(\mathbf{r})dv$. Note that the sensor and the emitter can be of any shape or size. It is the sensor's terminals that are small. When the j-th sensor is set to transmit, it is described by the source function $S_j(\mathbf{r}) = a_j \Delta v_j \delta(\mathbf{r}-\mathbf{r}_j)$, where a_j is its strength (or volume density) and Δv_j is its volume. Derive the adjoint source and the adjoint constant. State the integral equation for R_{jk} in terms of: (i) the actual incident field $U_j^{\text{inc}}(\mathbf{r}')$ due to the j-th sensor in transmitting mode, (ii) the total field due to the k-th emitter $U_k(\mathbf{r})$, and (iii) the scattering potential $K(\mathbf{r})$.

Solution: The sensor's mode, see Eq. (1.240), is $M_j(\mathbf{r}) = \delta(\mathbf{r} - \mathbf{r}_j)$; therefore, the adjoint source is $\check{S}_j(\mathbf{r}) = \delta(\mathbf{r} - \mathbf{r}_j)$, see Eq. (1.241). It is apparent that this adjoint source is a mathematical abstraction. Its strength is exactly equal to its vanishing volume Δv_j, so that $a_j \Delta v_j = 1$. At the same time, if the j-th sensor were to operate as an emitter, it would have some (known) strength a_j and volume Δv_j. It is described by $S_j(\mathbf{r}) = a_j \Delta v_j \delta(\mathbf{r} - \mathbf{r}_j)$. The adjoint constant is then $\check{A}_j = (a_j \Delta v_j)^{-1}$. Using Eq. (1.261), the integral model of the response R_{jk} is obtained as

$$R_{jk} = \frac{1}{a_j \Delta v_j} \iiint_V U_j^{\text{inc}}(\mathbf{r}') \cdot K(\mathbf{r}') U_k(\mathbf{r}') dv'. \tag{1.263}$$

In a slightly modified scenario, the response of interest could again be the field sampled at $\mathbf{r}_j$ but normalized to the strength of the transmitted signal a_k (assumed known), i.e., R_{jk} is the transmission coefficient $R_{jk} = \frac{1}{a_k} \iiint_V \delta(\mathbf{r} - \mathbf{r}_j) U_k(\mathbf{r}) dv$. The integral-equation model for this transmission coefficient is

$$R_{jk} = \frac{1}{a_k a_j \Delta v_j} \iiint_V U_j^{\text{inc}}(\mathbf{r}') \cdot K(\mathbf{r}') U_k(\mathbf{r}') dv'. \tag{1.264}$$

Example 2: The j-th sensor, centered on $\mathbf{r}_j = (0, 0, 0)$, samples the field along a finite line such that the observation domain is 1D and defined by $-L/2 \le z \le L/2$. Let the response be the average of the field values in this domain:

$$R_{jk} = \frac{1}{L} \int_{-L/2}^{L/2} U_k^{\text{sc}}(0, 0, z) dz. \tag{1.265}$$

Derive the adjoint source and the adjoint constant. State the integral equation for R_{jk} in terms of: (i) the actual incident field $U_j^{\text{inc}}(\mathbf{r}')$ due to the j-th sensor (when in transmitting mode), (ii) the total field due to the k-the emitter $U_k(\mathbf{r})$, and (iii) the scattering potential $K(\mathbf{r})$.

Solution: As per Eq. (1.238), the adjoint excitation is

$$\check{S}_j(z) = \frac{1}{L} \int_{-L/2}^{L/2} \delta(z - z') dz, \tag{1.266}$$

which results in

$$\check{S}(z) = \begin{cases} \frac{1}{L}, & |z| \le \frac{L}{2} \\ 0, & |z| > \frac{L}{2}. \end{cases} \tag{1.267}$$

This is a line source of finite length L centered at $\mathbf{r}_j = (0, 0, 0)$ and distributed along z. It has constant strength $\frac{1}{L}$ along its length.

On the other hand, if this averaging sensor were to operate as an emitter, its actual source distribution would be described by

$$\mathcal{S}_j(z) = \begin{cases} a_{\ell,j}, & |z| \leq \frac{L}{2} \\ 0, & |z| > \frac{L}{2}, \end{cases} \tag{1.268}$$

where $a_{\ell,j}$ is the source line density. Therefore, the adjoint constant is $\breve{A}_j = (a_{\ell,j}L)^{-1}$ and the integral-equation model for R_{jk} is

$$R_{jk} = \frac{1}{a_{\ell,j}L} \iiint_V U_j^{\text{inc}}(\mathbf{r}') \cdot K(\mathbf{r}')U_k(\mathbf{r}')dv'. \tag{1.269}$$

Let us now consider a time-domain example.

Example 3: The j-th sensor is centered on $\mathbf{r}_j = (0, 0, 0)$, and its excitation waveform (when operating as an emitter) is given by $w_j(t) = A_j \Pi(t - 0.5T_j)$, where $\Pi(t)$ is the box function

$$\Pi(t) = \begin{cases} 1, & |t| \leq 0.5T_j \\ 0, & \text{otherwise.} \end{cases} \tag{1.270}$$

The resulting incident field distribution is given by

$$u_j^{\text{inc}}(\mathbf{r}, t) = \frac{m_j A_j}{4\pi r} \Lambda(t - 0.5T_j - r/v), \tag{1.271}$$

where m_j is a factor accounting for the impact the sensor has on the signal strength, $r = |\mathbf{r}|$ is the distance between the observation point and the sensor's center, v is the phase velocity, and Λ is the triangular function

$$\Lambda(t) = \begin{cases} 1 - \frac{|t|}{T_j}, & |t| \leq T_j \\ 0, & \text{otherwise.} \end{cases} \tag{1.272}$$

State the integral-equation model of $r_{jk}(\mathbf{r}, t)$ in terms of the total field due to the k-th emitter $u_k(\mathbf{r}', t')$, and the normalized contrast function $\chi(\mathbf{r}')$.

Solution: The FT of the box function is

$$\tilde{\Pi}(\omega) = T_j \frac{\sin\left(\frac{\omega T_j}{2}\right)}{\frac{\omega T_j}{2}} = T_j \text{sinc}\left(f T_j\right), \tag{1.273}$$

where $f = \omega/(2\pi)$ is the frequency in *hertz* and the sinc function is $\text{sinc}(x) = \frac{\sin(\pi x)}{\pi x}$. The FT of $w_j(t)$ is then

$$\tilde{W}_j(\omega) = A_j T_j \text{sinc}\left(f T_j\right) e^{-i\pi f T_j}. \tag{1.274}$$

The FT of the triangular function is

$$\tilde{\Lambda}(\omega) = T_j \text{sinc}^2\left(f T_j\right). \tag{1.275}$$

Therefore, that of $u_j^{\text{inc}}(\mathbf{r}, t)$ is

$$\tilde{U}_j^{\text{inc}}(\mathbf{r}, \omega) = \frac{m_j A_j T_j}{4\pi r} \text{sinc}^2\left(f T_j\right) e^{-\text{i}\pi f T_j} e^{-\text{i}2\pi f r/v}. \tag{1.276}$$

The field "transfer function" is then

$$\tilde{H}_j^{\text{inc}}(\mathbf{r}, \omega) = \frac{\tilde{U}_j^{\text{inc}}(\mathbf{r}, \omega)}{\tilde{W}_j(\omega)} = \frac{m_j}{4\pi r} \text{sinc}(f T_j) e^{-\text{i}2\pi f r/v}. \tag{1.277}$$

The inverse FT of the Eq. (1.277) produces the field "impulse response":

$$h_j^{\text{inc}}(\mathbf{r}, t) = \frac{m_j}{4\pi r} \frac{1}{T_j} \Pi\left(t - r/v\right). \tag{1.278}$$

As per Eq. (1.257), the integral-equation model of $r_{jk}(\mathbf{r}, t)$ is obtained as

$$r_{jk}(t) = \frac{m_j}{T_j} \int_0^{T_{\max}} \iiint_V \frac{1}{4\pi r} \Pi\left(t - t' - r/v\right) \cdot \chi(\mathbf{r}') \nabla^2 u_k(\mathbf{r}', t') \, dv' \, dt'. \tag{1.279}$$

Using Eq. (1.270), Eq. (1.279) can be simplified to

$$r_{jk}(t) = \frac{m_j}{T_j} \int_{t_r - 0.5 T_j}^{t_r + 0.5 T_j} \iiint_V \frac{\chi(\mathbf{r}') \nabla^2 u_k(\mathbf{r}', t')}{4\pi r} \, dv' \, dt', \tag{1.280}$$

where $t_r = t - r/v$ is the retarded time.

If multiple sensors and emitters are used to perform the data acquisition, the procedure of finding Green's function as the incident field due to a sensor has to be repeated for all of them. However, if the medium surrounding the sensors is uniform and infinite, and if the same sensor is used at all observation locations, then the adjoint incident field distribution obtained for just one sensor location can be properly translated and/or rotated to obtain all other distributions. In MW imaging, the data are acquired over the acquisition surface, which surrounds the imaged region. In doing so, we sample the field by either mechanically scanning a sensor over the surface or by using an electronically switched array of sensors. In either case, the observation point moves over the acquisition surface while the sensor properties remain the same (or nearly so). In addition, the acquisition surfaces are typically chosen to have canonical shapes such as portions of planes, cylinders, or spheres. Such shapes make the reconstruction simpler. But they also allow for exploiting translational or rotational symmetries and can thus save significant time and effort when acquiring the system's Green functions through simulation or measurement.

Table 1.2 Fundamental Solutions to the Wave Equation

wavefront	*planar*	*spherical*	*cylindrical*
equation	$\frac{1}{v^2}\frac{\partial^2 f}{\partial t^2} - \frac{\partial^2 f}{\partial z^2} = \delta(z)\delta(t)$	$\frac{1}{v^2}\frac{\partial^2 f}{\partial t^2} - \frac{1}{r^2}\frac{\partial}{\partial r}\left(r^2\frac{\partial f}{\partial r}\right) = \delta_r^3\delta(t)$	$\frac{1}{v^2}\frac{\partial^2 f}{\partial t^2} - \frac{1}{\rho}\frac{\partial}{\partial \rho}\left(\rho\frac{\partial f}{\partial \rho}\right) = \delta_\rho^2\delta(t)$
range	$-\infty < z < \infty$	$0 \le r < \infty$	$0 \le \rho < \infty$
solution	$f_{\text{pw}}(z,t) = \frac{v}{2}H(t - \lvert z\rvert/v)$	$f_{\text{sw}}(r,t) = \frac{\delta(t - r/v)}{4\pi r}$	$f_{\text{cw}}(\rho,t) = \frac{H(t-\rho/v)}{2\pi\sqrt{t^2-(\rho/v)^2}}$

Notes: 1) $H(\cdot)$ is Heaviside's step function. 2) $\delta_r^3 = \delta(x)\delta(y)\delta(z)$. 3) $\delta_\rho^2 = \delta(x)\delta(y)$.

1.11 Fundamental Solutions to the Wave Equation

In this section we summarize and discuss the three fundamental solutions to the wave equation in the cases of planar, cylindrical, and spherical spatial symmetries. These solutions are critically important in constructing our forward models of scattering because they give us Green's functions in the case of a uniform unbounded background. The fundamental solutions are derived under the same symmetry assumptions used for the respective general solutions in a uniform medium; see Section 1.2. The fundamental solutions, however, are particular solutions derived under the following conditions: (i) the source is a δ-function *at the origin*, (ii) the boundaries are nonreflecting (unbounded medium), and (iii) the initial conditions are $u(\mathbf{r}, 0) = 0$, $\partial u(\mathbf{r}, 0)/\partial t = 0$.

Table 1.2 summarizes the wave equations in the three cases and their respective causal fundamental solutions [6, 31]. The reader will notice that the equations in Table 1.2 are based on the wave operator in the form $\mathcal{L} = \frac{1}{v^2}\frac{\partial^2}{\partial t^2} - \nabla^2$ whereas up until now we have worked under the convention that $\mathcal{L} = \nabla^2 - \frac{1}{v^2}\frac{\partial^2}{\partial t^2}$. The former convention is common in the mathematical studies on the fundamental solutions of the various partial differential equations. It is convenient because, for a positive impulse excitation, it yields a positive initial response at any point. The latter convention, however, is widely accepted in the engineering literature, and it is adopted throughout this text. The reader needs only to remember that when using a fundamental solution $f_{\text{w}}(\mathbf{r}, t)$ as Green's function $g(\mathbf{r}, t; \mathbf{r}', t)$, a minus sign must be applied, i.e.,

$$g(\mathbf{r}, t; \mathbf{r}', t) = -f_{\text{w}}(\mathbf{r} - \mathbf{r}', t - t'). \tag{1.281}$$

Fig. 1.22 illustrates the fundamental solution of the 1D wave equation (see Table 1.2, 2nd column). The source located at the origin $z = 0$ of the 1D space is also an impulse function of time. The source is constant in x and y (uniform planar source), and as a result, so is the wave it generates—a uniform plane wave propagating along z. Fig. 1.22 shows a "snapshot" of the fundamental plane wave solution $F_{\text{pw}}(z, t)$ at two time instants t_1 and t_2 where $t_1 < t_2$. This is a "step" wave, which expands with time maintaining a constant value of $0.5v$ all the way from $z = 0$ to $z = vt$, at which point it drops to zero. The left and right arrows in Fig. 1.22 point in the two directions $+z$ and $-z$ along which the "step" wave expands with time. The figure also shows a plot of the waveform as a function of time observed at the position $z_1 = vt_1$. Here t_1 is the time instant at which

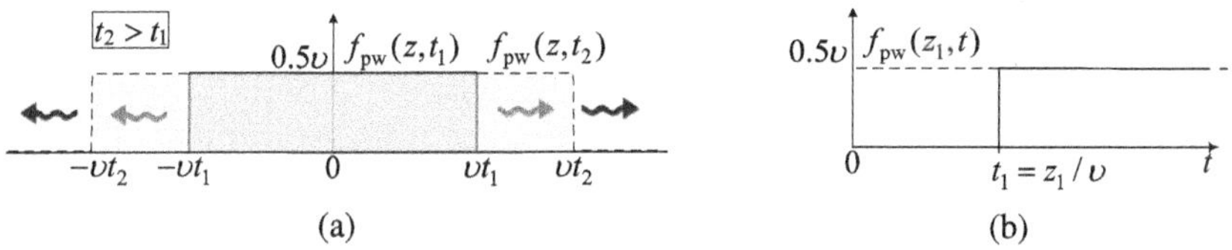

Figure 1.22 Illustration of the fundamental solution of the wave equation for a uniform plane wave; see Table 1.2, 2nd column. (a) The plot shows two "snapshots" at instants t_1 and t_2 where $t_2 > t_1$. The left and right wavy arrows point in the two directions $+z$ and $-z$ along which this "step" wave expands with time. (b) The plot shows the waveform observed at a position $z_1 = \upsilon t_1$.

the wave arrives at z_1. Prior to this instant, the waveform is zero. At t_1, the wave value "jumps" to 0.5υ and remains constant with time after that. Note that the 1D fundamental solution does not have singularities.

The fundamental solution in the case of spherical symmetry (see Table 1.2, 3rd column) is illustrated in Fig. 1.23. The wave has the shape of a δ-function in space and time. Its envelope in space is given by $\frac{1}{4\pi r}$, which is shown by the dotted line in the plot. To plot the δ-distribution, an approximation is used based on the expression [43]

$$\delta(r) = \frac{1}{\pi} \lim_{\varepsilon \to 0} \left(\frac{\varepsilon}{r^2 + \varepsilon^2} \right), \tag{1.282}$$

with the parameter ε set equal to 0.025. For a distance r from the origin, the temporal δ-distribution is centered at $t = r/\upsilon$.

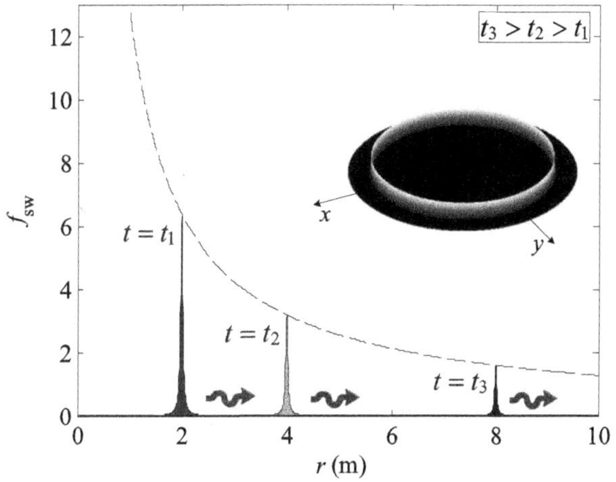

Figure 1.23 Illustration of the fundamental solution of the wave equation for a uniform spherical wave; see Table 1.2, 3rd column. The plot shows "snapshots" of this δ-wave as a function of r at three time instants, t_1, t_2, and t_3, where $t_1 < t_2 < t_3$. The wavy arrows point in the direction of expansion of the spherical wave. The wave "strength" decreases as $\sim 1/r$, a tendency shown by the dotted line. The wave propagates as an expanding, very thin spherical shell. A cut of the wave distribution in the xy plane at time $t = t_3$ is shown in the figure inset.

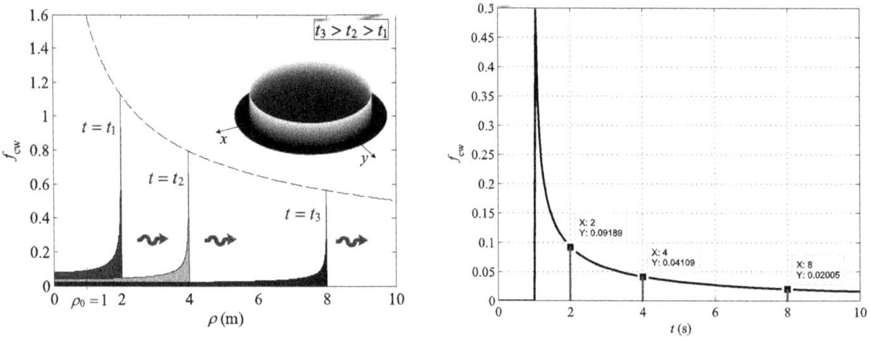

(a) (b)

Figure 1.24 Illustration of the fundamental solution of the wave equation for a uniform cylindrical wave; see Table 1.2, 4th column. (a) The plot shows "snapshots" of this wave as a function of ρ at three time instants, t_1, t_2, and t_3, where $t_1 < t_2 < t_3$. The wavy arrows point in the direction of expansion of the cylindrical wave. The wavefront "strength" decreases as $\sim 1/\sqrt{\rho}$, a tendency shown by the dotted line. A cut of the wave distribution in the xy plane at time $t = t_3$ is shown in the figure inset. (b) The plot shows the waveform observed at $\rho_0 = 1$ m.

Finally, the fundamental solution in the case of cylindrical symmetry (see Table 1.2, 4th column) is illustrated in Fig. 1.24. The front of the wave features a sharp ridge of infinite value, which corresponds to the singularity of Green's function at $t = \rho/v$. The singularity point is excluded from the curves plotted in Fig. 1.24. In contrast to the spherical fundamental solution (refer to Fig. 1.23), the cylindrical wave features a long, slowly decaying trail behind the wavefront ridge. This trail reminds us of the flat trail left behind by the plane wave (see Fig. 1.22-a). We could say that the cylindrical wave distribution is somewhat in between those of the plane wave and the spherical wave.

The reader is also reminded that we examined this very same waveform as an example in Section 1.2.C; see Eq. (1.15), Eq. (1.16), and Eq. (1.17). There, we already predicted the wave features that are observed in Fig. 1.24. First, we see the $\sim 1/\sqrt{\rho}$ behavior of the wavefront, i.e., the early stages of the wave. This behavior is shown by the dotted line in Fig. 1.24-a. Second, at a given observation point, we predicted a decay with time as $\sim 1/\sqrt{\Delta t}$ in the late stages of the wave (long after the wave ridge has passed). Here, Δt is the time interval that has elapsed since the wave impinged on that observation point; see Eq. (1.17). This manifests in Fig. 1.24-a as the long trail that the wave leaves behind. If we focus our attention on the wave behavior at $\rho_0 = 1$, which is well behind the wavefront at all three instants (t_1, t_2, and t_3), we observe the slow decay of the wave-trail value from the instant t_1 to t_2 and then on to t_3. To provide some reference values, we note that the wave plots in Fig. 1.24 correspond to a wave of velocity $v = 1$ m/s and to the time instances $t_1 = 2$ s, $t_2 = 4$ s, and $t_3 = 8$ s. The time the wave impinged on the point $\rho_0 = 1$ m is $t_0 = 1$ s (see Fig. 1.24-b); therefore, the wave-trail values are observed at $\Delta t_1 = 1$ s, $\Delta t_2 = 3$ s, and $\Delta t_3 = 7$ s. Thus, if the wave-trail value at Δt_1 is taken as a reference (this is the value of the $t = t_1$ curve at ρ_0), then the wave-trail values at Δt_2 and Δt_3 are about $\sqrt{3}$ and $\sqrt{7}$ times weaker, respectively. This trend with

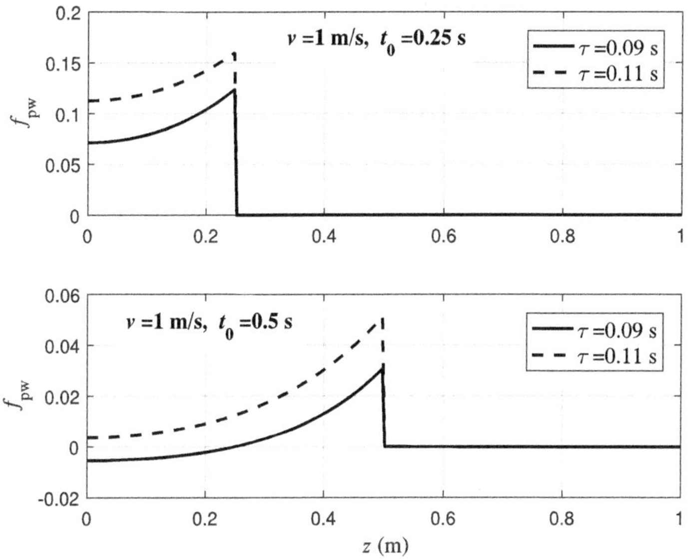

Figure 1.25 Illustration of the fundamental solution of the lossy-wave equation for a uniform plane wave (1D wave) based on Eq. (1.285). The phase velocity is set to $\upsilon = 1$ m/s. Two cases are shown where the relaxation time is set as $\tau = 0.11$ s and $\tau = 0.09$ s. The top and the bottom plots show the wave at two observation instants (t_0).

time is best observed in Fig. 1.24-b where the wave-trail values are shown by the cursor labels at $t_1 = 2$ s, $t_2 = 4$ s, and $t_3 = 8$ s.

The case of transients in lossy media is common both in acoustics and microwaves, especially when the latter are used for the nondestructive testing of materials, underground surveillance, or medical applications such as tissue imaging. Therefore, it would be useful to also summarize the fundamental solutions to the damped-wave equation,

$$\left(\frac{1}{\upsilon^2}\frac{\partial^2}{\partial t^2} + \frac{1}{\upsilon^2\tau}\frac{\partial}{\partial t} - \nabla^2\right) f(\mathbf{r}, t) = \delta_{\mathbf{r}}^n \delta(t), \tag{1.283}$$

where τ is the relaxation time and $\delta_{\mathbf{r}}^n$ is set to $\delta(z)$, $\delta_\rho^2 \equiv \delta(x)\delta(y)$, or $\delta_r^3 \equiv \delta(x)\delta(y)\delta(z)$, for a uniform plane, cylindrical, or spherical-wave solution, respectively. It is the relaxation time that controls the wave attenuation, with smaller τ values corresponding to faster attenuation. To relate τ to the EM constitutive parameters, we compare Eq. (1.283) to Eq. (1.1) to obtain the expression

$$\tau = \frac{\mu\epsilon}{\mu\sigma_{\mathrm{e}} + \epsilon\sigma_{\mathrm{m}}} = \left(\frac{\sigma_{\mathrm{e}}}{\epsilon} + \frac{\sigma_{\mathrm{m}}}{\mu}\right)^{-1}. \tag{1.284}$$

Larger conductivity values clearly correspond to shorter relaxation times and therefore faster attenuation of the wave as it propagates through the lossy medium.

The fundamental solutions to Eq. (1.283) are summarized next together with references where detailed derivations can be found.

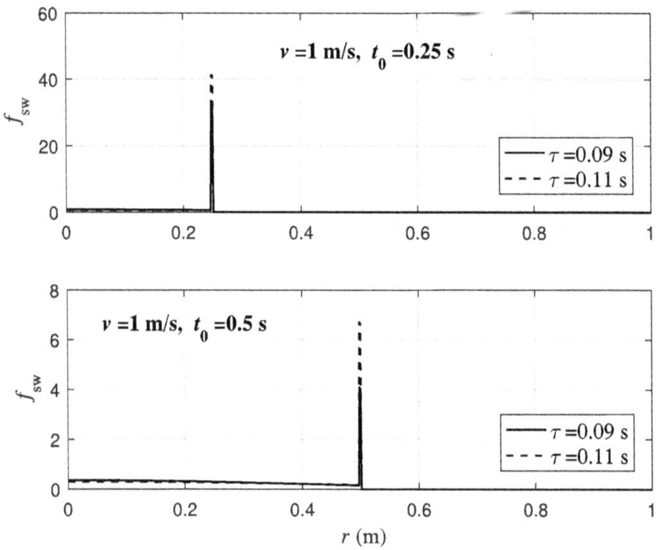

Figure 1.26 Illustration of the fundamental solution of the lossy-wave equation for a 3D spherical wave based on Eq. (1.286). The phase velocity is set to $v = 1$ m/s. Two cases are shown where the relaxation time is set as $\tau = 0.11$ s and $\tau = 0.09$ s. The top and the bottom plots show the wave at two observation instants (t_0).

- Uniform plane wave, $\nabla^2 = \frac{\partial^2}{\partial z^2}$, [3]:

$$f_{\mathrm{pw}}(z,t) = \frac{v}{2}\exp\left(-\frac{t}{2\tau}\right)H\left(t-\frac{|z|}{v}\right)J_0\left(\frac{1}{2\tau}\sqrt{t^2-\left(\frac{z}{v}\right)^2}\right). \tag{1.285}$$

Here, J_0 is the Bessel function of the first kind of order 0. Note that the Heaviside step function $H(\cdot)$ ensures a real argument for the Bessel function. When $\tau \to \infty$, $J_0 \to 1$, and Eq. (1.285) reduces to the loss-free fundamental solution in Table 1.2, 2nd column.

Fig. 1.25 illustrates the 1D lossy-wave fundamental solution when $v = 1$ m/s with two relaxation time constants, $\tau = 0.11$ s and $\tau = 0.09$ s. The wave is plotted as a function of position z at two instants indicated by the value of t_0. We observe that as the wavefront advances, it experiences significant attenuation. Also, the wave is no longer a "step" function as in the case of loss-free propagation.

- Uniform spherical wave, $\nabla^2 = \frac{1}{r^2}\frac{\partial}{\partial r}\left(r^2\frac{\partial}{\partial r}\right)$, [31, 44]:

$$f_{\mathrm{sw}}(r,t) = \frac{e^{-t/(2\tau)}}{4\pi t}\frac{1}{v}\frac{\partial}{\partial t}\left[H\left(t-\frac{r}{v}\right)I_0\left(\frac{1}{2\tau}\sqrt{t^2-\left(\frac{r}{v}\right)^2}\right)\right]. \tag{1.286}$$

Here, I_0 is the modified Bessel function of the first kind of order 0. If $\tau \to \infty$, then $I_0 \to 1$, $\frac{\partial I_0}{\partial t} \to 0$, and the time derivative of $H(\cdot)$ yields the δ-function. At the same time, since the δ-function is confined to space/time coordinates satisfying $r = vt$,

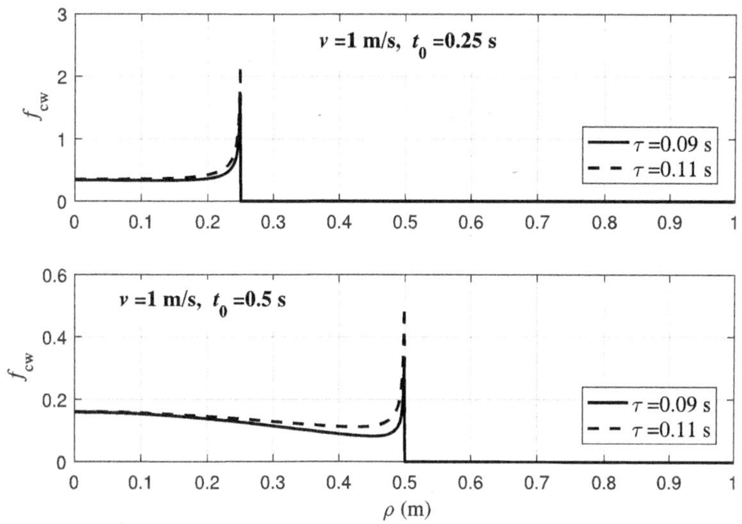

Figure 1.27 Illustration of the fundamental solution of the lossy-wave equation for a 2D cylindrical wave based on Eq. (1.287). The phase velocity is set to $v = 1$ m/s. Two cases are shown where the relaxation time is set as $\tau = 0.11$ s and $\tau = 0.09$ s. The top and the bottom plots show the wave at two observation instants (t_0).

the denominator $4\pi vt$ becomes $4\pi r$. Thus, for vanishingly small losses, Eq. (1.286) reduces to the fundamental solution in Table 1.2, 3rd column.

Fig. 1.26 illustrates the fundamental 3D lossy-wave solution with the settings for the speed v and the relaxation times as in the 1D case. The wavefront still exhibits singularity at $r = vt$ as in the loss-free case; however, we now observe deformation in the wave distribution in the form of a trail behind the wavefront. To make the wave trail visible, the wave singularity was removed in the plots in Fig. 1.26.

- Uniform cylindrical wave, $\nabla^2 = \frac{1}{\rho}\frac{\partial}{\partial\rho}\left(\rho\frac{\partial u}{\partial\rho}\right)$, [31]:

$$f_{\text{cw}}(\rho, t) = e^{-t/(2\tau)} \frac{H(t - \rho/v)}{2\pi\sqrt{t^2 - \left(\frac{\rho}{v}\right)^2}} \cosh\left(\frac{1}{2\tau}\sqrt{t^2 - \left(\frac{\rho}{v}\right)^2}\right). \quad (1.287)$$

As the cosh function tends to 1 for small arguments, again, we assert that this solution reduces to the respective loss-free solution; see Table 1.2, 4th column. Similarly to the loss-free case, the wavefront exhibits singularity at $t = \rho/v$. This is observed in Fig. 1.27.

1.12 Fundamental Solutions to the Helmholtz Equation

The three fundamental solutions to Helmholtz's equation—the phasors of the uniform planar, cylindrical, and spherical waves—are summarized in Table 1.3. Note that the

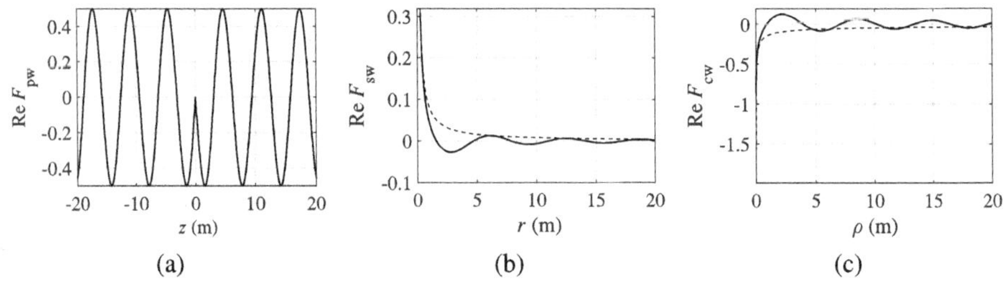

Figure 1.28 Illustration of the fundamental solutions of the Helmholtz equation (see Table 1.3) where the propagation constant is set as $\gamma = j\ (\mathrm{m}^{-1})$: (a) plane wave in terms of $\Re F_{\mathrm{pw}}(z)$, (b) spherical wave in terms of $\Re F_{\mathrm{sw}}(r)$, and (c) cylindrical wave in terms of $\Re F_{\mathrm{cw}}(\rho)$.

fundamental solutions accommodate both loss-free and lossy media. In the former case, $\gamma = \mathrm{i}\beta$, whereas in the latter $\gamma = \alpha + \mathrm{i}\beta$. Both α and β (the attenuation and phase constants, respectively) are nonnegative real.

Be aware that the equations in Table 1.3 are based on the Helmholtz operator in the form $\mathcal{L} = \gamma^2 - \nabla^2$, which is commonly used to describe the fundamental solutions. However, for reasons already explained in the previous section, throughout this text, we adhere to the convention of Helmholtz's operator being in the form $\mathcal{L} = \nabla^2 - \gamma^2$. Thus, when using a fundamental solution $F_{\mathrm{w}}(\mathbf{r})$ as Green's function $G(\mathbf{r}, \mathbf{r}')$, a minus sign must be applied, i.e.,

$$G(\mathbf{r}, \mathbf{r}') = -F_{\mathrm{w}}(\mathbf{r} - \mathbf{r}'). \tag{1.288}$$

The behavior of the frequency-domain fundamental solutions in the loss-free case of $\gamma = \mathrm{i}1\ (\mathrm{m}^{-1})$ is illustrated in Fig. 1.28. The real part of the respective phasor is plotted versus distance. As per Eq. (1.19), the real part of the phasor is the time-domain harmonic equivalent. Thus, Fig. 1.28 shows "snapshots" of the time-harmonic fundamental solutions.

It is important to comment on the asymptotic behavior of these solutions with increasing distance. The plane-wave solution in Fig. 1.28-a describes a guided (or confined) disturbance, as in a loss-free transmission line, where the wave is not allowed to spread and thus maintains constant strength with distance.

The spherical wave, on the other hand, describes a disturbance, the power of which diverges from the source uniformly in all directions. As a result, with distance increasing, the power spreads uniformly over an ever-expanding spherical surface, leading to the $\sim 1/r^2$ decay of its density. The factor $\frac{1}{4\pi r^2}$ is often referred to as the *free-space loss factor*.[30] This loss factor has nothing to do with damping due to dissipation. Dissipation, if present, leads to an additional exponential damping ($e^{-\alpha r}$). The free-space loss factor translates into the $\sim 1/r$ behavior of the wave intensity. This intensity (the signal

[30] In antenna engineering, the free-space loss factor is defined as $\frac{\lambda}{4\pi r^2}$, which also takes into account the frequency-dependent behavior of the antenna aperture in the Friis transmission equation [15].

Table 1.3 Fundamental Solutions to the Helmholtz Equation

wavefront	*planar*	*spherical*	*cylindrical*
equation	$\gamma^2 F - \frac{\partial^2 F}{\partial z^2} = \delta(z)$	$\gamma^2 F - \frac{1}{r^2}\frac{\partial}{\partial r}\left(r^2 \frac{\partial F}{\partial r}\right) = \delta_r^3$	$\gamma^2 F - \frac{1}{\rho}\frac{\partial}{\partial \rho}\left(\rho \frac{\partial F}{\partial \rho}\right) = \delta_\rho^2$
solution	$F_{\text{pw}}(z) = \dfrac{e^{-\gamma\|z\|}}{2\gamma}$	$F_{\text{sw}}(r) = \dfrac{e^{-\gamma r}}{4\pi r}$	$F_{\text{cw}}(\rho) = \dfrac{\mathrm{i}}{4} H_0^{(2)}(-\mathrm{i}\gamma\rho)$

envelope as a function of distance) is shown by the dotted line in Fig. 1.28-b. In contrast, the envelope of the guided wave in Fig. 1.28-a is constant.

The fundamental cylindrical wave spreads uniformly over an ever-expanding cylindrical surface as the travel distance ρ increases. This implies a 2D free-space loss factor of $1/\rho$ and signal decay of $\sim 1/\sqrt{\rho}$. Indeed, the asymptotic behavior of the Hankel function is given by

$$H_0^{(2)}(x) \underset{x\to\infty}{\longrightarrow} \sqrt{\frac{2\mathrm{i}}{\pi x}}\, e^{-\mathrm{i}x}. \tag{1.289}$$

The wave envelope that obeys this behavior is shown in Fig. 1.28-c by a dotted line. Note that the envelope's asymptotic behavior tracks the magnitude of Hankel's function accurately only for sufficiently large values of ρ ($\gtrsim 15$).

1.13 The Born Series in the Scalar Model of Scattering

So far, we have shown how to relate the scattered field to the contrast function in the imaged volume; see Eq. (1.216) or Eq. (1.218). This discussion emphasized the nonlinear nature of this relation in the inverse problem. However, under certain conditions, this relation can be linearized. The two best-known linear approximations of scattering are the Born and the Rytov approximations, which allow for reducing the inverse problem to the solution of a linear system of equations (typically rectangular) for the unknown contrast function. Linear-system solvers can provide solutions within seconds even for tens of thousands of unknowns, thus allowing for quasi-real-time image reconstruction, where the time for obtaining the image is comparable to or less than the time required to complete the measurements.

The Born approximation is best understood through the Born series. Both are named after Max Born, German physicist and mathematician, who developed the series for the inverse problems in quantum scattering.

Let us explicitly state the Helmholtz-equation integral solution for the total field through the scattering potential K. Using Eq. (1.216),

$$U(\mathbf{r}) = U^{\text{inc}}(\mathbf{r}) + \iiint_V G(\mathbf{r}, \mathbf{r}')K(\mathbf{r}')U(\mathbf{r}')\, dv'. \tag{1.290}$$

For the wave equation, as follows from Eq. (1.218), the total field in terms of the normalized contrast χ is

$$u(\mathbf{r},t) = u^{\mathrm{inc}}(\mathbf{r},t) + \int_{T_{\min}}^{T_{\max}}\iiint_V g(\mathbf{r},t;\mathbf{r}',t')\chi(\mathbf{r}')\nabla^2 u(\mathbf{r}',t')dv'dt'. \qquad (1.291)$$

In both cases, we are dealing with a linear integral equation with respect to the total field. Specifically, this is a Fredholm equation of the 2nd kind, the standard operator form of which is written as

$$u(\mathbf{x}) = u^{\mathrm{inc}}(\mathbf{x}) + \xi\mathcal{L}u(\mathbf{x}). \qquad (1.292)$$

Here, ξ is a constant and $\mathcal{L}$ is the integral operator. Let us define ξ as the normalized contrast of the largest absolute value, i.e., we set[31]

$$\xi = K_{\max},\ K_{\max} = \sup|K(\mathbf{r})|, \qquad (1.293)$$

in Eq. (1.290), whereas in Eq. (1.291),

$$\xi = \chi_{\max},\ \chi_{\max} = \sup|\chi(\mathbf{r})|. \qquad (1.294)$$

Then, the integral operator $\mathcal{L}$ in the case of Eq. (1.290) is

$$\mathcal{L}(\cdot) = \iiint_V \underbrace{G(\mathbf{r},\mathbf{r}')\bar{K}(\mathbf{r}')}_{\mathcal{K}(\mathbf{r},\mathbf{r}')}(\cdot)\,dv',\ \bar{K}(\mathbf{r}') = \frac{K(\mathbf{r}')}{K_{\max}}, \qquad (1.295)$$

and, in the case of Eq. (1.291), it is

$$\mathcal{L}(\cdot) = \int_{T_{\min}}^{T_{\max}}\iiint_V \underbrace{g(\mathbf{r},t;\mathbf{r}',t')\bar{\chi}(\mathbf{r}')\nabla^2}_{\mathcal{K}(\mathbf{r},t;\mathbf{r}',t')}(\cdot)\,dv'dt',\ \bar{\chi}(\mathbf{r}') = \frac{\chi(\mathbf{r}')}{\chi_{\max}}. \qquad (1.296)$$

Here, $\mathcal{K}$ is the kernel of the operator $\mathcal{L}$ and $(\cdot)$ denotes the scalar quantity on which $\mathcal{L}$ operates, i.e., the total field.

A well-known approach to solving Eq. (1.292) iteratively is the Neumann series expansion [11, 45]. The approach works provided a convergence criterion is met, which requires that the product of ξ and $\mathcal{K}$ is sufficiently small. Notice that the product $\xi\mathcal{K}$ is proportional to the contrast in both Eq. (1.295) and Eq. (1.296); thus, the convergence criterion translates into a requirement for small contrast. It is worth walking through the steps of the iterative procedure in order to appreciate the role of the contrast function in the convergence of the solution.

Assume that the normalized contrast is so small that, as a first approximation, we ignore completely the scattered field component in Eq. (1.292) and state our initial fairly crude solution as

$$u^{(0)}(\mathbf{x}) = u^{\mathrm{inc}}(\mathbf{x}). \qquad (1.297)$$

[31] The supremum is the least upper bound of a set.

We now substitute this solution in the right-hand side of the original equation, Eq. (1.292):

$$u^{(1)}(\mathbf{x}) = u^{\text{inc}}(\mathbf{x}) + \xi \mathcal{L} u^{(0)}(\mathbf{x}) = u^{\text{inc}}(\mathbf{x}) + \xi \mathcal{L} u^{\text{inc}}(\mathbf{x}), \tag{1.298}$$

where $u^{(1)}(\mathbf{x})$ is a better solution than $u^{(0)}(\mathbf{x})$ provided the convergence criterion is fulfilled. The second iteration results in

$$u^{(2)}(\mathbf{x}) = u^{\text{inc}}(\mathbf{x}) + \xi \mathcal{L} u^{(1)}(\mathbf{x}) = u^{\text{inc}}(\mathbf{x}) + \xi \mathcal{L} u^{\text{inc}}(\mathbf{x}) + \xi^2 \mathcal{L}\mathcal{L} u^{\text{inc}}(\mathbf{x}). \tag{1.299}$$

Explicitly, Eq. (1.299) is written out as

$$\begin{aligned} u^{(2)}(\mathbf{x}) &= u^{\text{inc}}(\mathbf{x}) \\ &+\xi \int_{\mathcal{D}} \mathcal{K}(\mathbf{x}, \mathbf{x}_1) u^{\text{inc}}(\mathbf{x}_1) d\Omega(\mathbf{x}_1) \\ &+\xi^2 \int_{\mathcal{D}} \mathcal{K}(\mathbf{x}, \mathbf{x}_1) \left[\int_{\mathcal{D}} \mathcal{K}(\mathbf{x}_1, \mathbf{x}_2) u^{\text{inc}}(\mathbf{x}_2) d\Omega(\mathbf{x}_2) \right] d\Omega(\mathbf{x}_1). \end{aligned} \tag{1.300}$$

The term $\mathcal{L}\mathcal{L}u^{\text{inc}}(\mathbf{x})$ is formally expressed as $\mathcal{L}^2 u^{\text{inc}}(\mathbf{x})$. Rearranging the expanded form of the third term in Eq. (1.300), this "squared" operator is expressed as

$$\mathcal{L}^2 u^{\text{inc}}(\mathbf{x}) = \int_{\mathcal{D}} d\Omega(\mathbf{x}_2) \underbrace{\int_{\mathcal{D}} d\Omega(\mathbf{x}_1)\, \mathcal{K}(\mathbf{x}, \mathbf{x}_1) \mathcal{K}(\mathbf{x}_1, \mathbf{x}_2)}_{\mathcal{K}_2(\mathbf{x}, \mathbf{x}_2)} u^{\text{inc}}(\mathbf{x}_2). \tag{1.301}$$

It has a kernel of its own $\mathcal{K}_2(\mathbf{x}, \mathbf{x}')$ such that

$$\mathcal{K}_2(\mathbf{x}, \mathbf{x}') = \int_{\mathcal{D}} \mathcal{K}(\mathbf{x}, \mathbf{x}_1) \mathcal{K}(\mathbf{x}_1, \mathbf{x}') d\Omega(\mathbf{x}_1). \tag{1.302}$$

We may continue the iterative process to obtain higher and higher order approximations to the solution of Eq. (1.292). The n-th order approximation is

$$u^{(n)}(\mathbf{x}) = u^{\text{inc}}(\mathbf{x}) + \sum_{m=1}^{n} \xi^m \mathcal{L}^m u^{\text{inc}}(\mathbf{x}). \tag{1.303}$$

The kernel of the operator $\mathcal{L}^m$ can be updated recursively using

$$\mathcal{K}_m(\mathbf{x}, \mathbf{x}') = \int_{\mathcal{D}} \mathcal{K}(\mathbf{x}, \mathbf{x}_1) \mathcal{K}_{m-1}(\mathbf{x}_1, \mathbf{x}') d\Omega(\mathbf{x}_1), \tag{1.304}$$

just the way we did in Eq. (1.302). This allows for the computation of the $\mathcal{L}^m u^{\text{inc}}$ term in the series Eq. (1.303) as

$$\mathcal{L}^m u^{\text{inc}}(\mathbf{x}) = \int_{\mathcal{D}} \mathcal{K}_m(\mathbf{x}, \mathbf{x}') u^{\text{inc}}(\mathbf{x}') d\Omega(\mathbf{x}'). \tag{1.305}$$

Eq. (1.303) is the Neumann (or power) series expansion. The solution to Eq. (1.292) is given by the infinite Neumann series, i.e.,

$$u(\mathbf{x}) = \lim_{n\to\infty} u^{(n)}(\mathbf{x}) = u^{\text{inc}}(\mathbf{x}) + \sum_{m=1}^{\infty} \xi^m \mathcal{L}^m u^{\text{inc}}(\mathbf{x}), \tag{1.306}$$

provided it converges. It can be shown that the series converges if [11, 45]

$$|\xi|^2 \int_{\mathcal{D}} d\Omega \int_{\mathcal{D}} d\Omega' |\mathcal{K}(\mathbf{x}, \mathbf{x}')|^2 < 1. \tag{1.307}$$

Whether the convergence condition above is fulfilled for the Born series expansions of Eq. (1.290) and Eq. (1.291) depends critically on the normalized contrast function since the respective kernels are proportional to it; see Eq. (1.295) and Eq. (1.296).

The expansion of the field solution in Neumann series provides some interesting insight into the forward model of scattering. From Eq. (1.306), it is clear that this solution can be written in terms of the incident field $u^{\text{inc}}(\mathbf{x})$ and a new operator $\mathcal{R}$ as

$$u(\mathbf{x}) = u^{\text{inc}}(\mathbf{x}) + \xi \mathcal{R} u^{\text{inc}}(\mathbf{x}). \tag{1.308}$$

This equation is very similar to our starting equation Eq. (1.292), but now it has only the known $u^{\text{inc}}(\mathbf{x})$ in the right-hand side. Using Eq. (1.305) and Eq. (1.306), the $\mathcal{R}$ operator is written out explicitly as

$$\mathcal{R}(\cdot) = \sum_{m=1}^{\infty} \xi^{(m-1)} \mathcal{L}^m(\cdot) = \sum_{m=1}^{\infty} \xi^{m-1} \int_{\mathcal{D}} \mathcal{K}_m(\mathbf{x}, \mathbf{x}')\, (\cdot)\, d\Omega(\mathbf{x}'). \tag{1.309}$$

This new operator has a kernel of its own,

$$\mathcal{K}_{\mathcal{R}}(\mathbf{x}, \mathbf{x}') = \sum_{m=1}^{\infty} \xi^{m-1} \mathcal{K}_m(\mathbf{x}, \mathbf{x}'), \tag{1.310}$$

with which we can write the solution Eq. (1.308) as

$$u(\mathbf{x}) = u^{\text{inc}}(\mathbf{x}) + \xi \int_{\mathcal{D}} \mathcal{K}_{\mathcal{R}}(\mathbf{x}, \mathbf{x}') u^{\text{inc}}(\mathbf{x}') d\Omega(\mathbf{x}'). \tag{1.311}$$

In the mathematics of integral equations, $\mathcal{K}_{\mathcal{R}}(\mathbf{x}, \mathbf{x}')$ is referred to as the *resolvent kernel*.

What is important here is that the resolvent kernel relates to Green's function for the medium *in the presence of the scatterer*[32] in exactly the same manner as $\mathcal{K}(\mathbf{x}, \mathbf{x}')$ relates to Green's function $\mathcal{G}(\mathbf{x}, \mathbf{x}')$ of the background. As a reminder, in Eq. (1.295), where $\mathcal{G}(\mathbf{x}, \mathbf{x}') \equiv G(\mathbf{r}, \mathbf{r}')$, we have

$$\mathcal{K}(\mathbf{r}, \mathbf{r}') = \bar{K}(\mathbf{r}') G(\mathbf{r}, \mathbf{r}'), \tag{1.312}$$

[32] Green's function in the presence of the scatterer is sometimes referred to as total (or full) Green's function [26].

whereas in Eq. (1.296), where $\mathcal{G}(\mathbf{x},\mathbf{x}') \equiv g(\mathbf{r},\mathbf{r}';t,t')$,

$$\mathcal{K}(\mathbf{r},t;\mathbf{r}',t') = g(\mathbf{r},t;\mathbf{r}',t')\bar{\chi}(\mathbf{r}')\nabla^2(\cdot). \tag{1.313}$$

Let us take as an example Eq. (1.290), written in terms of $\mathcal{K}(\mathbf{r},\mathbf{r}')$ as

$$U(\mathbf{r}) = U^{\text{inc}}(\mathbf{r}) + \xi \iiint_V \mathcal{K}(\mathbf{r},\mathbf{r}')U(\mathbf{r}')dv', \tag{1.314}$$

and compare it to Eq. (1.311) written in the case of Helmholtz's equation as

$$U(\mathbf{r}) = U^{\text{inc}}(\mathbf{r}) + \xi \iiint_V \mathcal{K}_{\mathcal{R}}(\mathbf{r},\mathbf{r}')U^{\text{inc}}(\mathbf{r}')dv'. \tag{1.315}$$

Both equations describe the same field. It follows that the scattered field can be described through either one of the following two expressions:

$$U^{\text{sc}}(\mathbf{r}) = \xi \iiint_V \bar{K}(\mathbf{r}')G(\mathbf{r},\mathbf{r}')U(\mathbf{r}')dv' = \xi \iiint_V \mathcal{K}_{\mathcal{R}}(\mathbf{r},\mathbf{r}')U^{\text{inc}}(\mathbf{r}')dv'. \tag{1.316}$$

Let us now consider a reversed scenario where what was the background in the previous case is now the unknown medium, whereas what was the unknown medium becomes the background. This changes the sign of the contrast K and, therefore, that of the U^{sc}. Also, what is $G(\mathbf{r},\mathbf{r}')$ in Eq. (1.316) is now $G_{\text{tot}}(\mathbf{r},\mathbf{r}')$—Green's function of the medium containing the scatterer. Meanwhile, what is $U(\mathbf{r}')$ in Eq. (1.316) is now $U^{\text{inc}}(\mathbf{r}')$. The scattered field in this alternative scenario can be written as

$$-U^{\text{sc}}(\mathbf{r}) = \xi \iiint_V \left(-\bar{K}(\mathbf{r}')\right) G_{\text{tot}}(\mathbf{r},\mathbf{r}')U^{\text{inc}}(\mathbf{r}')dv'. \tag{1.317}$$

Comparing Eq. (1.317) with the rightmost term in Eq. (1.316) reveals that

$$\mathcal{K}_{\mathcal{R}}(\mathbf{r},\mathbf{r}') = \bar{K}(\mathbf{r}')G_{\text{tot}}(\mathbf{r},\mathbf{r}'), \tag{1.318}$$

which proves the assertion that the resolvent kernel depends on Green's total function in the same manner as $\mathcal{K}(\mathbf{r},\mathbf{r}')$ depends on Green's background function. It follows that the scattered field can be described through either one of these two expressions:

$$U^{\text{sc}}(\mathbf{r}) = \iiint_V K(\mathbf{r}')G(\mathbf{r},\mathbf{r}')U(\mathbf{r}')dv' = \iiint_V K(\mathbf{r}')G_{\text{tot}}(\mathbf{r},\mathbf{r}')U^{\text{inc}}(\mathbf{r}')dv'. \tag{1.319}$$

The analysis in the case of the wave equation is analogous.

1.14 The Born Approximation in the Scalar Model of Scattering

The linear Born approximation, also known simply as the Born approximation, utilizes only the first term in the Neumann-series expansion, which is also the first iterate shown in Eq. (1.298). Explicitly, it is stated as

$$U(\mathbf{r}) \approx U^{\text{inc}}(\mathbf{r}) + \iiint_V G(\mathbf{r}',\mathbf{r})K(\mathbf{r}')U^{\text{inc}}(\mathbf{r}')\,dv', \tag{1.320}$$

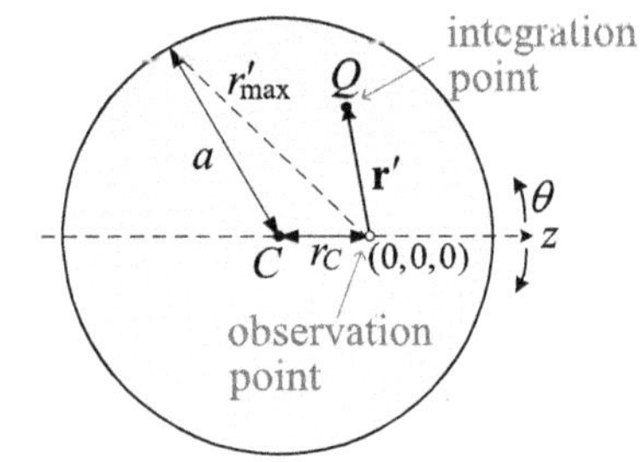

Figure 1.29 Illustration of the coordinate setup in deriving the error estimate for the Born approximation. The observation point is at the origin, $\mathbf{r} = (0, 0, 0)$, while the integration point Q is at $\mathbf{r}'$. The sphere of radius a (the volume over which the integration is performed) has its center C at a distance r_C from the origin.

for the frequency-domain scattering problem, whereas for the time-domain problem, it is

$$u(\mathbf{r}, t) \approx u^{\text{inc}}(\mathbf{r}, t) + \int_{t_{\min}}^{t_{\max}} \iiint_V g(\mathbf{r}', t'; \mathbf{r}, t)\chi(\mathbf{r}')\nabla^2 u^{\text{inc}}(\mathbf{r}', t')dv'dt'. \tag{1.321}$$

Clearly, the approximation lies in replacing the total field in the integral with the incident one, which is indeed admissible if the scattering is weak, i.e., if the scattered field is much weaker than the incident one. Weak scattering implies small contrast and/or small size of the scatterer. If valid, this approximation provides a simple linear relationship between the data (the left-hand side of Eq. (1.320) or Eq. (1.321)) and the contrast function (K in Eq. (1.320) or χ in Eq. (1.321)), allowing for very fast inversion.

Note that the Born approximation is applied to the computation of the field outside the imaged volume (the data equation). This is different from the application of the Neumann series in finding the total field *inside* the scatterer that was described in the previous section. To make the distinction, we refer to the field inside the scatterer as the *internal field* while that outside the scatterer is the *external field*.

The region of validity of the Born approximation is rather limited by the maximum allowed contrast and size of the scattering object, which determine the strength of the neglected scattered portion of the total internal field. The associated limitation is typically stated in the form [46, 47, 48, 49, 50]

$$k_b a \sup_{|\mathbf{r}| \leq a} |\mathcal{X}(\mathbf{r})| < 2\pi\zeta, \tag{1.322}$$

where $\mathcal{X}(\mathbf{r}) = \left[\mu_{r,s}(\mathbf{r})\varepsilon_{r,s}(\mathbf{r}) - \mu_{r,b}\varepsilon_{r,b}\right]/(\mu_{r,b}\varepsilon_{r,b})$, see Eq. (1.214), k_b is the background wavenumber, a is the smallest sphere that circumscribes the scatterer, and ζ is a sufficiently small number. Slaney *et al.* [48] derive that $\zeta \approx 0.25$ based on a requirement that the difference in phase between the incident field and the wave propagating through the scattering object be less than π. In [50], this value is derived as 0.16. A rigorous method of establishing the limitations under the Born approximation and the associated errors in the field solution is proposed in [47].

To understand better the limitations of the Born approximation, let us consider again the Born series expansion of the total field and inspect the most significant error term when the total internal field is approximated by the incident field. Eq. (1.300) shows that this term's supremum is

$$e_1(\mathbf{x}) = \sup \left| \xi \int_D \mathcal{K}(\mathbf{x}, \mathbf{x}_1) u^{\text{inc}}(\mathbf{x}_1) d\Omega(\mathbf{x}_1) \right|. \tag{1.323}$$

Consider e_1 in the case of the frequency-domain analysis where it is explicitly given by

$$e_1(\mathbf{r}) = \sup \left| \iiint_V G(\mathbf{r}, \mathbf{r}') K(\mathbf{r}') U^{\text{inc}}(\mathbf{r}')\, dv' \right|. \tag{1.324}$$

Using the inequality $\left|\int f(\mathbf{x}) d\mathbf{x}\right| \le \int |f(\mathbf{x})| d\mathbf{x}$ and Green's function of the form $\frac{\exp(-\mathrm{i}k_{\mathrm{b}}R)}{4\pi R}$, the maximum value of e_1 can be estimated from

$$e_1 \approx k_{\mathrm{b}}^2 \sup |\mathcal{X}(\mathbf{r})| \cdot \sup \left[\iiint_V \frac{1}{4\pi R} \left| U^{\text{inc}}(\mathbf{r}') \right| dv' \right], \tag{1.325}$$

where k_{b} is the background wavenumber and $R = |\mathbf{r} - \mathbf{r}'|$ is the distance between the scattering and the observation points.

Let us consider the case of an incident wave, which is uniform, plane, and of unit magnitude, i.e., $\left|U^{\text{inc}}(\mathbf{r}')\right| = 1$. This assumption produces an estimate of e_1, which is normalized to the strength of the incident field. As per Eq. (1.325), we need to solve the integral

$$I = \iiint_V \frac{1}{4\pi R}\, dv'. \tag{1.326}$$

To simplify the integral, let us assume that the volume V is the smallest sphere circumscribing the scattering object and this sphere has radius a. It is convenient to work in spherical coordinates where R is expressed as

$$R^2 = r^2 + r'^2 - 2rr' \left[\sin\theta \sin\theta' \cos(\phi - \phi') + \cos\theta \cos\theta' \right]. \tag{1.327}$$

With reference to Fig. 1.29, the observation location $\mathbf{r}$ is chosen at the origin of the spherical coordinate system, $\mathbf{r} = (0, 0, 0)$. This greatly simplifies the integral since $R = r'$. Let the center of the sphere C be on the z axis at some distance r_C from the origin. The point of integration has its radial coordinate r' limited between 0 and $r'_{\max} = \sqrt{a^2 + r_C^2 - 2ar_C \cos\theta'}$. Now, the integral in Eq. (1.326) can be evaluated as

$$I(r_C) = \iiint_{\text{sphere}} \frac{1}{4\pi |\mathbf{r}'|} dv' = \frac{1}{4\pi} \int_0^{2\pi} \int_0^{\pi} \int_0^{r'_{\max}(\theta')} r' \sin\theta' dr' d\theta' d\phi' = \frac{1}{2}\left(a^2 + r_C^2\right). \tag{1.328}$$

Since the largest value of r_C is $r_{C\max} = a$, we obtain that $\sup I(r) = a^2$, which, after substitution in Eq. (1.325), leads to an estimate of the maximum error as

$$e_1 \approx (ak_{\mathrm{b}})^2 \sup |\mathcal{X}(\mathbf{r})|. \tag{1.329}$$

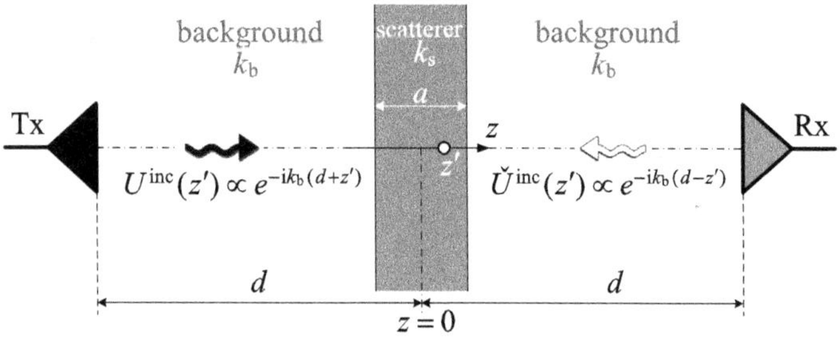

Figure 1.30 Illustration of the setup in deriving the constraint in Eq. (1.332). The incident field U^{inc} is assumed to be a uniform plane wave traveling in a straight line along z from the transmitter (Tx) to the receiver (Rx). The receiver, when in transmitting mode, generates a uniform plane wave as well ($\check{U}^{inc}$). The background's wavenumber is k_b, whereas that of the scatterer is k_s.

This error term gives the maximum possible strength of the neglected scattered component of the total internal field relative to the incident internal field. It is now clear that the Born approximation for the internal field is constrained by the condition

$$(ak_b)^2 \sup |\mathcal{X}(\mathbf{r})| \ll 1, \tag{1.330}$$

or, equivalently, (see also [31])

$$a^2 \sup |k_s^2(\mathbf{r}) - k_b^2| \ll 1. \tag{1.331}$$

It can be shown that Eq. (1.331) ensures that the Born approximation is accurate for the external field as well, although, in this case, it is overly restrictive. In [48], Slaney *et al.* state that "for the first-order Born approximation to be valid, a necessary condition is that the change in phase between the incident field and the wave propagating through the object be less than π." Mathematically,

$$a \sup |k_s(\mathbf{r}) - k_b| < \pi. \tag{1.332}$$

This condition is often referred to as the high-frequency limit of the Born approximation [49]. It addresses the accuracy of the Born approximation of the *external* field. This is in contrast to Eq. (1.331), which is a constraint for the *internal* field. The constraint in Eq.(1.332) is particularly relevant in cases such as computed tomography where the wave can be assumed to be traveling in a straight path and where the size of the object is large compared to the wavelength. It leads to an important observation: *the Born approximation of the external field is sensitive to errors in the overall phase of the total wave as it propagates through the object.*

Example 1: With reference to Fig. 1.30, assume that the imaged volume, centered at the origin, is illuminated by a uniform plane wave incident from the left, which is generated by an emitter positioned at a very large distance $d >> a$ from the origin. Here, a is the size of the imaged volume. Assume that the field travels along a straight path through the background medium and through the object. This allows for approximating the object as a slab of thickness a. Let the sensor be to the right of the imaged volume, also a distance d from the origin. Since the sensor is far from the imaged object, if it were to act as an emitter, it would also generate a uniform plane wave in the imaged volume. Verify the high-frequency constraint Eq. (1.332).

Solution: Green's function associated with the receiver (Rx) in Fig. 1.30 is $G(z', z) = \check{U}^{\text{inc}}(z') = e^{-\text{i}k_b(d-z')}$. The incident field due to the transmitter (Tx) is $U^{\text{inc}}(z') = e^{-\text{i}k_b(d+z')}$. Then, the linear Born approximation of the scattered field at the Rx is given by

$$U_B^{\text{sc}} = \int_{-a/2}^{a/2} Ke^{-\text{i}k_b(d-z')} \cdot e^{-\text{i}k_b(d+z')}dz', \tag{1.333}$$

where K is the scattering potential assumed to be constant in $z' \in [-a/2, a/2]$. This leads to

$$U_B^{\text{sc}} = aKe^{-\text{i}2k_b d}. \tag{1.334}$$

Let me point out some of the flaws of this solution. In addition to the ignored reflections at the interfaces (which we will continue ignoring to keep things simple), one apparent flaw in approximating the total field in the scatterer by the incident field in the background is the fact that the wavenumber in the scatterer is k_s, not k_b. Ignoring the reflections, a better approximation of the total field in the scatterer would be

$$U^{\text{tot}}(z') \approx e^{-\text{i}k_b(d-a/2)} \cdot e^{-\text{i}k_s(z'+a/2)}. \tag{1.335}$$

Now, the scattered field at the Rx is obtained as

$$\begin{aligned} U^{\text{sc}} &= \int_{-a/2}^{a/2} Ke^{-\text{i}k_b(d-z')} \cdot e^{-\text{i}k_b(d-a/2)} \cdot e^{-\text{i}k_s(z'+a/2)}dz' \\ &= aKe^{-\text{i}2k_b d} \cdot e^{-\text{i}\Delta ka/2} \cdot \frac{\sin(\Delta ka/2)}{\Delta ka/2} = U_B^{\text{sc}} \cdot e^{-\text{i}\Delta ka/2} \cdot \frac{\sin(\Delta ka/2)}{\Delta ka/2}. \end{aligned} \tag{1.336}$$

Here, $\Delta k = k_s - k_b$. We can now appreciate the meaning of the constraint in Eq. (1.332) with regard to the phase and magnitude errors in the Born approximation U_B^{sc}. Assuming that Eq. (1.336) is the accurate (reference) solution, the constraint in Eq. (1.332) ensures that: (i) the error $(\Delta ka/2)$ in the phase of the external U_B^{sc} is below 90°, and (ii) the error in the magnitude $|U_B^{\text{sc}}|$ is less than 4 dB. The latter statement comes from the ratio $\frac{|U^{\text{sc}}|}{|U_B^{\text{sc}}|} = |\frac{\sin(\Delta ka/2)}{\Delta ka/2}|$, which attains values between $\text{sinc}(0.5) \approx 0.6366$ and 1, provided $a\Delta k < \pi$.

To demonstrate the limitations of the linear Born approximation, let us consider the simple example of a uniform plane wave normally incident on a dielectric (loss-free, nonmagnetic) slab, which is infinite in the lateral directions and has a thickness of 60 mm. The rest of the propagation medium is vacuum. The wave is excited on the left of the slab (at Port 1) as a Gaussian pulse, the spectrum of which covers all frequencies from zero to about 5 GHz at a level above -3 dB. The scattered field is recorded at Port 1 and at Port 2, the latter being to the right of the slab. The simulations are performed with MEFiSTo-3D for various values of the slab's relative permittivity ϵ_r. Here, the object of interest is the dielectric slab, whereas the background medium is vacuum. Therefore, the incident field is simulated in vacuum with the slab absent.

According to Born's approximation, the incident field recorded at locations occupied by the slab in the actual scattering problem can represent the total internal field provided the constraint in Eq. (1.331) is observed. This is indeed the case if the slab's permittivity is low. Fig. 1.31-a shows the magnitude and phase distribution of the field inside the slab at the frequency of 1 GHz for three values of the slab permittivity. Fig. 1.31-b shows the analogous plots at 5 GHz. We observe that the incident and total internal field distributions are indeed very similar for $\epsilon_r = 1.1$, both in magnitude and phase. In fact, at 1 GHz, the phase error stays below 90° even for $\epsilon_r = 4.0$. Note that the constraint in Eq. (1.331) dictates that $\epsilon_r \ll 6.7$ (here, $\lambda \approx 30$ cm and $a = 6$ cm). It is worth noting that the magnitude in Born's approximation is in general more sensitive to the contrast increase than the phase because of the impact of the multiple reflection (multiple scattering) at interfaces. This is why the differences in the magnitude distribution are significant at all frequencies for $\epsilon_r = 4.0$.

Similar trends are observed for the exterior field (i.e., the scattered field recorded at Port 1 and Port 2). Fig. 1.32 shows a comparison of the actual scattered field values with those estimated by the Born and the Rytov approximations. The data refer to Port 2 and are plotted versus frequency in the cases of $\epsilon_r = 1.2$ and $\epsilon_r = 2$.

1.15 Convergence Criterion for the Born Series

As demonstrated earlier, the limitations of the linear Born approximation prevent its applications in the cases of strong scattering. On the other hand, the Neumann-series expansion considered in Section 1.13 provides the means to improve the estimate of the internal total field beyond the simple approximation of it being equal to the incident field. For that, the recursive kernel update in Eq. (1.304) needs to be carried out. The so obtained total-field solution would converge only if the condition in Eq. (1.307) holds. It is therefore important to obtain the condition that would ensure the Neumann-series convergence in terms of the size and maximum contrast of the imaged object. Employing the approach used to arrive at Eq. (1.330), this condition is derived as

$$a^2 \sup|K(\mathbf{r})| < 1.58, \tag{1.337}$$

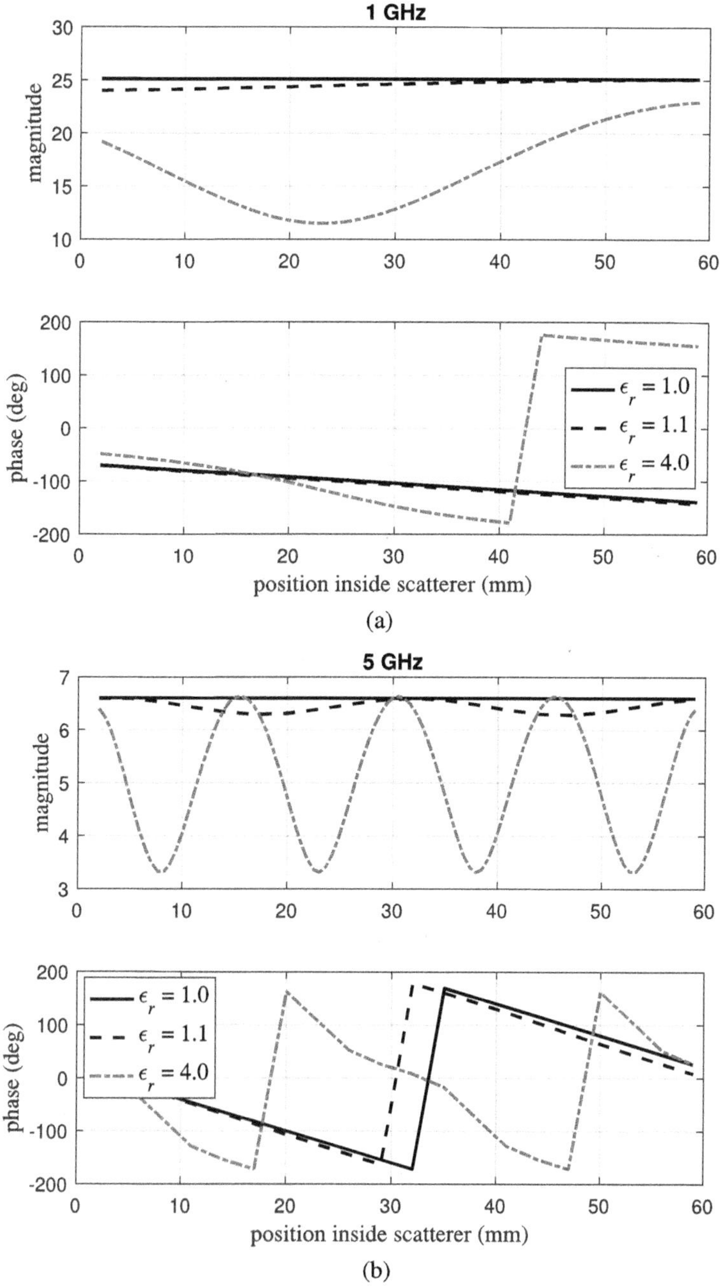

Figure 1.31 The field distribution inside the dielectric slab in the 1D wave example illustrating the limitations of Born's approximation: (a) at 1 GHz, (b) at 5 GHz. The distributions with $\epsilon_r = 1.0$ are those of the incident field.

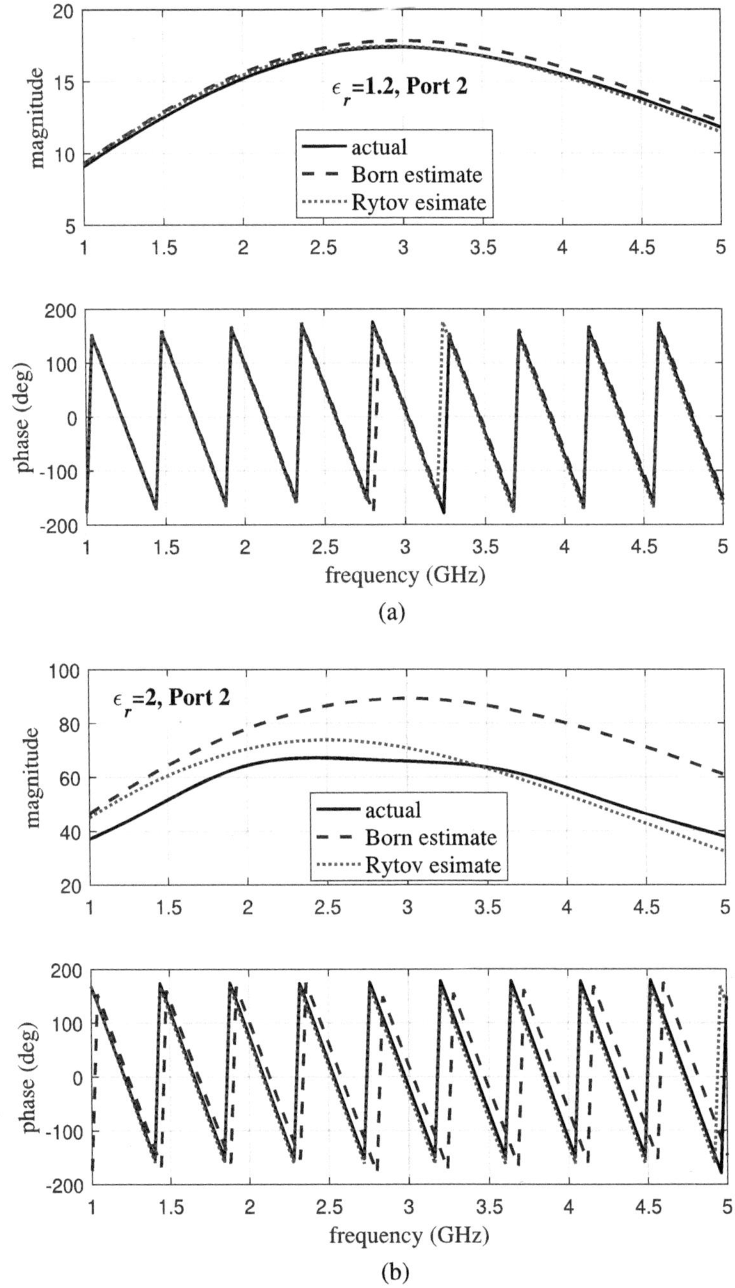

Figure 1.32 The scattered field recorded at Port 2 versus frequency in the 1D wave example illustrating the limitations of Born's approximation: (a) $\epsilon_r = 1.2$, (b) $\epsilon_r = 2$. An alternative estimate is also shown, which uses Rytov's approximation (discussed later).

where a is the radius of the sphere circumscribing the scatterer. Equivalently, Eq. (1.337) can be also stated in terms of the background wavenumber and the normalized contrast $\mathcal{X}$ (see Eq. (1.214)) as

$$(ak_{\mathrm{b}})^2 \sup|\mathcal{X}(\mathbf{r})| < 1.58. \tag{1.338}$$

We see that Eq. (1.338) is more relaxed than Eq. (1.330). This is expected since Eq. (1.330) states the requirement for the Born approximation to hold, whereas Eq. (1.338) states the requirement for the Born series to converge.

Example 2: Obtain the Born-series convergence condition Eq. (1.337). Work with the frequency-domain model of scattering and use the geometrical setup in Fig. 1.29.

Solution: The condition for the Born-series convergence, Eq. (1.307), requires the knowledge of the kernel $\mathcal{K}(\mathbf{r}, \mathbf{r}')$. In a uniform unbounded background, Green's function is $\frac{\exp(-ik_{\mathrm{b}}R)}{4\pi R}$. In this case, Eq. (1.307) is written as

$$E = [\sup |K(\mathbf{r})|]^2 \left(\frac{1}{4\pi}\right)^2 \iiint_V dv \iiint_V dv' \frac{1}{R^2} < 1, \tag{1.339}$$

where V is the volume of the scatterer, i.e., the sphere of radius a in Fig. 1.29, and R is evaluated as shown in Eq. (1.327). The integral over the primed coordinates in Eq. (1.339) is explicitly given by

$$\mathcal{I}' = \int_0^{2\pi}\int_0^{\pi}\int_0^{a} \frac{r'^2 \sin\theta' dr' d\theta' d\phi'}{r^2 + r'^2 - 2rr'\left[\sin\theta \sin\theta' \cos(\phi - \phi') + \cos\theta \cos\theta'\right]}, \tag{1.340}$$

assuming that the origin of the coordinate system is at the center of the sphere of volume V. Just like in the case of solving the integral in Eq. (1.326), we first adopt an auxiliary coordinate system, in which the observation point is at the origin, i.e., $\mathbf{r} = (0, 0, 0)$. The center C of the spherical volume V is a distance r_C away from the origin and along the z axis; refer to Fig. 1.29. Since the result of the integration $\mathcal{I}'$ depends only on the distance r_C between the volume's center and the observation point $\mathbf{r}$, we are free to choose the orientation of the z axis. In the auxiliary coordinate system, $\mathcal{I}'$ is

$$\mathcal{I}'(r_C) = \int_0^{2\pi}\int_0^{\pi}\int_0^{r'_{\max}(\theta')} \frac{r'^2 \sin\theta' dr' d\theta' d\phi'}{r'^2}, \tag{1.341}$$

which reduces to

$$\mathcal{I}'(r_C) = 2\pi \int_0^{\pi} \sin\theta' \sqrt{a^2 + r_C^2 - 2ar_C\cos\theta'}\, d\theta'. \tag{1.342}$$

This integral is solved exactly to obtain

$$\mathcal{I}'(r_C) = 4\pi\left(\frac{3a^2 + r_C^2}{3a}\right). \tag{1.343}$$

This result is further substituted in Eq. (1.339). Bearing in mind that $\mathcal{I}'$ depends only on r, which in the original coordinate system is $r = r_C$, we obtain

$$E = \left(\frac{\sup|K(\mathbf{r})|}{4\pi}\right)^2 \cdot 2\pi \cdot 2 \cdot 4\pi \int_0^a \left(\frac{3a^2 + r^2}{3a}\right) r^2 dr. \tag{1.344}$$

The convergence condition for the Neumann–series expansion follows as

$$E = (\sup|K(\mathbf{r})|)^2 \cdot 0.4a^4 < 1, \tag{1.345}$$

which can also be stated as

$$a^2 \sup|K(\mathbf{r})| < \frac{1}{\sqrt{0.4}} \approx 1.58. \tag{1.346}$$

1.16 The Rytov Approximation in the Scalar Model of Scattering

Another approach to linearizing the scalar inverse scattering problem is the Rytov approximation [14, 26, 48, 51]. Its region of applicability is somewhat different from that of the Born approximation, making it an important additional tool in inverse scattering. As discussed in [48], "the Born approximation produces a better estimate of the scattered field *amplitude* for large deviations in the refractive index for objects small in size. On the other hand, the Rytov approximation gives a more accurate estimate of the scattered field *phase* for large-sized objects with small deviations in refractive index." Here, the size of the object is considered in relation to the wavelength of the incident radiation; a "large-sized" object would measure several wavelengths or more. As discussed in Section 1.18, "small deviations in refractive index" are those of several percent or less.

The Rytov approximation of the total field is stated as

$$U(\mathbf{r}) \approx U^{\text{inc}}(\mathbf{r})e^{\psi_1(\mathbf{r})}, \tag{1.347}$$

where

$$\psi_1(\mathbf{r}) = \left[U^{\text{inc}}(\mathbf{r})\right]^{-1} \int_{\mathcal{D}} G(\mathbf{r}, \mathbf{r}')K(\mathbf{r}')U^{\text{inc}}(\mathbf{r}')d\Omega(\mathbf{r}'),$$

is a complex function. The domain $\mathcal{D}$ indicates the 2D or 3D spatial extent of the scattering object, and $K = \gamma_s^2 - \gamma_b^2$ is the scattering potential.

To understand the limitations associated with the Rytov approximation, it is worth walking through the steps leading to it. The essence of the Rytov approximation lies in the assumption that the scatterer affects mostly the phase of the incident radiation while the magnitude may experience some exponential decay. This is why it is the complex phase that is subjected to an expansion into a Neumann series. For that, the total field is cast into the form

$$U(\mathbf{r}) = e^{\psi(\mathbf{r})}, \tag{1.348}$$

where $\psi(\mathbf{r})$ is generally complex and is the subject of the series expansion. Substituting Eq. (1.348) into the Helmholtz equation for the total field,

$$\nabla^2 U(\mathbf{r}) - \gamma_b^2 U(\mathbf{r}) = K(\mathbf{r}) U(\mathbf{r}), \tag{1.349}$$

we arrive at[33]

$$\nabla^2 \psi(\mathbf{r}) + \nabla\psi(\mathbf{r}) \cdot \nabla\psi(\mathbf{r}) = \gamma_b^2(\mathbf{r}) \left[1 + \mathcal{X}(\mathbf{r})\right] = \gamma_s^2(\mathbf{r}), \tag{1.350}$$

where $\gamma_s = \mathrm{i}\omega\sqrt{\mu_s \varepsilon_s}$ is the complex propagation constant in the scatterer. Also, note that $1 + \mathcal{X} = \frac{\mu_s \varepsilon_s}{\mu_b \varepsilon_b}$. Eq. (1.350) is of the form of Ricatti's equation, which is nonlinear in its unknown function.

Ricatti's equation can be solved through an iterative procedure similar to the one used in solving the integral equation Eq. (1.290). We start with the function ψ_0, which describes the incident field through

$$U^{\text{inc}}(\mathbf{r}) = e^{\psi_0(\mathbf{r})}. \tag{1.351}$$

Since ψ_0 describes the incident field, it is assumed to be known. It must satisfy the Ricatti equation for the background medium,

$$\nabla^2 \psi_0(\mathbf{r}) + \nabla\psi_0(\mathbf{r}) \cdot \nabla\psi_0(\mathbf{r}) = \gamma_b^2. \tag{1.352}$$

We now expand the solution $\psi(\mathbf{r})$ of Eq. (1.350) in the vicinity of $\psi_0(\mathbf{r})$ using the Neumann series:

$$\psi(\mathbf{r}) = \lim_{n\to\infty} \psi^{(n)}, \quad \psi^{(n)}(\mathbf{r}) = \sum_{i=0}^{n} \psi_i(\mathbf{r}), \tag{1.353}$$

where the superscript (n) denotes the n-th iterate of the solution. Note that the convergence of the series requires that $|\psi_{i+1}| \le \xi\, |\psi_i|$ where $\xi < 1$. This expansion translates into an expansion for the field,

$$U(\mathbf{r}) = \exp\left[\psi_0(\mathbf{r}) + \psi_1(\mathbf{r}) + \psi_2(\mathbf{r}) + \cdots\right]. \tag{1.354}$$

This is Rytov's expansion. Notice that each phase term ψ_i adds a correction to the phase of the total field. The phase depends on the propagation constant γ of the medium through which the wave travels. In the simplest case of a traveling wave, the relationship is linear. Thus, each phase term ψ_i corresponds to a correction $\delta\gamma_i$ of the propagation constant. Then, the requirement $|\psi_{i+1}| \le \xi\, |\psi_i|$ translates into $|\delta\gamma_{i+1}| \le \xi |\delta\gamma_i|$ and $|\delta\gamma_1| \le \xi |\gamma_b|$, where $\xi < 1$.

[33] In this derivation, we have used the following identities: $\nabla^2(\cdot) = \nabla \cdot \nabla(\cdot)$ and $\nabla\left(e^{\psi}\right) = e^{\psi}\nabla\psi$.

The terms of the series are found iteratively as follows. The first solution iterate,

$$\psi^{(1)}(\mathbf{r}) = \psi_0(\mathbf{r}) + \psi_1(\mathbf{r}), \tag{1.355}$$

is substituted in Eq. (1.350), where Eq. (1.352) is further applied to cancel some terms, leading to

$$\nabla^2 \psi_1(\mathbf{r}) + 2\nabla\psi_0(\mathbf{r}) \cdot \nabla\psi_1(\mathbf{r}) + \underbrace{\nabla\psi_1(\mathbf{r}) \cdot \nabla\psi_1(\mathbf{r})}_{\approx 0} = K(\mathbf{r}). \tag{1.356}$$

The term $\nabla\psi_1 \cdot \nabla\psi_1$ is small, on the order of $O(\xi^2)$, and is neglected. Thus, ψ_1 is found as the solution to

$$\nabla^2 \psi_1(\mathbf{r}) + 2\nabla\psi_0(\mathbf{r}) \cdot \nabla\psi_1(\mathbf{r}) = K(\mathbf{r}) = \gamma_b^2 \mathcal{X}(\mathbf{r}), \tag{1.357}$$

where $\psi_0(\mathbf{r})$ and $K(\mathbf{r})$ are known.

The second solution iterate adds the second-order term of the Neumann series ψ_2:

$$\psi^{(2)}(\mathbf{r}) = \psi^{(1)}(\mathbf{r}) + \psi_2(\mathbf{r}) = \psi_0(\mathbf{r}) + \psi_1(\mathbf{r}) + \psi_2(\mathbf{r}). \tag{1.358}$$

To find the equation solved by ψ_2, Eq. (1.358) is substituted in Eq. (1.350). At the same time, the right side of Eq. (1.350), $\gamma_b^2(1+\mathcal{X})$, is replaced by

$$\gamma_b^2\left[1+\mathcal{X}(\mathbf{r})\right] = \nabla^2\left[\psi_0(\mathbf{r}) + \psi_1(\mathbf{r})\right] + \nabla\psi_0(\mathbf{r}) \cdot \nabla\psi_0(\mathbf{r}) + 2\nabla\psi_0(\mathbf{r}) \cdot \nabla\psi_1(\mathbf{r}), \tag{1.359}$$

which is obtained by adding Eq. (1.352) and Eq. (1.357). In the resulting equation, all small terms on the orders of $O(\xi^3)$ and $O(\xi^4)$ are neglected, leading to the equation for ψ_2:

$$\nabla^2 \psi_2(\mathbf{r}) + 2\nabla\psi_0(\mathbf{r}) \cdot \nabla\psi_2(\mathbf{r}) = -\nabla\psi_1(\mathbf{r}) \cdot \nabla\psi_1(\mathbf{r}). \tag{1.360}$$

Notice that the right side depends on ψ_1, which is already available from the previous iteration.

The iterative solution can continue, and it can be shown that the i-th term of the Neumann series Eq. (1.353), ψ_i, satisfies [14, 26]

$$\nabla^2 \psi_i(\mathbf{r}) + 2\nabla\psi_0(\mathbf{r}) \cdot \nabla\psi_i(\mathbf{r}) = -g_i(\mathbf{r}), \tag{1.361}$$

where

$$g_i(\mathbf{r}) = \begin{cases} -K(\mathbf{r}), & i = 1, \\ \sum_{k=1}^{i-1} \nabla\psi_k(\mathbf{r}) \cdot \nabla\psi_{i-k}(\mathbf{r}), & i \geq 2. \end{cases}$$

How is Eq. (1.361) solved? It can be transformed to a familiar equation by the substitution

$$\psi_i(\mathbf{r}) = W_i(\mathbf{r}) e^{-\psi_0(\mathbf{r})}, \tag{1.362}$$

where $\psi_0(\mathbf{r})$ is the incident-field "phase" that solves Eq. (1.352). It follows that $W_i(\mathbf{r})$ solves the familiar Helmholtz equation

$$\left(\nabla^2 - \gamma_b^2\right) W_i(\mathbf{r}) = -g_i(\mathbf{r}) e^{\psi_0(\mathbf{r})}. \tag{1.363}$$

In an integral form,

$$W_i(\mathbf{r}) = -\int_{\mathcal{D}} G(\mathbf{r}', \mathbf{r}) g_i(\mathbf{r}') e^{\psi_0(\mathbf{r}')} d\Omega(\mathbf{r}'), \tag{1.364}$$

where $G(\mathbf{r}', \mathbf{r})$ is Green's function of the background problem. According to Eq. (1.362), this translates into the integral solution for ψ_i as

$$\psi_i(\mathbf{r}) = -\left[U^{\mathrm{inc}}(\mathbf{r})\right]^{-1} \int_{\mathcal{D}} G(\mathbf{r}', \mathbf{r}) g_i(\mathbf{r}') U^{\mathrm{inc}}(\mathbf{r}') d\Omega(\mathbf{r}'), \tag{1.365}$$

where we have taken into account that $e^{\psi_0(\mathbf{r})} = U^{\mathrm{inc}}(\mathbf{r})$.

The Rytov approximation, which we initially stated as Eq. (1.347), is in effect the first-order Rytov expansion [see Eq. (1.354)]:

$$\begin{aligned} U_{\mathrm{R}}(\mathbf{r}) &= e^{[\psi_0(\mathbf{r})+\psi_1(\mathbf{r})]} = \\ & U^{\mathrm{inc}}(\mathbf{r}) \cdot \exp\left\{\left[U^{\mathrm{inc}}(\mathbf{r})\right]^{-1} \int_{\mathcal{D}} G(\mathbf{r}', \mathbf{r}) K(\mathbf{r}') U^{\mathrm{inc}}(\mathbf{r}') d\Omega(\mathbf{r}')\right\}. \end{aligned} \tag{1.366}$$

In deriving Eq. (1.366) we have taken into account that $g_1(\mathbf{r}) = -K(\mathbf{r})$; see Eq. (1.357) and Eq. (1.361).

1.17 Relationship between the Born and the Rytov Approximations

We recognize the integral in Eq. (1.366) as Born's approximation of the scattered field; see Eq. (1.320). Thus, Eq. (1.366) provides an explicit relationship between the Born and the Rytov approximations of the scattered field and of the total field:

$$\begin{aligned} U_{\mathrm{R}}(\mathbf{r}) &= U^{\mathrm{inc}}(\mathbf{r}) \cdot \exp\left\{\left[U^{\mathrm{inc}}(\mathbf{r})\right]^{-1} U_{\mathrm{B}}^{\mathrm{sc}}(\mathbf{r})\right\} \\ &= U^{\mathrm{inc}}(\mathbf{r}) \cdot \exp\left\{\left[U^{\mathrm{inc}}(\mathbf{r})\right]^{-1} U_{\mathrm{B}}(\mathbf{r}) - 1\right\}, \end{aligned} \tag{1.367}$$

where $U_{\mathrm{B}}^{\mathrm{sc}}(\mathbf{r})$ is the Born approximation of the scattered field, whereas $U_{\mathrm{B}}(\mathbf{r}) = U^{\mathrm{inc}}(\mathbf{r}) + U_{\mathrm{B}}^{\mathrm{sc}}(\mathbf{r})$ is the Born approximation of the total field. Notice that the Born approximation of the total field is a superposition of the incident field and the approximated scattered field. In contrast, the Rytov approximation of the total field is a correction to the incident field phase, which in itself is a function of the incident field.

The two approximations, clearly tend to converge together when the scattered field is much weaker than the incident field. If the exponent in Eq. (1.367) is expanded as $e^x = \sum_{k=0}^{\infty} \frac{x^k}{k!}$, and if the ratio $|U_{\mathrm{B}}^{\mathrm{sc}}/U^{\mathrm{inc}}|$ is sufficiently small to ignore the second and all subsequent terms, we obtain $U_{\mathrm{R}}(\mathbf{r}) \approx U^{\mathrm{inc}}(\mathbf{r})$, which is exactly the Born approximation for the internal total field. If the first two terms in the exponent expansion are taken, then Rytov's approximation reduces to $U^{\mathrm{inc}} + U_{\mathrm{B}}^{\mathrm{sc}}$, and it coincides with Born's approximation of the total external field. This is exactly the case with measurements far from the scatterer where the scattered field is sufficiently weaker than the incident one, making Rytov's and Born's approximations practically identical.

1.18 Limitations of the Rytov Approximation

We recall that the phase ψ_1 used in Rytov's expansion, Eq. (1.366), originated from Eq. (1.356)–(1.357) where the term $\nabla\psi_1 \cdot \nabla\psi_1$ was neglected. On one hand, we can view this term as a source term, which has been dismissed instead of adding it to the right-hand term $K(\mathbf{r})$. On the other hand, it can be viewed as a term quadratic in the unknown ψ_1 (and quadratic in the small parameter ξ), so that it is sufficiently small to be dismissed, rendering Eq. (1.357) linear in ψ_1. Recall that the convergence of the Rytov expansion requires that $|\psi_1| \le \xi|\psi_0|$, where $\xi < 1$. It then can be argued that the validity of Rytov's approximation depends on two conditions:

$$|\nabla\psi_1(\mathbf{r})|^2 << |\mathcal{X}(\mathbf{r})|\,|\gamma_b|^2 = |K(\mathbf{r})|\,, \tag{1.368}$$

and

$$|\nabla\psi_1| \le \xi|\nabla\psi_0|,\ \xi < 1. \tag{1.369}$$

Clearly, the smaller ξ is, the better the accuracy of Rytov's approximation. Assuming that the incident wave is a traveling wave of phase $\psi_0 = \varphi_0 - \gamma_b r$ (φ_0 is some initial phase and r is distance), we obtain that $|\nabla\psi_0| \approx |\gamma_b|$. The second condition can then be expressed as

$$|\nabla\psi_1| \le \xi|\gamma_b|,\ \xi < 1. \tag{1.370}$$

The two conditions, Eq. (1.368) and Eq. (1.370), are in fact very similar. We first note that they differ only in the factor $|\mathcal{X}(\mathbf{r})|$ in Eq. (1.368) versus ξ^2 in Eq. (1.370) (when squared). Under the somewhat crude assumption that the phase of the total field is also of the traveling-wave form, $\psi = \psi_0 + \psi_1 = \varphi_0 - \gamma_s r$, the requirement of Eq. (1.370) translates into $|\Delta\gamma_1| \le \xi|\gamma_b|$, where $\Delta\gamma_1 \approx \gamma_s - \gamma_b = \gamma_b\mathcal{X}$. Here, γ_s is the propagation constant in the scattering volume. Therefore,

$$|\mathcal{X}| \le \xi < 1, \tag{1.371}$$

is an inherent limitation of the Rytov approximation. It also tells us that observing Eq. (1.368) ensures that Eq. (1.370) is also observed. The combined conditions in Eq. (1.368) and Eq. (1.371) can now be stated as

$$\left[\frac{\lambda_b}{2\pi}\,|\nabla\psi_1(\mathbf{r})|\right]^2 << |\mathcal{X}(\mathbf{r})| < 1. \tag{1.372}$$

The physical meaning of these limitations is now more transparent: "the Rytov approximation is valid when the phase change over a single wavelength is small" [48]. Notice that, unlike the limitation of the Born approximation, see Eq. (1.332), the overall size of the scatterer does not play a role in Eq. (1.372). In [48], Slaney *et al.* conclude that while "the Born approximation is sensitive to the total phase shift in the object...", "...the Rytov approximation is sensitive to the change in refractive index."

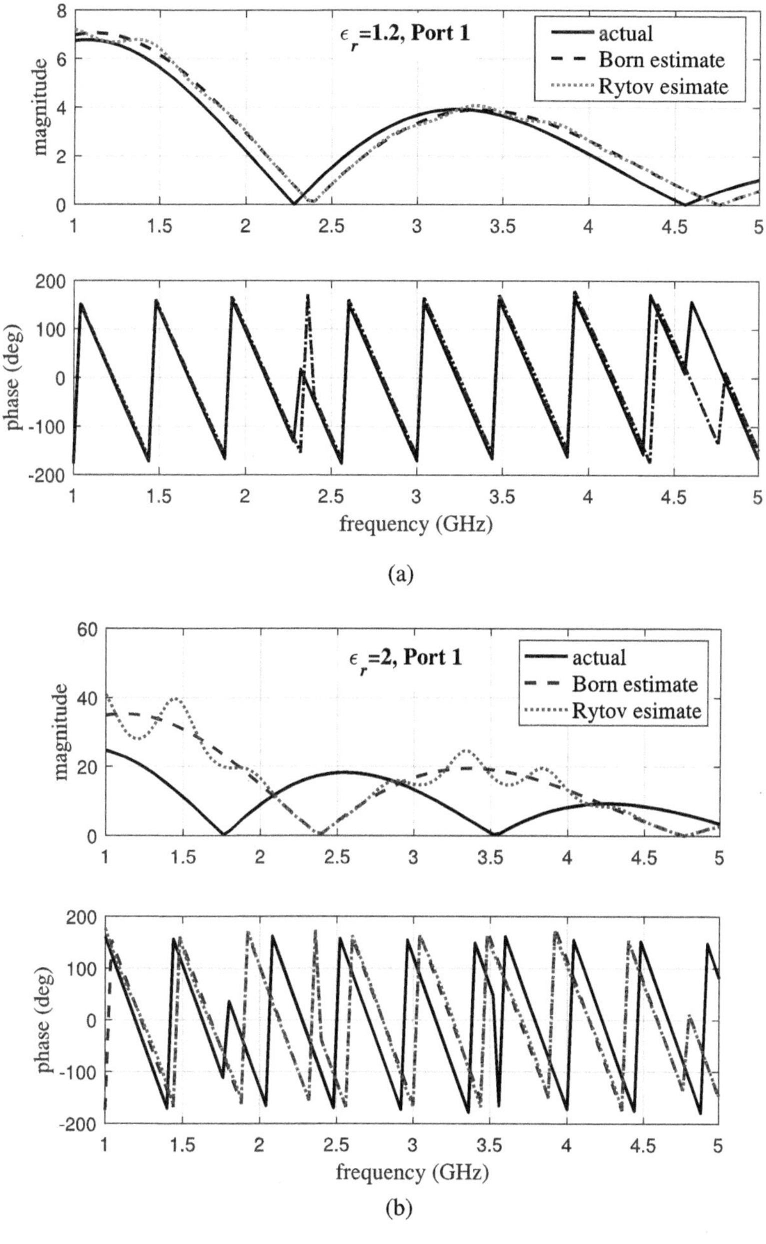

Figure 1.33 The scattered field recorded at Port 1 versus frequency in the 1D wave example illustrating the limitations of Born's and Rytov's approximations: (a) $\epsilon_r = 1.2$, (b) $\epsilon_r = 2$.

To illustrate the limitations of Rytov's approximation, we revisit the example of 1D scattering from a dielectric slab discussed in Section 1.14. Fig. 1.32 provides a comparison between the Born and Rytov approximations for the external field recorded at Port 2. Rytov's approximation provides a better estimate of the scattered field in both

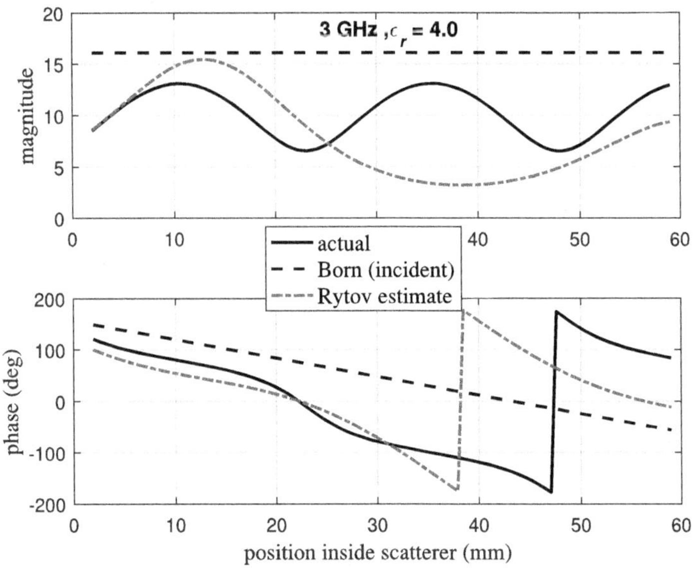

(a)

Figure 1.34 The field distribution inside the dielectric slab in the 1D wave example illustrating the limitations of the Born and Rytov approximations at $\epsilon_r = 4.0$. The Born approximation of the internal field is simply that of the incident field. Rytov's approximation tracks the phase of the total internal field better. This result is consistent across all frequencies.

magnitude and phase. Moreover, as predicted by theory, this estimate is largely frequency independent since Rytov's approximation is insensitive to the electrical size of the scatterer. It is instructive to look at the same comparison at Port 1, where the back-scattered field is recorded. This is shown in Fig. 1.33. This time, Rytov's approximation does not outperform Born's approximation. In general, Rytov's approximation performs best with forward-scattered (or transmitted) signals rather than with back-scattered ones. This is because it is introduced as a multiplicative correction to the incident field, which suits better the physics of forward-scattering than that of back-scattering.

Finally, we look at a comparison between the Born and Rytov approximations of the internal field in the same example. Fig. 1.34 shows a plot of the 3-GHz field distributions inside the dielectric slab when its permittivity is $\epsilon_r = 4.0$, the actual simulated distribution, along with its Born and Rytov approximations. Born's approximation is simply the incident field distribution. We observe that Rytov's approximation tracks the phase of the total internal field better. This result is consistent across all frequencies. With lower contrast, Rytov's approximations track better the magnitude of the total internal field as well.

2 Electromagnetic Scattering: The Vector Model

Electromagnetic (EM) scattering is intrinsically a vectorial wave-field problem. As discussed in Chapter 1, the scalar models can be very useful in uniform background regions where the EM field can be treated either through scalar-wave potentials or one rectangular component at a time. However, the full-wave description of the scattering arising at inhomogeneities is mathematically rigorous only through vectorial models.

2.1 Frequency-Domain Vector Solutions to Maxwell's Equations

The EM model is based on Maxwell's equations, which are summarized in Appendix A and aspects of which were already discussed in Chapter 1. There are quite a few numerical methods for solving Maxwell's equations directly either in the frequency or in the time domains. The literature on the subject is extensive, but a good start can be found in [29]. While the numerical solutions are widely employed in imaging as forward models, they are not in the scope of this text. Here, we focus on the formulation of the integral solutions to Maxwell's equation for the **E** or the **H** field, starting with the analysis in the frequency domain. These integral equations are not only valuable in understanding the forward problem of scattering but they are also necessary to understand how the inverse problem is solved in the frequency domain.

As shown in Section B.2 of Appendix B, the second-order differential equations for the electric and magnetic field phasors in a uniform medium are derived from Maxwell's equations as:

$$\nabla \times \nabla \times \mathbf{E} - \omega^2 \mu\varepsilon \mathbf{E} = -\mathrm{i}\omega\mu \mathbf{J}_\mathrm{i} - \nabla \times \mathbf{M}_\mathrm{i}, \tag{2.1}$$

$$\nabla \times \nabla \times \mathbf{H} - \omega^2 \mu\varepsilon \mathbf{H} = -\mathrm{i}\omega\varepsilon \mathbf{M}_\mathrm{i} + \nabla \times \mathbf{J}_\mathrm{i}, \tag{2.2}$$

where $\mathbf{J}_\mathrm{i}$ is the impressed electric current density, $\mathbf{M}_\mathrm{i}$ is the (fictitious) impressed magnetic current density, and μ and ε are the complex permeability and permittivity, respectively. In the following development, we will also use the complex propagation constant $\gamma = \mathrm{i}\omega\sqrt{\mu\varepsilon}$. Note that the medium's propagation properties are described equally well by the complex wave number such that $k^2 = -\gamma^2$, i.e., $k = \omega\sqrt{\mu\varepsilon}$. In a loss-free medium, the wave number is real whereas the propagation constant is imaginary, often making the wave number a more convenient choice.

2.2 The Vector Helmholtz Equation and Green's Dyadic

The linear partial differential operator $\mathcal{L} = \nabla \times \nabla \times +\gamma^2$ is known as the vector Helmholtz operator. Just like the scalar Helmholtz equation, integral-equation solutions can be derived, provided we know the impulse response associated with $\mathcal{L}$ and the respective boundary conditions. The difference here is that the impulse excitation must be considered in the three directions corresponding to the basis of the coordinate system. For example, in a Cartesian coordinate system, three impulse excitations must be considered: $\hat{\mathbf{x}}\delta(\mathbf{r} - \mathbf{r}')$, $\hat{\mathbf{y}}\delta(\mathbf{r} - \mathbf{r}')$, and $\hat{\mathbf{z}}\delta(\mathbf{r} - \mathbf{r}')$. The impulse response to each one of these excitations is a vector itself. For example, $\mathbf{G}_x = \hat{\mathbf{x}}G_{xx} + \hat{\mathbf{y}}G_{yx} + \hat{\mathbf{z}}G_{zx}$ is the vectorial impulse response to the x-oriented source $\hat{\mathbf{x}}\delta(\mathbf{r} - \mathbf{r}')$. For a complete description of the system behavior, we need to know all three vectorial impulse responses:

$$\mathbf{G}_\xi(\mathbf{r}, \mathbf{r}') = \hat{\mathbf{x}}G_{x\xi}(\mathbf{r}, \mathbf{r}') + \hat{\mathbf{y}}G_{y\xi}(\mathbf{r}, \mathbf{r}') + \hat{\mathbf{z}}G_{z\xi}(\mathbf{r}, \mathbf{r}'), \; \xi \equiv x, y, z. \tag{2.3}$$

Each vector function $\mathbf{G}_\xi(\mathbf{r}, \mathbf{r}')$, $\xi \equiv x, y, z$, solves the vector Helmholtz equation with the respective impulse excitation:

$$\mathcal{L}\mathbf{G}_\xi(\mathbf{r}, \mathbf{r}') = \nabla \times \nabla \times \mathbf{G}_\xi(\mathbf{r}, \mathbf{r}') + \gamma^2(\mathbf{r})\mathbf{G}_\xi(\mathbf{r}, \mathbf{r}') = \hat{\xi}\delta(\mathbf{r} - \mathbf{r}'), \; \xi \equiv x, y, z. \tag{2.4}$$

Let us now form a 3×3 matrix such that its columns are the vector functions $\mathbf{G}_\xi(\mathbf{r}, \mathbf{r}')$:

$$\bar{\mathbf{G}}(\mathbf{r}, \mathbf{r}') = \begin{bmatrix} G_{xx}(\mathbf{r}, \mathbf{r}') & G_{xy}(\mathbf{r}, \mathbf{r}') & G_{xz}(\mathbf{r}, \mathbf{r}') \\ G_{yx}(\mathbf{r}, \mathbf{r}') & G_{yy}(\mathbf{r}, \mathbf{r}') & G_{yz}(\mathbf{r}, \mathbf{r}') \\ G_{zx}(\mathbf{r}, \mathbf{r}') & G_{zy}(\mathbf{r}, \mathbf{r}') & G_{zz}(\mathbf{r}, \mathbf{r}') \end{bmatrix}, \tag{2.5}$$

and a 3×3 matrix such that its columns are the three vectorial impulse excitations oriented along x, y, and z:

$$\bar{\mathbf{I}}\delta(\mathbf{r} - \mathbf{r}') = \begin{bmatrix} \delta(\mathbf{r} - \mathbf{r}') & 0 & 0 \\ 0 & \delta(\mathbf{r} - \mathbf{r}') & 0 \\ 0 & 0 & \delta(\mathbf{r} - \mathbf{r}') \end{bmatrix}. \tag{2.6}$$

Here, $\bar{\mathbf{I}}$ is the identity 3×3 matrix. Now the three vector equations in Eq. (2.4) can be compactly written as

$$\mathcal{L}\bar{\mathbf{G}}(\mathbf{r}, \mathbf{r}') = \nabla \times \nabla \times \bar{\mathbf{G}}(\mathbf{r}, \mathbf{r}') + \gamma^2(\mathbf{r})\bar{\mathbf{G}}(\mathbf{r}, \mathbf{r}') = \bar{\mathbf{I}}\delta(\mathbf{r} - \mathbf{r}'), \tag{2.7}$$

provided $\mathcal{L}$ transforms $\bar{\mathbf{G}}$ one column at a time, so that the resulting matrix is

$$\mathcal{L}\bar{\mathbf{G}} = \left[\mathcal{L}\mathbf{G}_x, \mathcal{L}\mathbf{G}_y, \mathcal{L}\mathbf{G}_z\right]. \tag{2.8}$$

The impulse response described by Eq. (2.5) is termed *Green's dyadic*. Dyadics are second-rank tensors (here 3×3 matrices) that describe a linear relationship between two vectors.[1] Green's dyadic is simply Green's function for vector fields. More precisely, it is a matrix of nine scalar Green's functions, where $G_{\xi\zeta}$ provides the ξ ($\xi = x, y, z$)

[1] Tensors are a particular type of matrix. They describe physical quantities (force, inertia, stress, permeability, etc.) which must remain unchanged (invariant) upon change of the coordinate system such as rotation,

component of a response to the ζ ($\zeta = x, y, z$) oriented δ-function excitation. An excellent treatment on dyadic analysis, with all pertinent properties enabling EM solutions, can be found (as an appendix) in [4].

Green's dyadic of the vector Helmholtz equation in the case of a uniform unbounded medium[2] is important in MW imaging because it is often the assumption for the background. This dyadic is in essence the vector Helmholtz equation fundamental solution. It is well known and can be expressed in terms of the fundamental solution of the scalar Helmholtz equation as [4, 31]:

$$\bar{\mathbf{G}}(\mathbf{r}, \mathbf{r}') = \left(\bar{\mathbf{I}} - \frac{1}{\gamma^2}\nabla\nabla\right) F_{\text{SW}}(r), \tag{2.9}$$

where $F_{\text{SW}} = \frac{\exp(-\gamma r)}{4\pi r}$ and $r = |\mathbf{r} - \mathbf{r}'|$ (see Table 1.3). Explicitly, in a spherical coordinate system, Eq. (2.9) is written as

$$\bar{\mathbf{G}}(\mathbf{r}, \mathbf{r}') = \frac{e^{-\gamma r}}{4\pi\gamma^2 r^3}\left[(1 + \gamma r + \gamma^2 r^2)\bar{\mathbf{I}} - (3 + 3\gamma r + \gamma^2 r^2)\hat{\mathbf{r}}\hat{\mathbf{r}}\right], \tag{2.10}$$

where $\hat{\mathbf{r}} = (\mathbf{r} - \mathbf{r}')/r$. This is a diagonal tensor, with its nonzero elements being $G_{rr} = f(a-b)$, $G_{\theta\theta} = fa$, $G_{\phi\phi} = fa$, where $f = \frac{\exp(-\gamma r)}{4\pi\gamma^2 r^3}$, $a = 1 + \gamma r + \gamma^2 r^2$, and $b = 3 + 3\gamma r + \gamma^2 r^2$.

2.3 Integral Solutions to the Vector Helmholtz Equation

A. Integral Solution in an Unbounded Region

The construction of integral solutions to the vector Helmholtz equation for a field **F** with source **S**,

$$\mathcal{L}\mathbf{F} = \nabla \times \nabla \times \mathbf{F}(\mathbf{r}) + \gamma^2(\mathbf{r})\mathbf{F}(\mathbf{r}) = \mathbf{S}(\mathbf{r}), \tag{2.11}$$

can be viewed as an extension to the respective scalar solution; see Section 1.7. Explicitly, the interior solution describing the impact of volume sources in an unbounded region is [4],

$$F_x(\mathbf{r}) = \iiint_V \left[G_{xx}(\mathbf{r}, \mathbf{r}')S_x(\mathbf{r}') + G_{xy}(\mathbf{r}, \mathbf{r}')S_y(\mathbf{r}') + G_{xz}(\mathbf{r}, \mathbf{r}')S_z(\mathbf{r}')\right] dv' \tag{2.12}$$

$$F_y(\mathbf{r}) = \iiint_V \left[G_{yx}(\mathbf{r}, \mathbf{r}')S_x(\mathbf{r}') + G_{yy}(\mathbf{r}, \mathbf{r}')S_y(\mathbf{r}') + G_{yz}(\mathbf{r}, \mathbf{r}')S_z(\mathbf{r}')\right] dv' \tag{2.13}$$

$$F_z(\mathbf{r}) = \iiint_V \left[G_{zx}(\mathbf{r}, \mathbf{r}')S_x(\mathbf{r}') + G_{zy}(\mathbf{r}, \mathbf{r}')S_y(\mathbf{r}') + G_{zz}(\mathbf{r}, \mathbf{r}')S_z(\mathbf{r}')\right] dv'. \tag{2.14}$$

translation, or type (e.g., switching between Cartesian and spherical coordinates). Tensors offer a generalized framework that accommodates scalars (these are tensors of rank 0), vectors (these are tensors of rank 1), matrices (tensors of rank 2), etc. A succinct yet easy to understand introduction to the subject of tensors can be found in [52].

2 The unbounded medium solution requires that Green's dyadic describes waves that tend to zero as the distance from the δ-function source grows to infinity, i.e., $\bar{\mathbf{G}}(\mathbf{r}, \mathbf{r}') \to 0$ for $|\mathbf{r} - \mathbf{r}'| \to \infty$. This requirement sets the boundary condition for the vector Helmholtz equation, thereby ensuring uniqueness.

Eqs. (2.12) to (2.14) can be written more concisely in a vector form as

$$\mathbf{F}(\mathbf{r}) = \iiint_V \left[S_x(\mathbf{r}')\mathbf{G}_x(\mathbf{r},\mathbf{r}') + S_y(\mathbf{r}')\mathbf{G}_y(\mathbf{r},\mathbf{r}') + S_z(\mathbf{r}')\mathbf{G}_y(\mathbf{r},\mathbf{r}')\right] dv', \tag{2.15}$$

where $\mathbf{G}_\xi$ ($\xi = x, y, z$) is the ξ-th column of Green's dyadic. The field is now shown to be a vectorial sum of three contributions, each due to one component of the source (x, y, and z). Employing Green's dyadic $\bar{\mathbf{G}}$ in Eq. (2.5), Eq. (2.15) is now written as

$$\mathbf{F}(\mathbf{r}) = \iiint_V \bar{\mathbf{G}}(\mathbf{r},\mathbf{r}') \cdot \mathbf{S}(\mathbf{r}')dv'. \tag{2.16}$$

Eq. (2.16) is in a form analogous to that obtained in the scalar frequency-domain model of radiation, see Eq. (1.198), provided its Green's dyadic is causal.

The reader is reminded that, in the unbounded case, the interior region occupies the entire space with its surface S_∞ expanding into infinity. Just like in the scalar analysis, in general, the interior solution contains a volume integral accounting for the presence of internal sources as well as a surface integral over the field boundary values. However, here the surface integral over S_∞ is set to zero in accordance with our understanding that it represents contributions from sources *outside* S_∞ and that such sources do not exist.[3] The vanishing of the surface integral over S_∞ is ensured if $\mathbf{F}$ satisfies the *vector-field radiation condition* [4]

$$\hat{\mathbf{r}} \times \nabla \times \mathbf{F}(\mathbf{r}) - \gamma \mathbf{F}(\mathbf{r}) = o\left(r^{-1}\right), \tag{2.17}$$

where $r = |\mathbf{r}|$ is the radius of S_∞. When the $\mathbf{E}$ and $\mathbf{H}$ vectors in Maxwell's equations satisfy this condition, they also relate to each other through the *EM radiation boundary conditions* (also known as Silver-Müller boundary conditions):

$$\hat{\mathbf{r}} \times \mathbf{E} = \eta\mathbf{H} + o\left(r^{-1}\right),$$
$$\mathbf{H} \times \hat{\mathbf{r}} = \eta\mathbf{E} + o\left(r^{-1}\right),$$

where $\eta = \sqrt{\frac{\mu}{\epsilon}}$ is the intrinsic impedance of the medium.

B. Integral Solution in a Bounded Region

Just like in the scalar analysis, see Eq. (1.129), the integral solution of the vector Helmholtz equation may involve not only impressed volume sources in V but also surface sources on the surface S that bounds V. The analogy continues in stating that these surface sources are defined by the boundary values of the vector field, as is shown next.

The derivation of the general *interior* integral solution employs Green's vector identity,

$$\iiint_V (\mathbf{B} \cdot \nabla \times \nabla \times \mathbf{A} - \mathbf{A} \cdot \nabla \times \nabla \times \mathbf{B})\, dv = \oiint_S (\mathbf{A} \times \nabla \times \mathbf{B} - \mathbf{B} \times \nabla \times \mathbf{A}) \cdot d\mathbf{s}, \tag{2.18}$$

[3] Sometimes in scattering theory it is assumed that far-away sources (sources at infinity) exist and that their contribution is represented by a known field $\mathbf{E}_{\text{inc}}^{\text{far}}(\mathbf{r})$, typically in the form of a plane wave.

and it is similar to the derivation in the scalar case (see Section 1.5.2). As usual, the unit normal $\hat{\mathbf{n}}$ of $d\mathbf{s}$ points out of the volume V.

Consider now Eq. (2.11) together with the equation satisfied by one of Green's vectors:

$$\nabla \times \nabla \times \mathbf{G}_\xi(\mathbf{r}, \mathbf{r}') + \gamma^2(\mathbf{r})\mathbf{G}_\xi(\mathbf{r}, \mathbf{r}') = \hat{\xi}\delta(\mathbf{r} - \mathbf{r}'), \tag{2.19}$$

where ξ could be x, or y, or z. Take the dot product of Eq. (2.11) with $\mathbf{G}_\xi$ as well as the dot product of Eq. (2.19) with $\mathbf{F}$, subtract the two, and apply the identity Eq. (2.18). After formally exchanging $\mathbf{r}$ and $\mathbf{r}'$, the interior solution for the field's ξ-th component is obtained as

$$F_\xi(\mathbf{r}) = \iiint_V \mathbf{S}(\mathbf{r}') \cdot \mathbf{G}_\xi(\mathbf{r}', \mathbf{r})dv' + \oiint_S \{\mathbf{G}_\xi(\mathbf{r}', \mathbf{r}) \times [\nabla' \times \mathbf{F}(\mathbf{r}')] - \mathbf{F}(\mathbf{r}') \times [\nabla' \times \mathbf{G}_\xi(\mathbf{r}', \mathbf{r})]\} \cdot d\mathbf{s}', \mathbf{r} \in V, \xi = x, y, z, \tag{2.20}$$

where $\nabla' \times$ operates on $\mathbf{r}'$. First, we use the triple-product vector identities and rewrite Eq. (2.20) as

$$F_\xi(\mathbf{r}) = \iiint_V \mathbf{S}(\mathbf{r}') \cdot \mathbf{G}_\xi(\mathbf{r}', \mathbf{r})dv' + \oiint_S \{[\nabla' \times \mathbf{F}(\mathbf{r}')] \times \hat{\mathbf{n}} \cdot \mathbf{G}_\xi(\mathbf{r}', \mathbf{r}) - [\hat{\mathbf{n}} \times \mathbf{F}(\mathbf{r}')] \cdot [\nabla' \times \mathbf{G}_\xi(\mathbf{r}', \mathbf{r})]\} ds', \mathbf{r} \in V, \xi = x, y, z. \tag{2.21}$$

Second, we cast this component-wise solution into a form that uses Green's dyadic as

$$\mathbf{F}(\mathbf{r}) = \iiint_V \mathbf{S}(\mathbf{r}') \cdot \bar{\mathbf{G}}(\mathbf{r}', \mathbf{r})dv' + \oiint_S \{[\nabla' \times \mathbf{F}(\mathbf{r}')] \times \hat{\mathbf{n}} \cdot \bar{\mathbf{G}}(\mathbf{r}', \mathbf{r}) - [\hat{\mathbf{n}} \times \mathbf{F}(\mathbf{r}')] \cdot [\nabla' \times \bar{\mathbf{G}}(\mathbf{r}', \mathbf{r})]\} ds', \mathbf{r} \in V. \tag{2.22}$$

In Eq. (2.22), $\mathbf{F}(\mathbf{r})$ and all vectors in the integral that are pre-multiplying Green's dyadic, and its curl can be viewed as rows. Third, we make use of the EM reciprocity theorem [31], which states that

$$\bar{\mathbf{G}}^{\mathrm{T}}(\mathbf{r}', \mathbf{r}) = \bar{\mathbf{G}}(\mathbf{r}, \mathbf{r}'). \tag{2.23}$$

Here, $\bar{\mathbf{G}}^{\mathrm{T}}$ is the transpose (but *not* conjugated) of Green's dyadic. From reciprocity, it also follows that $\nabla' \times \bar{\mathbf{G}}(\mathbf{r}', \mathbf{r}) = [\nabla \times \bar{\mathbf{G}}(\mathbf{r}, \mathbf{r}')]^{\mathrm{T}}$ [34]. Note that $\nabla\times$ operates on $\mathbf{r}$ whereas $\nabla'\times$ operates on $\mathbf{r}'$. Finally, taking the transpose of both sides of Eq. (2.22) and employing Eq. (2.23) yields

$$\mathbf{F}(\mathbf{r}) = \iiint_V \bar{\mathbf{G}}(\mathbf{r}, \mathbf{r}') \cdot \mathbf{S}(\mathbf{r}')dv' + \oiint_S \{\bar{\mathbf{G}}(\mathbf{r}, \mathbf{r}') \cdot [\nabla' \times \mathbf{F}(\mathbf{r}')] \times \hat{\mathbf{n}} - [\nabla \times \bar{\mathbf{G}}(\mathbf{r}, \mathbf{r}')] \cdot [\hat{\mathbf{n}} \times \mathbf{F}(\mathbf{r}')]\} ds', \mathbf{r} \in V. \tag{2.24}$$

Here, $\mathbf{F}(\mathbf{r})$ and all vectors in the integral that are post-multiplying Green's dyadic and its curl can be viewed as columns. Eq. (2.24) is the interior solution to Helmholtz's vector equation. The *exterior* solution (for $\mathbf{r} \notin V$) has the same right side as Eq. (2.24), but the left side is zero.

2.4 Electric and Magnetic Field Integral Equations

2.4.1 EM Integral Equations in Terms of Green's Dyadic

Eq. (2.24) and the respective exterior integral solution can now be applied to the case of the $\mathbf{E}$ or $\mathbf{H}$ field; see Eq. (2.1) and Eq. (2.2). The electric field integral equation (EFIE) is obtained as

$$\iiint_V \bar{\mathbf{G}}(\mathbf{r},\mathbf{r}') \cdot \mathbf{S}_\mathrm{e}(\mathbf{r}')dv' + \mathrm{i}\omega \oiint_S \bar{\mathbf{G}}(\mathbf{r},\mathbf{r}') \cdot \mu \left[\hat{\mathbf{n}} \times \mathbf{H}(\mathbf{r}')\right] ds' + \oiint_S \left[\nabla \times \bar{\mathbf{G}}(\mathbf{r},\mathbf{r}')\right] \cdot \left[\mathbf{E}(\mathbf{r}') \times \hat{\mathbf{n}}\right] ds' = \begin{cases} \mathbf{E}(\mathbf{r}) & \mathbf{r} \in V, \\ 0 & \mathbf{r} \notin V, \end{cases} \tag{2.25}$$

where $\mathbf{r} \in V$ is the case of the interior solution and $\mathbf{r} \notin V$ is that of the exterior solution. As before, $\hat{\mathbf{n}}$ on S points outward. Note that in obtaining the first term in the surface integral we have used Faraday's law, $\nabla \times \mathbf{E} = -\mathrm{i}\omega\mu\mathbf{H}$. Also, in accordance with Eq. (2.1), the source is described as $\mathbf{S}_\mathrm{e} = -\mathrm{i}\omega\mu\mathbf{J}_\mathrm{i} - \nabla \times \mathbf{M}_\mathrm{i}$. Thus, the volume integral in Eq. (2.25) is explicitly written as

$$\mathcal{I}_\mathrm{vol}(\mathbf{r}) = -\mathrm{i}\omega \iiint_V \bar{\mathbf{G}}(\mathbf{r},\mathbf{r}') \cdot \mu\mathbf{J}_\mathrm{i}(\mathbf{r}')dv' - \iiint_V \bar{\mathbf{G}}(\mathbf{r},\mathbf{r}') \cdot \left[\nabla' \times \mathbf{M}_\mathrm{i}(\mathbf{r}')\right] dv'. \tag{2.26}$$

It is worth noting that the second term in Eq. (2.26) can be written in another mathematically equivalent form (provided $\mathbf{M}_\mathrm{i} = 0$ at S), namely,

$$\mathcal{I}_\mathrm{vol}(\mathbf{r}) = -\mathrm{i}\omega \iiint_V \bar{\mathbf{G}}(\mathbf{r},\mathbf{r}') \cdot \mu\mathbf{J}_\mathrm{i}(\mathbf{r}')dv' - \iiint_V \left[\nabla \times \bar{\mathbf{G}}(\mathbf{r},\mathbf{r}')\right] \cdot \mathbf{M}_\mathrm{i}(\mathbf{r}')dv'. \tag{2.27}$$

If we compare the volume integrals in Eq. (2.27) to the surface ones in Eq. (2.25), we notice that the field tangential values at S, $\hat{\mathbf{n}} \times \mathbf{H}$ and $\mathbf{E} \times \hat{\mathbf{n}}$, have the same effect as surface electric and magnetic currents, respectively, on the $\mathbf{E}$-field solution. Indeed, the EM surface equivalence principle shows that such equivalent (fictitious) currents can be very useful in solving scattering problems. We discuss this topic in more detail in the next two sections.

We now return to the definition of the general form of the magnetic field integral equation (MFIE), which appears in a form dual to that of the EFIE, namely,

$$\iiint_V \bar{\mathbf{G}}(\mathbf{r},\mathbf{r}')\cdot\mathbf{S}_\mathrm{m}(\mathbf{r}')dv' + \mathrm{i}\omega\oiint_S \bar{\mathbf{G}}(\mathbf{r},\mathbf{r}')\cdot\varepsilon\left[\mathbf{E}(\mathbf{r}')\times\hat{\mathbf{n}}\right]ds' + \oiint_S \left[\nabla\times\bar{\mathbf{G}}(\mathbf{r},\mathbf{r}')\right]\cdot\left[\mathbf{H}(\mathbf{r}')\times\hat{\mathbf{n}}\right]ds' = \begin{cases}\mathbf{H}(\mathbf{r}) & \mathbf{r}\in V,\\ 0 & \mathbf{r}\notin V,\end{cases} \tag{2.28}$$

Here, the source is $\mathbf{S}_\mathrm{m} = -\mathrm{i}\omega\varepsilon\mathbf{M}_\mathrm{i} + \nabla\times\mathbf{J}_\mathrm{i}$. For a definition of duality in electromagnetism and a list of the duality correspondences, the reader is referred to Section A.3 in Appendix A.

When solving the integral equations above, a difficult problem is encountered: Green's dyadic is singular when the observation and source points coincide, $\mathbf{r}\to\mathbf{r}'$; see Eq. (2.10). The integrals themselves are finite, but must be treated with care. A rigorous treatment can be found in [4]. Here, it suffices to introduce the relevant terminology and state the final result for the volume integrals.

The integration volume V must be split into two parts. The first part is a vanishingly small volume δ_V (e.g., a sphere or a cube) that is centered on the source point, thereby isolating it from the remaining volume where there is no singularity. As it turns out, the integral over δ_V tends to a finite value as the volume shrinks to a point. Using Green's dyadic in Eq. (2.10), it can be shown that [4]

$$\lim_{\delta_V\to 0}\iiint_{\delta_V}\bar{\mathbf{G}}(\mathbf{r},\mathbf{r}')\cdot\mathbf{S}(\mathbf{r}')dv' = \frac{1}{3\gamma^2}\mathbf{S}(\mathbf{r}). \tag{2.29}$$

The integral over the remaining volume, $V-\delta_V$, is termed the *principal value integral*:

$$\lim_{\delta_V\to 0}\iiint_{V-\delta_V}\bar{\mathbf{G}}(\mathbf{r},\mathbf{r}')\cdot\mathbf{S}(\mathbf{r}')dv' = \mathrm{PV}\iiint_V \bar{\mathbf{G}}(\mathbf{r},\mathbf{r}')\cdot\mathbf{S}(\mathbf{r}')dv'. \tag{2.30}$$

Thus, the rigorous expression for the field at an observation point that contains a source is written in terms of the singularity term in Eq. (2.29) plus the principal value integral. For example, in the case of the **E**-field in a volume of electric-current distribution, the volume integral is expressed as

$$\iiint_V \bar{\mathbf{G}}(\mathbf{r},\mathbf{r}')\cdot\left[-\mathrm{i}\omega\mu\mathbf{J}_\mathrm{i}(\mathbf{r}')\right]dv' = \mathrm{PV}\iiint_V \bar{\mathbf{G}}(\mathbf{r},\mathbf{r}')\cdot\left[-\mathrm{i}\omega\mu\mathbf{J}_\mathrm{i}(\mathbf{r}')\right]dv' - \frac{\mathbf{J}_\mathrm{i}(\mathbf{r})}{3\mathrm{i}\omega\varepsilon}. \tag{2.31}$$

2.4.2 EM Integral Equations in Terms of Green's Scalar Function

Poggio and Miller [53, chapter 4] give the EFIE and the MFIE in a vector form, which does not use Green's dyadic. Moreover, their development accommodates in a very simple form the possibility of the observation point being on the surface S. This is important in the analysis of the scattering from perfect-electric-conductor (PEC) objects where the EFIE and/or the MFIE are solved for observation points on the surface of the metallic scatterers. Since this surface is also the place where the induced surface sources of scattering reside (these are surface electric currents), singularities arise in Green's function. It is emphasized that these forms of the EFIE and the MFIE are equivalent mathematically to Eq. (2.25) and Eq. (2.28).

The EFIE of 3D scattering is now given as

$$- T \iiint_V \left[i\omega\mu \mathbf{J}_{\mathrm{i}}(\mathbf{r}') G(\mathbf{r}, \mathbf{r}') + \mathbf{M}_{\mathrm{i}}(\mathbf{r}') \times \nabla' G(\mathbf{r}, \mathbf{r}') - \frac{\rho_{\mathrm{e,i}}}{\varepsilon} \nabla' G(\mathbf{r}, \mathbf{r}') \right] dv'$$
$$- T \cdot \mathrm{PV} \oiint_S \{ i\omega\mu \left[\hat{\mathbf{n}}' \times \mathbf{H}(\mathbf{r}') \right] G(\mathbf{r}, \mathbf{r}') + \left[\mathbf{E}(\mathbf{r}') \times \hat{\mathbf{n}}' \right] \times \nabla' G(\mathbf{r}, \mathbf{r}')$$
$$- \left[\hat{\mathbf{n}}' \cdot \mathbf{E}(\mathbf{r}') \right] \nabla' G(\mathbf{r}, \mathbf{r}') \} \, ds' = \begin{cases} \mathbf{E}(\mathbf{r}) & \mathbf{r} \in V, \\ 0 & \mathbf{r} \notin V, \end{cases} \tag{2.32}$$

where $\rho_{\mathrm{e,i}}$ is the impressed electric charge density and $\hat{\mathbf{n}}'$ is the surface unit normal, which points *inward*, i.e., $\hat{\mathbf{n}}' = -\hat{\mathbf{n}}$. In the scattering from PEC objects and in the surface equivalence principle (see Section 2.7), it is customary to use $\hat{\mathbf{n}}'$ instead of $\hat{\mathbf{n}}$ in defining the surface sources of scattering. Therefore, it is prudent to adhere to this convention every time induced surface sources are going to be employed. Further, in Eq. (2.32), $G(\mathbf{r}, \mathbf{r}')$ is Green's function of Helmholtz's scalar equation and ∇' operates on $\mathbf{r}'$. In a uniform unbounded medium, $G(\mathbf{r}, \mathbf{r}') = \frac{\exp(-\gamma|\mathbf{r}-\mathbf{r}'|)}{4\pi|\mathbf{r}-\mathbf{r}'|}$. Note that here Green's scalar function relates to the fundamental solution *without* a minus sign. The constant T allows for treating the singularity of Green's function in the surface integral when $\mathbf{r} \to S$. If $\mathbf{r}$ does not belong to S, there is no singularity, $T = 1$, and there is no need to take the surface integral as a principal value (PV). However, if $\mathbf{r} \in S$, then

$$T = \left(1 - \frac{\Omega}{4\pi} \right)^{-1}, \tag{2.33}$$

where Ω is the solid angle subtended by S with its vertex at $\mathbf{r}$ as $\mathbf{r}$ approaches S. For example, if the surface is smooth (i.e., locally flat) at $\mathbf{r}$, then $\Omega = 2\pi$ and $T = 2$. If the scatterer is a very sharp long object (e.g., the end of a wire), $\Omega = 4\pi$ and $T \to \infty$, i.e., the field is singular.

The MFIE in terms of Green's scalar function is in a form dual to Eq. (2.32).

2.4.3 Electric Field Equations vs. Magnetic Field Equations

In principle it suffices to solve either the EFIE or the MFIE (not both). **E** and **H** are related through Maxwell's equations, and knowing one enables the computation of the other. However, the reader should be aware that in some instances it is advantageous numerically to use both. This is particularly true in the analysis of the scattering from

PEC or penetrable but *uniform* scatterers. In these cases, both the EFIE and the MFIE can be written for the observation points on the scatterer's surface. Through discretization, each can provide a linear system of equations for the unknown tangential field components. Once these components are computed, the field can be determined everywhere. Yet, spurious numerical solutions are avoided by combining both equations in what is known as the combined field integral equation (CFIE) method [54]. A detailed discussion and useful insights can be found in [34].

In the inverse scattering involving complex heterogeneous dielectric objects, it is common to use the EFIE model only. The reason lies in the forms of the Helmholtz equation for the **E** and **H** vectors in a heterogeneous medium; see Appendix B:

$$\mu\nabla \times \frac{1}{\mu}\nabla \times \mathbf{E} - \omega^2\mu\varepsilon\mathbf{E} = \mathbf{S}_{\mathrm{e}}, \tag{2.34}$$

$$\varepsilon\nabla \times \frac{1}{\varepsilon}\nabla \times \mathbf{H} - \omega^2\mu\varepsilon\mathbf{H} = \mathbf{S}_{\mathrm{m}}. \tag{2.35}$$

It is clear that in a magnetically homogeneous medium Eq. (2.34) reduces to Eq. (2.1), which served as the basis for deriving the EFIE in Eq. (2.25) and Eq. (2.32). Thus, with the proper choice of Green's dyadic, Eq. (2.25) would hold for heterogeneous dielectric media as well. Also, as we see later, the analysis of dielectric scattering in terms of the **E**-field allows for a simple expression of the induced source in the form $K\mathbf{E}$, where K is the scattering potential.

In contrast, Eq. (2.35) is not solved by the MFIE Eq. (2.28) if the dielectric medium is heterogeneous. More importantly, a contrast in ε results in a very complicated expression for the induced scattering source. Yet, an inverse-problem solution in terms of **H** and the MFIE would be advantageous in the case of scattering from a magnetically heterogeneous object.

2.5 Field Boundary Values and Uniqueness of the EM Forward Model

As an application of Eq. (2.25), let us consider the scenario depicted in Fig. 2.1, where the region of interest V does not contain sources. The sources, $\mathbf{J}_{\mathrm{i}}$ and $\mathbf{M}_{\mathrm{i}}$, are in the exterior region V^{e}. Both regions, however, have the same constitutive parameters and Green's dyadics. Thus, the boundary S between them is fictitious. In this case, the interior Eq. (2.25) contains only the surface integral:

$$\mathbf{E}(\mathbf{r}) = \oiint_S \left\{\mathrm{i}\omega\bar{\mathbf{G}}(\mathbf{r},\mathbf{r}')\cdot\mu\left[\hat{\mathbf{n}}\times\mathbf{H}(\mathbf{r}')\right] + \left[\nabla\times\bar{\mathbf{G}}(\mathbf{r},\mathbf{r}')\right]\cdot\left[\mathbf{E}(\mathbf{r}')\times\hat{\mathbf{n}}\right]\right\} ds', \ \mathbf{r}\in V. \tag{2.36}$$

Clearly, **E** and **H** at S represent implicitly the contribution of the exterior sources ($\mathbf{J}_{\mathrm{i}}$ and $\mathbf{M}_{\mathrm{i}}$) to the interior field solution; otherwise, the solution would be trivial.

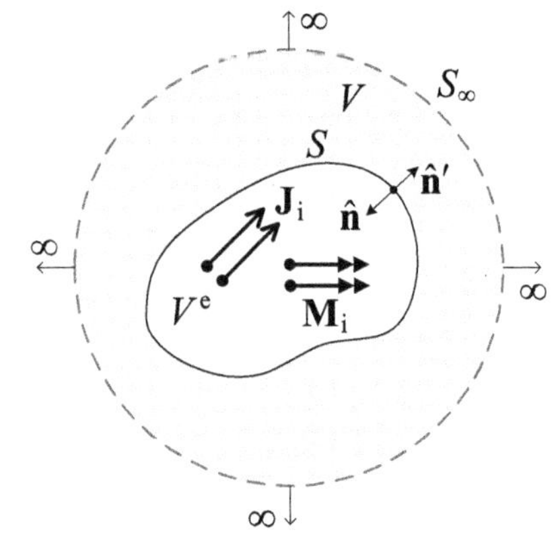

Figure 2.1 Illustration of a volume V, which excludes the EM sources $\mathbf{J}_i$ and $\mathbf{M}_i$. The sources reside in the complementary volume V^e such that $V \cup V^e$ is all space and $V \cap V^e$ is zero. V and V^e share a common boundary S where $\hat{\mathbf{n}}$ points out of V.

At the same time, the exterior Eq. (2.25) written with respect to the region V_e (i.e., observation point is in V while integration is in V_e) yields

$$0 = \iiint_{V_e} \bar{\mathbf{G}}(\mathbf{r}, \mathbf{r}') \cdot \mathbf{S}_i(\mathbf{r}')dv' - \oint\!\!\!\oint_S \left\{ i\omega\bar{\mathbf{G}}(\mathbf{r}, \mathbf{r}') \cdot \mu \left[\hat{\mathbf{n}} \times \mathbf{H}(\mathbf{r}')\right] + \left[\nabla \times \bar{\mathbf{G}}(\mathbf{r}, \mathbf{r}')\right] \cdot \left[\mathbf{E}(\mathbf{r}') \times \hat{\mathbf{n}}\right] \right\} ds', \; \mathbf{r} \in V, \tag{2.37}$$

where $\mathbf{S}_i(\mathbf{r}') = -i\omega\mu\mathbf{J}_i(\mathbf{r}') - \nabla \times \mathbf{M}_i(\mathbf{r}')$. The minus sign before the surface integral arises here because we use the same unit normal $\hat{\mathbf{n}}$ as in Eq. (2.36), which is inward with respect to V_e; see Fig. 2.1. Notice that adding Eq. (2.36) and Eq. (2.37) leads to the relation $\mathbf{E}(\mathbf{r}) = \iiint \bar{\mathbf{G}}(\mathbf{r}, \mathbf{r}') \cdot \mathbf{S}_i(\mathbf{r}')dv'$, which is expected since S is a fictitious boundary in the otherwise unbounded region $V \cup V_e$.

The exterior solution Eq. (2.37) shows that the boundary values of $\mathbf{E}$ and $\mathbf{H}$ are not independent. For a given source $\mathbf{S}_i(\mathbf{r}')$ in V^e, if $\mathbf{E} \times \hat{\mathbf{n}}$ is known at S, then, in principle, $\hat{\mathbf{n}} \times \mathbf{H}$ can be found and *vice versa*. This result is analogous to the scalar Helmholtz solution where we found that the boundary values of the field U_B and its normal derivative $\frac{\partial U_B}{\partial n}$ are not independent; see Section 1.7.3.

The above observation regarding the $\mathbf{E}$ and $\mathbf{H}$ boundary values is in accord with the *uniqueness theorem of electromagnetism* [1, 5], stated below.[4]

[4] The uniqueness theorem stated here is what is usually found in the textbooks. A more careful look at the mathematical proof [55] reveals that this statement is not complete. Uniqueness of the EM solution is also ensured by specifying exactly one pair of two collinear $\mathbf{E}$ and $\mathbf{H}$ field components: either $(\hat{\tau}_1 \cdot \mathbf{E}, \hat{\tau}_1 \cdot \mathbf{H})$, or $(\hat{\tau}_2 \cdot \mathbf{E}, \hat{\tau}_2 \cdot \mathbf{H})$, or $(\hat{\mathbf{n}} \cdot \mathbf{E}, \hat{\mathbf{n}} \cdot \mathbf{H})$. Here, $\hat{\mathbf{n}}$ is the surface unit normal whereas $\hat{\tau}_1$ and $\hat{\tau}_2$ are the surface tangential unit vectors. These types of boundary conditions are rarely used in EM analysis based on the field vectors $\mathbf{E}$ and $\mathbf{H}$; however they are relevant in choosing the boundary conditions for the EM scalar-wave functions.

For a unique solution to exist in a region of space V, in addition to the sources and the constitutive parameters in V, the following conditions need to be specified at the boundary of V:

(1) the tangential **E**-field components; or
(2) the tangential **H**-field components; or
(3) the tangential **E**-field components on some portions of the surface, and the tangential **H**-field components on the remainder.

The uniqueness theorem is telling us that, at each point on the boundary surface, it is sufficient to know either the tangential electric field or the tangential magnetic field but not both. Specifying both over-determines the forward problem and if the two boundary conditions (for **E** and for **H**) are inconsistent, the solution will be compromised.

Next, we see how the equivalent surface sources are derived from the boundary values of **E** and **H**.

2.6 Boundary Conditions and Surface Sources in Electromagnetism

Surface sources are used often in EM modeling. They are described by the following surface densities: (a) electric current surface density $\mathbf{J}_s$ (measured in A·m^{-1}), (b) electric charge surface density ρ_s (C·m^{-2}), (c) magnetic current surface density $\mathbf{M}_s$ (V·m^{-1}), and (d) magnetic charge surface density m_s (Wb·m^{-2}).

The surface densities appear naturally when applying Maxwell's equations at a surface. As a starting point, Maxwell's frequency-domain equations (see Appendix A) are re-stated here in their integral form:

$$\oint_C \mathbf{H} \cdot d\mathbf{c} = \iint_{S_C} \mathrm{i}\omega\mathbf{D} \cdot d\mathbf{s} + \iint_{S_C} \mathbf{J}_\mathrm{i} \cdot d\mathbf{s}, \tag{2.38}$$

$$-\oint_C \mathbf{E} \cdot d\mathbf{c} = \iint_{S_C} \mathrm{i}\omega\mathbf{B} \cdot d\mathbf{s} + \iint_{S_C} \mathbf{M}_\mathrm{i} \cdot d\mathbf{s}, \tag{2.39}$$

where S_C can be any surface bounded by the contour C. $\mathbf{J}_\mathrm{i}$ and $\mathbf{M}_\mathrm{i}$ are the impressed current densities, electric and magnetic, respectively.

When Eq. (2.38) and Eq. (2.39) are applied to a surface—the boundary between two regions—they lead to what is known as the *boundary conditions* of electromagnetism. An illustration in Fig. 2.2 shows how S_C and C are chosen at an interface, when deriving the boundary conditions. The regions 1 and 2 share a boundary (the interface), the unit normal $\hat{\mathbf{n}}$ of which points from region 1 into region 2 by convention. An orthogonal coordinate system can be associated with the interface with the basis $(\hat{\tau}_1, \hat{\tau}_2, \hat{\mathbf{n}})$, such that $\hat{\tau}_1 \times \hat{\tau}_2 = \hat{\mathbf{n}}$. The surface S_C can be defined so that it contains the unit vector $\hat{\mathbf{n}}$ and one of the tangential unit vectors, e.g., $\hat{\tau}_1$. To define the direction of the contour C,

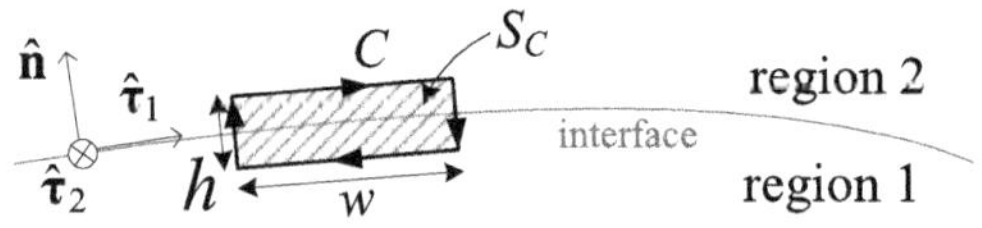

Figure 2.2 Illustration of the interface between two regions, region 1 and region 2, and the choice of the contour C and its respective surface S_C. The unit normal $\hat{\mathbf{n}}$ of the interface points from region 1 into region 2. The two unit vectors tangential to the surface are denoted as $\hat{\tau}_1$ and $\hat{\tau}_2$. They are mutually orthogonal and must satisfy $\hat{\tau}_1 \times \hat{\tau}_2 = \hat{\mathbf{n}}$.

the unit normal of S_C is chosen along the positive direction of the other tangential unit vector, e.g., $\hat{\tau}_2$. The so obtained direction of C is illustrated in Fig. 2.2.

Consider the application of Eq. (2.38) to the surface S_C illustrated in Fig. 2.2. To simplify the notations, the subscript in the source notation is omitted hereafter. To reduce Eq. (2.38) to a relationship at the interface, its limit is taken when the height h of the contour C tends to zero:

$$\lim_{h\to 0} \oint_C \mathbf{H} \cdot d\mathbf{c} = \lim_{h\to 0} \iint_{S_C} \mathrm{i}\omega \mathbf{D} \cdot d\mathbf{s} + \lim_{h\to 0} \iint_{S_C} \mathbf{J} \cdot d\mathbf{s}. \tag{2.40}$$

Assuming that there are no field singularities, the left side of Eq. (2.40) reduces to

$$\lim_{h\to 0} \oint_C \mathbf{H} \cdot d\mathbf{c} = \int_w \left(\mathbf{H}^{(2)} - \mathbf{H}^{(1)}\right) \cdot \hat{\tau}_1 d\tau_1, \tag{2.41}$$

Here, the superscript denotes the region to which the field belongs. Note that while $\mathbf{H}^{(1)}$ and $\mathbf{H}^{(2)}$ are both evaluated at the interface, they belong to two distinct regions and may not be equal, i.e., the field may be discontinuous.

The first term in the right side of Eq. (2.40) is considered next. The assumption of no field singularity at the interface leads to the conclusion that it is zero since S_C tends to zero at the rate of h. Explicitly,

$$\lim_{h\to 0} \left(h \int_w \mathrm{i}\omega \mathbf{D} \cdot \hat{\tau}_2 \, d\tau_1 \right) = 0. \tag{2.42}$$

Here, w is the width of the contour C.

Finally, we turn to the last term in Eq. (2.40). If the source distribution $\mathbf{J}$ is not singular at the interface, then that term is zero as well. However, if a "sheet" of current is present, the source is singular as it is a δ-function along $\hat{\mathbf{n}}$:

$$\mathbf{J}(\tau_1, \tau_2, n) = \mathbf{J}_\mathrm{s}(\tau_1, \tau_2) \cdot \delta(n), \tag{2.43}$$

where τ_1, τ_2, and n are the surface coordinates. $\mathbf{J}_\mathrm{s}$ is exactly the surface current density of the source. It is a mathematical abstraction but it is commonly used in electrical engineering to describe currents on metallic sheets the thickness of which is negligible. Such sheets are commonplace in the technology of printed circuit boards, microwave and antenna engineering. The physical meaning of the surface current is more obvious when the relationship in Eq. (2.43) is inverted:

$$\mathbf{J}_s(\tau_1, \tau_2) = \lim_{h \to 0} \int_{-h/2}^{h/2} \mathbf{J}(\tau_1, \tau_2, n) dn = \lim_{h \to 0} [h\mathbf{J}(\tau_1, \tau_2)] . \tag{2.44}$$

In the scenario illustrated in Fig. 2.2, if surface currents are present, the last term in Eq. (2.40) reduces to

$$\lim_{h \to 0} \iint_{S_C} \mathbf{J} \cdot d\mathbf{s} = \int_{w} \mathbf{J}_s \cdot \hat{\tau}_2 d\tau_1 . \tag{2.45}$$

The results in Eq. (2.41), Eq. (2.42) and Eq. (2.45) are substituted in Eq. (2.40) to obtain

$$\left(\mathbf{H}^{(2)} - \mathbf{H}^{(1)}\right) \cdot \hat{\tau}_1 = \mathbf{J}_s \cdot \hat{\tau}_2 . \tag{2.46}$$

In an analogous manner (using a contour in the plane of $\hat{\mathbf{n}}$ and $\hat{\tau}_2$), a second boundary condition is obtained as

$$\left(\mathbf{H}^{(1)} - \mathbf{H}^{(2)}\right) \cdot \hat{\tau}_2 = \mathbf{J}_s \cdot \hat{\tau}_1 . \tag{2.47}$$

Eq. (2.46) and Eq. (2.47) can be unified in the vector expression

$$\hat{\mathbf{n}} \times \left(\mathbf{H}^{(2)} - \mathbf{H}^{(1)}\right) = \mathbf{J}_s . \tag{2.48}$$

In essence, this is the surface version of Ampère's law.

Following the same procedure, we find the boundary condition that follows from Faraday's law, Eq. (2.39):

$$\hat{\mathbf{n}} \times \left(\mathbf{E}^{(1)} - \mathbf{E}^{(2)}\right) = \mathbf{M}_s . \tag{2.49}$$

The boundary conditions that relate the field to the surface charge densities, ρ_s and m_s, follow from Maxwell's divergence equations in a straightforward manner; see, for example, [5]. Their derivation is not discussed here but they are stated in Table 2.1, which summarizes all four EM boundary conditions together with the Maxwell equation that they are derived from. Note that the surface charges are not independent of the surface currents since they obey the continuity law; e.g., $\nabla_\tau \cdot \mathbf{J}_s = -\mathrm{i}\omega\rho_s$, where $\nabla_\tau \cdot$ denotes the 2D (surface) divergence operator at the interface.

We notice from Eq. (2.48) and Eq. (2.49) that in the absence of surface currents, the tangential **E** and **H** field components are continuous, i.e.,

$$\hat{\mathbf{n}} \times \mathbf{H}^{(2)} = \hat{\mathbf{n}} \times \mathbf{H}^{(1)}, \tag{2.50}$$

$$\hat{\mathbf{n}} \times \mathbf{E}^{(2)} = \hat{\mathbf{n}} \times \mathbf{E}^{(1)} . \tag{2.51}$$

Table 2.1 EM Boundary Conditions

Maxwell's Equation	Boundary Condition
$\nabla \times \mathbf{H} = \mathrm{i}\omega\mathbf{D} + \mathbf{J}$	$\hat{\mathbf{n}} \times \left(\mathbf{H}^{(2)} - \mathbf{H}^{(1)}\right) = \mathbf{J}_s$
$-\nabla \times \mathbf{E} = \mathrm{i}\omega\mathbf{B} + \mathbf{M}$	$\hat{\mathbf{n}} \times \left(\mathbf{E}^{(1)} - \mathbf{E}^{(2)}\right) = \mathbf{M}_s$
$\nabla \cdot \mathbf{D} = \rho$	$\hat{\mathbf{n}} \cdot \left(\mathbf{D}^{(2)} - \mathbf{D}^{(1)}\right) = \rho_s$
$\nabla \cdot \mathbf{B} = m$	$\hat{\mathbf{n}} \cdot \left(\mathbf{B}^{(2)} - \mathbf{B}^{(1)}\right) = m_s$

This boundary condition is common as it applies to dielectric and magnetic interfaces where surface currents do not exist.

On the other hand, at the interface between a PEC object and a dielectric medium, surface electric currents do exist. Let region 2 be a dielectric medium, which borders a PEC object[5] (region 1). Then, Eq. (2.48) dictates that a surface current density is induced by a nonzero tangential magnetic field such that

$$\hat{\mathbf{n}} \times \mathbf{H}^{(2)} = \mathbf{J}_s. \tag{2.52}$$

This is because the PEC object is perfectly opaque to the EM field and $\mathbf{H}^{(1)} = 0$. Thus, the surface of the PEC object becomes the place where real high-frequency electric currents are induced by the microwave field. Note that $\hat{\mathbf{n}}$ points *out of* the PEC region.

A mathematically similar situation arises at the surface of perfect-magnetic-conductor (PMC) objects where surface magnetic currents are induced such that

$$-\hat{\mathbf{n}} \times \mathbf{E}^{(2)} = \mathbf{M}_s. \tag{2.53}$$

The difference with the PEC case is that PMC objects do not exist in nature at microwave frequencies. Artificial surfaces can be designed to behave approximately as PMC surfaces in a narrow frequency band. In any case, the surface magnetic current, just like its volume counterpart, is a fictitious quantity, which, as we will see next, proves useful in the surface equivalence principle of electromagnetism.

2.7 Surface Equivalence Principle in Electromagnetism

The boundary conditions enable the expression of the field boundary values as surface sources, which replace equivalently the actual exterior sources as far as the field in the interior volume is concerned. Fig. 2.3-a shows an actual problem where the sources generate the field $(\mathbf{E}, \mathbf{H})$ in the overall region $V \cup V^e$. The division of the overall region into two parts, V, where the field is observed, and V^e, where the sources reside, is artificial, i.e., the boundary S does not need to indicate an actual change in the properties of the medium.

Fig. 2.3-b shows an equivalent (fictitious) problem constructed so that the field in V^e has been set to zero. Yet, the field in V remains $(\mathbf{E}, \mathbf{H})$ as in the original problem.

[5] Practically all metals in the electronic industry (copper, aluminum, gold, etc.) can be considered as PEC at MW frequencies due to an extremely high specific conductivity σ_e on the order of $\sim 10^7$ S/m.

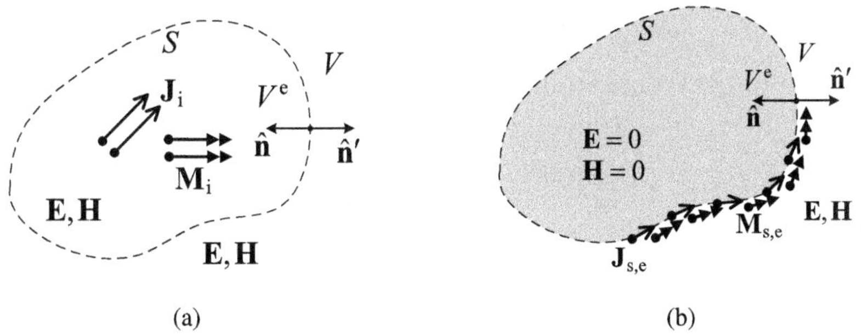

Figure 2.3 Illustration of Love's equivalence principle. (a) The actual problem in which the field $(\mathbf{E}, \mathbf{H})$ is generated by sources described mathematically by electric and magnetic current densities $\mathbf{J}_i$ and $\mathbf{M}_i$. The surface S encloses the sources by traversing through source-free spatial points. S is fictitious and need not be associated with any physical interface. (b) The equivalent problem in which the region V^e features zero field whereas the field in V remains the same as in the actual problem. The field discontinuity at S demands the presence of surface sources. These surface sources are equivalent to the actual sources in V^e in the sense that they generate exactly the same field $(\mathbf{E}, \mathbf{H})$ in V. Note, however, that they generate zero field in V^e—this is the EM extinction theorem.

The uniqueness theorem tells us that in order to preserve $(\mathbf{E}, \mathbf{H})$ in V as in the original problem, we must set its tangential values at S as those in the original problem. Now the field has a discontinuity at S. The boundary conditions, see Table 2.1, assert that such discontinuity can exist only if surface sources are present at S. The convention in setting the surface equivalent sources is that the unit normal, here $\hat{\mathbf{n}}'$, points out of the source region V^e and into the source-free region V; refer to Fig. 2.3. Since the field in V^e is set to zero, these equivalent surface sources are obtained from Eq. (2.48) and Eq. (2.49) as:

$$\hat{\mathbf{n}}' \times \mathbf{H} = \mathbf{J}_{s,e}, \tag{2.54}$$

$$-\hat{\mathbf{n}}' \times \mathbf{E} = \mathbf{M}_{s,e}. \tag{2.55}$$

Note the added subscript $(\cdot)_e$, which emphasizes that these are equivalent surface sources, i.e., they do not exist physically.

The method of deriving the equivalent surface currents by setting the field to zero in the region of the original sources (here V^e) is referred to as *Love's equivalence principle* [56]. This method is common in antenna theory, especially the analysis of aperture antennas [15], where it is advantageous to exclude the complicated antenna structure from the analyzed volume and replace it with (approximately) known surface currents on a surface (the *aperture*) surrounding the antenna. Love's equivalence can be viewed as an expression of the interior and exterior EFIE and MFIE, Eq. (2.25) and Eq. (2.28), in terms of the equivalent surface currents instead of the field boundary values. Just like the EFIE and the MFIE indicate, the field in the exterior (source) region is zero whereas the field in the interior (source-free) region is due to the surface sources.

2.8 Surface Equivalence Principle in EM Radiation

We can now substitute Love's equivalent sources, Eqs. (2.54)–(2.55), into the interior source-free solution for the **E**-field; see Eq. (2.25). Bearing in mind that $\hat{\mathbf{n}} = -\hat{\mathbf{n}}'$, the **E**-field in V is obtained as

$$\mathbf{E}(\mathbf{r}) = -\oiint_S \mathrm{i}\omega\mu\bar{\mathbf{G}}(\mathbf{r},\mathbf{r}')\cdot\mathbf{J}_{\mathrm{s,e}}(\mathbf{r}')ds' - \oiint_S \left[\nabla\times\bar{\mathbf{G}}(\mathbf{r},\mathbf{r}')\right]\cdot\mathbf{M}_{\mathrm{s,e}}(\mathbf{r}')ds',\ \mathbf{r}\in V. \tag{2.56}$$

Notice that the surface integrals in the above EFIE have the same form as their volume counterparts in Eq. (2.27).

The expression for the **H** field is the dual of Eq. (2.56):

$$\mathbf{H}(\mathbf{r}) = -\oiint_S \mathrm{i}\omega\varepsilon\bar{\mathbf{G}}(\mathbf{r},\mathbf{r}')\cdot\mathbf{M}_{\mathrm{s,e}}(\mathbf{r}')ds' + \oiint_S \left[\nabla\times\bar{\mathbf{G}}(\mathbf{r},\mathbf{r}')\right]\cdot\mathbf{J}_{\mathrm{s,e}}(\mathbf{r}')ds',\ \mathbf{r}\in V. \tag{2.57}$$

The above EFIE and MFIE show that the radiation from an EM source can be replaced equivalently by the radiation from (fictitious) surface electric and magnetic current sources on any surface (the aperture) that encloses the actual source. These equivalent surface sources are determined from the EM field boundary values through Eq. (2.54) and Eq. (2.55).

Example: A uniform plane EM wave is due to sources in region 1 ($z < 0$). The **E**-field is measured at the planar surface at $z = 0$ as $\mathbf{E}_0 = E_0\hat{\mathbf{x}}$ where $E_0 = 5$ V/m. The **H**-field at $z = 0$ is given by $\mathbf{H}_0 = H_0\hat{\mathbf{y}}$ where $H_0 = E_0/\eta_0$ and η_0 is the medium's wave impedance. The medium is vacuum and $\eta_0 = \sqrt{\frac{\mu_0}{\epsilon_0}} \approx 377\ \Omega$. The frequency is $f = 3$ GHz. Find the expression for the **E** field in region 2 using Love's equivalence principle.

Solution: Our first task is to find Green's dyadic. As per Eq. (2.9), it can be found from the fundamental solution of the scalar Helmholtz equation F_{sw}. A plane wave traveling in the z direction is a 1D problem, for which $F_{\mathrm{sw}} = \frac{\exp(-\gamma|z|)}{2\gamma}$ (see Table 1.3), where γ is the propagation constant. Here, the medium is vacuum, the frequency is 3 GHz; therefore,

$$\gamma = \mathrm{i}\omega\sqrt{\mu_0\epsilon_0} = \mathrm{i}2\pi\frac{f}{c} \approx \mathrm{i}2\pi\frac{3\cdot 10^9}{3\cdot 10^8} = \mathrm{i}20\pi\ \mathrm{m}^{-1}.$$

Green's scalar function $g(z, z')$ in both regions is explicitly written as

$$g(z,z') = \begin{cases} \dfrac{e^{-\gamma z}}{2\gamma}, z \geq 0, \\ \dfrac{e^{\gamma z}}{2\gamma}, z < 0. \end{cases} \tag{2.58}$$

The gradient of $g(z, z')$ is then found as

$$\nabla g(z, z') = \begin{cases} -\hat{\mathbf{z}}\dfrac{e^{-\gamma z}}{2}, z \geq 0, \\ \hat{\mathbf{z}}\dfrac{e^{\gamma z}}{2}, z < 0. \end{cases} \tag{2.59}$$

Further, $\nabla\nabla g(z, z')$ is found as:[6]

$$\nabla\nabla g(z, z') = \begin{bmatrix} 0 & 0 & 0 \\ 0 & 0 & 0 \\ 0 & 0 & \frac{\partial^2 g}{\partial z^2} \end{bmatrix}, \tag{2.60}$$

where $\frac{\partial^2 g}{\partial z^2} = \frac{\gamma}{2}\exp(-\gamma|z - z'|) = \gamma^2 g$. Finally, Green's dyadic is found as

$$\bar{\mathbf{G}}(z, z') = \bar{\mathbf{I}}g(z, z') - \frac{1}{\gamma^2}\nabla\nabla g(z, z') = \begin{bmatrix} g & 0 & 0 \\ 0 & g & 0 \\ 0 & 0 & 0 \end{bmatrix}, \tag{2.61}$$

where g is defined in Eq. (2.58).

To calculate $\mathbf{E}$ in region 2 via the EFIE Eq. (2.56) we also need $\nabla \times \bar{\mathbf{G}}$. Bearing in mind that g depends only on z, we obtain

$$\nabla \times \bar{\mathbf{G}} = \begin{bmatrix} 0 & -\frac{\partial g}{\partial z} & 0 \\ \frac{\partial g}{\partial z} & 0 & 0 \\ 0 & 0 & 0 \end{bmatrix}, \tag{2.62}$$

where $\frac{\partial g}{\partial z}$ in region 2 is given by

$$\left.\frac{\partial g}{\partial z}\right|_{z'=0} = -\frac{e^{-\gamma z}}{2}, z \geq 0. \tag{2.63}$$

In the 1D case, the EFIE Eq. (2.56) reduces to

$$\mathbf{E}(z) = -\mathrm{i}\omega\mu\bar{\mathbf{G}}(z, z') \cdot \mathbf{J}_{\mathrm{s,e}}(z' = 0) - \left[\nabla \times \bar{\mathbf{G}}(z, z')\right] \cdot \mathbf{M}_{\mathrm{s,e}}(z' = 0), z \geq 0. \tag{2.64}$$

As per the boundary conditions Eq. (2.54) and Eq. (2.55), the equivalent surface sources are found as

$$\mathbf{J}_{\mathrm{s,e}} = \hat{\mathbf{z}} \times \hat{\mathbf{y}}\frac{E_0}{\eta_0} = -\hat{\mathbf{x}}\frac{E_0}{\eta_0}, \tag{2.65}$$

$$\mathbf{M}_{\mathrm{s,e}} = -\hat{\mathbf{z}} \times \hat{\mathbf{x}}E_0 = -\hat{\mathbf{y}}E_0. \tag{2.66}$$

Substituting these in the 1D EFIE above with $E_0 = 5$ V/m, we obtain

$$\mathbf{E}(z) = \hat{\mathbf{x}}\frac{1}{2}\underbrace{\left(\frac{\mathrm{i}\omega\mu_0}{\gamma\eta_0} + 1\right)}_{2} E_0 e^{-\gamma z} = \hat{\mathbf{x}}5e^{-\mathrm{i}20\pi z}\ \mathrm{V/m}, z \geq 0. \tag{2.67}$$

[6] Here, we use the tensor formula $\nabla\mathbf{a} = (\nabla a_x)\hat{\mathbf{x}} + (\nabla a_y)\hat{\mathbf{y}} + (\nabla a_z)\hat{\mathbf{z}}$.

This is an expected result as it shows the uninterrupted propagation of the plane EM wave into region 2 through the fictitious boundary at $z' = 0$.

2.9 Surface Equivalence Principle in EM Scattering

In Section 1.7.4, we showed how to employ the exterior integral solution of the scalar Helmholtz equation to the problem of scattering from a body of uniform constitutive parameters. There, the equivalent surface sources appeared as single-layer and double-layer scalar distributions on the object's surface. The same approach can be applied to the vector EM scattering from a uniform object. This time, the equivalent surface sources appear in the form of electric and magnetic surface current distributions.

The problem is illustrated in Fig. 2.4-a. Region 1 is the background of constitutive parameters μ_1 and ε_1. It hosts a scatterer (region 2), the constitutive parameters (μ_2 and ε_2) of which are different from those of region 1. In the absence of the scatterer, the antenna in region 1 generates the incident field $\mathbf{E}_{\text{inc}}$, which is assumed known. The scatterer is responsible for the scattered field $\mathbf{E}_{\text{sc}}^{(1)}$ in region 1. The goal is to determine $\mathbf{E}_{\text{sc}}^{(1)}$ along with $\mathbf{E}_{\text{sc}}^{(2)}$ (the field in region 2). Similarly to the radiation problem described in the previous section, this goal is achieved if the equivalent surface currents are known. Unlike the radiation problem, however, here these sources cannot be found directly from the incident field because the total field at the object's surface S is different from $\mathbf{E}_{\text{inc}}$.

To determine the surface equivalent sources, a system of equations is formed based on the EFIE. Using the same approach as the one in deriving Eq. (2.56), the interior and exterior EFIE solutions for regions 1 and 2 are written using Green's dyadics of the two regions ($\bar{\mathbf{G}}_1$ and $\bar{\mathbf{G}}_2$):

- region 1

$$\mathbf{E}_{\text{inc}} + \oiint_S i\omega\mu\bar{\mathbf{G}}_1(\mathbf{r},\mathbf{r}')\cdot\mathbf{J}_{s,e}(\mathbf{r}')ds' + \oiint_S [\nabla\times\bar{\mathbf{G}}_1(\mathbf{r},\mathbf{r}')]\cdot\mathbf{M}_{s,e}(\mathbf{r}')ds' = \begin{cases}\mathbf{E}^{(1)}(\mathbf{r}) & \mathbf{r}\in V_1\\ 0 & \mathbf{r}\in V_2\end{cases} \tag{2.68}$$

- region 2

$$-\oiint_S i\omega\mu\bar{\mathbf{G}}_2(\mathbf{r},\mathbf{r}')\cdot\mathbf{J}_{s,e}(\mathbf{r}')ds' - \oiint_S [\nabla\times\bar{\mathbf{G}}_2(\mathbf{r},\mathbf{r}')]\cdot\mathbf{M}_{s,e}(\mathbf{r}')ds' = \begin{cases}\mathbf{E}^{(2)}(\mathbf{r}) & \mathbf{r}\in V_2\\ 0 & \mathbf{r}\in V_1.\end{cases} \tag{2.69}$$

In the above EFIEs, we have exploited the continuity of the tangential field components at the interface S in order to express the field in both regions in terms of the same surface equivalent currents. This continuity follows from the fact that in the actual problem, there are no surface sources at S. Therefore, according to Eq. (2.48) and Eq. (2.49), the

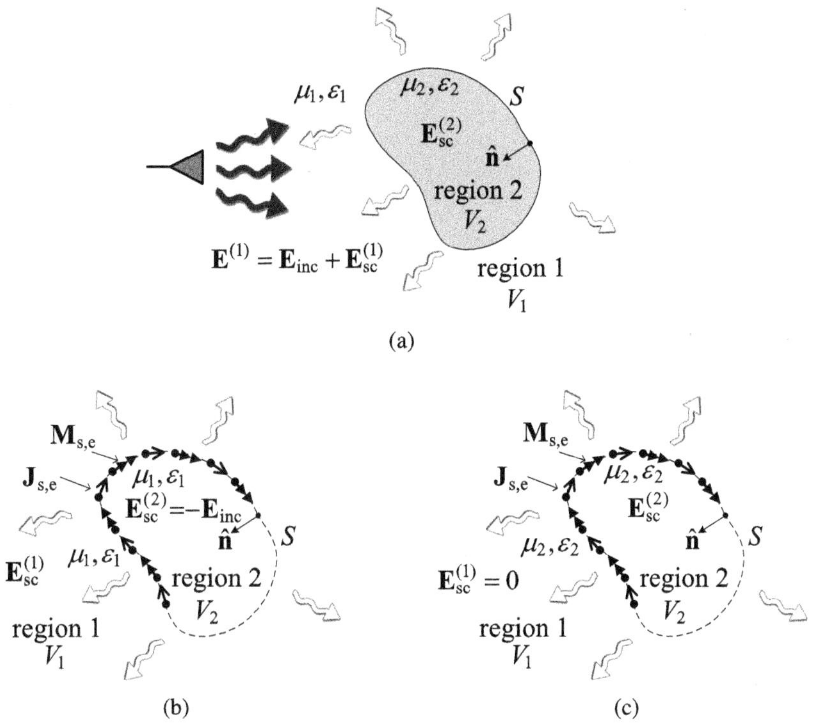

Figure 2.4 Illustration of the application of the equivalence principle in EM scattering. (a) The actual problem where an antenna in region 1 generates the incident field $\mathbf{E}_{\text{inc}}$, which is the field that would have existed if the scatterer were not present. The scattering object (region 2) is the reason for the additional (scattered) field term $\mathbf{E}_{\text{sc}}^{(1)}$. The total field in region 1, $\mathbf{E}^{(1)}$, is a superposition of the incident and scattered field terms. The field in region 2, $\mathbf{E}_{\text{sc}}^{(2)}$, has no explicit incident term because region 2 does not enclose sources. (b) The equivalent problem for determining the scattered field in region 1; see Eq. (2.68). The equivalent surface sources reside on the surface S and they radiate in open space where the constitutive parameters are those of region 1. The scattered field in region 2 equals the incident field with a minus sign because the total field in region 2 is zero. (c) The analogous equivalent problem for determining the scattered field in region 2; see Eq. (2.69).

tangential **E** and **H** field components are continuous, i.e.,

$$\hat{\mathbf{n}} \times \mathbf{H}^{(2)} = \hat{\mathbf{n}} \times \mathbf{H}^{(1)} = \hat{\mathbf{n}} \times \mathbf{H}_{\text{B}}, \tag{2.70}$$

$$\hat{\mathbf{n}} \times \mathbf{E}^{(2)} = \hat{\mathbf{n}} \times \mathbf{E}^{(1)} = \hat{\mathbf{n}} \times \mathbf{E}_{\text{B}}, \tag{2.71}$$

where $\mathbf{E}_{\text{B}}$ and $\mathbf{H}_{\text{B}}$ are the field boundary values at S. At the same time, as per Love's equivalence principle, Eqs. (2.54)–(2.55), the surface equivalent sources for region 1 are

$$\mathbf{J}_{\text{s,e}}^{(1)} = \hat{\mathbf{n}}_{21} \times \mathbf{H}_{\text{B}} = -\hat{\mathbf{n}} \times \mathbf{H}_{\text{B}}, \tag{2.72}$$

$$\mathbf{M}_{\text{s,e}}^{(1)} = \mathbf{E}_{\text{B}} \times \hat{\mathbf{n}}_{21} = -\mathbf{E}_{\text{B}} \times \hat{\mathbf{n}}, \tag{2.73}$$

whereas those for region 2 are

$$\mathbf{J}_{s,e}^{(2)} = \hat{\mathbf{n}}_{12} \times \mathbf{H}_B = \hat{\mathbf{n}} \times \mathbf{H}_B, \tag{2.74}$$

$$\mathbf{M}_{s,e}^{(2)} = \mathbf{E}_B \times \hat{\mathbf{n}}_{12} = \mathbf{E}_B \times \hat{\mathbf{n}}. \tag{2.75}$$

Here, $\hat{\mathbf{n}}_{12}$ points from region 1 to region 2 and $\hat{\mathbf{n}}_{21} = -\hat{\mathbf{n}}_{12}$. When expressing the surface sources above, we have followed the rule of Love's equivalence, which sets the unit normal so that it points into the region where the field is observed. It is customary in scattering analyses, to set a common unit normal $\hat{\mathbf{n}}$ at S so that it points from the region where the actual source resides into the volume of the scatterer (see Fig 2.4). Thus, $\hat{\mathbf{n}} = \hat{\mathbf{n}}_{12} = -\hat{\mathbf{n}}_{21}$, which sets the surface sources that appear in Eqs. (2.68)–(2.69) as

$$\mathbf{J}_{s,e} = \mathbf{J}_{s,e}^{(2)} = -\mathbf{J}_{s,e}^{(1)}, \tag{2.76}$$

$$\mathbf{M}_{s,e} = \mathbf{M}_{s,e}^{(2)} = -\mathbf{M}_{s,e}^{(1)}. \tag{2.77}$$

The interior solutions in Eqs. (2.68)–(2.69) (region 1 for $\mathbf{r} \in V_1$ and region 2 for $\mathbf{r} \in V_2$) provide the means of computing the field everywhere as long as the equivalent surface sources are known. To find these sources, on the other hand, we use the exterior solutions (region 1 for $\mathbf{r} \in V_2$ and region 2 for $\mathbf{r} \in V_1$). There are two integral equations for the two unknown surface source distributions $\mathbf{J}_{s,e}$ and $\mathbf{M}_{s,e}$. $\mathbf{E}_{inc}$ is considered known and so are Green's dyadics $\bar{\mathbf{G}}_1$ and $\bar{\mathbf{G}}_2$. In an unbounded problem, these dyadics are in the form of Eq. (2.9), with the constitutive parameters (or γ) being those of the respective medium. A system of two surface integral equations is formed from the two exterior integral equations by letting the observation point be on S. This system can be efficiently solved using the method of moments (MoM); see, for example, [34].

2.10 Induced Sources of EM Scattering

Let us now consider the construction of induced (or contrast) sources in the context of the vector model of scattering. The procedure is outlined in the case of the scattering from dielectric targets, which are magnetically the same as the background. This procedure can be easily modified for the dual case of magnetic targets, which have no dielectric contrast with the background.

The starting point is the **E**-field Helmholtz equation Eq. (2.34) when written for the source-free scattering region:

$$\mu_b(\mathbf{r})\nabla \times \frac{1}{\mu_b(\mathbf{r})}\nabla \times \mathbf{E}(\mathbf{r}) - \omega^2\mu_b(\mathbf{r})\varepsilon_s(\mathbf{r})\mathbf{E}(\mathbf{r}) = 0. \tag{2.78}$$

Note that the permeability μ_b is that of the background whereas the permittivity ε_s of the scatterer may be different from that of the background. For reasons that will become clear shortly, Eq. (2.78) is written in the equivalent form

$$\mu_o\nabla \times \frac{1}{\mu_b(\mathbf{r})}\nabla \times \mathbf{E}(\mathbf{r}) - \omega^2\mu_o\varepsilon_s(\mathbf{r})\mathbf{E}(\mathbf{r}) = 0. \tag{2.79}$$

In the context of scattering, Eq. (2.79) is the equation for the *total* **E** field. Similarly to the case of scalar scattering, the total field is represented as the superposition of the incident and scattered components,

$$\mathbf{E}(\mathbf{r}) = \mathbf{E}_{\text{inc}}(\mathbf{r}) + \mathbf{E}_{\text{sc}}(\mathbf{r}), \tag{2.80}$$

where $\mathbf{E}_{\text{inc}}$ satisfies the equation written for the background medium,

$$\mu_0 \nabla \times \frac{1}{\mu_{\text{b}}(\mathbf{r})} \nabla \times \mathbf{E}_{\text{inc}}(\mathbf{r}) - \omega^2 \mu_0 \varepsilon_{\text{b}}(\mathbf{r}) \mathbf{E}_{\text{inc}}(\mathbf{r}) = 0. \tag{2.81}$$

Subtracting Eq. (2.81) from Eq. (2.78) and making use of Eq. (2.80), we arrive at the equation for $\mathbf{E}_{\text{sc}}$:

$$\mu_0 \nabla \times \frac{1}{\mu_{\text{b}}(\mathbf{r})} \nabla \times \mathbf{E}_{\text{sc}}(\mathbf{r}) - \omega^2 \mu_0 \varepsilon_{\text{b}}(\mathbf{r}) \mathbf{E}_{\text{sc}}(\mathbf{r}) = \omega^2 \mu_0 \Delta\varepsilon(\mathbf{r}) \mathbf{E}(\mathbf{r}) \tag{2.82}$$

where

$$\Delta\varepsilon(\mathbf{r}) = \varepsilon_{\text{s}}(\mathbf{r}) - \varepsilon_{\text{b}}(\mathbf{r}) \tag{2.83}$$

is the *permittivity contrast.*

Eq. (2.82) indicates two fundamental features of the scattered field. First, $\mathbf{E}_{\text{sc}}$ can be viewed as the result of the action of the induced source of scattering,

$$\mathbf{S}_{\text{ind}}(\mathbf{r}) = \omega^2 \mu_0 \Delta\varepsilon(\mathbf{r}) \mathbf{E}(\mathbf{r}), \tag{2.84}$$

which depends on the total field **E** and the permittivity contrast $\Delta\varepsilon$.[7] Note that the dependence on $\Delta\varepsilon$ is not only explicit but also implicit through $\mathbf{E}(\mathbf{r})$. In other words, *the induced source of scattering is a nonlinear function of the permittivity contrast.* Occasionally, we will cast the induced source of scattering in the form of electric current density $\mathbf{J}_{\text{ind}}$. Bearing in mind that, for the linear operator of Eq. (2.79), $\mathbf{S}_{\text{e}} = -\mathrm{i}\omega\mu_0 \mathbf{J}$, $\mathbf{J}_{\text{ind}}$ is expressed explicitly in terms of the permittivity contrast as

$$\mathbf{J}_{\text{ind}}(\mathbf{r}) = \mathrm{i}\omega \Delta\varepsilon(\mathbf{r}) \mathbf{E}(\mathbf{r}). \tag{2.85}$$

Second, $\mathbf{E}_{\text{sc}}$ can be viewed as the field generated by the induced source *in the background.* This is important because it enables the use of Green's background dyadic $\bar{\mathbf{G}}_{\text{b}}$, which we assume is known.

The so-obtained induced source $\mathbf{S}_{\text{ind}}$ can be substituted in the EFIE Eq. (2.25) to obtain an integral-equation model of scattering. In particular, in an unbounded medium,

[7] The induced source of scattering in the dual case of magnetic targets that have no dielectric contrast with the background is obtained from the **H**-field vector Helmholtz equation using the same procedure. This source is $\mathbf{S}^{\text{m}}_{\text{ind}}(\mathbf{r}) = \omega^2 \epsilon_0 \Delta\mu(\mathbf{r}) \mathbf{H}(\mathbf{r})$.

the surface integral in Eq. (2.25) is zero, and we obtain the total field as

$$\mathbf{E}(\mathbf{r}) = \mathbf{E}_{\text{inc}}(\mathbf{r}) + \underbrace{\omega^2 \iiint_V \bar{\mathbf{G}}_{\text{b}}(\mathbf{r}, \mathbf{r}') \cdot \left[\mu_0 \Delta\varepsilon(\mathbf{r}')\mathbf{E}(\mathbf{r}')\right] dv'}_{\mathbf{E}_{\text{sc}}(\mathbf{r})}, \tag{2.86}$$

where the incident field $\mathbf{E}_{\text{inc}}$ represents the illumination due to the known impressed sources radiating in the absence of scatterers. In contrast, the second term (the scattered field component) is due to the unknown induced source of scattering.[8]

Let me note that the surface integral in Eq. (2.25) is zero not only when S expands to infinity. If Green's dyadic $\bar{\mathbf{G}}_{\text{b}}$ is chosen so that its constituent vectors $\mathbf{G}_\xi$, $\xi = x, y, z$, satisfy Eq. (2.79) with an impulse source in the right-hand side, along with the same boundary conditions at S as $\mathbf{E}$, it can be shown that the surface integral vanishes regardless of whether S is finite or infinite [31]. The reader is encouraged to verify this statement by revisiting Eq. (2.21) wherein one requires that $\mathbf{F}$ and $\mathbf{G}_\xi$ have proportional values at S. Note that if Green's dyadic $\bar{\mathbf{G}}_{\text{b}}$ is to satisfy an equation with a left-hand side as that in Eq. (2.78), then its impulse excitation must be scaled by a factor of $\mu_{\text{r,b}}$.

If the background is uniform, then $\bar{\mathbf{G}}_{\text{b}}$ is available in the closed form of Eq. (2.9). This is why the assumption of a uniform background is often made when solving inverse scattering problems.

Note that the derivation of Eq. (2.86) tacitly assumed that the medium is isotropic and the induced source of scattering would not have any impact on the scattered wave's polarization. If the medium is anisotropic, the same approach can be applied, i.e., the field is split into incident and scattered components along with representing the permittivity tensor as the sum of the background and the contrast tensors. The scattering from anisotropic dielectric and magnetic targets is an advanced topic and is not pursued here. Moreover, the approximation of isotropic scattering from penetrable dielectric/magnetic targets is acceptable in most MW imaging applications.

Finally, note that the scattering from metallic (or PEC) objects is described by induced surface electric current density. The polarization of such surface currents strongly depends on the shape of the object's surface and its orientation with respect to the illuminating source. Thus, the scattered field generated by the PEC-induced sources is practically *always* of a polarization that is different from that of the incident field. A thorough treatment of the scattering from PEC objects can be found, for example, in [57].

2.11 Green's Function as an Incident Field in the Vector Data Equation

Closed-form analytical expressions for Green's dyadic of Helmholtz's equation are available only for simple cases such as uniform or stratified unbounded regions. This

[8] The MFIE for the **H**-field scattering from magnetic targets has a form dual to Eq. (2.86); i.e., the MFIE can be obtained from the EFIE in Eq. (2.86) using the duality substitutions found in Table A.2 in Appendix A.

limits substantially the applicability of the integral-equation scattering models such as the EFIE-based model in Eq. (2.86). Another practical limitation arises from the fact that the data provided by MW measurements in the frequency domain are scalar phasors, not the **E** or **H** vectors.[9] As discussed in the next chapter, the standard frequency-domain responses in microwave network analysis are the scattering parameters, or, *S*-parameters, which are scalar phasors. Consequently, there is a need to link the measured scalar responses to the vectorial field distribution at the receiving antenna. This link is often oversimplified, which degrades the accuracy of the forward model. That is why it is important to have a general methodology of casting the forward model of scattering into a data equation where the data are in the form of the actual measured scalar responses.

This methodology is presented next. It is shown that: (i) a form of *Green's vector function* exists and can be computed for any scalar response defined as a functional of the field distribution, and (ii) the so-found Green's vector function is valid in a heterogeneous (but reciprocal) medium bounded by any type of boundary conditions. Green's vector function is a contraction of Green's dyadic because the data point is now a scalar, not a vector. Most importantly, just like in the scalar model of scattering (see Section 1.10), Green's vector function is directly obtainable through simulations or measurements performed in the scatterer-free environment.

2.11.1 Scalar Responses from Vector Fields

Let us consider the background medium as a complex structure, which includes not only the propagation medium, which is possibly heterogeneous, but also the antenna structures, enclosures, and so on. Thus, it may include metallic, dielectric, and/or absorbing components. We impose the restriction that the medium is reciprocal so that Green's dyadic satisfies the condition in Eq. (2.23).[10]

Let us assume that the **E** field observed at **r** is due to the volume source $\mathbf{S}(\mathbf{r}')$. It can then be expressed in terms of Green's background dyadic as

$$\mathbf{E}(\mathbf{r}) = \iiint_V \bar{\mathbf{G}}_b(\mathbf{r}, \mathbf{r}') \cdot \mathbf{S}(\mathbf{r}') d\mathbf{r}'. \tag{2.87}$$

If the volume V is bounded by a finite closed surface S, Green's dyadic is required to satisfy the same boundary conditions on S as **E** so that the integral solution does not involve surface integration.

Next, consider a sensor (e.g., an antenna with its transmission lines, matching circuitry, connectors, etc.), that produces a scalar response R at the receiver. Because the

[9] The full vectorial measurement of the MW field is difficult and time consuming. It is accomplished using field probes, which are in essence small linearly polarized antennas. The field probe must be small in order to minimize the distortion of the measured field. Yet, it consists of not only an antenna but also a connector and a coaxial cable, which as a whole do cause distortions. Moreover, each of the three field vector components must be acquired through a separate measurement where the field probe is properly oriented.

[10] Most materials are electromagnetically reciprocal [31, 34]. Materials described by symmetric constitutive tensors are reciprocal. Anisotropic nonmagnetic dielectrics that exhibit loss, however minor, are also reciprocal [58]. A nonlinear medium is *not* reciprocal.

field may be sampled at multiple locations by multiple sensors, the particular response of interest is indexed by the subscript i. If the sensor is a linear device, then R_i is a linear functional of the field:

$$R_i = \mathcal{R}_i \left\{ \mathbf{E}(\mathbf{r}) \right\}, \; R_i \in \mathbb{C}. \tag{2.88}$$

The functional $\mathcal{R}_i$ is applied to both sides of Eq. (2.87) to obtain the integral equation for the scalar response R_i. Since $\mathcal{R}_i$ operates on $\mathbf{r}$, it can be inserted in the integral and applied to $\bar{\mathbf{G}}_\text{b}(\mathbf{r}, \mathbf{r}')$. This results in

$$R_i = \iiint_V \mathbf{E}_i^\dagger(\mathbf{r}') \cdot \mathbf{S}(\mathbf{r}') dv', \tag{2.89}$$

where $\mathbf{E}_i^\dagger(\mathbf{r}')$ is an auxiliary vector field associated with the i-th sensor, the components of which are explicitly given by

$$\mathbf{E}_i^\dagger(\mathbf{r}') = \left[\mathcal{R}_i \left\{ \mathbf{G}_x(\mathbf{r}, \mathbf{r}') \right\}, \mathcal{R}_i \left\{ \mathbf{G}_y(\mathbf{r}, \mathbf{r}') \right\}, \mathcal{R}_i \left\{ \mathbf{G}_z(\mathbf{r}, \mathbf{r}') \right\} \right]. \tag{2.90}$$

Here, $\mathbf{G}_\xi(\mathbf{r}, \mathbf{r}')$, $\xi = x, y, z$, is the ξ-th column of Green's dyadic $\bar{\mathbf{G}}_\text{b}(\mathbf{r}, \mathbf{r}')$.

It is apparent that $\mathbf{E}_i^\dagger(\mathbf{r}')$ serves as the kernel for the scalar-response integral equation. As such it can be termed *Green's vector function* (in analogy to Green's dyadic function).[11] It is shown next that it is the background's vectorial response to a particular impulse excitation that is uniquely determined by the i-th sensor, i.e., by its functional $\mathcal{R}_i$.

We now return to the problem of finding the mathematical form of the impulse excitation generating Green's i-th vector $\mathbf{E}_i^\dagger(\mathbf{r}')$. Using reciprocity, $\bar{\mathbf{G}}_\text{b}(\mathbf{r}, \mathbf{r}') = \bar{\mathbf{G}}_\text{b}^\text{T}(\mathbf{r}', \mathbf{r})$, Green's vector function can also be written as

$$\mathbf{E}_i^\dagger(\mathbf{r}') = \left[\mathcal{R}_i \left\{ \mathbf{G}^x(\mathbf{r}', \mathbf{r}) \right\}, \mathcal{R}_i \left\{ \mathbf{G}^y(\mathbf{r}', \mathbf{r}) \right\}, \mathcal{R}_i \left\{ \mathbf{G}^z(\mathbf{r}', \mathbf{r}) \right\} \right], \tag{2.91}$$

where $\mathbf{G}^\xi(\mathbf{r}', \mathbf{r})$, $\xi = x, y, z$, is the ξ-th row of Green's dyadic $\bar{\mathbf{G}}_\text{b}(\mathbf{r}', \mathbf{r}) = \bar{\mathbf{G}}_\text{b}^\text{T}(\mathbf{r}, \mathbf{r}')$. $\bar{\mathbf{G}}_\text{b}(\mathbf{r}', \mathbf{r})$ satisfies

$$\mathcal{L}_\text{b}' \bar{\mathbf{G}}_\text{b}(\mathbf{r}', \mathbf{r}) = \bar{\mathbf{I}} \delta(\mathbf{r}' - \mathbf{r}), \tag{2.92}$$

where $\mathcal{L}_\text{b}' = \mu_0 \nabla' \times \mu_\text{b}^{-1} \nabla' \times + \gamma_\text{b}^2$, $\gamma_\text{b} = \text{i}\omega\sqrt{\mu_0 \varepsilon_\text{b}}$, operates on $\mathbf{r}'$. Note that $\mathcal{L}'$ has the form of the operator in Eq. (2.79). Applying $\mathcal{R}_i$ to both sides of Eq. (2.92) and taking into account that it operates on $\mathbf{r}$, the following is obtained:

$$\mathcal{L}_\text{b}' \mathcal{R}_i \left\{ \bar{\mathbf{G}}_\text{b}(\mathbf{r}', \mathbf{r}) \right\} = \mathcal{R}_i \left\{ \bar{\mathbf{I}} \delta(\mathbf{r}' - \mathbf{r}) \right\}. \tag{2.93}$$

Since $\mathcal{R}_i$ operates over the rows of $\bar{\mathbf{G}}_\text{b}(\mathbf{r}', \mathbf{r})$, the governing equation for Green's vector function is explicitly given by

$$\mathcal{L}_\text{b}' \mathbf{E}_i^\dagger(\mathbf{r}') = \mathcal{R}_i \left\{ \bar{\mathbf{I}} \delta(\mathbf{r}' - \mathbf{r}) \right\}. \tag{2.94}$$

[11] It would appear that the term *Green's vector function* was introduced first in [59], where it was derived for the case of common-impedance S-parameters of antennas operating in a homogeneous open medium.

There are two important conclusions to be drawn from Eq. (2.94). First, Green's vector function is an auxiliary field that satisfies the EM equation describing the propagation in the background. In fact, due to the requirement imposed earlier that the boundary conditions satisfied by Green's dyadic are the same as those imposed on the actual field solution in Eq. (2.87), this auxiliary field also complies with the boundary conditions of the actual EM problem. Second, the auxiliary source $\mathbf{S}_{\text{a}}^{(i)}$ generating Green's i-th vector is of the form

$$\mathbf{S}_i^{\dagger}(\mathbf{r}') = \mathcal{R}_i\left\{\bar{\mathbf{I}}\delta(\mathbf{r}'-\mathbf{r})\right\}. \tag{2.95}$$

It is shown next that the source defined in Eq. (2.95) is in fact proportional (through a complex constant) to the actual source exciting the i-th (receiving) antenna, when this antenna is to operate as a transmitter.

2.11.2 Scattering Model in the Case of Field Point Sampling

Let us consider the receiving antenna structure as part of the heterogeneous propagation medium. This structure guides the signal to the antenna terminals, which we assume are sufficiently small to be considered a point of field sampling. The response, which results from this sampling, could be voltage or current or any other network parameter that would represent the antenna as a signal generator at the input of the receiver.

In the most basic terms, we define the i-th scalar response resulting from the sampled field as

$$R_i = \hat{\mathbf{p}}_i \cdot \mathbf{E}(\mathbf{r}_i). \tag{2.96}$$

Here, $\hat{\mathbf{p}}_i$ is the unit polarization vector of the field component that is being sampled, and $\mathbf{r}_i$ is the sampling location. This representation is useful in a number of scenarios. In simulation-based studies, point-like sources and sampling are often preferred because they simplify the problem's geometry and accelerate the computations. Eq. (2.96) is also directly related to a receiving antenna parameter known as *effective height* [15], which is particularity relevant for electrically small antennas (e.g., dipoles or monopoles). The open-circuit voltage V^{oc} at the terminals of such antennas is conveniently expressed in terms of the effective height $\mathbf{h} = \hat{\mathbf{p}}h$ as [60, chapter 4]

$$V^{\text{oc}} = \mathbf{h} \cdot \mathbf{E}, \tag{2.97}$$

where $\mathbf{E}$ is the field at the position of the antenna's feed point. The effective height $\mathbf{h}$ (measured in m) describes both the antenna polarization and the strength of its voltage response. The i-th response R_i in Eq. (2.96) can be easily modified to represent V_i^{oc} by replacing $\hat{\mathbf{p}}_i$ with $\mathbf{h}_i$.

To obtain the particular form of the response functional $\mathcal{R}_i$ associated with the point-sampling response in Eq. (2.96), the latter is reexpressed as

$$R_i = \mathcal{R}_i\{\mathbf{E}\} = \iiint_V (\hat{\mathbf{p}}_i \cdot \mathbf{E})\,\delta(\mathbf{r}-\mathbf{r}_i)d\mathbf{r}, \tag{2.98}$$

Comparing Eq. (2.98) and Eq. (2.95) reveals that the auxiliary source is in the form

$$S^{\dagger}_{i,\xi}(\mathbf{r}') = \iiint_V (\hat{\mathbf{p}}_i \cdot \hat{\xi})\delta(\mathbf{r} - \mathbf{r}_i)\delta(\mathbf{r}' - \mathbf{r})d\mathbf{r}, \ \xi \equiv x, y, z, \tag{2.99}$$

which further leads to

$$S^{\dagger}_{i,\xi}(\mathbf{r}') = (\hat{\mathbf{p}}_i \cdot \hat{\xi})\delta(\mathbf{r}' - \mathbf{r}_i), \ \xi \equiv x, y, z. \tag{2.100}$$

In deriving Eq. (2.100) from Eq. (2.99), the identity [14]

$$\int_{-\infty}^{\infty} \delta(a - x)\delta(x - b)dx = \delta(a - b). \tag{2.101}$$

has been used. It is clear from Eq. (2.100) that the auxiliary source is a point source polarized along $\hat{\mathbf{p}}_i$, i.e.,

$$\mathbf{S}^{\dagger}_i(\mathbf{r}') = \hat{\mathbf{p}}_i\delta(\mathbf{r}' - \mathbf{r}_i). \tag{2.102}$$

For instance, if a sensor samples the **E**-field x-component at $\mathbf{r}_i$, then the auxiliary source is $\mathbf{S}^{\dagger}_i(\mathbf{r}') = \hat{\mathbf{x}}\delta(\mathbf{r}' - \mathbf{r}_i)$ in units of m^{-3}. If the response of interest is the respective V_i^{oc}, then the auxiliary source is $\mathbf{S}^{\dagger}_i(\mathbf{r}') = \hat{\mathbf{x}}h_i\delta(\mathbf{r}' - \mathbf{r}_i)$ in units of m^{-2}.

To derive Green's vector function generated by $\mathbf{S}^{\dagger}_i$, let us now assume that the i-th antenna operates in a transmitting mode. Since its terminals act as a point-like receiver responding to the **E**-field component along $\hat{\mathbf{p}}_i$, they also act as a point-like current source along $-\hat{\mathbf{p}}_i$. The minus sign reflects the fact that the **E** field associated with a transmitting current element is in a direction opposite to that of the current; see [1, Section 3-1]. Let us think of the antenna terminals as a current element at $\mathbf{r}_i$. A current density source representing a current element of polarization $-\mathbf{p}_i$ is expressed as

$$\mathbf{J}_i(\mathbf{r}') = -\hat{\mathbf{p}}_i\delta(\mathbf{r}' - \mathbf{r}_i)J_i\Delta\Omega_i. \tag{2.103}$$

Here, $J_i\Delta\Omega_i$ is the current element (in A·m), wherein J_i (in $\text{A}\cdot\text{m}^{-2}$) is the density strength and $\Delta\Omega_i$ (m^3) is the volume occupied by the source. In a simulation, $\Delta\Omega_i$ is the discretization mesh element where the source resides. In more practical terms, the current element for an antenna of effective height $\mathbf{h}_i$ and current I_i at its feed point is $I_i\mathbf{h}_i$; therefore, its current density is expressed as $\mathbf{J}_i(\mathbf{r}') = \mathbf{h}_iI_i\delta(\mathbf{r}' - \mathbf{r}_i)$. In MW and antenna engineering, it is customary to work with the input power P_i of the antenna, from which we can calculate I_i, provided we know the antenna impedance $Z_{i,\text{A}}$.

The current element in Eq. (2.103) that describes the actual excitation of the i-th antenna, when this antenna operates in a transmitting mode, corresponds to the Helmholtz-equation source function (for the linear operator in Eq. (2.79)) in the form

$$\mathbf{S}_i(\mathbf{r}') = \hat{\mathbf{p}}_i\text{i}\omega\mu_0J_i\Delta\Omega_i\delta(\mathbf{r}' - \mathbf{r}_i). \tag{2.104}$$

The comparison between the source function of the actual excitation in Eq. (2.104) and that of the auxiliary one in Eq. (2.102) shows simple proportionality, which allows for deriving Green's vector function from the actual incident field $\mathbf{E}_i^{\text{inc}}$ that the i-th antenna would generate when in transmitting mode. Explicitly,

$$\mathbf{E}_i^{\dagger}(\mathbf{r}') = \frac{\mathbf{E}_i^{\text{inc}}(\mathbf{r}')}{\mathrm{i}\omega\mu_0 J_i \Delta\Omega_i}. \tag{2.105}$$

Green's vector function from Eq. (2.105) is now substituted in Eq. (2.89), which yields the integral equation for the point-wise response for a general source $\mathbf{S}(\mathbf{r}')$:

$$R_i = \frac{1}{\mathrm{i}\omega\mu_0(\mathbf{r}_i) J_i \Delta\Omega_i} \iiint_V \mathbf{E}_i^{\text{inc}}(\mathbf{r}') \cdot \mathbf{S}(\mathbf{r}') dv'. \tag{2.106}$$

In the context of scattering, $\mathbf{S}(\mathbf{r}')$ includes both impressed and induced sources that generate the incident field $\mathbf{E}_{\text{inc}}$ and the scattered field $\mathbf{E}_{\text{sc}}$, respectively. Since the total field is a superposition $\mathbf{E} = \mathbf{E}_{\text{inc}} + \mathbf{E}_{\text{sc}}$, see Eq. (2.86), and the operator $\mathcal{R}_i$ is linear, the response R_i, too, can be expressed as a superposition:

$$R_i = \mathcal{R}_i \{\mathbf{E}_{\text{inc}}\} + \mathcal{R}_i \{\mathbf{E}_{\text{sc}}\} = R_{i,\text{b}} + R_{i,\text{sc}}. \tag{2.107}$$

$R_{i,\text{b}}$ is the response acquired in the absence of scatterers and is referred to as the *background response* or *baseline response*. $R_{i,\text{sc}}$ is the *scattering response*, and it is the one that carries the information about the scatterer. When the expression of the induced source of scattering Eq. (2.84) is substituted in Eq. (2.106), we obtain the data integral equation for the i-th point-sampling scattering response as:

$$R_{i,\text{sc}} = \hat{\mathbf{p}}_i \cdot \mathbf{E}_{\text{sc}}(\mathbf{r}_i) = \frac{-\mathrm{i}\omega}{J_i \Delta\Omega_i} \iiint_V \Delta\varepsilon(\mathbf{r}') \mathbf{E}_i^{\text{inc}}(\mathbf{r}') \cdot \mathbf{E}(\mathbf{r}') dv', \tag{2.108}$$

where $\mathbf{E}_i^{\text{inc}}(\mathbf{r}')$ is the field generated by the current element of Eq. (2.103) in the scatterer-free background. This result is adjusted to the case of open-circuit voltage response by simply multiplying by h_i:

$$V_{i,\text{sc}}^{\text{oc}} = \frac{-\mathrm{i}\omega h_i}{J_i \Delta\Omega_i} \iiint_V \Delta\varepsilon(\mathbf{r}') \mathbf{E}_i^{\text{inc}}(\mathbf{r}') \cdot \mathbf{E}(\mathbf{r}') dv'. \tag{2.109}$$

At the same time, however, the current element $J_i \Delta\Omega_i$ of the actual antenna of effective height h_i is $J_i \Delta\Omega_i = I_i h_i$, as discussed earlier. As a reminder, I_i is the current at the i-th antenna terminals when this antenna transmits. Thus,

$$V_{i,\text{sc}}^{\text{oc}} = \frac{-\mathrm{i}\omega}{I_i} \iiint_V \Delta\varepsilon(\mathbf{r}') \mathbf{E}_i^{\text{inc}}(\mathbf{r}') \cdot \mathbf{E}(\mathbf{r}') dv'. \tag{2.110}$$

Let me note that V_i^{oc} is a useful response because it enables the representation of the receiving antenna as an equivalent voltage generator using Thevenin's circuit theorem [15]. In the next chapter, we will derive a similar scalar data equation for the case of the scattering parameters, which are widely used in MW measurements and network analysis.

To appreciate the importance of the results in Eq. (2.108) and Eq (2.110), it should be re-iterated that they are valid for a very wide class of environments that can be

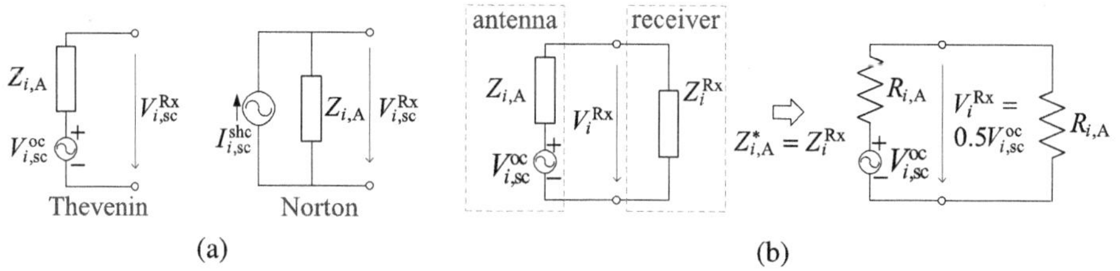

Figure 2.5 (a) Thevenin and Norton equivalents of the receiving antenna representing it as a signal generator. (b) Conjugate impedance match results in a simple voltage divider formed by the antenna internal resistance and that of the receiver.

heterogeneous, bounded or open. The limitation on the propagation medium is that it cannot be nonreciprocal.

Finally, it should be noted that $\mathbf{E}_i^{\mathrm{inc}}(\mathbf{r}')$ in Eq. (2.108) and Eq (2.110) can be acquired via simulation or field-probe measurements in the scatterer-free setup wherein the i-th antenna operates as a transmitter. Such measurements are often time consuming and require relatively expensive equipment such as field probes and positioners or scanners, but they are feasible and need to be done only once as part of the system calibration.

Example 1: Using Eq. (2.110), find the data integral equation for the i-th antenna (of effective height $\mathbf{h}_i$) if the scattering response of interest is: (a) the short-circuit current $I_{i,\mathrm{sc}}^{\mathrm{shc}}$ and (b) the voltage at the receiver's input $V_{i,\mathrm{sc}}^{\mathrm{Rx}}$ under conditions of maximum power transfer from the antenna to the receiver. The antenna impedance is $Z_{i,\mathrm{A}}$.

Solution: To solve part (a), we use the relationship between the Thevenin and Norton equivalents of the antenna (see Fig. 2.5-a), according to which

$$I_i^{\mathrm{shc}} = V_i^{\mathrm{oc}}/Z_{i,\mathrm{A}}. \tag{2.111}$$

Using Eq. (2.110), we obtain that

$$I_{i,\mathrm{sc}}^{\mathrm{shc}} = \frac{-\mathrm{i}\omega}{I_i Z_{i,\mathrm{A}}} \iiint_V \Delta\varepsilon(\mathbf{r}')\mathbf{E}_i^{\mathrm{inc}}(\mathbf{r}') \cdot \mathbf{E}(\mathbf{r}')dv', \tag{2.112}$$

where $\mathbf{E}_i^{\mathrm{inc}}(\mathbf{r}')$ is the incident field distribution due to the i-th antenna when it transmits.

To solve part (b), we recall that maximum power transfer occurs when there is a conjugate impedance match between the generator (the i-th antenna) and the load (the i-th receiver), i.e., when $Z_{i,\mathrm{A}}^* = Z_i^{\mathrm{Rx}}$; see Fig. 2.5-b. Under these conditions, the antenna and the receiver input form a simple voltage divider, so that $V_{i,\mathrm{sc}}^{\mathrm{Rx}} = 0.5V_{i,\mathrm{sc}}^{\mathrm{oc}}$. Therefore, using Eq. (2.110), we obtain that

$$V_{i,\mathrm{sc}}^{\mathrm{Rx}} = \frac{-\mathrm{i}\omega}{2I_i} \iiint_V \Delta\varepsilon(\mathbf{r}')\mathbf{E}_i^{\mathrm{inc}}(\mathbf{r}') \cdot \mathbf{E}(\mathbf{r}')dv'. \tag{2.113}$$

In the next example, we consider the scattering model for the transmission and reflection coefficients acquired with two dipole antennas. It illustrates the influence of the

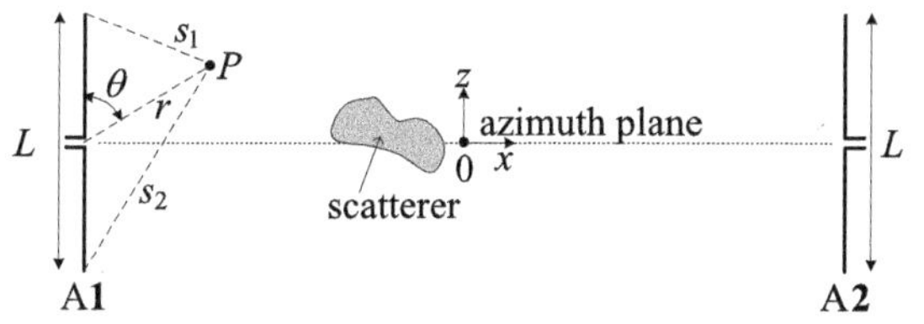

Figure 2.6 Illustration of a setup where two dipoles are aligned with the z axis with their centers lying in a common azimuth plane. The z-component of the **E** field of a dipole oriented along z can be determined analytically at *any* position using Eq. (2.114) wherein r is the distance from the dipole's center while s_1 and s_2 are those from the dipole's edges.

various parameters associated with the network of the imaging setup that comprises the antennas and the propagation medium.

Example 2: Two half-wavelength dipole antennas are aligned vertically along the z axis with their centers being in a common azimuth plane; see Figure 2.6 where the dipoles are denoted as Antenna 1 (**A1**) and Antenna 2 (**A2**). Assume that the antennas are loss-free and impedance matched to their respective circuits. The background medium is air, i.e., $\mu_b = \mu_0$ and $\varepsilon_b = \epsilon_0$. Assume that the scatterer is far from the antennas and its extent in the z direction is much smaller than its distance to any of the antennas. Assume also that the scatterer is weak and Born's approximation applies i.e., $\mathbf{E} \approx \mathbf{E}^{\text{inc}}$ inside the scatterer. Derive the scalar model of scattering for: (a) the scattering component of the transmission coefficient T_{21}^{sc} between the voltage at the terminals of **A1** (the transmitting antenna) and that at the terminals of the receiver at **A2**; (b) the scattering component of the reflection coefficient R_{11}^{sc} at **A1**.

Solution: To proceed toward the solution, we need to recall some well-known characteristics of the half-wavelength dipole:

(i) the radiation resistance is $R_{\text{A}} \approx 73\ \Omega$ [61, chapter 5];
(ii) in the azimuth plane, the **E** field has only a vertical, or z, component, i.e., $\mathbf{E} = \hat{\mathbf{z}}E_z$;
(iii) there is a closed-form expression for E_z at *any* point in space [61, chapter 5]:

$$E_z(P) = -\frac{\mathrm{i}I_0\eta_0}{4\pi}\left[\frac{e^{-\mathrm{i}k_0 s_1}}{s_1} + \frac{e^{-\mathrm{i}k_0 s_2}}{s_2} - 2\cos\left(\frac{k_0 L}{2}\right)\frac{e^{-\mathrm{i}k_0 r}}{r}\right]. \tag{2.114}$$

Here, $k_0 = \omega\sqrt{\mu_0\epsilon_0} = \frac{2\pi}{\lambda_0}$ is the wavenumber, whereas the distances r, s_1, and s_2 are from the observation point P to the center of the dipole and to its top and bottom ends, respectively (see Fig. 2.6). $\eta_0 \approx 377\ \Omega$ is the intrinsic impedance of air, and I_0 is the phasor of the current at the dipole's feed point.

First, we apply Eq. (2.114) to our particular setup. For a half-wavelength dipole ($L = \lambda_0/2$), $\cos\left(\frac{k_0 L}{2}\right) = 0$, which sets the last term to zero. Further, the scatterer is set to be far from the dipoles. This allows the use of the so-called far-zone approximations [15]. The far-zone "phase" approximations are $s_1 \approx r - \frac{L}{2}\cos\theta$ and $s_2 \approx r + \frac{L}{2}\cos\theta$; they are applied to the arguments of the exponents in Eq. (2.114). Here, θ is the angle between the dipole's axis and the vector **r** pointing from the dipole's center toward the observation point P (see Fig. 2.6). The far-zone "amplitude" approximations state that $\frac{1}{s_1} \approx \frac{1}{s_2} \approx \frac{1}{r}$. Thus, we obtain the approximate expression for E_z as

$$E_z(P) \approx -\frac{\mathrm{i} I_0 \eta_0}{2\pi} \frac{e^{-\mathrm{i}k_0 r}}{r} \cos\left(\frac{\pi}{2}\cos\theta\right). \tag{2.115}$$

Moreover, since the scatterer's extent along z is much smaller than the distance to any of the antennas, θ is close to 90° and $\cos\left(\frac{\pi}{2}\cos\theta\right) \approx 1$. Thus, the E_z far-field component due to a half-wavelength dipole antenna for a target at and around the azimuth plane is obtained as

$$E_z(P) \approx -\frac{\mathrm{i} I_0 \eta_0}{2\pi} \frac{e^{-\mathrm{i}k_0 r}}{r}. \tag{2.116}$$

To solve case (a) in Example 2, we derive the integral equation for the scattering component T_{21}^{sc} of the transmission coefficient. For that, we use the solution of part (b) in Example 1, Eq. (2.113). This solution is applied to the voltage at the A**2** receiver, i.e., $i = 2$. Also, we take into account that: (i) the **E** field of a dipole at and around the azimuth plane is polarized along z, and (ii) the scatterer is weak and we can use Born's linear approximation. Thus, the scattering component of the voltage at the terminals of the A**2** receiver is obtained as

$$V_{2,\mathrm{sc}}^{\mathrm{A}} = \frac{-\mathrm{i}\omega}{2I_2} \iiint_V \Delta\varepsilon(\mathbf{r}') E_{z,2}^{\mathrm{inc}}(\mathbf{r}') E_{z,1}^{\mathrm{inc}}(\mathbf{r}') dv'. \tag{2.117}$$

Here, I_2 is the current at the terminals of A**2** when this antenna operates in a transmitting mode so as to generate $E_{z,2}^{\mathrm{inc}}(\mathbf{r}')$. Since T_{21}^{sc} is the ratio of $V_{2,\mathrm{sc}}^{\mathrm{A}}$ to the voltage V_1^{A} fed to the terminals of A**1**,

$$T_{21}^{\mathrm{sc}} = \frac{V_{2,\mathrm{sc}}^{\mathrm{A}}}{V_1^{\mathrm{A}}} = \frac{V_{2,\mathrm{sc}}^{\mathrm{A}}}{R_{\mathrm{A}} I_1}, \tag{2.118}$$

its integral equation is

$$T_{21}^{\mathrm{sc}} = \frac{-\mathrm{i}\omega}{2R_{\mathrm{A}} I_1 I_2} \iiint_V \Delta\varepsilon(\mathbf{r}') E_{z,2}^{\mathrm{inc}}(\mathbf{r}') E_{z,1}^{\mathrm{inc}}(\mathbf{r}') dv'. \tag{2.119}$$

We can now substitute the expressions for $E_{z,1}^{\mathrm{inc}}$ and $E_{z,2}^{\mathrm{inc}}$ as obtained from Eq. (2.116) to arrive at

$$T_{21}^{\mathrm{sc}} \approx \frac{\mathrm{i}\omega\eta_0^2}{8\pi^2 R_{\mathrm{A}}} \iiint_V \Delta\varepsilon(\mathbf{r}') \frac{e^{-\mathrm{i}k_0[r_1(\mathbf{r}')+r_2(\mathbf{r}')]}}{r_1(\mathbf{r}') r_2(\mathbf{r}')} dv', \tag{2.120}$$

where $r_1(\mathbf{r}')$ and $r_2(\mathbf{r}')$ are the distances from the integration point to the centers of A**1** and A**2**, respectively. Observe that the transmission coefficient does not depend on the actual antenna excitations, I_1 or I_2. It depends only on the medium constitutive parameters as well as the antenna resistance and geometry, all of which determine the distributions $\mathbf{E}_1^{\mathrm{inc}}(\mathbf{r}')$ and $\mathbf{E}_2^{\mathrm{inc}}(\mathbf{r}')$. This is expected. Another expected result is that the transmission coefficient does not change if we exchange the transmission and reception roles of the antennas, i.e., make A**2** transmitting and A**1** receiving. The exchange of the subscripts in Eq. (2.120) shows that $T_{21}^{\text{sc}} = T_{12}^{\text{sc}}$, which is in essence a statement of reciprocity.

The solution for the scattering component of the reflection coefficient R_{11}^{sc} at **A1** (case b) follows the same steps, leading to

$$R_{11}^{\text{sc}} \approx \frac{\mathrm{i}\omega\eta_0^2}{8\pi^2 R_{\text{A}}} \iiint_V \Delta\varepsilon(\mathbf{r}') \frac{e^{-\mathrm{i}2k_0 r_1'(\mathbf{r}')}}{\left[r_1'(\mathbf{r}')\right]^2} dv'. \tag{2.121}$$

2.12 The Linear Born and Rytov Approximations in Vector Scattering

The Born and Rytov approximations have already been introduced and discussed in detail in the context of scalar scattering in Sections 1.13 through 1.18. The underlying mathematical model there was Helmholtz's scalar equation.

The Born linear approximation follows from the expansion of the field into a Neumann power series. This expansion is generic in the sense that it can be applied to any field (scalar or vector) provided its solution can be written as a Fredholm equation of the second kind. The integral equation of vector scattering from dielectric targets, Eq. (2.86), is in this form because it can be written in very general terms as

$$\mathbf{E}(\mathbf{r}) = \mathbf{E}^{\text{inc}}(\mathbf{r}) + \xi \mathcal{L}\mathbf{E}(\mathbf{r}), \tag{2.122}$$

where $\mathcal{L}$ is the vector integral operator defined by Green's dyadic of the background and the normalized permittivity contrast. This is an extension of the scalar Fredholm equation Eq. (1.292) to vector fields. The methodology used to arrive at Born's linear approximation in the scalar model holds in the case of the vector operators as well, with the entailing limitation for weak scattering. Thus, *the linear Born approximation can be applied to vector scattering*. When applied to Eq. (2.86), for example, it yields

$$\mathbf{E}(\mathbf{r}) = \mathbf{E}^{\text{inc}}(\mathbf{r}) + \underbrace{\omega^2 \iiint_V \bar{\mathbf{G}}_{\text{b}}(\mathbf{r}, \mathbf{r}') \cdot \left[\mu_0 \Delta\varepsilon(\mathbf{r}')\mathbf{E}^{\text{inc}}(\mathbf{r}')\right] dv'}_{\mathbf{E}^{\text{sc}}(\mathbf{r})}. \tag{2.123}$$

The approximation of the total internal field by its incident counterpart can be applied to all other vector models of scattering that we have considered so far, including the data integral equations for the scalar EM responses considered in Section 2.11.2.

The Rytov approximation is derived on the basis of Helmholtz's scalar equation (see Section 1.16). Therefore, its application to the vector model is not straightforward. To appreciate the particulars of the extension of the Rytov and the *scalar* Born linearizing approximations to the vector model of scattering, we first need to scalarize it, i.e., to reduce it to a set of scalar equations. Let us revisit Eq. (2.78), which governs the total **E** field inside the volume of a dielectric scatterer. If we impose the restriction of a magnetically uniform background, $\mu_b(\mathbf{r}) = \text{const.}$, Eq. (2.78) is reduced to

$$\nabla \times \nabla \times \mathbf{E}(\mathbf{r}) - \omega^2 \mu_b \varepsilon_s(\mathbf{r})\mathbf{E}(\mathbf{r}) = 0. \tag{2.124}$$

Using the identity $\nabla \times \nabla \times = \nabla\nabla \cdot -\nabla^2$, Eq. (2.124) is expressed as

$$\nabla^2 \mathbf{E}(\mathbf{r}) + \omega^2 \mu_b \varepsilon_s(\mathbf{r})\mathbf{E}(\mathbf{r}) = \nabla\nabla \cdot \mathbf{E}(\mathbf{r}). \tag{2.125}$$

The right side of Eq. (2.125) is not zero in general. Gauss law of electricity dictates that $\nabla \cdot [\varepsilon_s(\mathbf{r})\mathbf{E}(\mathbf{r})] = 0$ in the source-free volume of the scatterer but here ε_s is a function of position, which implies that $\nabla \cdot \mathbf{E}(\mathbf{r}) \neq 0$.

The incident field equation is obtained in the same manner:

$$\nabla^2 \mathbf{E}^{\text{inc}}(\mathbf{r}) + \omega^2 \mu_b \varepsilon_b(\mathbf{r})\mathbf{E}^{\text{inc}}(\mathbf{r}) = \nabla\nabla \cdot \mathbf{E}^{\text{inc}}(\mathbf{r}). \tag{2.126}$$

If the background is uniform in its permittivity, then the right side of Eq. (2.126) is zero.

From Eq. (2.125) and Eq. (2.126), it follows that the equation for the scattered field $\mathbf{E}^{\text{sc}} = \mathbf{E} - \mathbf{E}^{\text{inc}}$ is

$$\nabla^2 \mathbf{E}^{\text{sc}}(\mathbf{r}) + \omega^2 \mu_b \varepsilon_b(\mathbf{r})\mathbf{E}^{\text{sc}}(\mathbf{r}) = \nabla\nabla \cdot \mathbf{E}^{\text{sc}}(\mathbf{r}) - \omega^2 \mu_b \Delta\varepsilon(\mathbf{r})\mathbf{E}(\mathbf{r}). \tag{2.127}$$

In a Cartesian coordinate system, the Laplacian operates on the three vector components independently, and we can rewrite Eq. (2.127) as three scalar equations:

$$\nabla^2 E_\xi^{\text{sc}}(\mathbf{r}) + \omega^2 \mu_b \varepsilon_b(\mathbf{r}) E_\xi^{\text{sc}}(\mathbf{r}) = \frac{\partial}{\partial \xi}\left[\nabla \cdot \mathbf{E}^{\text{sc}}(\mathbf{r})\right] - \omega^2 \mu_b \Delta\varepsilon(\mathbf{r}) E_\xi(\mathbf{r}),\ \xi = x, y, z. \tag{2.128}$$

We observe that the three field components couple through the term $\nabla \cdot \mathbf{E}^{\text{sc}}$ in the right-hand side, which describes the induced scattering source for each component:

$$\mathcal{S}_\xi^{\text{ind}}(\mathbf{r}) = \frac{\partial}{\partial \xi}\left[\nabla \cdot \mathbf{E}^{\text{sc}}(\mathbf{r})\right] - \omega^2 \mu_b \Delta\varepsilon(\mathbf{r}) E_\xi(\mathbf{r}),\ \xi = x, y, z. \tag{2.129}$$

Accordingly, the scattered field integral solution (in an unbounded region) is

$$E_\xi^{\text{sc}}(\mathbf{r}) = \iiint_V G_b(\mathbf{r}, \mathbf{r}')\mathcal{S}_\xi^{\text{ind}}(\mathbf{r}')dv', \tag{2.130}$$

where $G_b(\mathbf{r}, \mathbf{r}')$ is Green's scalar function. The induced source here differs substantially from the source $\mathcal{S}^{\text{ind}} = K(\mathbf{r})U(\mathbf{r})$ ($U \equiv E_\xi$) that we encountered in the scalar case (see Eq. (1.213) in Section 1.8) in that it is no longer simply proportional to the ξ-th component of the total internal field E_ξ. It is now a linear functional of all three components of $\mathbf{E}$. In fact, using Gauss law, $\nabla \cdot \mathbf{E}^{\text{sc}}$ can be expressed explicitly in terms of the incident and total fields as

$$\nabla \cdot \mathbf{E}^{\text{sc}}(\mathbf{r}) = \left[\frac{\nabla\varepsilon_b(\mathbf{r})}{\varepsilon_b(\mathbf{r})}\right] \cdot \mathbf{E}^{\text{inc}}(\mathbf{r}) - \left[\frac{\nabla\varepsilon_s(\mathbf{r})}{\varepsilon_s(\mathbf{r})}\right] \cdot \mathbf{E}(\mathbf{r}). \tag{2.131}$$

The fact that the induced source of scattering $\mathcal{S}_\xi^{\text{ind}}$ now depends not only on the contrast $\Delta\varepsilon$ but also on its gradient (through $\nabla\varepsilon_s$), complicates the derivation of the constraints for both the Born and the Rytov approximations. Such analysis is beyond the scope of an introductory-level text and will not be pursued here. Yet it should be noted that Eq. (2.131) suggests that if the background is uniform ($\nabla\varepsilon_b = 0$) and if the permittivity profile of the scatterer is a slowly varying function of position ($\frac{\nabla \varepsilon_s}{\varepsilon_s} \to 0$), then $\nabla \cdot \mathbf{E}^{sc}$ can be neglected and the scalar Born and Rytov approximations can be used to analyze the vector scattering one rectangular field component at a time.

2.13 TM$_z$ Approximation: The EM Model of Tomography

The 3D vector EM model is extremely complex not only because the scattering object is intrinsically heterogeneous but also because the acquisition setup in MW imaging is usually large and comprises many components, such as enclosures, scanners, antennas, connectors, and cables. This complexity translates into prohibitive computational requirements if a full-wave simulation is used in lieu of a forward model. Simulation-based models are common in the iterative reconstruction methods and they are the determining factor for the time requirements of the reconstruction. If the simulation model can be reduced to a 2D one, that would mean a drastic reduction of the time requirements.

Indeed, a reduction to a 2D simulation can be done. The procedure is illustrated next through the frequency-domain analysis, but in the time domain the approach is the same. The trick is to introduce a plane of symmetry in the field distribution.

Let us assume that the background propagation medium is magnetically uniform, in which case we know that the total **E** field in the scattering volume is governed by Eq. (2.124). Next, a field symmetry is enforced at the $z = 0$ plane so that $\frac{\partial \mathbf{E}}{\partial z} = 0$. The Cartesian component-wise form of Eq. (2.124) becomes

$$\frac{\partial}{\partial y}\left[\frac{\partial E_y(\mathbf{r})}{\partial x} - \frac{\partial E_x(\mathbf{r})}{\partial y}\right] - \omega^2 \mu_b \varepsilon_s(\mathbf{r}) E_x(\mathbf{r}) = 0, \qquad (2.132)$$

$$-\frac{\partial}{\partial x}\left[\frac{\partial E_y(\mathbf{r})}{\partial x} - \frac{\partial E_x(\mathbf{r})}{\partial y}\right] - \omega^2 \mu_b \varepsilon_s(\mathbf{r}) E_y(\mathbf{r}) = 0, \qquad (2.133)$$

$$-\left[\frac{\partial^2 E_z(\mathbf{r})}{\partial x^2} + \frac{\partial^2 E_z(\mathbf{r})}{\partial y^2}\right] - \omega^2 \mu_b \varepsilon_s(\mathbf{r}) E_z(\mathbf{r}) = 0. \qquad (2.134)$$

We observe that the E_z field component is now decoupled from E_x and E_y. In addition, we require that $E_x = E_y = 0$ at $z = 0$.[12] This leaves E_z as the only nonzero **E** field component in the $z = 0$ plane along with two magnetic field components, H_y and H_z, which can be determined from E_z using Maxwell's equations as

$$\mathrm{i}\omega\mu_b H_x = -\frac{\partial E_z}{\partial y}, \ \mathrm{i}\omega\mu_b H_y = \frac{\partial E_z}{\partial x}. \qquad (2.135)$$

The three-component EM field that we just described is *transverse magnetic* with respect to z (or TM$_z$) field since its **H** vector is transverse to z. It is completely described

[12] This requirement imposes the so-called *electric wall* boundary conditions. At an electric wall, the tangential **E** field components must vanish.

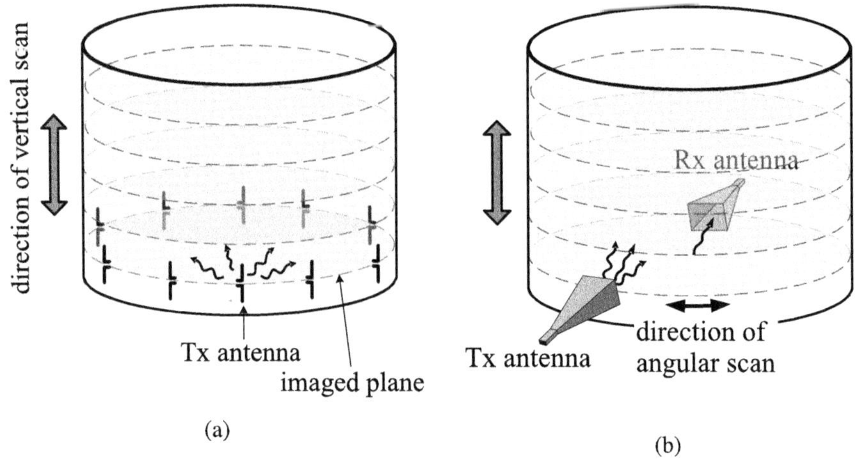

Figure 2.7 Illustration of a tomographic acquisition setup. The data are acquired and a 2D image is reconstructed one plane at a time. To generate a 3D image, this is repeated at multiple vertical locations and the resulting image is a collection of slices. (a) Circular array of electronically switched antennas (here shown as dipoles): at each measurement one antenna transmits and all others receive. The scan is fast but the angular sampling rate is limited by the minimum possible center-to-center separation between the antennas. (b) One or more antennas (here shown as horns) are scanned in a circular fashion in addition to the vertical scanning. The mechanical circular scan allows for a very fine angular sampling but may be slow.

by E_z alone, the governing equation of which, as per Eq. (2.134), is the 2D scalar Helmholtz equation

$$\nabla_{\perp}^2 E_z(\mathbf{r}) + \omega^2 \mu_b \varepsilon_s(\mathbf{r}) E_z(\mathbf{r}) = 0, \;\; z = 0. \tag{2.136}$$

Here, $\nabla_{\perp}^2$ is the 2D Laplace operator in the plane transverse to z; e.g., its Cartesian form is $\nabla_{\perp}^2 = \frac{\partial^2}{\partial x^2} + \frac{\partial^2}{\partial y^2}$. The numerical solution of this equation is fast as the problem is two-dimensional. Moreover, its associated inverse scattering problem is simpler in comparison with a 3D reconstruction as $\varepsilon_s(\mathbf{r})$ is reconstructed in one plane only.

The approach to imaging where the data acquisition and the reconstruction are done one plane at a time is referred to as *tomography*.[13] Typical tomographic arrangements are shown in Fig. 2.7. Fig. 2.7-a shows a circular array of antennas at the desired vertical position and the data are acquired in the respective plane. The transmission and possibly the reflection coefficients are acquired where in each acquisition one antenna transmits and all others receive. Using RF switches, this is repeated for all antennas in the array, so that illumination from all angles is achieved. To obtain a complete 3D image of the target, the array is scanned vertically obtaining the data and reconstructing the image one slice at a time. The data acquisition with electronically switched arrays is fast but

[13] The word *tomography* is a term from the Greek *tomos*, meaning a "cut" or a "slice", and *graphos*, also derived from Greek, meaning "write".

the angular sampling rate is limited by the minimum possible center-to-center separation between the antennas.

Fig. 2.7-b shows another possibility where one or more antennas are scanned in a circular fashion, instead of employing a dense switched array. Employing one antenna allows for acquiring the reflection coefficient only whereas with more antennas the measurement of the transmission coefficients becomes also possible. Mechanical scanning is slower than electronic switching; however, the system is easier to design because the center-to-center distance between the involved antennas (if more than one) can be large. This is why it is often preferred in initial tests and algorithm development. Moreover, mechanical scanning allows for very fine angular sampling. Often electronic switching is combined with mechanical scanning to take advantage of the speed of electronic switching and the superior sampling rate of the mechanical scan.

It is critical to design the antennas together with the whole setup so that the field TM_z symmetry is ensured. Of course, the symmetry can be strictly ensured only in a scatterer-free environment, i.e., only for the incident field. The presence of scatterers inevitably distorts this symmetry unless the scatterers have a shape which varies very slowly in the vertical direction, e.g., they are shaped as long cylinders. To reduce this distortion, the antenna beamwidth in the vertical direction should be made as small as possible in order to reduce the scattering from the vertical positions above and below the plane of symmetry.

An antenna, which generates a TM_z field in one of its *principal planes*,[14] is a linearly polarized antennas such that: (i) the direction of the **E** field it generates (i.e., its polarization vector $\hat{\mathbf{p}}$) is orthogonal to the plane of symmetry (the acquisition plane of the tomographic setup), and (ii) its E-plane radiation pattern is symmetric with respect to this plane. Examples of such antennas include dipoles, monopoles, bow-tie antennas, Vivaldi antennas, rectangular horns, and many others.

2.14 Simplifying Approximations in the Vector State Equation

2.14.1 Fundamental Solutions Used as Green's Functions

In Section 2.11, we showed how to represent Green's dyadic in the **E** field data equation when this equation is associated with a scalar response such as an open-circuit voltage, a transmission or a reflection coefficient, etc. We showed that Green's dyadic contracts to a vector (Green's vector function), which can be easily obtained from the **E** field distribution generated by the receiving antenna, when this antenna operates as a transmitter in the scatterer-free setup. Thus, Green's vector function in the data equation can

[14] An antenna's principal plane is a plane that contains the direction of maximum radiation. There are two mutually orthogonal principal planes associated with the 3D radiation pattern of an antenna. For a linearly polarized antenna, one principal plane is chosen so that it contains the **E** field vector of the radiated field plus the direction of maximum radiation. This is the E-plane and the associated 2D cut of the antenna radiation pattern is referred to as the E-plane principal pattern. The other plane contains the **H** field vector (as $\mathbf{H} \perp \mathbf{E}$ in the far zone) and the direction of maximum radiation. This is the H-plane and the 2D radiation pattern in this plane is the H-plane principal pattern.

be obtained fairly accurately using simulations or measurements wherein the receiving antenna is set to transmit.

In the state equation, however, this approach is inapplicable. To appreciate the difference with the data equation, let us restate the state equation in the case of scattering from a dielectric target:

$$\mathbf{E}(\mathbf{r}) = \mathbf{E}_{\rm inc}(\mathbf{r}) + \omega^2 \iiint_V \bar{\mathbf{G}}_{\rm b}(\mathbf{r}, \mathbf{r}') \cdot \left[\mu_{\rm b}(\mathbf{r}')\Delta\varepsilon(\mathbf{r}')\mathbf{E}(\mathbf{r}')\right] dv', \; \mathbf{r} \in V. \tag{2.137}$$

In the state equation above, the response $\mathbf{E}(\mathbf{r})$ is the *vector* field *inside* the scattering region, therefore, Green's dyadic cannot be contracted to a vector. To fully characterize $\bar{\mathbf{G}}_{\rm b}(\mathbf{r}, \mathbf{r}')$, an impulse current-density excitation needs to be placed at each $\mathbf{r}' \in V$ and the $\mathbf{E}$ vector needs to be computed at $\forall \mathbf{r} \in V$. While this is in principle possible to accomplish through simulation in an arbitrary heterogeneous medium, the associated computational burden would be prohibitive. This is why it is common to assume, as a starting point, that the background is uniform. If, in addition, we also assume that the background is unbounded, then the fundamental solutions can be used.

For example, in the tomographic scenario considered in the previous Section 2.13, the scalar 2D Green function in a uniform open region employs the 2D fundamental solution to Helmholtz's equation (see Table 1.3):

$$G_{\rm 2D}(\mathbf{r}, \mathbf{r}') = -\frac{\rm i}{4} H_0^{(2)}\left[-{\rm i}\gamma_{\rm b}(\rho - \rho')\right], \tag{2.138}$$

where $\gamma_{\rm b} = {\rm i}\omega\sqrt{\mu_{\rm b}\varepsilon_{\rm b}}$ and $\mathbf{r} = (\rho, \phi)$ denotes the position in the imaged plane S. The corresponding state equation is

$$E_z(\mathbf{r}) = E_z^{\rm inc}(\mathbf{r}) - \frac{{\rm i}\omega^2\mu_{\rm b}}{4} \iint_S \Delta\varepsilon(\mathbf{r}') \cdot H_0^{(2)}\left[-{\rm i}\gamma_{\rm b}(\rho - \rho')\right] \cdot E_z(\mathbf{r}')ds', \; \mathbf{r} \in S. \tag{2.139}$$

On the other hand, the 3D vector scattering requires Green's 3D dyadic, which in a uniform open medium is given by Eq. (2.9).

What are the implications of the assumption that the background propagation medium is uniform and unbounded?

First, if the background is not quite uniform but contains low-contrast heterogeneities, these heterogeneities become part of the unknown imaged object. This is acceptable if we know in advance what these heterogeneities are so that we can identify them in the reconstructed image.

Second, if the background contains high-contrast objects and especially if these are large (e.g., the background is stratified), then the assumption is likely to result in reconstruction failure. It is best to employ more adequate Green's dyadics such as those for layered media; see, for example, [31].

Third, even if the propagation medium is uniform, the antenna structures, positioners, and enclosures of the measurement setup inevitably make the overall background heterogeneous. Moreover, they cause reflections and multiple scattering which contradicts the assumption of open (reflection-free) boundaries. How is then the presence of such objects compatible with the assumption of Green's dyadic being that of a uniform unbounded region? Recall that we are dealing here with the state equation, which is written for observation points inside the scatterer. The antennas and the enclosures reside outside the scatterer and influence its internal field only if their own scattering couples back into its volume. If the imaged object is far from the antennas and the enclosures and if these enclosures are lined with MW absorbers, then as far as the scattering region is concerned, it does indeed comply with the condition of open boundaries. However, if the measurements are done in the near zone (nondestructive testing, tissue imaging, etc.), then this assumption would not hold and is likely to lead to strong image artifacts or complete reconstruction failure.

2.14.2 Fundamental Solutions Used as Approximations of the Interior Field

In Section 2.12, we discussed the application of the Born and Rytov approximations to the total internal field $\mathbf{E}(\mathbf{r}')$, $\mathbf{r}' \in V$. Both of these involve the knowledge of the incident field within the imaged volume. We discussed the possibility of measuring with field probes or simulating the incident field generated by each transmitting antenna in the setup. These are indeed rigorous methods that can provide high-fidelity estimates of the incident-field distributions. In near-field measurements, these methods are strongly recommended because the near field of antennas is not amenable to closed-form analysis, it contains strong reactive components as well as evanescent modes. However, these methods are rather time-consuming and require specific equipment or simulation software neither of which is cheap.

On the other hand, the far field of an antenna in an open environment can be analytically approximated. Common approximations include: (1) the *locally plane wave*, $\sim \exp(-\mathrm{i}k_\mathrm{b}r)$, (2) the *spherical wave*, $\sim \frac{\exp(-\mathrm{i}k_\mathrm{b}r)}{r}$, or (3) in the 2D case of tomography, the *cylindrical wave* $\sim H_0^{(2)}(k_\mathrm{b}\rho)$. Here, k_b is the (possibly complex) wave number in the background[15] whereas r and ρ are the distances from the antenna to the observation point in 3D and 2D problems, respectively. The spherical-wave approximation is sometimes referred to as the *isotropic-wave approximation*.

The locally plane-wave approximation works best in the far zone where the size of the target is much smaller than the distance to the illuminating antenna. It ignores the impact of the amplitude factor, $1/r$. This factor is indeed much less sensitive to small variations in r than the phase factor $\exp(-\gamma_\mathrm{b}r)$. Consider an example in a loss-free medium, where $r \approx 100\lambda$ is the distance to the center of a target (λ is the wavelength) whereas the distance between two points inside the target and along $\mathbf{r}$ is $\approx \lambda/2$. The amplitude factors at the two points would be $\sim \frac{1}{99.75\lambda}$ and $\sim \frac{1}{100.25\lambda}$; clearly, of very similar values. The phase factors, however, would be $\exp[-\mathrm{i}(k_\mathrm{b}r + \pi/2)]$ (for the more distant point) and $\exp[-\mathrm{i}(k_\mathrm{b}r - \pi/2)]$ (for the closer point). This is a significant difference of π in the phase that would reverse the sign of the scattered signal.

The locally plane-wave approximation is widely used in millimeter-wave reflection holography [62, 63], which exploits frequencies anywhere from about 30 GHz to about

[15] Recall that the wave propagation in a lossy medium can be represented either through the complex propagation constant γ_b or through the complex wavenumber k_b, which relate as $\gamma_\mathrm{b} = \mathrm{i}k_\mathrm{b}$.

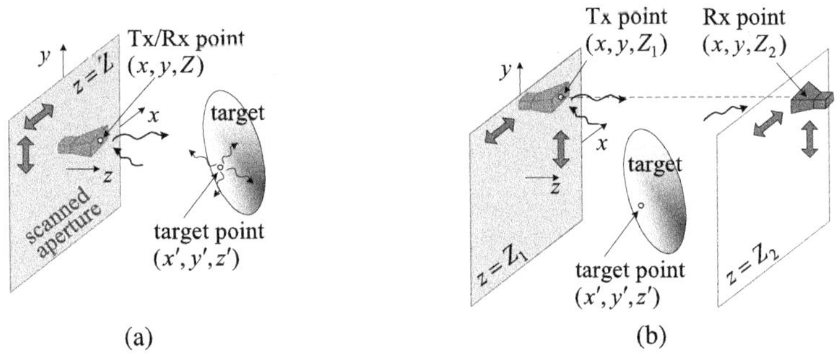

Figure 2.8 Illustration of a planar acquisition setup. (a) The data are collected through a scan of a single aperture (the acquisition plane) by recording the back-scattered signal. The system is monostatic, meaning that the points of transmission and reception coincide. (b) The data are collected on two mutually parallel planes on both sides of the imaged object (the target). Two antennas, aligned along each other's boresight, scan the acquisition planes together, allowing for recording both back-scattered and forward-scattered signals.

300 GHz. This is a *qualitative* imaging method,[16] which is employed in commercial imaging systems, most notably, the whole-body scanners in airports. It is worthwhile stating the forward model of reflection holography in the case of data acquired in a plane [63],

$$s(x, y, \omega) = \iiint_V f(x', y', z') e^{-i2k\sqrt{(x-x')^2+(y-y')^2+(z-Z)^2}} dx'dy'dz', \tag{2.140}$$

and commenting on the underlying approximations. In Eq. (2.140), k is the background's wave number. Fig. 2.8-a shows the setup of planar reflection holography and the coordinate variables involved in Eq. (2.140). In Eq. (2.140), $s(x, y, \omega)$ is the scattered signal picked up by the millimeter-wave transceiver at the position (x, y) on the acquisition plane $z = Z$ (the scanned aperture) at the frequency ω, and $f(x', y', z')$ is referred to as the *reflectivity function*. The reflectivity function, which is the subject of reconstruction, is a qualitative representation of the amount of scattering (or reflection) from a point within the imaged volume. As such it is influenced as a whole by the permittivity and conductivity contrasts (at these frequencies, most targets are nonmagnetic and the permeability contrast is usually neglected). We should also point out the assumption of the reflectivity being frequency-independent. This assumption does not hold generally due to the dispersion in the permittivity and permeability; however, it is often acceptable in a relatively narrow frequency band and/or for materials which do nor exhibit dispersion in the microwave and millimeter-wave bands (e.g., ceramics and metals). Extending the

[16] Qualitative imaging refers to a type of methods where the goal is to locate and identify the shape of objects, which exhibit contrast with the background. The actual value of the contrast is not estimated although the images are often a good representation of the *relative* magnitude of the contrast. Thus, the qualitative images capture the shape and the location of the scattering objects. In contrast, quantitative methods aim at the complete reconstruction of the complex-valued permittivity (or permeability) distributions in the imaged volume.

approach to frequency-dependent reflectivity is not difficult [64] but we will discuss this later in conjunction with the inversion approach employed in holographic methods.

The reflectivity is also dependent on the ability of the target to reflect waves of the particular polarization of the transmitting antenna. A long very thin metallic object oriented along the **E** vector of the incident wave, will reflect strongly, resulting in a bright line in the image. This very same object, however, if oriented perpendicularly to $\mathbf{E}_{inc}$, will reflect nothing (its radar cross-section becomes zero) and will be invisible in the reconstructed image. Another polarization consideration must address the possibility of the target reflecting a wave which is *not* aligned with the receiving antenna. For example, if our long very thin metallic object is oriented at an angle of 45° with respect to $\mathbf{E}_{inc}$, it will capture and reflect back only one-half of the scattered power that it could have reflected under optimal conditions (perfect alignment with $\mathbf{E}_{inc}$). Moreover, its reflected wave is polarized at an angle of 45° with respect to $\mathbf{E}_{inc}$. If the receiving antenna is the same as the transmitting antenna, it will be polarizationally mismatched with the back-scattered wave by 45° and it will capture only one-half of the maximum possible scattered power (the power that the receiving antenna would capture if its polarization vector were aligned with the wave). Thus, the overall captured back-scattered power would be only one-quarter of the maximum possible scattered signal power that would have been realized under optimal polarization conditions.

In summary, the target reflectivity depends not only on the permittivity/conductivity contrast but also on the polarization of the transmitting and receiving antennas in relation to the target's orientation and shape. To minimize the impact of the polarization effects, it is common to use circularly polarized antennas, the field of which contains both vertically and horizontally polarized components.

The comparison between Eq. (2.140) and the scalar-response models in Eq. (2.110) and Eq. (2.108) shows that the qualitative holography model makes the following approximations:

i. the total field is equal to the incident field (Born's linear approximation);
ii. subsequently to the first approximation, the total field is made equal to e^{-ikr}, where $r = \sqrt{(x-x')^2 + (y-y')^2 + (z-Z)^2}$ (plane-wave approximation);
iii. Green's vector function (the incident field of the receiving antenna when it operates in a transmitting mode) is made equal to e^{-ikr} as well, because reflection holography uses a *monostatic radar* arrangement where the distances traveled by the incident and reflected waves are the same.[17]

The plane-wave approximation is sufficiently accurate when the following conditions are fulfilled: (i) the target is in the far zone of the transmitting and receiving antennas, (ii) the distance between the target and the scanned aperture is larger than the largest

[17] In *monostatic radar*, the points of transmission and reception coincide. Often the same antenna is used for both transmission and reception. In contrast, in *bistatic radar*, there are two antennas, one transmitting and one receiving, residing at different locations. Their transmission coefficient is the measured radar signal. Finally, in *multistatic radar*, one antenna transmits and a multitude of other antennas receive, all at different locations. Several transmission coefficients are acquired simultaneously.

extent of the scanned aperture, and (iii) this distance is at least an order of magnitude larger than the wavelength.

In near-zone imaging, however, some or all of these conditions may not hold. The forward model needs to be improved in such cases. In [65], the application of holography to near-field imaging is proposed where the possibility of acquiring both reflection and transmission coefficients has been considered. The scattering component of the reflection coefficient represents the back-scattered signal whereas that of the transmission coefficient represents the forward-scattered signal. This setup is illustrated in Fig. 2.8-b. The acquisition surfaces are again planar but this time two antennas are needed. The two antennas are aligned along each other's *boresight*[18] so that the strength of the forward-scattered signal is maximized. This is necessary since this near-field setup envisions imaging in lossy media, where an increased signal-path length translates into signal attenuation. Such alignment implies that the points of transmission and reception have the same lateral coordinates (x, y).

The forward model in the near-field holography method of [65] is improved in two ways. First, the incident field distribution due to the transmitting antenna is simulated so that evanescent and fast-varying field components in the antenna near zone are well represented. The linear Born approximation is still employed, i.e., the total internal field is approximated by the simulated incident field. This is necessary to enable real-time reconstruction as in the original far-zone holography. Second, Green's vector function is approximated as a spherical wave, $\sim \frac{\exp(-ikr)}{r}$. Now, the forward model in terms of the reflectivity function becomes [65]:

$$s(x, y, Z_i; \omega) = \iiint_V f(x', y', z') \cdot \frac{e^{-ikr}}{r} \cdot E_0^{\text{inc}}(x' - x, y' - y, z'; Z_j; \omega) dx' dy' dz', i, j = 1, 2. \quad (2.141)$$

Notice that Z_i $(i = 1, 2)$ denotes the location of the plane of reception whereas Z_j $(j = 1, 2)$ denotes that of transmission so that the above equation accommodates both back-scattered and forward-scattered signals. Green's function depends on $r = [(x' - x)^2 + (y' - y)^2 + (z' - Z_i)^2]^{1/2}$ and it represents an assumption that the receiving antenna acts as a point source if it were to operate as a transmitter. This is acceptable in the particular study since there the responses are indeed field point samples. $E_0^{\text{inc}}(x', y', z')$ is the simulated field distribution due to the transmitting antenna when this antenna is at the center of the scanned plane $(x = y = 0, z = Z_j)$. Here, the fact that the background medium is open and unbounded (which is yet another approximation) is used to obtain the incident field distribution in the case when the transmitting antenna is at any other location on the Z_j plane, $x \neq 0 \vee y \neq 0$, by a simple coordinate translation of the E_0^{inc} distribution. The assumption that the imaged region is uniform and unbounded also enables the use of the spherical wave approximation of Green's function. The examples presented in [65] confirm the significant

[18] The *boresight* of an antenna is the direction of its maximum radiation.

improvement in the reconstructed images of various targets as a result of the improved forward model.

Yet another improvement of the linearized model of scattering has been proposed in [66]. There, the imaging method is far-zone planar reflection holography where the same antenna is used as a transmitter and a receiver. This time, however, the radiation pattern of the antenna is taken into account. The motivation behind this work is the fact that the antenna pattern does impact considerably the strength of the back-scattered signal when the scattering occurs off boresight. Taking the antenna pattern into account is especially beneficial when focused (high-gain) antennas are used because their radiated power density drops off fast as the direction moves away from boresight. Besides, the phaseless radiation pattern of antennas is always known. For example, commercially available antennas must be accompanied by a data sheet providing this pattern along with many other antenna parameters. The pattern is termed *phaseless* because it describes the real-valued normalized power density of the radiation as a function of the direction of radiation (the angular position of the observation point) but it does not include information about the signal phase as a function of this direction. It is customary to use the spherical angular coordinates, θ and ϕ, to describe the direction of radiation, where the antenna is at the origin and $\theta = 0$ is the antenna boresight. Thus, the power pattern can be expressed as

$$P(\theta, \phi, \omega) = \left|\bar{E}_{\text{inc}}(\theta, \phi, \omega)\right|^2, \tag{2.142}$$

where $\left|\bar{E}_{\text{inc}}\right|$ is the normalized radiated field strength in the far zone.

At the same time, the linearized forward model of reflection holography in the far zone can be expressed as

$$s(\mathbf{r}, \omega) = \iiint_V f(x', y', z')\mathbf{E}_{\text{inc}}(\mathbf{r}') \cdot \mathbf{E}_{\text{inc}}(\mathbf{r}')dx'dy'dz', \tag{2.143}$$

where $\mathbf{r} = (x, y, Z)$, $\mathbf{r}' = (x', y', z')$ and $\mathbf{E}_{\text{inc}} = \hat{\mathbf{p}}E_{\text{inc}}$ is the incident field distribution of polarization $\hat{\mathbf{p}}$ and a complex-valued phasor E_{inc}. Thus, $\mathbf{E}_{\text{inc}} \cdot \mathbf{E}_{\text{inc}} = E_{\text{inc}}^2$. The magnitude of E_{inc}^2, when normalized, is nothing but the antenna power pattern as per Eq. (2.142). Its phase on the other hand can be closely approximated in the far zone using the delay factor $e^{-\text{i}2kr}$, where $r = |\mathbf{r}' - \mathbf{r}|$. Thus, when the transceiver is at $x = y = 0$ $(z = Z)$, the kernel of Eq. (2.143) is obtained as

$$E_{0,\text{inc}}^2(x', y', z'; \omega) \equiv E_{\text{inc}}^2(x', y', z'; 0, 0, Z; \omega) = P(\theta, \phi, \omega) \cdot e^{-\text{i}2kr}. \tag{2.144}$$

Here, the angular position (θ, ϕ) can be readily obtained from (x', y', z') through rectangular-to-spherical coordinate transformation. In order to obtain the kernel for any other transceiver position $(x \neq 0 \vee y \neq 0)$, coordinate translation is applied, i.e.,

$$E_{\text{inc}}^2(x', y', z'; x, y, Z; \omega) = E_{0,\text{inc}}^2(x' - x, y' - y, z'; \omega). \tag{2.145}$$

Again, the improvement over the simplified plane-wave approximation is demonstrated through various simulation and experimental results.

2.15 Time-Domain Solutions to Maxwell's Equations

There are quite a few methods for the direct numerical solution of Maxwell's equations in the time domain [29, 67] such as the finite-difference time-domain (FDTD) method [68, 69, 27], the transmission-line matrix (TLM) method [22, 24], the finite-element time-domain (FETD) method [70, 71, 72], the time-domain integral-equation (TDIE) methods [73], the time-domain marching-on-in-degree (TD-MOD) method [74], etc. These methods are often employed as forward models to solve the inverse scattering problem. However, just like in the frequency domain, they do not provide an explicit link between the contrast function and the observed scattered field or the respective measured response. This is why stating the time-domain vector integral equations of scattering is worthwhile. For the sake of brevity, only the final form of the time-domain EFIE is stated below. The MFIE can be obtained via duality substitutions. The derivation along with many references can be found, for example, in [53, Chapter 4] and [75] (in [76]).

Before we proceed, we remark that the direct inversion methods based on pulsed (or time-domain) responses, such as those acquired with ultra-wideband (UWB) radar, usually exploit the scalar models of scattering, which have been discussed in detail in Chapter 1. However, stating the full vectorial model will make the reader aware of the approximations made in the scalar models and the entailing potential loss of accuracy.

The time-domain EFIE in an open loss-free unbounded medium is derived by taking the inverse Fourier transform of Eq. (2.32). The contribution of the volume sources (if present) is obtained as [53, Chapter 4]:

$$\mathcal{E}_V(\mathbf{r},t) = -\frac{1}{4\pi}\iiint_V \left\{\frac{\mu}{r}\cdot\frac{\partial \mathcal{J}_\mathrm{i}(\mathbf{r}',t')}{\partial t'} + \mathcal{M}_\mathrm{i}(\mathbf{r}',t')\times\frac{(\mathbf{r}-\mathbf{r}')}{r^3}\right.$$
$$\left. + \frac{\partial \mathcal{M}_\mathrm{i}(\mathbf{r}',t')}{\partial t'}\times\frac{(\mathbf{r}-\mathbf{r}')}{vr^2} - \frac{\rho_{\mathrm{e,i}}(\mathbf{r}',t')}{\epsilon}\cdot\frac{(\mathbf{r}-\mathbf{r}')}{r^3} - \frac{1}{\epsilon}\frac{\partial\rho_{\mathrm{e,i}}(\mathbf{r}',t')}{\partial t'}\cdot\frac{(\mathbf{r}-\mathbf{r}')}{vr^2}\right\}dv', \tag{2.146}$$

where $r = |\mathbf{r}-\mathbf{r}'|$, $t' = t - \frac{r}{v}$ is the retarded time, and $v = 1/\sqrt{\mu\epsilon}$ is the speed of light.

The contribution of sources that are excluded from V by a surface S is represented by the surface integral over S, which has essentially the same form as the volume integral above except that the surface equivalent sources are invoked, namely, $\mathcal{J}_{\mathrm{s,e}} = \hat{\mathbf{n}}'\times\mathcal{H}$, $\mathcal{M}_{\mathrm{s,e}} = \mathcal{E}\times\hat{\mathbf{n}}'$, and $\rho_{\mathrm{s,e}} = \epsilon\hat{\mathbf{n}}'\cdot\mathcal{E}$. Here, $\hat{\mathbf{n}}'$ points *into* V. Explicitly, the surface-integral contribution in terms of the field vectors at S is given by

$$\mathcal{E}_S(\mathbf{r},t) = -\frac{1}{4\pi}\oiint_S \left\{\frac{\mu}{r}\cdot\frac{\partial}{\partial t'}\left[\hat{\mathbf{n}}'\times\mathcal{H}(\mathbf{r},t')\right] + \left[\mathcal{E}(\mathbf{r}',t')\times\hat{\mathbf{n}}'\right]\times\frac{(\mathbf{r}-\mathbf{r}')}{r^3}\right.$$
$$+\left[\frac{\partial\mathcal{E}(\mathbf{r}',t')}{\partial t'}\times\hat{\mathbf{n}}'\right]\times\frac{(\mathbf{r}-\mathbf{r}')}{vr^2} - \left[\hat{\mathbf{n}}'\cdot\mathcal{E}(\mathbf{r}',t')\right]\frac{(\mathbf{r}-\mathbf{r}')}{r^3}$$
$$\left. - \left[\hat{\mathbf{n}}'\cdot\frac{\partial\mathcal{E}(\mathbf{r}',t')}{\partial t'}\right]\frac{(\mathbf{r}-\mathbf{r}')}{vr^2}\right\}ds'. \tag{2.147}$$

The singularities arising in the cases when $\mathbf{r} \to \mathbf{r}'$ are treated with the same approaches as in the frequency domain; see Section 2.4.

It is instructive to consider the simple scattering case of a loss-free dispersion-free isotropic dielectric scatterer of permittivity contrast $\Delta\epsilon(\mathbf{r}) = \epsilon_s(\mathbf{r}) - \epsilon_b$. Since the model in Eq. (2.146) is valid in a uniform background only, we have to assume that ϵ_b is constant. The time-domain version of the frequency-domain induced source due to the permittivity contrast, see Eq. (2.85), is

$$\mathcal{J}_{\text{ind}}(\mathbf{r}, t) = \Delta\epsilon(\mathbf{r})\frac{\partial \mathcal{E}(\mathbf{r}, t)}{\partial t}. \tag{2.148}$$

There is no magnetic-current induced source, i.e., $\mathcal{M}_{\text{ind}} = 0$, since the scatterer does not have magnetic contrast with the background. The induced electric charge-density source is determined from $\mathcal{J}_{\text{ind}}$ via charge conservation,

$$-\frac{\partial \rho_{\text{ind}}(\mathbf{r}, t)}{\partial t} = \nabla \cdot \mathcal{J}_{\text{ind}}(\mathbf{r}, t) = \frac{\partial}{\partial t}\nabla \cdot [\Delta\epsilon(\mathbf{r})\mathcal{E}(\mathbf{r}, t)]. \tag{2.149}$$

It follows that

$$\rho_{\text{ind}}(\mathbf{r}, t) = -\nabla \cdot \{[\epsilon_s(\mathbf{r}) - \epsilon_b]\,\mathcal{E}(\mathbf{r}, t)\} = \nabla \cdot [\epsilon_b\mathcal{E}(\mathbf{r}, t)]. \tag{2.150}$$

Here, we have used the fact that the scatterer does not contain impressed charge sources, therefore, as per Gauss law, $\nabla \cdot (\epsilon_s\mathcal{E}) = 0$. On the other hand, the rightmost term in Eq. (2.150) is not likely to be zero because the total electric field exists in the volume of the scattering object, the permittivity of which differs from that of the background.

Next, we consider the impact of Born's linear approximation, $\mathcal{E} \approx \mathcal{E}_{\text{inc}}$, on the induced sources. The approximated $\mathcal{J}_{\text{ind}}$ is proportional to $\frac{\partial \mathcal{E}_{\text{inc}}}{\partial t}$ and the permittivity contrast. The approximated ρ_{ind}, however, vanishes under this approximation since $\nabla \cdot (\epsilon_b\mathcal{E}_{\text{inc}}) = 0$. Thus, under Born's approximation, only one induced source, $\mathcal{J}_{\text{ind}}$, is retained. Substituting it in Eq. (2.146) yields the linearized scattering equation as

$$\mathcal{E}_{\text{sc}}(\mathbf{r}, t) \approx -\iiint_V \frac{\mu_b\Delta\epsilon(\mathbf{r}')}{4\pi r} \cdot \frac{\partial^2\mathcal{E}_{\text{inc}}(\mathbf{r}', t')}{\partial t'^2} dv'. \tag{2.151}$$

We observe that the assumption of isotropic scattering along with Born's approximation leads to a vectorial expression that appears in the same form as the scalar relation that we already established in Section 1.10.1; see Eq. (1.232). In fact, if one assumes that the polarization of $\mathcal{E}_{\text{inc}}$ is constant throughout the scatterer's volume, then this is also the polarization of $\mathcal{E}_{\text{sc}}$ and the field waveforms $\mathscr{E}_{\text{sc}}$ and $\mathscr{E}_{\text{inc}}$ relate through the scalar relation of Eq. (1.232).

3 Scattering Parameters in Microwave Imaging

The scattering parameters, also known as S-parameters, are the most common network parameters used to describe the performance of the microwave (MW) devices, circuits, and networks in the frequency domain. They are also the measured quantities in frequency-sweep measurements. The most widely used MW test instrument is the *vector network analyzer* (VNA), which measures accurately and efficiently the S-parameters of N-port networks. The most common VNAs have 2 ports, i.e., $N = 2$, but VNAs with 3, 4, 8, and even 24 ports are commercially available. Moreover, radio-frequency (RF) switches are available that can increase the number of ports that a VNA can handle.

In MW imaging, the illuminating and receiving antennas together with the measurement setup and the imaged object form a MW network. This is illustrated in Fig. 3.1. If the setup consists of N antennas, the VNA (possibly with an additional RF switch) must have N ports to which the antennas are connected with precision coaxial cables and connectors. The cables and the connectors come in many varieties and have widely varying performance characteristics. Note that using high-quality cables and connectors is critical for a good (repeatable) measurement.[1] It is this network's S-parameters that comprise the data. Since the VNA measures one frequency at a time, we have a subset of S-parameter data at each frequency. At each frequency, one port (one antenna) at a time is excited while all other antennas (marked as Rx in Fig. 3.1) receive. This allows for acquiring the respective transmission S-parameters. The reflection S-parameter at the terminals of the transmitting (Tx) antenna is also acquired. This is repeated for each antenna (each port) in the setup. Thus, at each frequency we have a set of $N \times N$ S-parameters, which form the so-called *scattering matrix* or S-matrix of the network.

3.1 Basics of S-Parameters

Electrical engineers are familiar with the various parameters used to characterize an N-port ($2N$-terminal) network such as the Z-parameters (the impedance matrix), the

[1] Test-instrument grade cables and connectors must feature low loss and a good impedance match to the system impedance. The latter is usually 50 Ω in the MW range. In addition, the cables must ensure low phase distortion when flexed.

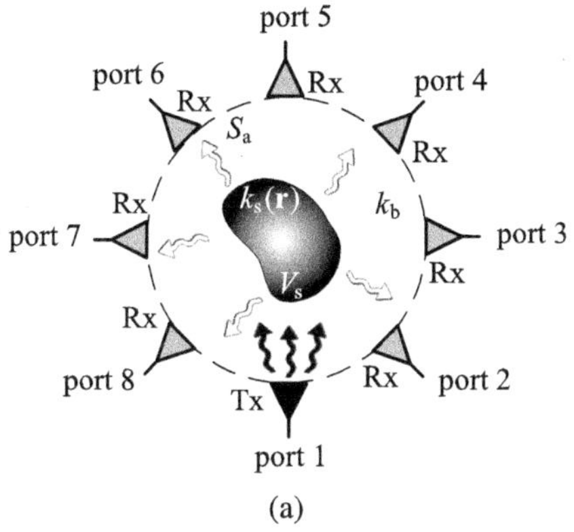

Figure 3.1 Illustration of a MW network formed by eight antennas arranged in a circular fashion surrounding the imaged object (the region V_s). To acquire the full S-matrix at each frequency, the antennas are excited one at a time with all other antennas receiving. In this illustration antenna 1 transmits and all others receive. This measurement yields the elements of the first column of the network's S-matrix, $[S_{11}\ S_{21} \cdots S_{81}]^T$.

Y-parameters (the admittance matrix), the H-parameters (the hybrid matrix), and the $ABCD$ parameters of 2-ports (the transmission matrix). All of these parameter sets relate the voltages and the currents at the network's ports.

Fig. 3.2 shows the common symbols for a single-port, a 2-port, and a 3-port network with the convention for the directions of the port voltages and currents. Knowing the network parameters of a circuit or a device allows for treating it as a "black box" and for incorporating it in the analysis of a larger network. As seen from Fig. 3.2, every port is a pair of terminals where the module interconnects with other modules or with a measurement instrument.

3.1.1 Power Waves

In the MW networks, the interconnects are not simple two-terminal wires; they are waveguides (the name is self-explanatory). Strictly speaking, all transmission lines are waveguides, as long as they are electrically long. The waveguides can take the form of coaxial cables, printed traces in an integrated circuit, metallic waveguides, or dielectric (optical) fibers. While in the first two cases (a coaxial cable or a pair of printed traces)

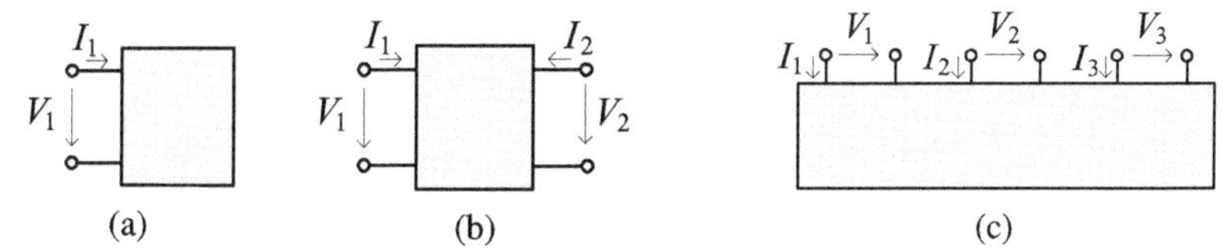

Figure 3.2 Examples of electrical networks and their symbol definitions: (a) 1-port, (b) 2-port, and (c) 3-port.

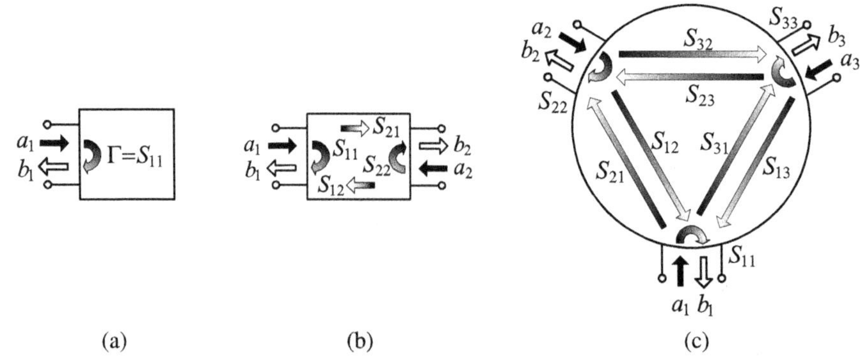

Figure 3.3 Examples of MW networks and their symbol definitions: (a) 1-port device with incident and reflected waves and its reflection coefficient Γ, which is identical to the only element of its scattering matrix (S_{11}); (b) 2-port device with its four wave quantities and four scattering parameters, and (c) 3-port device with its six wave quantities and nine scattering parameters.

we still could define and measure voltages and currents (if the frequency is sufficiently low), this is impossible in the case of metallic waveguides and optical fibers. Thus, all network parameters based on the current/voltage port variables are inadequate in such cases.

Moreover, even if we could define and measure the voltages and currents for certain waveguides, we still would face a significant limitation, namely, the electrical length from the port cross-section (the device connector) to the actual high-frequency structure enclosed by the module. Adjusting this length—which is often necessary in the design of the system—would change the voltage and current port quantities in a fairly complicated manner because they represent the superposition of incident and reflected voltage and current waves along the interconnect. Thus, the network parameters change in a complicated way as functions of the lengths of the interconnects.

For these reasons, MW engineers prefer to work with wave quantities: the port incident and reflected waves. By convention, the incident port waves are denoted by the symbol a, whereas the reflected ones are denoted by b. Fig. 3.3 illustrates the conventions in describing 1-port, 2-port, and 3-port MW networks in terms of the wave quantities. The wave quantities are in essence traveling waves: the incident waves travel toward the device, whereas the reflected waves travel away from the device. That is why the incident waves are also called *incoming*, whereas the reflected waves are also called *outgoing* or *scattered*. In fact, the terms "outgoing" or "scattered" are often preferred because a wave traveling away from the device may not be due to reflection at the port, but to transmission through the device.

The physical meaning of the incoming and outgoing port waves is easy to understand. Any given port, say the n-th port, of a MW device, is connected to its respective (n-th) waveguide (or a transmission line). Assume first that a MW generator is connected at the other end of this n-th waveguide. It will generate a wave traveling toward the device.

Provided our device presents a matched load to the waveguide, there will be no reflected wave, and the waveguide will carry exactly and only the incoming wave a_n.

The outgoing wave b_n can be understood in a similar manner. Assume that MW power is fed into the device through another port, say the m-th port, where $m \neq n$. Then we can expect that, in general, some power will emerge at the n-th port and will travel down the n-th waveguide away from the device. Now assume that this waveguide is infinitely long or is loaded with a matched load at the other end. Then that outgoing traveling wave is the only wave carried by the waveguide. This is exactly the wave b_n.

The wave quantities are complex numbers. The wave-quantity magnitude is the square root of the power carried by the traveling wave. Thus, if we denote the power carried by the incoming wave at the n-th port by P_n^+ and that carried by the outgoing wave by P_n^-, we have

$$|a_n| = \sqrt{P_n^+}, \ |b_n| = \sqrt{P_n^-} \quad \left(\mathrm{W}^{1/2}\right). \tag{3.1}$$

This is why the wave quantities are also referred to as *root-power waves* (or simply *power waves*).

The angles (or the phases) of the power waves correspond to the phases of the respective traveling waves, which, in EM terms, are the phases of their electric fields at the port:

$$\angle a_n = \angle E_n^+, \ \angle b_n = \angle E_n^-. \tag{3.2}$$

There is a straightforward relationship between the power waves and the voltage and current waves that exist in TEM transmission lines such as the familiar coaxial or twin-lead cables. If the transmission line at the n-th port has real characteristic impedance $Z_{c,n}$ (loss-free line), which is normally the case in measurements, the incident and the outgoing power waves a_n and b_n relate to the root-mean-square (RMS) phasors of the incident voltage wave V_n^+ and the reflected voltage wave V_n^-, respectively, as

$$a_n = \frac{V_n^+}{\sqrt{Z_{c,n}}}, \ b_n = \frac{V_n^-}{\sqrt{Z_{c,n}}} \quad \left(\mathrm{W}^{1/2}\right). \tag{3.3}$$

It can be shown that in the general case of a complex characteristic impedance, the denominators in the expressions in Eq. (3.3) should be $\sqrt{\Re Z_{c,n}}$ ($\Re Z_{c,n} > 0$).

Finally, taking into account the expressions for the incident and reflected voltage waves in terms of the total voltage and current at the n-th port device terminals, $V_n^+ = 0.5(V_n + Z_{c,n}I_n)$ and $V_n^- = 0.5(V_n - Z_{c,n}^* I_n)$, we obtain the well-known Penfield-Kurokawa expressions [77, 78]:

$$a_n = \frac{V_n + Z_{c,n}I_n}{2\sqrt{\Re Z_{c,n}}}, \ b_n = \frac{V_n - Z_{c,n}^* I_n}{2\sqrt{\Re Z_{c,n}}}. \tag{3.4}$$

Here, $Z_{c,n}$ is the characteristic impedance of the n-th port. These expressions are important because they allow for obtaining the relationships between the scattering parameters of a network on one hand and its equivalent voltage-current parameters (Z, Y, etc.) on the other. For a summary of these relationships, see, for example, [13].

3.1.2 Scattering Matrix

The scattering matrix (or S-matrix) relates the vector of incident waves $\boldsymbol{a}$ to the vector of scattered waves $\boldsymbol{b}$ as

$$\boldsymbol{S}\boldsymbol{a} = \boldsymbol{b}, \tag{3.5}$$

where $\boldsymbol{a} = [a_1\ a_2 \cdots a_N]^T$, $\boldsymbol{b} = [b_1\ b_2 \cdots b_N]^T$, N is the number of ports, and

$$\boldsymbol{S} = \begin{bmatrix} S_{11} & \cdots & S_{1N} \\ \vdots & & \vdots \\ S_{N1} & \cdots & S_{NN} \end{bmatrix}. \tag{3.6}$$

The elements of the scattering matrix $\boldsymbol{S}$ are the scattering parameters (or S-parameters). They are in essence the reflection and transmission root-power coefficients of the MW network.

The diagonal elements of $\boldsymbol{S}$ are the reflection coefficients. As follows from Eq. (3.5), the reflection coefficient at the i-th port S_{ii} ($i = 1, \ldots, N$) can be written in terms of the incoming and outgoing power waves at the i-th port as

$$S_{ii} = \left.\frac{b_i}{a_i}\right|_{a_k=0}, \quad k \neq i. \tag{3.7}$$

The off-diagonal elements of $\boldsymbol{S}$ are the transmission coefficients. Similarly to Eq. (3.7), the transmission coefficient from the i-th port to the j-th port S_{ji} is expressed in terms of the outgoing wave b_j at the j-th port and the incoming wave at the i-th port a_i as

$$S_{ji} = \left.\frac{b_j}{a_i}\right|_{a_k=0}, \quad k \neq i. \tag{3.8}$$

Notice the condition $a_k = 0, k \neq i$, in Eq. (3.7) and Eq. (3.8). It tells us that in order to measure S_{ji} or S_{ii}, the excitation must be provided at the i-th port (so that $a_i \neq 0$) while at all other ports' incident waves must not be present. This imposes two requirements on these ports: (1) they must not be connected to generators, and (2) they must be loaded with a matched impedance so that the power waves leaving the device do not get reflected back toward the device.

The S-parameters are complex numbers. The magnitude of the S_{ji} parameter shows the proportion of the root-power wave (the signal) incident on port i that reaches port j and leaves the device. If $i = j$, the same principle applies—$|S_{ii}|$ shows the proportion of the signal incident on port i that, as a result of the interaction with the network, is injected back into the i-th port and travels away as a reflected wave toward the generator.

An important property of the S-matrix of a reciprocal network is that it is symmetric, i.e., $S_{ij} = S_{ji}$ [13]. Networks consisting of linear components and linear media are reciprocal, and this is often the case in MW imaging. A MW imaging setup, such as the one illustrated in Fig. 3.3, consists of the antennas together with the imaged object. The imaged object is usually a composition of linear matter, i.e., its constitutive parameters do not depend on the field strength. The antennas, too, are usually linear, unless they are integrated with nonlinear detectors and/or amplifiers.[2] Thus, in an N-port S-parameter measurement, only $N(N+1)/2$ S-parameters are independent. For example, the S-matrix of a 2-port linear network has four elements but, because $S_{12} = S_{21}$, only three of them are independent.

In passive networks (i.e., networks that do not contain amplifiers and/or generators), the S-parameters satisfy $|S_{ij}| \leq 1, i, j = 1, \ldots, N$. This reflects the fact that none of the outgoing waves can carry more power than the incoming one. An S-parameter of unity magnitude indicates full transmission/reflection, whereas zero magnitude indicates full power absorption by the device.

Consider as an example a loss-free piece of transmission line (a cable). This is a 2-port device. Let the line's characteristic impedance Z_c ($Z_c \in \mathbb{R}$) be the same as the internal impedance of the generator at port 1 ($Z_c = Z_G$). Such a line is said to be impedance-matched to the generator. An impedance match is always desirable since it ensures maximum power delivery from the generator to the line in the somewhat abstract scenario of this line being infinitely long [13]. Moreover, it ensures unobstructed (reflection-free) propagation of the wave from the generator through the interconnect and into the transmission line. In reality, the cable is finite in length. However, if it is loaded at port 2 with a matched load, $Z_L = Z_c$, it behaves just like an infinitely long transmission line since it delivers the wave power to the load fully, i.e., without reflection. Thus, a piece of cable that is impedance-matched at both ends exhibits zero reflection and full transmission. In terms of S-parameters, $|S_{11}| = 0$ and $|S_{21}| = 1$, the power incident on port 1 has been fully absorbed and has been fully transmitted onto port 2. A piece of cable is a linear device; therefore, it is reciprocal and $|S_{12}| = 1$. In addition, due to symmetry, $|S_{22}| = 0$.

Let us now assume that our loss-free cable is shorted at port 2. Now we have a 1-port network, which is characterized by one S-parameter, S_{11}, the magnitude of which is $|S_{11}| = 1$. This is because the power initially absorbed at port 1 travels down the cable to the short where it is fully reflected back toward port 1 (toward the generator).

The phase of the S_{ji} parameter shows the change in the wave's phase as it emerges at the j-th port relative to the phase of the incident wave at the i-th port. For example, the phase of the S_{11} parameter of a very short (zero-length) transmission line shorted at the end is $\angle S_{11} = 180^\circ$, indicating the phase reversal of the incident electric field (or the incident voltage) at the short. Combining this with our knowledge that $|S_{11}| = 1$ (full reflection), we obtain that $S_{11} = -1$. If, on the other hand, the line's length is a quarter of a wavelength, and the line is still shorted at the end, $S_{11} = +1$, because in addition to the phase reversal at the short, the wave experiences a phase delay of 180° by traversing the line twice (to and from the short-circuit load).

There is an ambiguity in the phase of the S-parameters that is often referred to as *phase wrapping*. This is important in MW imaging because it adds a degree of nonuniqueness to the solution of the inverse problem. Phase wrapping refers to the fact that the angle of an S-parameter in the complex plane varies only between $-\pi$ and π (or 0 and 2π), when in reality the phase delay can be anywhere between 0 and infinity. Phase unwrapping corrects the radian phase angles by adding multiples of $\pm 2\pi$ when

[2] Such antennas are referred to as *active* antennas.

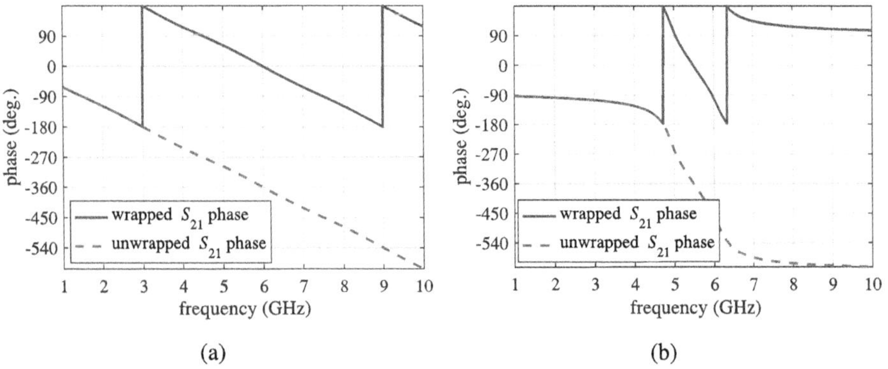

Figure 3.4 Examples of phase wrapping in the transmission S_{21} parameter compared with the actual (unwrapped) phase: (a) 50-cm piece of coaxial cable with air insulation; (b) bandpass (5 GHz to 6 GHz) filter.

absolute jumps between consecutive elements are greater than or equal to π. Phase unwrapping is such a common problem in high-frequency engineering that there is a MATLAB function for it called *unwrap* [79]. Fig. 3.4 illustrates the wrapped phases of the transmission S-parameters of two 2-port devices, a piece of cable, and a bandpass Chebyshev filter, together with the actual phase curves obtained with the unwrapping function in MATLAB.

3.1.3 Measurements with a Vector Network Analyzer

The network analyzer generates a sinusoidal signal (the incident wave) that is applied to each port of the device, one port at a time. When the i-th port is excited, the outgoing waves are measured simultaneously on all (matched) ports, and this measurement yields the i-th column of the S-matrix. A scalar network analyzer measures only the amplitude ratio of the incident and the outgoing waves, i.e., it provides only the magnitudes of the scattering parameters. A vector network analyzer (VNA)[3] is a much more complicated instrument, which can also provide the phase difference between the incident and the outgoing waves, thus measuring the S-parameter phases as well. The availability of vectorial information about the MW circuit opens many possibilities in addition to the S-parameter output, including: (a) plotting the input impedance of a 1-port device on a Smith chart and (b) emulating time-domain measurements through an inverse Fourier transform of a very wideband measurement. The latter application is referred to as the *time-domain option* of the VNA operation, and typically, it makes use of the whole bandwidth of the instrument.

The architecture and the basic principles of operation of the VNA are the subject of MW engineering courses and they are beyond the scope of this text. Here, it is only emphasized that unlike low-frequency test instruments, the VNA needs to be calibrated before each measurement. This is necessary to compensate for systematic measurement

[3] The name is derived from the representation of the complex S-parameters as vectors in the complex plane.

errors in the instrument with the greatest possible precision. Also, the calibration allows compensation for the effects of the connecting cables, connectors, and adapters, so that the measured S-parameters describe the device itself, not the device together with the cables connecting it to the VNA.

For 1-port (S_{11}) measurements, 1-port calibration must be performed. The method is termed the OSM (open-short-matched) *cal* technique (also known as OSL or SOL, for open-short-load or short-open-load). This is because the calibration consists of three separate measurements on known precision terminations: an *open*, a *short*, and a *matched load* (50-Ω load). These terminations are attached one at a time to the connector to which the measured device will be attached. The results of the three *cal* measurements are stored. Based on them the instrument can calculate the measurement errors. These include not only its internal errors but also the influence of the cable and its two connectors on the VNA side and on the device side. These measurement errors are then de-embedded from each subsequent device measurement.

In 2-port measurements, the same OSM procedure has to be repeated at both ports. In addition, a *thru* and an *isolation* measurements are needed. The *thru* measurement consists of connecting the two ports (or their respective cable/adapter extensions) directly, i.e., using a proper precision adapter (for example, a female-to-female SMA[4] adapter). The VNA then acquires both the reflection and transmission coefficients. This *cal* step removes internal errors such as port-matching errors and transmission-tracking errors. Also, it is used by the VNA to set automatically the reference positions of the two ports.

The *isolation* measurement consists of connecting precision 50-Ω loads to both ports and acquiring the transmission coefficient. It measures the crosstalk between the test ports in a transmission measurement (S_{21} or S_{12}). Ideally, the *crosstalk* transmission coefficients are zero. In practice, in most network analyzers, the crosstalk is in the range of −100 dB to −120 dB. The *isolation* calibration step is often skipped (and the VNAs allow for it) because it is important only when measuring very weak transmission coefficients such as those in the stop-band of a good filter where the rejection is better than −60 dB.

A 2-port VNA can be used to measure the scattering parameters of devices that have more than 2 ports, but this requires more work. This is because one has to manually connect the desired pair of device ports to the VNA while at the same time matched (50-Ω) loads are connected to all remaining ports. This has to be repeated until all port pairs are measured. Of course, such an approach is not practical if the network has 4 or more ports, which is often the case with MW imaging systems.

The 2-port VNAs are most common, but multi-port VNAs are available commercially for as many as 3, 4, 8, and even 24 ports. Multi-port VNAs are very expensive and fairly large instruments. This is why MW imaging setups are often custom made that allow for the S-parameter measurements of N-port networks where N is in the range of 10 to 20 in tomographic circular arrays [80, 81] and as high as 60 [82] in hemispherical arrays.

Readers interested in learning more about MW measurements with VNAs are directed to two excellent resources, namely, [83] and [84].

[4] SMA stands for *subminiature version A*.

3.2 S-Parameter Sensitivities

The sensitivity of a response (e.g., a scattering parameter) represents its gradient in the space of the system variables, e.g., the shape and/or the constitutive parameters of the inspected object or device. Response sensitivity analysis is an extensive branch of interdisciplinary science developed mostly in the mathematical and engineering disciplines [85, 86, 87, 88]. Response sensitivities provide crucial information in engineering tasks such as design optimization, modeling, tolerance and yield analyses, uncertainty analysis, stability analysis, detection, and imaging, because they describe the rate at which the response changes as a result of variations in the system. Many books on engineering design optimization have chapters dedicated to the subject; see, for example, [89, 90, 91, 92]. This is because a whole class of nonlinear optimization methods, the *gradient-based methods*, rely on the availability of response sensitivities.

The sensitivities (or the gradients) of all responses of interest are represented in their entirety by a matrix, referred to as the response *Jacobian*,[5] from which the gradient of any desired cost function can be easily determined. The Jacobian is at the heart of the iterative update of the system variables driving the gradient-based optimization procedures toward an optimal solution.

If a minute variation in a variable of the system results in large changes in its performance (its responses), the system is said to be sensitive to this variable. Large sensitivity is desirable when designing sensors to detect defects or abnormalities in an inspected object. On the other hand, large sensitivity is *not* desirable when it comes to fabrication tolerances because in that situation tiny inaccuracies in the fabrication process may result in a large number of defective units.

Gradient-based nonlinear optimization is widely employed in the solution of the inverse problem of scattering when linearizing approximations such as Born's and Rytov's cannot be applied. Such optimization strategies employ algorithms such as the conjugate-gradient, the Gauss–Newton, the Newton–Kantorovich, the quasi–Newton methods, and modifications of these. For a review with abundant references, the reader is referred to [46]. In all of these strategies, response gradients are used to compute an update of the contrast function or the induced sources of scattering such that an underlying EM model (e.g., a simulation) would yield responses that match better the data. This is why such methods are also termed *model-based imaging* [93].

In image reconstruction based on frequency-sweep measurements, the responses are the scattering parameters, whereas the system variables are the values of the complex contrast function, see Eq. (1.212), at each voxel[6] of the imaged volume. The number of scattering parameters involved is large because they are acquired at many sampling locations and at many frequencies. The number of system variables is even larger because it corresponds to the number of the voxels comprising the imaged volume. The number of variables is determined by the overall imaged volume and the voxel size, which,

[5] The Jacobian matrix for a vector of responses $\boldsymbol{r} = [r_1\ r_2 \cdots r_N]^T$ in the parameter space defined by $\boldsymbol{p} = [p_1\ p_2 \cdots p_M]^T$ is given by $\boldsymbol{J}_{M\times N} = [\nabla_{\boldsymbol{p}} \boldsymbol{r}]^T$. Here, $\nabla_{\boldsymbol{p}} r_n = \left[\frac{\partial r_n}{\partial p_1}\ \frac{\partial r_n}{\partial p_2} \cdots \frac{\partial r_n}{\partial p_M}\right]$, $n = 1, \ldots, N$.

[6] The term *voxel* refers to the smallest portion of a volume within which the medium properties are assumed constant. It is analogous to the meaning of *pixel* in a 2D image.

in general, is chosen to be comparable to the expected spatial resolution limit. In 2D imaged domains such as those in tomography, this number is on the order of thousands, whereas in 3D domains it is on the order of tens of thousands. The Jacobian is thus a very large matrix.

Model-based reconstruction using nonlinear optimization aims to minimize a cost function, $F(\boldsymbol{r}(\boldsymbol{p}))$, $F \in \mathbb{R} \mid F \geq 0$, that measures the difference between the data $\bar{\boldsymbol{r}}$ (the measured responses) and the responses $\boldsymbol{r}$ produced by the simulation model with the current iterate of the contrast distribution $\boldsymbol{p}$. It is the gradient of this cost function $\nabla_{\boldsymbol{p}} F$ that is used in updating $\boldsymbol{p}$. When $\boldsymbol{p}$ is composed of the values of the contrast function within the imaged volume, this gradient is referred to as the *Fréchet derivative*. The Fréchet derivative is obtained easily from the response Jacobian using the chain rule of differentiation. The term "Fréchet derivative" is also used to describe the gradient $\nabla_{\boldsymbol{p}} r$ of any response r of interest, not only the cost function.

Fréchet derivatives are also used in qualitative reconstruction with the sensitivity-based method [94, 95, 96]. This method performs in real time because it uses a linearized Born model and a direct reconstruction scheme that does not involve the solution of a linear system of equations. Later, it was shown in [97] that this real-time qualitative technique can be easily modified to produce quantitative images, provided the S-parameter sensitivities are obtained experimentally using known electrically small dielectric scatterers. These techniques are discussed in more detail in Chapter 4.

In view of the importance of the Fréchet derivatives of the S-parameters in MW imaging, it would be beneficial to provide their *analytical sensitivity formulas* here. A *sensitivity formula* is an expression of the desired response derivative in terms of the system state variables. In EM frequency-domain analysis, the state variables are the EM field components distributed over the 3D region of interest. Frequency-domain EM analysis is relevant here because the S-parameters are intrinsically frequency-domain responses, i.e., they do not have direct time-domain meaning. If the state variables (the EM field) are known, any desired response, the S-parameters included, can be computed from them because, by definition, a response has an a priori defined relationship (analytical or discrete) with the state variables. If, in addition, we have an analytical sensitivity formula for this response, we can compute its derivative with respect to any desired constitutive or shape structural variable. This computation is very fast because it amounts to integration over the field state variables.

Note that if a sensitivity formula for a given response is not available, one has to resort to parameter perturbation and employ a finite-difference approximation or another higher-order response-level estimation. For example, let us assume that a state of the scattering object has been simulated and has produced a response r_0, but we also want to know its derivative $\frac{\partial r}{\partial p}\big|_{p=p_0}$ at this state. Here, p is one of the optimizable variables, e.g., the permittivity of a structural component in the model. In the current state, $p = p_0$. We could perturb p in the forward direction (i.e., increase its value) by $0.5\Delta p$, simulate this perturbed state, and obtain p_0^+. This step is then repeated in the backward direction to obtain p_0^-. Using central finite difference, $\frac{\partial r_0}{\partial p}$ can now be approximated as $\frac{\Delta r_0}{\Delta p_0}$, where $\Delta r_0 = r_0^+ - r_0^-$ and $\Delta p_0 = p_0^+ - p_0^-$. There are two significant drawbacks to this approach. First, it requires three simulations to produce the response and its derivative. These are two simulations in addition to the one producing r_0. If we have M optimizable variables, the number of additional simulations increases to $2M$. In MW imaging, the Fréchet derivative is taken with respect to thousands of variables ($M \sim 10^3$), and this computational overhead is prohibitive. Second, choosing the perturbation Δp is tricky. It has to be sufficiently small to overcome the often strongly nonlinear dependence $r(p)$; yet, it has to be sufficiently large to ensure that Δp_0 rises well above the numerical errors of the simulation. The choice is often ad hoc; therefore the derivative estimate is often of questionable accuracy. In summary, perturbative response-level estimates of the Fréchet derivative are not feasible in MW imaging.

S-parameter sensitivity analysis is an integral part of the EM forward model and as such has been studied extensively. S-parameters have been analyzed in the context of equivalent-circuit networks as early as the late 1960s [98, 99, 100]. Comprehensive reviews of these early developments can be found in [101, 102, 103]. With the advent of EM numerical methods and simulation tools, the focus has shifted toward field-based S-parameter sensitivity analysis. A plethora of so-called exact sensitivity formulas for S-parameters have been reported that use the full-wave field solution to efficiently compute the desired derivatives; see, for example [94, 103, 104, 105, 106, 107, 108]. The reason why there is such a wide variety of sensitivity expressions for the same type of response is because each one of these expressions is valid only for the particular numerical method used by the simulation: frequency-domain or time-domain, finite-element or finite-difference and so on. To this day, exact sensitivity computations, mostly in the framework of the finite-element methods, are widely used in MW imaging [109, 110, 111, 112].

Here, we are particularly interested in the *analytical sensitivity formulas* that give the analytical relationship between the S-parameter derivative with respect to a constitutive or shape object parameter and the field distribution [113, 114]. The advantage of an analytical sensitivity expression over the exact formulas is that it is general, i.e., it does not depend on the particulars of the numerical method or the simulator. As long as the field values within the simulated object can be exported (which is a utility available in every modern simulation software), the user can compute the Jacobian independently and can use the same formula with any simulation tool that becomes available. The whole process requires less than a second on a PC even if tens of thousands of derivatives need to be computed.

The sensitivity formulas relevant to MW imaging are outlined and illustrated next.

3.2.1 General S-Parameter Sensitivity Formula

The general analytical sensitivity formula for S-parameters was first derived in [113]. In the case of a linear reciprocal medium, which is predominant in MW imaging, this formula appears as

$$\frac{dS_{ij}}{dp_n} = \frac{\mathrm{i}\omega}{2a_i a_j} \iiint_V \left\{ \left[\frac{d\bar{\mu}(\mathbf{r}')}{dp_n} \cdot \mathbf{H}_j(\mathbf{r}') \right] \cdot \mathbf{H}_i(\mathbf{r}') - \left[\frac{d\bar{\varepsilon}(\mathbf{r}')}{dp_n} \cdot \mathbf{E}_j(\mathbf{r}') \right] \cdot \mathbf{E}_i(\mathbf{r}') \right\} dv', \tag{3.9}$$

where $\bar{\mu}$ and $\bar{\varepsilon}$ are the (symmetric) permeability and permittivity tensors, respectively; a_ξ ($\xi = i, j$) is the power wave at the ξ-th port; and p_n ($n = 1, \ldots, N_p$) is any constitutive or shape parameter in the set of N_p optimizable variables. The subscripts in the **E** and **H** field notations indicate which port is excited when the respective field is obtained. The field phasors in Eq. (3.9) are RMS phasors, i.e., their absolute values correspond to the RMS values of the respective time-harmonic fields. Throughout this chapter, RMS phasors are used, unless stated otherwise.

To understand the application of Eq. (3.9), one should appreciate the fact that when the shape or constitutive parameters of an object change, $\bar{\mu}(\mathbf{r}')$ and $\bar{\varepsilon}(\mathbf{r}')$ change locally at its surface or volume, respectively. Thus, the field values are needed only where the local constitutive parameters depend on p_n. Often, the object is much smaller than the overall computational volume V, which leads to nonzero values of the permittivity and permeability derivatives at very few locations, and the computations via Eq. (3.9) are practically instantaneous. Even if the integration volume V comprises the whole region of the simulation, the computation is still performed within a fraction of a second.

Most importantly, due to the fact that Eq. (3.9) is an analytical result, the sensitivities it yields are very accurate. Their accuracy is limited only by the accuracy of the field (**E** and **H**) solution.

Eq. (3.9) has one important limitation. It does not take into account changes at the ports, i.e., the volume V must *not* include the port cross-sections. In practical applications, this is usually not a problem since the ports are at the antenna terminals, which are well removed from the imaged object.

3.2.2 S-Parameter Sensitivity to Constitutive Parameters

Let us consider the simpler case of isotropic materials where the constitutive tensors are diagonal 3×3 matrices: $\bar{\mu} = \mu\bar{\mathbf{I}}$ and $\bar{\varepsilon} = \varepsilon\bar{\mathbf{I}}$. Eq. (3.9) then reduces to

$$\frac{dS_{ij}}{dp_n} = \frac{\mathrm{i}\omega}{2a_i a_j} \iiint\limits_V \left[\frac{d\mu(\mathbf{r}')}{dp_n}\mathbf{H}_j(\mathbf{r}') \cdot \mathbf{H}_i(\mathbf{r}') - \frac{d\varepsilon(\mathbf{r}')}{dp_n}\mathbf{E}_j(\mathbf{r}') \cdot \mathbf{E}_i(\mathbf{r}')\right] dv'. \quad (3.10)$$

Let us further consider an object of volume V_O of any shape, wherein μ and ε are constant: $\mu = \mu_O$ and $\varepsilon = \varepsilon_O$. Let $p_n \equiv \varepsilon_O$, i.e., ε_O is an optimizable variable. The S-parameter sensitivity to changes in ε_O is then quantified by the derivative

$$\frac{dS_{ij}}{d\varepsilon_O} = \frac{-\mathrm{i}\omega}{2a_i a_j} \iiint\limits_{V_O} \mathbf{E}_j(\mathbf{r}') \cdot \mathbf{E}_i(\mathbf{r}') dv'. \quad (3.11)$$

Notice that the derivative does not depend explicitly on the particular value of ε_O. It is entirely determined by the i-th and j-th **E**-field distributions inside the object. These do depend on ε_O.

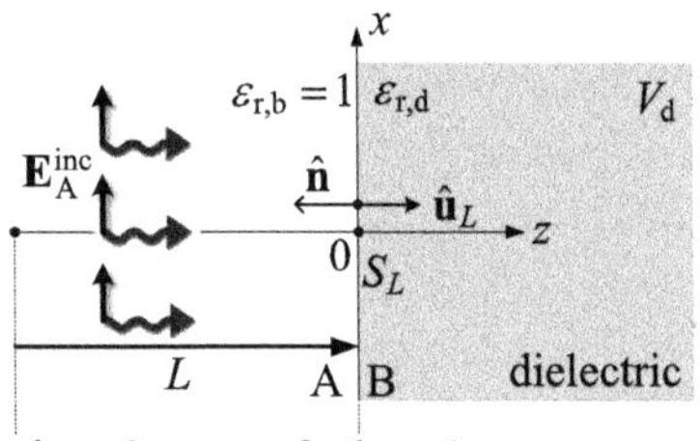

Figure 3.5 Illustration of the problem setup in finding the sensitivities of the reflection coefficient in the dielectric-interface examples. A plane wave is normally incident from region A (vacuum) upon a dielectric interface with region B. Region B has relative permittivity $\varepsilon_{\rm r,d}$. We seek the sensitivity of the back-scattered reflection S-parameter S_{11} with respect to $\varepsilon_{\rm r,d}$ (in Example 1) and with respect to the distance L between port 1 (where the plane wave is launched) and the interface (Example 3).

The S-parameter sensitivity with respect to $\mu_{\rm O}$ is analogous and entirely determined by the **H**-field distribution inside the object:

$$\frac{dS_{ij}}{d\mu_{\rm O}} = \frac{{\rm i}\omega}{2a_i a_j} \iiint_{V_{\rm O}} \mathbf{H}_j(\mathbf{r}') \cdot \mathbf{H}_i(\mathbf{r}') dv'. \tag{3.12}$$

Let us first consider a simple example that illustrates the implementation of Eq. (3.11) and validates it against a known analytical solution.

Example 1: Fig. 3.5 shows a plane wave in region A normally incident upon a dielectric interface with region B. This arrangement can be viewed as a 2-port network, where ports 1 and 2 are located in regions A and B, respectively. The RMS **E**-field phasor of the incident wave is $\mathbf{E}_{\rm A}^{\rm inc}(z) = \hat{\mathbf{x}} \exp(-{\rm i}k_0 z)$, which sets the phase reference of port 1 at $z = 0$; see reference plane 1 in Fig. 3.5. The background is vacuum, $\varepsilon_{\rm b} = \epsilon_0$, $\mu_{\rm b} = \mu_0$. The relative permittivity of region B is $\varepsilon_{\rm r,d}$. Find the analytical expression for the derivative of S_{11} with respect to the relative permittivity $\varepsilon_{\rm r,d}$.

Solution: The sensitivity formula in this case is

$$\frac{dS_{11}}{d\varepsilon_{\rm r,d}} = \frac{-{\rm i}\omega\epsilon_0}{2a_1^2} \iiint_{V_{\rm d}} E_{\rm B1}^2(\mathbf{r}') dv', \tag{3.13}$$

where $E_{\rm B1}(\mathbf{r}')$ is the total field in the volume $V_{\rm d}$ of the dielectric (see region B in Fig. 3.5) due to the excitation at port 1. The power carried by the incident wave through a cross-section S is

$$a_1^2 = \frac{|\mathbf{E}_{\rm A}^{\rm inc}|^2}{\eta_0} \cdot S = \frac{S}{\eta_0} \text{ W}, \tag{3.14}$$

where $\eta_0 = \sqrt{\frac{\mu_0}{\epsilon_0}}$ is the intrinsic impedance of the background (vacuum). Also, because the field is constant in the lateral directions, Eq. (3.13) can be expressed as

$$\frac{dS_{11}}{d\varepsilon_{\rm r,d}} = \frac{-{\rm i}\omega\epsilon_0\eta_0}{2S} \cdot S \cdot \int_0^\infty E_{\rm B1}^2(z')dz' = \frac{-{\rm i}k_0}{2} \int_0^\infty E_{\rm B1}^2(z')dz', \tag{3.15}$$

where $k_0 = \omega\sqrt{\mu_0\epsilon_0}$ is the vacuum wavenumber.

We need the expression for the **E**-field in region B, $E_{B1}(z')$. This is the transmitted field, which is expressed in terms of the incident field as $E_{B1}(z') = t \cdot E_A^{inc}(z' = 0) \cdot \exp(-ik_d z')$, where $k_d = k_0\sqrt{\varepsilon_{r,d}}$ is the wavenumber in the dielectric and t is the interface transmission coefficient, $t = \frac{2\eta_d}{\eta_d + \eta_0}$. Here, $\eta_d = \eta_0/\sqrt{\varepsilon_{r,d}}$ is the intrinsic impedance of the dielectric medium. Thus,

$$E_{B1}(z') = \frac{2\eta_d}{\eta_d + \eta_0} \cdot \exp(-ik_d z'). \tag{3.16}$$

Substituting Eq. (3.16) into Eq. (3.15) yields

$$\frac{dS_{11}}{d\varepsilon_{r,d}} = \frac{-ik_0}{2}\left(\frac{2\eta_d}{\eta_d + \eta_0}\right)^2 \int_0^\infty e^{-i2k_d z'} dz'. \tag{3.17}$$

The solution to the improper integral $\int_0^\infty e^{-i2k_d z'} dz'$ can be obtained as the limit of its solution when the condition $\Re(i2k_d) > 0$ holds. In physical terms, this condition means that the dielectric medium is lossy ($\Re\gamma_d = \Re(ik_d) = \alpha_d > 0$). However minute the dielectric loss is, the integral solves as

$$\int_0^\infty e^{-i2k_d z'} dz' = \frac{1}{i2k_d}. \tag{3.18}$$

The integral in the loss-free case is simply the limit of Eq. (3.18) when $\Im k_d \to 0$. Substituting Eq. (3.18) into Eq. (3.17) and making use of $\eta_d = \eta_0/\sqrt{\varepsilon_{r,d}}$ lead to the final result:

$$\frac{dS_{11}}{d\varepsilon_{r,d}} = -\frac{1}{\sqrt{\varepsilon_{r,d}}\left(1 + \sqrt{\varepsilon_{r,d}}\right)^2}. \tag{3.19}$$

This result can be validated against an analytical formula that can be easily obtained from the well-known expression for the interface reflection coefficient [12]:

$$r = \frac{\eta_d - \eta_0}{\eta_d + \eta_0} = \frac{1 - \sqrt{\varepsilon_{r,d}}}{1 + \sqrt{\varepsilon_{r,d}}}. \tag{3.20}$$

In the context of this example, the physical meaning of S_{11} coincides with that of r. Taking the derivative of Eq. (3.20) with respect to $\varepsilon_{r,d}$ leads to a result identical to that in Eq. (3.19).

Example 2: In a monostatic arrangement (S_{11} measurement), a large dielectric slab of thickness D is illuminated by a normally incident plane wave $\mathbf{E}_A^{inc}(z) = \hat{\mathbf{x}}\exp(-ik_0 z)$ (see Fig. 3.6). The background is vacuum. The phase reference $z = 0$ is set at the left end of the slab. Find the derivative $\frac{dS_{11}}{d\varepsilon_{r,d}}$ with respect to the relative permittivity $\varepsilon_{r,d}$ of the slab for a sweep of D from 0 to λ_d, where λ_d is the wavelength in the dielectric. Consider three values of $\varepsilon_{r,d}$: (a) 2, (b) 5, and (c) 10.

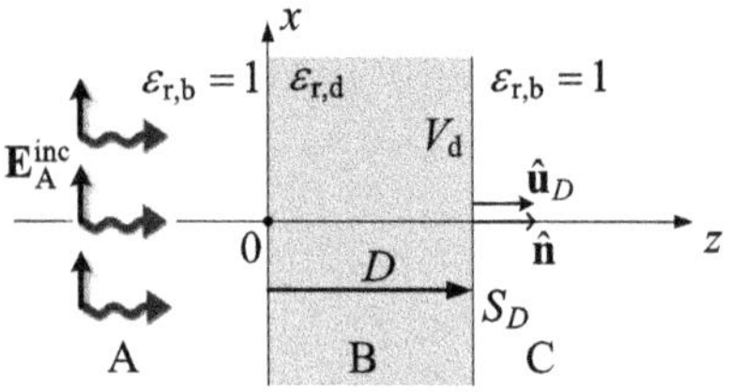

Figure 3.6 Illustration of the problem setup in finding the sensitivities of the reflection coefficient in the dielectric-slab examples. A plane wave illuminates a dielectric slab of thickness D and relative permittivity $\varepsilon_{r,d}$ (see region B). The background (regions A and C) is vacuum. We seek the sensitivity of the back-scattered reflection S-parameter S_{11} with respect to $\varepsilon_{r,d}$ (in Example 2) and with respect to the electrical thickness $\bar{D} = \frac{D}{\lambda_d}$ (in Example 4) for various thicknesses and relative permittivities.

Solution: The derivation of the sensitivity formula here follows the same initial steps as in the previous example that lead to

$$\frac{dS_{11}}{d\varepsilon_{r,d}} = \frac{-ik_0}{2} \int_0^D E_{B1}^2(z')dz'. \tag{3.21}$$

The expression for $E_{B1}(z')$ is found using the field boundary conditions, i.e., the continuity of the **E** and **H** tangential components at the interfaces $z = 0$ and $z = D$ (see, e.g., [13, problem 1.5]) as

$$E_{B1}(z') = Ae^{-ik_d z'} + Be^{ik_d z'}. \tag{3.22}$$

The constants A and B are found as

$$A = \frac{t}{1 - r^2 e^{-i2k_d D}}, \tag{3.23}$$

$$B = \frac{t \cdot r}{r^2 - e^{i2k_d D}}, \tag{3.24}$$

where t and r are the transmission and reflection coefficients, respectively, from region A into region B,

$$t = \frac{2\eta_d}{\eta_d + \eta_0}, \tag{3.25}$$

$$r = \frac{\eta_d - \eta_0}{\eta_d + \eta_0}. \tag{3.26}$$

Substituting Eq. (3.22) in Eq. (3.21) leads to the sensitivity expression

$$\frac{dS_{11}}{d\varepsilon_{r,d}} = -\frac{ik_0 D}{2}\left[2AB + \frac{\sin(k_d D)}{k_d D}\left(A^2 e^{-ik_d D} + B^2 e^{ik_d D}\right)\right]. \tag{3.27}$$

We reexpress Eq. (3.27) in terms of the slab's electrical thickness $\bar{D} = \frac{D}{\lambda_d}$ to obtain

$$\frac{dS_{11}}{d\varepsilon_{r,d}} = -\frac{i\pi \bar{D}}{\sqrt{\varepsilon_{r,d}}}\left[2AB + \frac{\sin(2\pi \bar{D})}{2\pi \bar{D}}\left(A^2 e^{-i2\pi \bar{D}} + B^2 e^{i2\pi \bar{D}}\right)\right]. \tag{3.28}$$

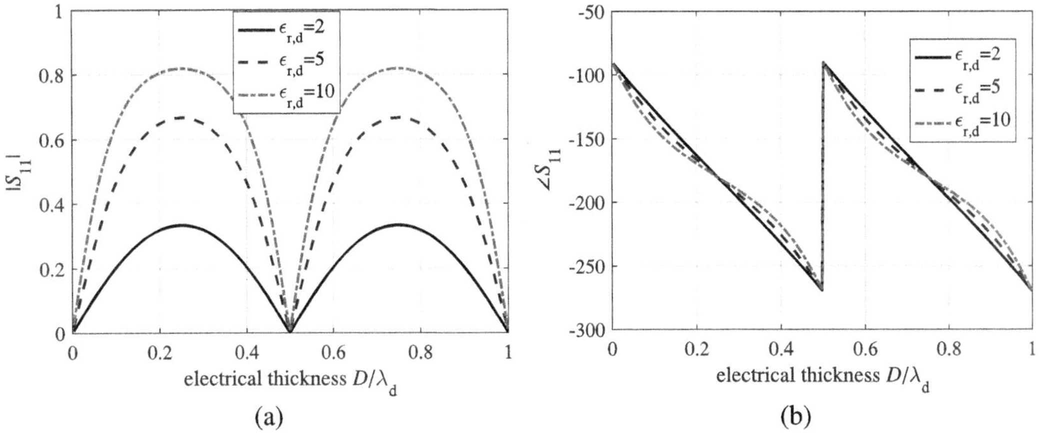

Figure 3.7 Plots of S_{11} for the dielectric-slab example versus a sweep of values for the slab's electrical thickness $\bar{D} = \frac{D}{\lambda_d}$ and three values of the slab's relative permittivity ($\varepsilon_{r,d} = 2$, 5, and 10): (a) magnitude, (b) unwrapped phase in degrees.

Fig. 3.7 shows the magnitude and phase of S_{11} plotted versus a sweep of $\bar{D}$ from 0 to 1, which corresponds to the required sweep of D from 0 to λ_d. Here, S_{11} has been computed from the coefficients A and B in Eqs. (3.23) and (3.24) as $S_{11} = A + B - 1$. As required, three values of the slab's relative permittivity are considered: $\varepsilon_{r,d} = 2$, 5, and 10. As expected, we see zero reflection when the slab has a thickness of exactly half-wavelength.

Fig. 3.8 shows the real and imaginary parts of $\frac{dS_{11}}{d\varepsilon_{r,d}}$ computed with Eq. (3.28) for the same normalized thickness values and the same permittivities of the slab. We observe that the derivative depends strongly on the slab's permittivity and length. This dependence is implicit through the field distribution; see Eq. (3.11).

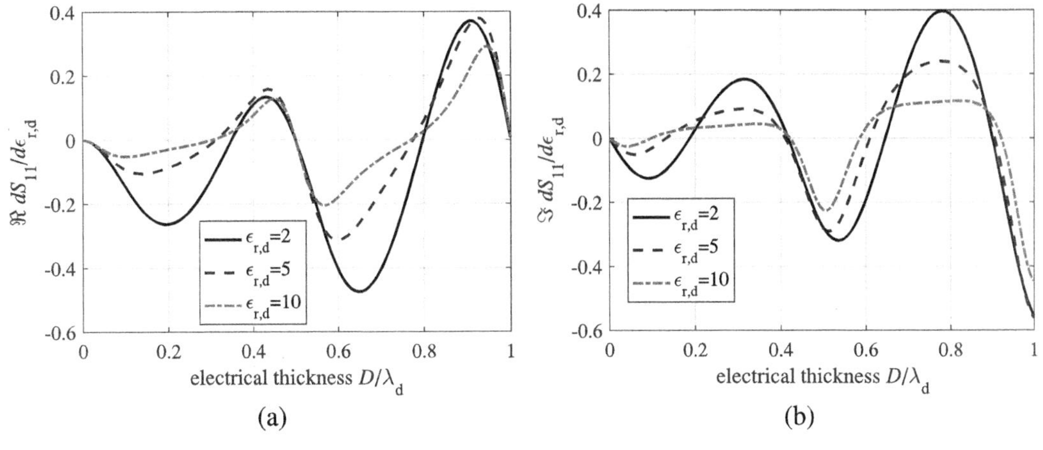

Figure 3.8 S_{11} sensitivity in the dielectric-slab example, $\frac{dS_{11}}{d\varepsilon_{r,d}}$, obtained with Eq. (3.28): (a) real part, (b) imaginary part. The derivative is plotted versus a sweep of $\bar{D}$ and for $\varepsilon_{r,d} = 2$, 5, and 10.

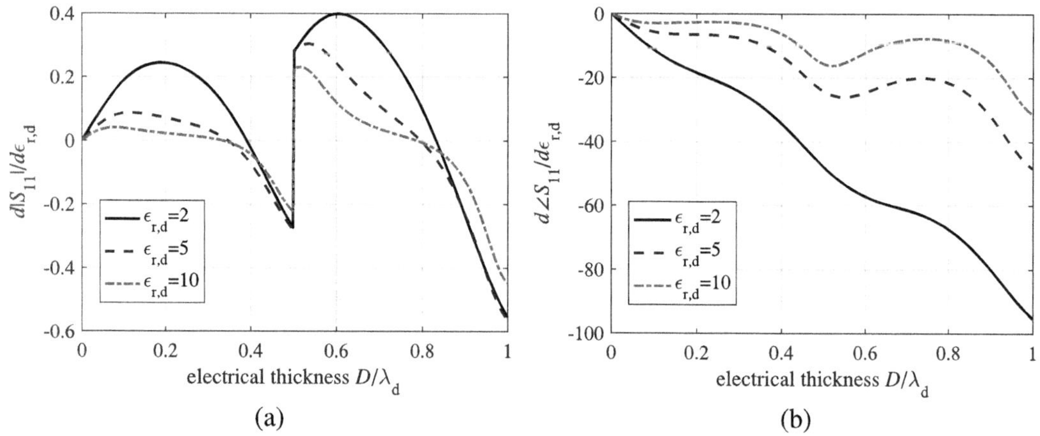

Figure 3.9 S_{11} sensitivity in the dielectric-slab example, $\frac{dS_{11}}{d\varepsilon_{r,d}}$, obtained with Eq. (3.28) and Eqs. (3.29) and (3.30): (a) magnitude sensitivity, $\frac{d|S_{11}|}{d\varepsilon_{r,d}}$, (b) phase sensitivity, $\frac{d\angle S_{11}}{d\varepsilon_{r,d}}$, in degrees.

Often, we are interested in the derivatives of the magnitude and phase of the S-parameter. These can be obtained from its complex-valued derivative. Let $F = F_R + \mathrm{i}F_I = |F| \exp\phi$ be a complex response, the derivative of which is $\frac{dF}{dp} = \frac{dF_R}{dp} + \mathrm{i}\frac{dF_I}{dp}$. The derivative of $|F|$ is found as

$$\frac{d|F|}{dp} = |F|^{-1} \cdot \left(F_R \frac{dF_R}{dp} + F_I \frac{dF_I}{dp} \right), \tag{3.29}$$

whereas the derivative of the phase is found as

$$\frac{d\phi}{dp} = |F|^{-2} \cdot \left(F_R \frac{dF_I}{dp} - F_I \frac{dF_R}{dp} \right). \tag{3.30}$$

Do not make the mistake of assuming that $\frac{d|F|}{dp}$ is the same as $\left|\frac{dF}{dp}\right|$! The result produced by $\left|\frac{dF}{dp}\right|$ is different from Eq. (3.29). Fig. 3.9 shows the derivatives of the magnitude and the phase of S_{11} computed using Eq. (3.29) and Eq. (3.30).

As discussed earlier, the Fréchet derivative in imaging is taken with respect to the voxel's constitutive parameters. In MW imaging, usually the parameter of interest is the complex relative permittivity $\varepsilon_r(\mathbf{r}')$. Within a voxel, the permittivity is constant by definition. Since a voxel is electrically very small—it is a discretization mesh element in numerical simulations—it can also be assumed that the field is constant within its volume. Thus, the S-parameter sensitivity formula for the relative permittivity $\varepsilon_{r,n}$ of the n-th voxel is derived as a special case of Eq. (3.11) as

$$\frac{dS_{ij}}{d\varepsilon_{r,n}} = \frac{-\mathrm{i}\omega\epsilon_0}{2a_i a_j} \mathbf{E}_j(\mathbf{r}_n) \cdot \mathbf{E}_i(\mathbf{r}_n) \Delta\Omega_n, \tag{3.31}$$

where $\mathbf{r}_n$ is the position of the n-th voxel and $\Delta\Omega_n$ is its volume. It is clear from Eq. (3.31) that obtaining the S-parameter Jacobian and the respective Fréchet derivative from the field solution is simple and fast even if the number of voxels is on the order of tens of thousands.

Occasionally, especially in the inverse scattering from lossy dielectric objects, the real and the imaginary parts of the voxel's complex permittivity may be treated as separate optimizable variables. For example, it may be known beforehand that the scattering object has only conductivity contrast with the background. Bearing in mind that $\varepsilon_{\mathrm{r},n} = \varepsilon'_{\mathrm{r},n} - \mathrm{i}\varepsilon''_{\mathrm{r},n}$, separate sensitivity expressions can be obtained from Eq. (3.31) for $\varepsilon'_{\mathrm{r},n}$ and $\varepsilon''_{\mathrm{r},n}$:

$$\frac{dS_{ij}}{d\varepsilon'_{\mathrm{r},n}} = \frac{dS_{ij}}{d\varepsilon_{\mathrm{r},n}} = \frac{-\mathrm{i}\omega\epsilon_0}{2a_ia_j}\mathbf{E}_j(\mathbf{r}_n)\cdot\mathbf{E}_i(\mathbf{r}_n)\Delta\Omega_n, \tag{3.32}$$

and

$$\frac{dS_{ij}}{d\varepsilon''_{\mathrm{r},n}} = -\mathrm{i}\frac{dS_{ij}}{d\varepsilon_{\mathrm{r},n}} = \frac{-\omega\epsilon_0}{2a_ia_j}\mathbf{E}_j(\mathbf{r}_n)\cdot\mathbf{E}_i(\mathbf{r}_n)\Delta\Omega_n. \tag{3.33}$$

Eq. (3.33) can be equivalently stated in terms of the voxel conductivity σ_n as (see Eq. (1.24))

$$\frac{dS_{ij}}{d\sigma_n} = \frac{1}{\omega\epsilon_0}\frac{dS_{ij}}{d\varepsilon''_{\mathrm{r},n}} = \frac{-1}{2a_ia_j}\mathbf{E}_j(\mathbf{r}_n)\cdot\mathbf{E}_i(\mathbf{r}_n)\Delta\Omega_n. \tag{3.34}$$

3.2.3 S-Parameter Sensitivity to Shape Parameters of Dielectric Objects

A special class of inversion problems is concerned with the scattering from a homogeneous penetrable (nonmetallic) object. The object may have dielectric and/or magnetic contrast with the background. Let us consider the case of dielectric contrast. Objects with magnetic contrast are treated in an analogous manner. The task in such problems is to find the shape and/or the permittivity of the scattering object. The sensitivity formula for the permittivity has already been considered in the previous section. Here, we focus on the shape sensitivity.

The S-parameter sensitivity formula for shape parameters of isotropic dielectric objects has been derived in [114] starting from the general formula Eq. (3.9). Here, we briefly outline the derivation and discuss its implementation.

With reference to Fig. 3.10, consider a background medium of constitutive parameters μ_{b} and ε_{b}, where a dielectric object resides. The object's permeability is the same as that of the background, but its permittivity ε_{d} is different, $\varepsilon_{\mathrm{d}} \neq \varepsilon_{\mathrm{b}}$. We consider a shape parameter p such that its changes would deform the object's surface or a portion of it. Fig. 3.10 shows a smooth portion S of the object's surface, which is affected by a change in p. The outward unit normal of S is $\hat{\mathbf{n}}$.

Let the position of an interface point be denoted as $\mathbf{r}_{\mathrm{s}}$ and the position of an observation point in its vicinity be $\mathbf{r}$. The permittivity changes abruptly at the material interface

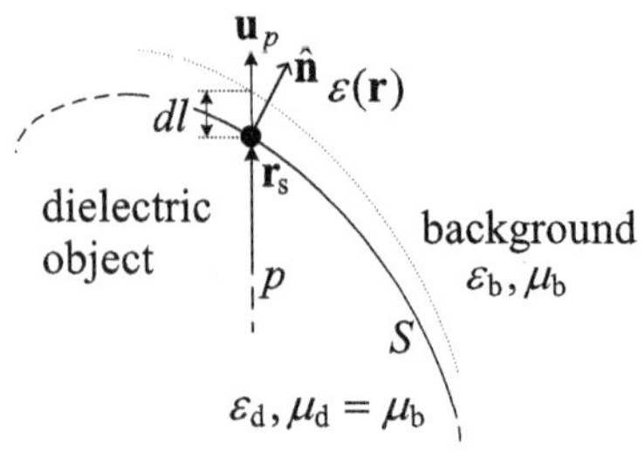

Figure 3.10 Illustration of a portion of the surface S of a dielectric object immersed in a background medium of permeability μ_b and permittivity ε_b. The object has the same permeability as the background but different permittivity, $\varepsilon_d \neq \varepsilon_b$. The surface is smooth and can be locally approximated as a flat interface. Its outward unit normal is $\hat{\mathbf{n}}$. A change in the shape parameter p results in a deformation or a displacement of the surface (see dotted line). The direction of the surface displacement at a point $\mathbf{r}_s \in S$ that corresponds to an increase in p is indicated by the unit vector $\hat{\mathbf{u}}_p$.

S according to

$$\varepsilon(\mathbf{r}) = \varepsilon_d + (\varepsilon_b - \varepsilon_d) H\left[(\mathbf{r} - \mathbf{r}_s) \cdot \hat{\mathbf{n}}\right], \tag{3.35}$$

where $H(\cdot)$ is Heaviside's step function. When p changes, in general, the position $\mathbf{r}_s$ of the interface point may change along with the orientation $\hat{\mathbf{n}}$ of its surface element.

We next recall the general S-parameter sensitivity formula Eq. (3.10), which dictates that when the dielectric properties are affected by a parameter p, the respective sensitivity is

$$\frac{dS_{ij}}{dp} = \frac{-i\omega}{2a_i a_j} \iiint_V \left[\frac{d\varepsilon(\mathbf{r}')}{dp} \mathbf{E}_j(\mathbf{r}') \cdot \mathbf{E}_i(\mathbf{r}')\right] dv'. \tag{3.36}$$

Therefore, we need the permittivity derivative with respect to the shape parameter, $\frac{d\varepsilon(\mathbf{r}')}{dp}$. As per Eq. (3.35), this is obtained as

$$\frac{d\varepsilon(\mathbf{r}')}{dp} = (\varepsilon_b - \varepsilon_d)\, \delta(\zeta) \frac{d\zeta}{dp}, \tag{3.37}$$

where $\delta(\cdot)$ is the 1D Dirac delta function (measured in m^{-1}) and its argument $\zeta(p) = \left[\mathbf{r}' - \mathbf{r}_s(p)\right] \cdot \hat{\mathbf{n}}(p)$ is an implicit function of p through $\mathbf{r}_s$ and $\hat{\mathbf{n}}$. Its derivative is found as

$$\frac{d\zeta(\mathbf{r}')}{dp} = -\frac{d\mathbf{r}_s}{dp} \cdot \hat{\mathbf{n}} + (\mathbf{r}' - \mathbf{r}_s) \cdot \frac{d\hat{\mathbf{n}}}{dp}. \tag{3.38}$$

Assuming that the derivative $\frac{d\hat{\mathbf{n}}}{dp}$ is finite—this is true if the surface is smooth—one obtains for points belonging to the object's surface that

$$\frac{d\zeta(\mathbf{r}' \equiv \mathbf{r}_s)}{dp} = -\frac{d\mathbf{r}_s}{dp} \cdot \hat{\mathbf{n}} = -\frac{d\nu(\mathbf{r}')}{dp}, \tag{3.39}$$

where $\nu(\mathbf{r}') = \mathbf{r}_s \cdot \hat{\mathbf{n}}$.

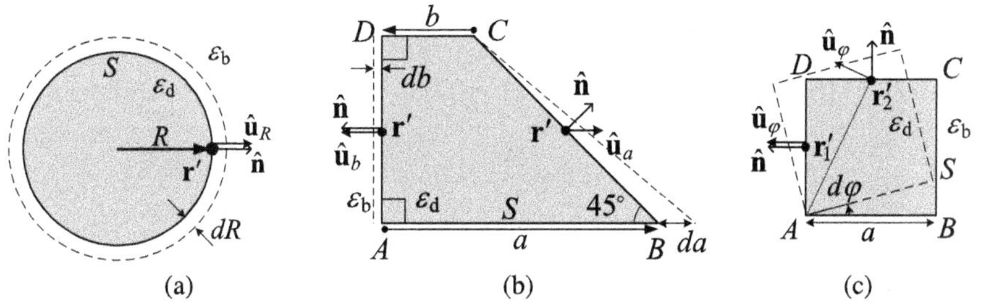

Figure 3.11 Examples illustrating the geometrical meaning of the surface derivative $\frac{dv}{dp}$ in Eq. (3.39) and Eq. (3.40): (a) cross-section in the shape of a circle with the radius being the shape parameter, $p \equiv R$, where $\frac{dv}{dR} = 1$ for $\forall \mathbf{r}' \in S$; (b) cross-section in the shape of a rectangular trapezoid with two shape parameters, a and b, where $\frac{dv}{db} = 1$ for $\forall \mathbf{r}' \in \overline{AD}$, while $\frac{dv}{da}$ changes from $\frac{1}{\sqrt{2}}$ to 0 depending on the position of $\mathbf{r}' \in \overline{BC}$; and (c) cross-section in the shape of a square of side a with the shape parameter being the rotation angle φ.

Substituting Eq. (3.37) into Eq. (3.36) and taking into account the electric field boundary conditions at the interface lead to [114]

$$\frac{dS_{ij}}{dp_n} = \frac{-i\omega}{2a_i a_j} \iint_S \frac{dv(\mathbf{r}')}{dp} \cdot \Big[(\varepsilon_d - \varepsilon_b)\,\mathbf{E}_{i,\tan}(\mathbf{r}') \cdot \mathbf{E}_{j,\tan}(\mathbf{r}') - \left(\frac{1}{\varepsilon_d} - \frac{1}{\varepsilon_b}\right) D_{i,n}(\mathbf{r}') D_{j,n}(\mathbf{r}')\Big]\, ds', \tag{3.40}$$

where $\mathbf{E}_{\xi,\tan}$ ($\xi = i, j$) is the tangential component of the $\mathbf{E}$ field at the surface due to the excitation at the ξ-th port, and $D_{\xi,n} = \hat{\mathbf{n}} \cdot \mathbf{D}_\xi$ is the normal component of the electric flux density at the surface due to the excitation at the ξ-th port.

To grasp the geometrical meaning of $\frac{dv}{dp}$ in Eq. (3.39), one could reexpress it as

$$\frac{dv}{dp} = \frac{dl}{dp} \hat{\mathbf{u}}_p \cdot \hat{\mathbf{n}}, \tag{3.41}$$

where dl is the differential linear displacement of the surface element corresponding to the perturbation dp and $\hat{\mathbf{u}}_p$ is its direction; see Fig 3.10. Let us consider some simple examples.

Fig. 3.11-a shows the cross-section of a cylinder, the radius R of which is a shape parameter, $p \equiv R$. An increase in R affects the surface S over which the integration of Eq. (3.40) is taken. The changed surface S resulting from an increment in R is indicated by a dashed line. The direction of the surface displacement at any point $\mathbf{r}' \in S$ is in the radial direction and is shown by the unit vector $\hat{\mathbf{u}}_R$. This is also the direction of the surface unit normal, $\hat{\mathbf{n}} = \hat{\mathbf{u}}_R$. The value of this displacement dl is equal to the increment in R, i.e., $dl = dR$; therefore $\frac{dl}{dR} = 1$. Thus, $\frac{dv}{dR} = 1$.

A similar scenario arises with the shape parameter $p \equiv b$ in Fig. 3.11-b. There, b is the top base of a rectangular trapezoid, and its increase pushes the left leg ($\overline{AD}$) to the left as indicated by $\hat{\mathbf{u}}_b$ and the dashed line showing its displacement. Again, for $\mathbf{r}' \in \overline{AD}$, $dl = db$ and $\hat{\mathbf{n}} = \hat{\mathbf{u}}_b$, so that $\frac{dv}{db} = 1$.

The situation is a bit more complicated with the shape parameter $p \equiv a$. An increase in a pushes the vertex B to the right, but the vertex C is unaffected. If we consider the surface point $\mathbf{r}'$ at the midpoint of $\overline{BC}$, the linear displacement is $dl(\mathbf{r}') = \frac{1}{2}da$. Therefore, $\frac{dl(\mathbf{r}')}{da} = \frac{1}{2}$. Moreover, $\hat{\mathbf{n}}$ and $\hat{\mathbf{u}}_a$ are at a 45° angle, and $\hat{\mathbf{n}} \cdot \hat{\mathbf{u}}_a = \frac{1}{\sqrt{2}}$. Thus, $\frac{dv(\mathbf{r}')}{da} = \frac{1}{2\sqrt{2}}$. Note that as $\mathbf{r}'$ varies from B to C, $\frac{dl(\mathbf{r}')}{da}$ changes from 1 to 0, and $\frac{dv(\mathbf{r}')}{da}$ changes from $\frac{1}{\sqrt{2}}$ to 0 accordingly.

Fig. 3.11-c illustrates the case of an angular (or rotational) shape parameter. The angle φ determines the rotation of the square about the vertex A. In the case of the surface point $\mathbf{r}'_1$, which is the midpoint of the side $\overline{AD}$, $dl = 0.5ad\varphi$. Also, $\hat{\mathbf{u}}_\varphi = \hat{\mathbf{n}}$. Therefore, $\frac{dv(\mathbf{r}'_1)}{d\varphi} = 0.5a$. The reader is encouraged to show that at the midpoint of $\overline{CD}$, $\mathbf{r}'_2$, $\frac{dv(\mathbf{r}'_2)}{d\varphi}$ is also equal to $0.5a$.

To validate Eq. (3.40) and to illustrate it implementation, let us consider again the dielectric-interface example shown in Fig. 3.5.

Example 3: A dielectric interface at $z = 0$ is illuminated by the plane wave $\mathbf{E}_A^{\text{inc}}(z) = \hat{\mathbf{x}} \exp[-\mathrm{i}k_0(z + L)]$. Now, the phase reference for port 1 is set at $z = -L$; see reference plane 2 in Fig. 3.5. Find the analytical expression for the derivative of S_{11} with respect to the distance L from port 1 to the interface. The surface unit normal $\hat{\mathbf{n}}$ is in the opposite direction of $\hat{\mathbf{u}}_L$ as shown in Fig. 3.5.

Solution: Here, $\hat{\mathbf{n}} = -\hat{\mathbf{u}}_L$; therefore $\frac{dv}{dD} = -1$. Moreover, the $\mathbf{E}$-field is tangential to the interface; therefore $D_{1n} = 0$. Thus, Eq. (3.40) simplifies to

$$\frac{dS_{11}}{dL} = \frac{\mathrm{i}\omega\epsilon_0}{2a_1^2} \iint_{S_L} (\varepsilon_{\mathrm{r,d}} - 1) E_{\mathrm{B1}}^2(\mathbf{r}') ds'. \tag{3.42}$$

The power carried by the plane wave is $a_1^2 = \frac{S}{\eta_0}$, where S is a cross-section in region A. The integrand in Eq. (3.42) is constant across S_L, the area of which equals S. Thus,

$$\frac{dS_{11}}{dL} = \frac{\mathrm{i}k_0}{2} (\varepsilon_{\mathrm{r,d}} - 1) E_{\mathrm{B1}}^2(z' = 0), \tag{3.43}$$

where $E_{\mathrm{B1}}(z' = 0)$ is the total field at the interface. This is exactly the field transmitted into region B for $z' = 0$, i.e.,

$$E_{\mathrm{B1}}(z' = 0) = t \cdot E_{\mathrm{A}}^{\mathrm{inc}}(z' = 0) = t \cdot e^{-\mathrm{i}k_0 L} = \left(\frac{2}{1 + \sqrt{\varepsilon_{\mathrm{r,d}}}}\right) \cdot e^{-\mathrm{i}k_0 L}, \tag{3.44}$$

where t is the interface transmission coefficient as defined in Eq. (3.25). Substituting Eq. (3.44) into Eq. (3.43) yields the desired analytical expression for the sensitivity:

$$\frac{dS_{11}}{dL} = -\mathrm{i}2k_0 \underbrace{\left(\frac{1 - \sqrt{\varepsilon_{\mathrm{r,d}}}}{1 + \sqrt{\varepsilon_{\mathrm{r,d}}}}\right)}_{r} e^{-\mathrm{i}2k_0 L}. \tag{3.45}$$

The term in the brackets is the interface reflection coefficient r as per Eq. (3.26).

This result can be validated against an analytical formula that can be obtained from the well-known expression for the transformation of the S-parameters upon changing the length of a port [13]. In the case of S_{11}, the expression is

$$S_{11}(z_1 = L) = \underbrace{S_{11}(z_1 = 0)}_{r}\, e^{-\mathrm{i}2k_0 L}, \tag{3.46}$$

where z_1 is the position of port 1. In the current setup, $S_{11}(z_1 = 0) \equiv r$. Differentiating Eq. (3.46) with respect to L yields the same result as in Eq. (3.45).

Next, a slightly more complicated scenario is presented where the thickness of a dielectric slab is the shape variable.

Example 4: Consider again the dielectric-slab arrangement shown in Fig. 3.6. S_{11} is measured in the presence of a large dielectric slab of relative permittivity $\varepsilon_{\mathrm{r,d}}$ and thickness D. The background is vacuum. The slab is illuminated by $\mathbf{E}_{\mathrm{A}}^{\mathrm{inc}}(z) = \hat{\mathbf{x}} \exp(-\mathrm{i}k_0 z)$. Find the derivative $\frac{dS_{11}}{d\bar{D}}$ with respect to the electrical thickness $\bar{D} = \frac{D}{\lambda_{\mathrm{d}}}$ of the slab for a sweep of $\bar{D}$ from 0 to 1. The interface at $z = 0$ is fixed, whereas the interface at $z = D$ defines the perturbation face S_D, that is affected by a change in D. The surface unit normal $\hat{\mathbf{n}}$ is aligned with the unit vector of the parameter perturbation $\hat{\mathbf{u}}_D$ as shown in Fig. 3.6. Consider three values of $\varepsilon_{\mathrm{r,d}}$: (a) 2, (b) 5, and (c) 10.

Solution: Following steps identical to those in the previous example, we arrive at

$$\frac{dS_{11}}{dD} = \frac{-\mathrm{i}k_0}{2}(\varepsilon_{\mathrm{r,d}} - 1)\, E_{\mathrm{B1}}^2(z' = D), \tag{3.47}$$

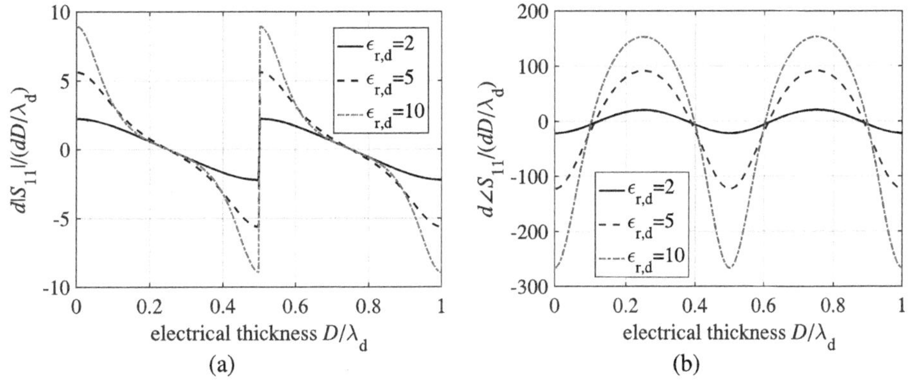

Figure 3.12 Example 4 results: S_{11} sensitivities with respect to $\bar{D}$ (the slab's electrical thickness) versus a sweep of $\bar{D}$ and for $\varepsilon_{\mathrm{r,d}} = 2$, 5, and 10: (a) magnitude sensitivity, $\frac{d|S_{11}|}{d\bar{D}}$; (b) phase sensitivity, $\frac{d\angle S_{11}}{d\bar{D}}$, in degrees.

where $E_{B1}(z'=D)$ can be calculated using Eq. (3.22). Since $\frac{\bar{D}}{D}=\frac{k_0\sqrt{\varepsilon_{r,d}}}{2\pi}$,

$$\frac{dS_{11}}{d\bar{D}}=-i\pi\left(\frac{\varepsilon_{r,d}-1}{\sqrt{\varepsilon_{r,d}}}\right)E_{B1}^2(z'=D). \tag{3.48}$$

Fig. 3.12 shows the sensitivity of the magnitude and the phase of $\frac{dS_{11}}{d\bar{D}}$ as computed with Eq. (3.48).

3.2.4 S-Parameter Sensitivity to Shape Parameters of Metallic Objects

The analytical sensitivity formula for the S-parameter sensitivity to shape parameters of metallic objects has been derived in [113]. Here, we only state the final result and discuss its applications.

Fig. 3.13 illustrates the problem. A shape parameter p is associated with the surface S of a metallic object (or a portion of it). An increase in p leads to a displacement of the surface element at $\mathbf{r}'$ in the direction $\hat{\mathbf{u}}_p$, which, in general, does not coincide with the surface unit normal $\hat{\mathbf{n}}$ (pointing out of the metallic region). The conductivity σ_m of the metallic body is very large and is assumed infinite, $\sigma_m \to \infty$. The conductivity of the background is negligibly small and is assumed zero, $\sigma_b = 0$. Thus, the background is an ideal dielectric of permittivity ε_b and it hosts a perfect-electric-conductor (PEC) object, the permittivity of which is $\varepsilon_m = \epsilon_0$. Both, the PEC object and the background are nonmagnetic, $\mu_b = \mu_m = \mu_0$.

The S-parameter shape sensitivity formula involves the integration of the field quantities on the object's surface [113]:

$$\frac{dS_{ij}}{dp_n}=\frac{-i\omega}{2a_ia_j}\iint_S \frac{dv(\mathbf{r}')}{dp}\left[\varepsilon_b(\mathbf{r}')E_{i,n}(\mathbf{r}')\cdot E_{j,n}(\mathbf{r}')+\mu_0\mathbf{H}_{i\,\tan}(\mathbf{r}')\cdot\mathbf{H}_{j\,\tan}(\mathbf{r}')\right]ds', \tag{3.49}$$

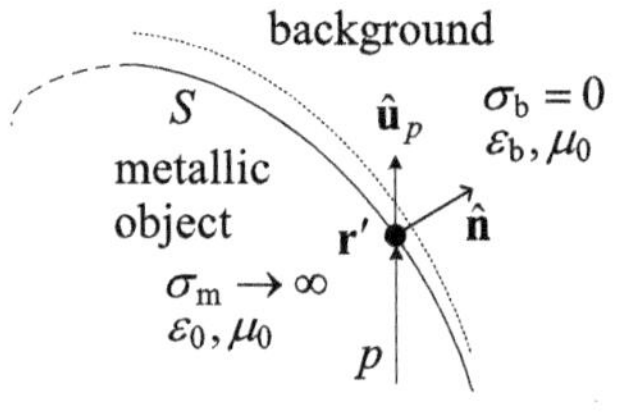

Figure 3.13 Illustration of a portion of the surface S of a nonmagnetic metallic object immersed in a nonmagnetic background medium of permittivity ε_b. The object is assumed to be PEC, i.e., its conductivity is infinite, $\sigma_m \to \infty$. The surface is assumed smooth. Its outward unit normal is $\hat{\mathbf{n}}$. A change in the shape parameter p results in a deformation or a displacement of the surface (see dashed line). The direction of the surface displacement at a point $\mathbf{r}' \in S$ that corresponds to an increase in p is indicated by the unit vector $\hat{\mathbf{u}}_p$.

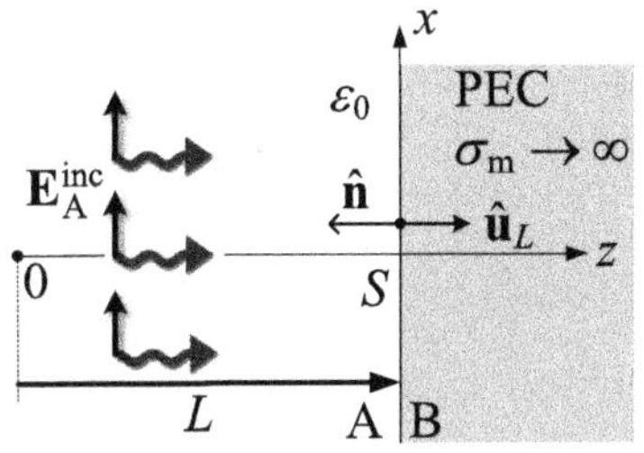

Figure 3.14 Illustration of the problem in the example with the metallic plate. A plane wave illuminates a large PEC interface (region B). The background (region A) is vacuum. We seek the sensitivity of S_{11} with respect to L, the distance between port 1 (where the incident wave is launched) and the PEC surface.

where $\frac{dv}{dp}$ is the same derivative as the one defined in Eq. (3.41). As in the case of a dielectric object, the subscripts i and j denote the port of excitation.

We notice that the formula involves the normal **E**-field components and the tangential **H**-field components. According to the EM field boundary conditions (see Table 2.1), these are directly related to the induced (electric) surface charge and current densities, respectively. Eq. (3.49) can then be expressed in terms of these densities as [113]

$$\frac{dS_{ij}}{dp_n} = \frac{-\mathrm{i}\omega}{2a_i a_j} \iint\limits_S \frac{dv(\mathbf{r}')}{dp} \left[\frac{\rho_{s,i}(\mathbf{r}') \cdot \rho_{s,j}(\mathbf{r}')}{\varepsilon_b(\mathbf{r}')} + \mu_0 \mathbf{J}_{s,i}(\mathbf{r}') \cdot \mathbf{J}_{s,j}(\mathbf{r}') \right] ds'. \tag{3.50}$$

Example 5: Consider a uniform plane wave $\mathbf{E}_A^{inc}(z) = \hat{\mathbf{x}} \exp(-\mathrm{i}k_0 z)$ normally incident upon a large PEC plate (see Fig. 3.14). The background (region A) is vacuum. The phase reference ($z = 0$) is the location of port 1 (the port of excitation). Find the derivative $\frac{dS_{11}}{dL}$ with respect to the separation length L between port 1 and the PEC plate. The surface unit normal $\hat{\mathbf{n}}$ is in the opposite direction than the unit vector of the parameter perturbation $\hat{\mathbf{u}}_L$ as shown in Fig. 3.14.

Solution: The sensitivity formula Eq. (3.50) is written for the S_{11} sensitivity with respect to L as

$$\frac{dS_{11}}{dL} = \frac{-\mathrm{i}\omega}{2a_1^2} \iint\limits_S \frac{dv}{dL} \left\{ \frac{\left[\rho_{s,1}(\mathbf{r}')\right]^2}{\epsilon_0} + \mu_0 \left[J_{s,1}(\mathbf{r}')\right]^2 \right\} ds'. \tag{3.51}$$

Here, we have also taken into account that region A is vacuum. The incident wave power is given by $a_1^2 = \frac{S}{\eta_0}$. At the same time, the integrand in Eq. (3.51) is constant in the cross-section S. Since $\hat{\mathbf{n}} = -\hat{\mathbf{u}}_L$, $\frac{dv}{dL} = -1$. Also, since the **E** field does not have a normal component at S, $\rho_{s,1} = 0$. Thus, Eq. (3.51) reduces to

$$\frac{dS_{11}}{dL} = \frac{\mathrm{i}\omega\eta_0\mu_0}{2} \left[J_{s,1}(z' = L)\right]^2. \tag{3.52}$$

The current density $\mathbf{J}_{s,1}$ induced on the PEC surface relates to the total **H**-field at the surface through $\hat{\mathbf{n}} \times \mathbf{H}_A = \mathbf{J}_{s,1}$ as per the EM boundary conditions. Since the reflection coefficient at the PEC surface is $\Gamma = -1$, $\mathbf{H}_A = 2\mathbf{H}_A^{\text{inc}}(z = L) = \hat{\mathbf{y}}2\exp(-ik_0 L)/\eta_0$. Substituting this result in Eq. (3.52) yields the sensitivity we seek:

$$\frac{dS_{11}}{dL} = i2k_0 e^{-i2k_0 L}. \tag{3.53}$$

We can easily check the validity of this result because we know the analytical form of the reflection coefficient in this arrangement: $S_{11} = -\exp(-i2k_0 L)$. Here, the minus sign is due to the **E**-field phase reversal upon reflection from the PEC wall, whereas the $(2k_0 L)$ phase delay is due to the wave traversing the distance L to and from the PEC wall. Differentiating this expression with respect to L yields the same result as the one in Eq. (3.53).

3.3 S-Parameter Data Equation

In Section 2.11, the concept of Green's vector function was introduced in order to construct integral data equations for scalar responses such as voltages, currents, and their respective reflection and transmission coefficients. It was pointed out that when the response is a vector function (e.g., the **E** field), Green's function is a dyadic; however, when the response is scalar, Green's dyadic contracts to a vector. The scattering parameters are the most common scalar responses used in MW imaging with frequency-domain measurements. Therefore, deriving their respective Green's vector function would be useful. It enables the construction of the S-parameter data equation, which is critically important in both direct (linear) and optimization-based inversion schemes.

Let us consider Green's vector function $\mathbf{E}_i^\dagger$ associated with the outgoing power wave b_i at the i-th (receiving) port of a MW imaging network formed by the inspected object and the antennas. As per Eq. (2.89), if Green's vector function is known, b_i can be found as the response to a source function $\mathbf{S}$ according to

$$b_i = \iiint_V \mathbf{E}_i^\dagger(\mathbf{r}') \cdot \mathbf{S}(\mathbf{r}') dv'. \tag{3.54}$$

In imaging, we are interested in the scattering component of the outgoing wave, $b_{i,\text{sc}}$, due to the induced sources of scattering $\mathbf{S}_{\text{ind}}$. In the scattering from dielectric objects, $\mathbf{S}_{\text{ind}} = \omega^2 \mu_0 \Delta\varepsilon \mathbf{E}$; see Eq. (2.84). Therefore,

$$b_{i,\text{sc}} = \omega^2 \mu_0 \iiint_V \Delta\varepsilon(\mathbf{r}') \mathbf{E}_i^\dagger(\mathbf{r}') \cdot \mathbf{E}(\mathbf{r}') dv', \tag{3.55}$$

where $\mathbf{E}(\mathbf{r}')$ is the total internal field.

In the scattering from PEC objects, the induced source is the surface current density on the object's surface $\mathbf{J}_{s,e}$, to which the induced surface source relates as

$\mathbf{S}_{\text{ind,s}} = -\mathrm{i}\omega\mu_0\mathbf{J}_{\text{ind,s}}$; see Eqs. (2.68) and (2.69). Therefore,

$$b_{i,\text{sc}} = -\mathrm{i}\omega\mu_0 \iint_S \mathbf{E}_i^{\dagger}(\mathbf{r}') \cdot \mathbf{J}_{\text{ind,s}}(\mathbf{r}')ds'. \tag{3.56}$$

Using the generic formula Eq. (2.95) for the auxiliary source generating Green's vector function as well as the known analytical relationship between the outgoing wave and the field solution, Green's vector function for b_i is obtained as [115]:

$$\mathbf{E}_i^{\dagger}(\mathbf{r}') = \frac{\mathbf{E}_i^{\text{inc}}(\mathbf{r}')}{\mathrm{i}\omega\mu_0 2a_i}, \tag{3.57}$$

where $\mathbf{E}_i^{\text{inc}}(\mathbf{r}')$ is the field due to the i-th port when it is excited by the power wave a_i.

To obtain the integral-equation solutions for the S_{ij} parameter ($j = 1, \ldots, N$), Eq. (3.57) is substituted into Eq. (3.55) and Eq. (3.56), and the result is divided by a_j as per the S-parameter definition Eqs. (3.7) and (3.8). The result in the case of dielectric scattering is

$$S_{ij}^{\text{sc}} = \frac{-\mathrm{i}\omega}{2a_i a_j} \iiint_V \Delta\varepsilon(\mathbf{r}')\mathbf{E}_i^{\text{inc}}(\mathbf{r}') \cdot \mathbf{E}_j(\mathbf{r}')dv', \tag{3.58}$$

where $\mathbf{E}_j$ is the total internal field when port j is excited. In the case of scattering from a PEC object,

$$S_{ij}^{\text{sc}} = \frac{-1}{2a_i a_j} \iint_S \mathbf{E}_i^{\text{inc}}(\mathbf{r}') \cdot \mathbf{J}_{\text{ind,s}}(\mathbf{r}')dv'. \tag{3.59}$$

3.3.1 S-Parameter Data Equation with Common System Impedance

It is instructive to compare the S-parameter data equation Eq. (3.58) with Eq. (2.113), which has been derived in Section 2.11.2 as the data equation for the voltage delivered from the i-th receiving antenna to its receiver under conditions of maximum power transfer.

As a reminder, maximum power transfer occurs when the impedances of the antenna and the receiver are the conjugates of each other. A conjugate match of complex impedances is relatively easy to achieve, especially in a narrow frequency band, if the two devices are connected directly. But, in MW systems, there is also a cable (with connectors on both ends) between the two. Thus, to ensure maximum power transfer, an impedance match is needed between the antenna and the cable on one end and between the cable and the receiver on the other. Moreover, under these conditions, no reflections occur at either end of the cable, and it carries a purely traveling wave. The cable is usually very low-loss, which ensures that its characteristic impedance Z_c is practically real-valued. This means that to achieve the best performance both the antenna and the receiver must be designed so that their internal impedances are real and equal to Z_c.

Modular system design and fabrication are greatly simplified if all MW devices and interconnects are designed for an impedance match to an impedance of the same value, often referred to as the *system impedance*. The MW industry has adopted a standard system impedance of $Z_0 = 50\ \Omega$. Thus, it is common that all ports in a MW network have the same characteristic impedance, $Z_{c,i} = Z_0$, $i = 1, \ldots, N$.

To return to our task of comparing Eq. (3.58) with Eq. (2.113), let us divide both sides of Eq. (2.113) by $\sqrt{Z_{c,i}}$. The result is

$$\frac{V_{i,\mathrm{sc}}^{\mathrm{Rx}}}{\sqrt{Z_{c,i}}} = \frac{-\mathrm{i}\omega}{2I_i\sqrt{Z_{c,i}}} \iiint_V \Delta\varepsilon(\mathbf{r}')\mathbf{E}_i^{\mathrm{inc}}(\mathbf{r}') \cdot \mathbf{E}(\mathbf{r}')dv'. \tag{3.60}$$

The left side of Eq. (3.60) is the square root of the power delivered to the receiver, which, when matched, has an internal impedance $Z_{c,i}$. Moreover, $V_{i,\mathrm{sc}}^{\mathrm{Rx}}$ is the pure traveling voltage wave in the cable incident upon the receiver and absorbed by it. This wave is also the outgoing voltage wave at the antenna terminals, minus a phase-delay factor due to the cable, which can be taken care of by choosing the port reference plane at the receiver terminals. Thus, as per the definition of the power waves Eq. (3.3), the left side of Eq. (3.60) is exactly $b_{i,\mathrm{sc}}$, the scattering portion of the root-power wave delivered by the i-th antenna to its receiver. At the same time, $I_i\sqrt{Z_{c,i}}$ is the root-power wave a_i exciting the (matched) antenna when the receiver is replaced by a (matched) transmitter. Thus, Eq. (3.60) can be restated as

$$b_{i,\mathrm{sc}} = \frac{-\mathrm{i}\omega}{2a_i} \iiint_V \Delta\varepsilon(\mathbf{r}')\mathbf{E}_i^{\mathrm{inc}}(\mathbf{r}') \cdot \mathbf{E}(\mathbf{r}')dv', \tag{3.61}$$

which is identical to $b_{i,\mathrm{sc}}$ in Eq. (3.55) (when Eq. (3.57) is substituted). Eq. (3.61) leads to Eq. (3.58) for S_{ij}^{sc} upon division by a_j and setting $\mathbf{E}(\mathbf{r}') \equiv \mathbf{E}_j(\mathbf{r}')$.

In fact, the S-parameter data equation Eq. (3.58) can be stated in terms of the incident voltages V_ξ^+ ($\xi = i, j$) using Eq. (3.3) as

$$S_{ij}^{\mathrm{sc}} = \frac{-\mathrm{i}\omega\sqrt{Z_{c,i}Z_{c,j}}}{2V_i^+V_j^+} \iiint_V \Delta\varepsilon(\mathbf{r}')\mathbf{E}_i^{\mathrm{inc}}(\mathbf{r}') \cdot \mathbf{E}_j(\mathbf{r}')dv'. \tag{3.62}$$

With a common system impedance Z_0, this becomes

$$S_{ij}^{\mathrm{sc}} = \frac{-\mathrm{i}\omega Z_0}{2V_i^+V_j^+} \iiint_V \Delta\varepsilon(\mathbf{r}')\mathbf{E}_i^{\mathrm{inc}}(\mathbf{r}') \cdot \mathbf{E}_j(\mathbf{r}')dv'. \tag{3.63}$$

The expression in Eq. (3.63) was derived first in [59][7] using a field expansion in spherical modes with the assumption of a uniform background medium and a common system impedance.

[7] There appears to be a missing minus sign in the final result, Eq. (43), in [59].

3.3.2 Relation between the Data Equation and the Fréchet Derivative of the S-Parameters

The S-parameter Fréchet derivative was derived in Eq. (3.31). Let us assume that this derivative is associated with a nominal state of the inspected object that represents the scatterer-free background, i.e.,

$$\frac{dS_{ij}^{\mathrm{b}}}{d\varepsilon_{\mathrm{r},n}} = \frac{-i\omega\epsilon_0}{2a_i a_j}\mathbf{E}_j^{\mathrm{inc}}(\mathbf{r}_n)\cdot\mathbf{E}_i^{\mathrm{inc}}(\mathbf{r}_n)\Delta\Omega_n, \tag{3.64}$$

where $\mathbf{E}_\xi^{\mathrm{inc}}$ ($\xi = i, j$) is the incident field due to the excitation of the ξ-th port, $\varepsilon_{\mathrm{r},n}$ is the permittivity of the n-th voxel, and $\Delta\Omega_n$ is its volume. In the context of scattering, dS_{ij}^{b} represents the change in the S-parameter due to the variation $\varepsilon_{\mathrm{r},n}$ in the relative permittivity of the n-th voxel in the background, i.e.,

$$\frac{dS_{ij}^{\mathrm{b}}}{d\varepsilon_{\mathrm{r},n}} = \lim_{\Delta\varepsilon_{\mathrm{r},n}\to 0}\left(\frac{\Delta S_{ij}^{\mathrm{b}}}{\Delta\varepsilon_{\mathrm{r},n}}\right) = \lim_{\Delta\varepsilon_{\mathrm{r},n}\to 0}\left(\frac{S_{ij}^{\mathrm{sc},n}}{\Delta\varepsilon_{\mathrm{r},n}}\right), \tag{3.65}$$

where $S_{ij}^{\mathrm{sc},n}$ is the scattering portion of S_{ij} arising due to the presence of the contrast $\Delta\varepsilon_{\mathrm{r},n}$.

On the other hand, the data equation Eq. (3.58) can be applied to the case of scattering from a single voxel (the n-th voxel) to obtain

$$S_{ij}^{\mathrm{sc},n} = \frac{-i\omega}{2a_i a_j}\iiint_{\Delta\Omega_n}\Delta\varepsilon_n\mathbf{E}_i^{\mathrm{inc}}(\mathbf{r}_n)\cdot\mathbf{E}_j(\mathbf{r}_n)dv_n, \tag{3.66}$$

where $\Delta\varepsilon_n = \epsilon_0[\varepsilon_{\mathrm{r},n} - \varepsilon_{\mathrm{r,b}}(\mathbf{r}_n)]$ is the contrast, which is nonzero only in the n-th voxel. Since the voxel is electrically very small, it can also be assumed that the field distributions $\mathbf{E}_i^{\mathrm{inc}}(\mathbf{r}_n)$ and $\mathbf{E}_j(\mathbf{r}_n)$ are constant in Ω_n. Moreover, the scattering from this single voxel is weak, which allows the linear Born approximation to be used, i.e., $\mathbf{E}_j(\mathbf{r}_n) \approx \mathbf{E}_j^{\mathrm{inc}}(\mathbf{r}_n)$. Thus, Eq. (3.66) can be written as

$$S_{ij}^{\mathrm{sc},n} \approx \frac{-i\omega\epsilon_0}{2a_i a_j}\Delta\varepsilon_{\mathrm{r},n}\mathbf{E}_i^{\mathrm{inc}}(\mathbf{r}_n)\cdot\mathbf{E}_j^{\mathrm{inc}}(\mathbf{r}_n)\Omega_n, \tag{3.67}$$

where $\Delta\varepsilon_{\mathrm{r},n} = \varepsilon_{\mathrm{r},n} - \varepsilon_{\mathrm{r,b}}(\mathbf{r}_n)$ is the contrast in the relative permittivity. Dividing both sides of Eq. (3.67) by $\Delta\varepsilon_{\mathrm{r},n}$ and taking the limit as in Eq. (3.65) lead to the sensitivity formula in Eq. (3.64). Thus, the S-parameter Fréchet derivative can be viewed as a special application of the S-parameter data equation to the scattering from very small (voxel-sized) dielectric targets.

4 Linear Inversion in Real Space

This chapter presents a group of methods for image reconstruction, both qualitative and quantitative, that solve linearized models of scattering such as those based on the Born or Rytov approximations. These methods operate on position variables in real space (e.g., x, y, and z) directly. This is in contrast to the group of methods described in the next chapter, which perform the inversion in Fourier space.

Inversion in real space can be done with both frequency-domain data (e.g., S-parameters) and time-domain data (e.g., voltage waveforms). In both cases, the reconstruction is very fast and is done in real time, i.e., an image is produced with practically no delay after the data are acquired.

The reconstruction is particularly fast when using qualitative methods because they do not solve large linear systems of equations. The qualitative images uncover the shape and location of contrast features in the background without quantifying their electrical properties. Despite the lack of quantitative output, these methods are widely used in the detection of defects or abnormalities because the feature's shape and location are often sufficient to reach a diagnostic decision.

4.1 Notations and Terminology

The reconstruction methods are described in the context of the imaging of penetrable objects, where the object of reconstruction is a normalized form of the dielectric contrast $\Delta\varepsilon(\mathbf{r}) = \varepsilon_{\text{O}}(\mathbf{r}) - \varepsilon_{\text{b}}(\mathbf{r})$. Here, ε_{O} is the complex permittivity of the inspected object, and ε_{b} is that of the background. We assume that, in the bandwidth of interest, $\Delta\varepsilon(\mathbf{r})$ is frequency-independent. The extension to dispersive contrasts is straightforward provided the frequency dependence is known and separable from the spatial position, i.e., if we can express the dispersive contrast as $\Delta\varepsilon(\mathbf{r},\omega) = \Delta\varepsilon(\mathbf{r})\phi(\omega)$, where $\phi(\omega)$ is a known function. This is because $\phi(\omega)$ can be made part of the kernel of the data equation [116].

We start by stating the data equation, Eq. (2.89), as

$$R_i^{\text{sc}}(\omega) = \kappa_i(\omega) \iiint_{V_{\text{O}}} \Delta\varepsilon(\mathbf{r}')\mathbf{E}_{i,\text{Rx}}^{\text{inc}}(\mathbf{r}',\omega) \cdot \mathbf{E}_{i,\text{Tx}}(\mathbf{r}',\omega)dv', \ i = 1,\ldots,N_R, \tag{4.1}$$

where i denotes one of the N_R experiments, ω is the frequency, κ_i is a known complex constant dependent on the type of response, and V_{O} is the volume of the inspected object.

$\mathbf{E}_{i,\mathrm{Rx}}^{\mathrm{inc}}$ is the incident field due to the receiving (Rx) antenna in the i-th experiment if this antenna were to operate as a transmitter. $\mathbf{E}_{i,\mathrm{Tx}}$ is the total field generated by the transmitting (Tx) antenna in the i-th experiment. If R_i^{sc} is the parameter S_{jk}^{sc}, then $\kappa_i = \frac{-\mathrm{i}\omega}{2a_j a_k}$; see Eq. (3.58). Notice that Eq. (4.1) refers to the scattering portion of the total response R_i. In the context of the Born model, every response is a superposition of its baseline and scattered portions, $R_i = R_i^{\mathrm{sc}} + R_i^{\mathrm{b}}$. For shorter notations, hereafter, the symbols $S_i \equiv R_i^{\mathrm{sc}}$ and $B_i \equiv R_i^{\mathrm{b}}$ are used for the scattered and the baseline responses, respectively, i.e., $R_i = S_i + B_i$.

Also, each (i-th) experiment is associated with a pair of Tx and Rx positions along with the respective antennas. There may be more than one pair of Tx/Rx antennas associated with a pair of Tx/Rx positions. A typical example is the acquisition of co-polarized (co-pol) and cross-polarized (cross-pol) signals where, in general, four experiments could be associated with each Tx/Rx position pair: VV, HH, VH, and HV. Here, V and H stand for vertical and horizontal polarizations, respectively. The first letter in the ordered pair XY (X,Y =V or H) stands for the polarization of the Rx antenna, whereas the second one denotes that of the Tx antenna.

Finally, note that frequency-sweep measurements are performed one frequency at a time; therefore, ω denotes one of the frequencies in a discrete set of N_ω frequencies. Hereafter, when necessary, the frequency is denoted as ω_m, $m = 1, \ldots, N_\omega$.

With the Born approximation, Eq. (4.1) is linearized with respect to the unknown $\Delta\varepsilon(\mathbf{r})$. Then the i-th signal at the m-th frequency is expressed as

$$S_i^{(m)} \approx \kappa_i^{(m)} \iiint_{V_{\mathrm{O}}} \Delta\varepsilon(\mathbf{r}') \left[\mathbf{E}_{i,\mathrm{Rx}}^{\mathrm{inc}}(\mathbf{r}') \cdot \mathbf{E}_{i,\mathrm{Tx}}^{\mathrm{inc}}(\mathbf{r}')\right]^{(m)} dv', \ i = 1, \ldots, N_R, \tag{4.2}$$

where the superscript indicates the frequency. We assume we know the field distributions $\mathbf{E}_{i,\mathrm{Rx}}^{\mathrm{inc}}(\mathbf{r}', \omega_m)$ and $\mathbf{E}_{i,\mathrm{Tx}}^{\mathrm{inc}}(\mathbf{r}', \omega_m)$ or their combined dot-product distribution. The latter is referred to as the *resolvent kernel* (or simply *kernel*) of the data equation in Eq. (4.2). With far-zone measurements, the incident fields can be approximated as locally plane waves or isotropic spherical waves. For improved accuracy, such analytical distributions can be multiplied by the respective antenna patterns, if available; see Section 2.14.2 and Eq. (2.142). With near-zone data, such approximations are not adequate, and simulated field distributions or calibration measurements are often used.

4.2 Acquiring the Resolvent Kernel: Simulation or Measurement

The acquisition of the kernel in Eq. (4.2) is part of the system calibration. It is worth briefly discussing the advantages and disadvantages of this calibration step when using simulations or measurements.

Acquiring the field dot-product distribution through simulations requires simulating the measurements in the scatterer-free environment for all N_R responses at all N_ω frequencies. Note that at each (i-th) response, two simulations are required—one for the excitation with the Tx antenna and one for the excitation with the Rx antenna. This can amount to quite a few simulations. With the complexity of an imaging setup being fairly high, the computational time may easily last more than a day.

The advantage of the simulations is in obtaining the two distributions $\mathbf{E}^{\text{inc}}_{i,\text{Rx}}(\mathbf{r}', \omega_m)$ and $\mathbf{E}^{\text{inc}}_{i,\text{Tx}}(\mathbf{r}', \omega_m)$ separately with full vectorial information. This is important when pursuing image improvement beyond the qualitative result. Quantitative reconstruction usually involves iterative algorithms, where the two field distributions[1] are updated separately.

Also, strictly speaking, the simulations are free of stochastic noise. However, numerical errors are present, which are deterministic in nature. Mesh-convergence errors in the simulated responses result from a coarse discretization mesh. Other numerical errors include imperfect absorbing boundary conditions at the radiating boundaries or at the ports. The numerical errors are usually not a major concern because they can be reduced by mesh refinement and stricter convergence criteria. Unfortunately, this refinement may also lead to a prohibitive computational burden.

The major disadvantage, however, is that simulations often fail to properly model the actual setup, even in a known background. This failure is associated with the so-called *modeling errors* [117], which are much more difficult to reduce than compared to the numerical errors. The modeling errors are rooted in the inability to predict all influencing factors arising in the practical implementation of the acquisition setup. These include fabrication tolerances of the antennas and the positioning components, uncertainties in the constitutive parameters of the materials (especially the absorbers) used to build the measurement chamber, aging of the materials and the connectors, and deformations due to temperature or humidity. In addition, the simulation models often ignore complexities in the cables, the connectors, and the fine structural components such as screws, brackets, and thin supporting plates.

The experimental acquisition of the kernel $\mathbf{E}^{\text{inc}}_{i,\text{Rx}} \cdot \mathbf{E}^{\text{inc}}_{i,\text{Tx}}$ offers the best fidelity [97, 118]. The concept is simple—an electrically small scatterer of aspect ratio close to unity (e.g., a cube or a sphere) and of known contrast ($\Delta\varepsilon_{\text{co}} = \varepsilon_{\text{co}} - \varepsilon_{\text{b}}$) is placed in the background and is measured. Together with the background setup, this scatterer forms what is referred to as the *calibration object* (CO). An aspect ratio close to unity is desirable so that the small scatterer is insensitive to the polarization of the illumination. For the scatterer to be electrically small, its size in any given direction must be smaller than $\lambda_{\text{b}}/8$, where λ_{b} is the wavelength in the background. This ensures that it is illuminated by a relatively uniform wave field. The scatterer's permittivity ε_{co} should be on the order of that of the background, so that the limits of Born's approximation Eq. (1.332) are observed.

The small scatterer is placed at the center of the imaged volume ($\mathbf{r}' = 0$), and the antennas are scanned across the full extent of the acquisition surfaces.[2] Fig. 4.1 shows

[1] Remember that $\mathbf{E}^{\text{inc}}_{i,\text{Rx}}(\mathbf{r}', \omega_m)$ represents Green's function, whereas $\mathbf{E}^{\text{inc}}_{i,\text{Tx}}(\mathbf{r}', \omega_m)$ approximates the total internal field.

[2] Sometimes, with small imaged volumes, the antennas are fixed, whereas it is the imaged object that is carried along the scan path by supporting plates attached to the scanning mechanism. Fixing the antennas reduces the uncertainty in MW measurements. If the antennas move, their connectors and cables may flex or strain, causing minute signal variations that may well exceed the noise floor of the instrument.

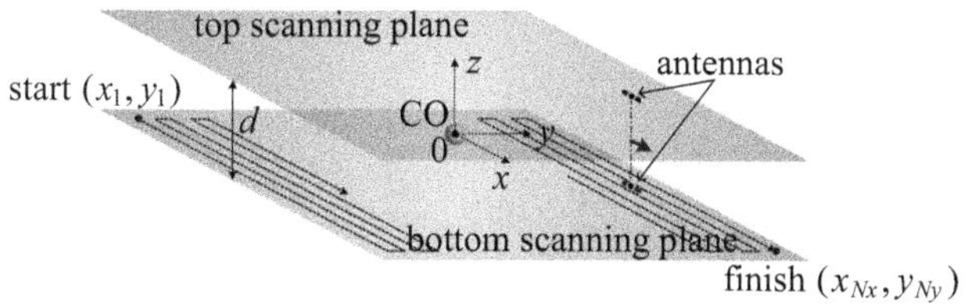

Figure 4.1 Sketch of a planar raster scanning setup and the acquisition of the CO data. A pair of dipole antennas scan together two parallel planes on both sides of the imaged volume. Their scanning path follows the raster route shown with a dotted line. The small scattering object is placed at the center of the imaged volume (the origin of the coordinate system). The sampling involves N_x samples along x and N_y samples along y. If both reflection and transmission coefficients are acquired, the total number of independent responses is $3N_xN_y$.

a CO measurement in a planar acquisition setup. Using Eq. (4.2) and bearing in mind that the scatterer is electrically very small, the kernel at $\mathbf{r}' = 0$ is estimated as

$$\mathcal{K}_i^{(m)}(\mathbf{r}' = 0) \approx \left[\mathbf{E}_{i,\mathrm{Rx}}^{\mathrm{inc}} \cdot \mathbf{E}_{i,\mathrm{Tx}}^{\mathrm{inc}}\right]_{\mathbf{r}'=0}^{(m)} \approx \frac{S_{0i}^{(m)}}{\kappa_i^{(m)} \Delta\varepsilon_{\mathrm{co}} \Omega_{\mathrm{co}}}, \quad i = 1, \ldots, N_R, \tag{4.3}$$

where Ω_{co} is the volume of the scatterer and $S_{0i}^{(m)}$ $(i = 1, \ldots, N_R,\ m = 1, \ldots, N_\omega)$ are the collected CO responses. These responses form the *point-spread function* (PSF) of the imaging system. The PSF describes the system's response to a point scatterer as a function of the observation position.[3] The PSF description of an imaging system inherently assumes that it is linear with respect to the scattering source, which is indeed the case in the framework of the linear Born approximation. The PSF itself is nothing but the impulse response of the linear system.

In a uniform background medium, measuring the PSF with the point scatterer at the center of the imaged volume $(\mathbf{r}' = 0)$ is often sufficient since the CO responses for all other positions $\mathbf{r}' \in V_{\mathrm{O}}$ can be obtained from it via coordinate translation. Take as an example the planar-scanning scenario depicted in Fig. 4.1 where the Tx and Rx antennas move together. Each response is then a function of the lateral position (x_i, y_i) of the Tx/Rx antenna pair in the i-th experiment. In addition, the position of the Rx antenna could be either $z_{\mathrm{Rx}} = d/2$ or $z_{\mathrm{Rx}} = -d/2$, depending on whether it is located at the bottom or the top acquisition planes. In the following discussion, the index i is associated with the observation position (x_i, y_i) only. Thus, in the planar scan shown in Fig. 4.1, the PSFs are now denoted as $S_{0\xi}^{(m)}(x_i, y_i, z_{\mathrm{Rx}})$. Here, ξ indicates that there may be different types of PSFs (e.g., a reflection and a transmission coefficient for a co-pol and a cross-pol measurement) associated with (x_i, y_i).

[3] The term "point-spread function" comes from optical (e.g., photographic) imaging systems where it describes the image obtained with a point source. Depending on how well the image is focused, it does not show a perfect point at the location of the source but rather a smeared spot; hence the term "point-spread."

The CO response, in general, depends not only on the observation location (x_i, y_i, z_{Rx}) but also on the position of the point scatterer $\mathbf{r}' = (x', y', z')$. This can be expressed as

$$S_{co,\xi}^{(m)}(x_i, y_i, z_{Rx}; x', y', z'), m = 1, \ldots, N_\omega, \xi = 1, \ldots, N_T; z_{Rx} = \pm d/2, \quad (4.4)$$

where N_T is the number of types of responses acquired at a given point. The CO response with the point scatterer at the origin is the PSF, i.e.,

$$S_{co,\xi}^{(m)}(x_i, y_i, z_{Rx}; 0, 0, 0) = S_{0\xi}^{(m)}(x_i, y_i, z_{Rx}). \quad (4.5)$$

If the background is uniform, it is invariant to lateral translations, and the CO response for any position $\mathbf{r}' = (x', y', 0)$ of the point scatterer in the plane $z' = 0$ is obtained as

$$S_{co,\xi}^{(m)}(x_i, y_i, z_{Rx}; x', y', 0) = S_{0\xi}^{(m)}(x_i - x', y_i - y', z_{Rx}). \quad (4.6)$$

Therefore, the resolvent kernel at $(x', y', 0)$ for the response at (x_i, y_i, z_{Rx}) is estimated from the respective PSF as

$$\mathcal{K}_{i,\xi}^{(m)}(x', y', 0) \approx \frac{S_{0\xi}^{(m)}(x_i - x', y_i - y', z_{Rx})}{\kappa_\xi^{(m)} \Delta\varepsilon_{co}\Omega_{co}}, \quad i = 1, \ldots, N_R. \quad (4.7)$$

Obtaining the CO responses from the PSFs at various range (or depth) positions of the point scatterer ($z' \neq 0$) is more complicated because the planar system scans only laterally and translations along range do not apply. One obvious approach is to acquire the PSFs at several range positions of the point scatterer ($x' = 0, y' = 0, z'$ varies) where slices of the image are to be produced. To obtain image slices at other range positions, the PSFs can be interpolated with respect to z.

Another approach uses the *angular spectrum representation* of the EM field, which is a widely used method of analysis in radiation and scattering [119, 14, 120, 121, chapter 7]. This approach is described next. If the medium is homogeneous, isotropic, linear, and source-free, then the field satisfies the vector Helmholtz equation,

$$\nabla^2\mathbf{E} + k^2\mathbf{E} = 0, \quad (4.8)$$

where $k = -\mathrm{i}\gamma$ is the complex wavenumber. As shown in [121, chapter 7], the field $\mathbf{E}(x, y, z)$ at any range location $z = const.$ can be determined from that at some reference location $z = 0$, using the relationship in Fourier space,

$$\tilde{\mathbf{E}}(k_x, k_y; z) = \tilde{\mathbf{E}}(k_x, k_y; 0)e^{\mp \mathrm{i}k_z z}, \quad (4.9)$$

where $\tilde{\mathbf{E}}(k_x, k_y; z)$ is the 2D Fourier transform (FT) of the field at the $z = const.$ plane, k_x and k_y are the Fourier variables corresponding to x and y, respectively, and k_z is the z wavenumber obtained from the eigenvalue equation

$$k_z^2 = k^2 - k_x^2 - k_y^2. \quad (4.10)$$

There is arbitrariness in the sign of k_z obtained from Eq. (4.10). This can be resolved by simple physical constraints. For example, if the wave propagates in a loss-free medium strictly along $+z$, then convention requires that $k_z = k > 0$ ($k_z \in \mathbb{R}\,|k_z > 0$).

This implies the constraint $\Re k_z \geq 0$. This convention also requires a minus sign in the exponent of Eq. (4.9):[4]

$$\tilde{\mathbf{E}}(k_x, k_y; z) = \tilde{\mathbf{E}}(k_x, k_y; 0)e^{-\mathrm{i}k_z z}, z \geq 0. \tag{4.11}$$

This choice of sign must be retained even if the wave propagates at some acute angle with respect to $+z$. It also has to hold when the medium is lossy, which now imposes the constraint $\Im k_z \leq 0$ in order to eliminate waves that grow in amplitude as they propagate away from the origin $z = 0$. Similarly, for waves propagating along $-z$, Eq. (4.9) is written as

$$\tilde{\mathbf{E}}(k_x, k_y; z) = \tilde{\mathbf{E}}(k_x, k_y; 0)e^{\mathrm{i}k_z z}, z \leq 0, \tag{4.12}$$

where k_z is subject to the same constraints as before, i.e.,

$$\Re k_z \geq 0 \;,\; \Im k_z \leq 0. \tag{4.13}$$

Now, we can state Eq. (4.10) as

$$k_z \equiv \pm\sqrt{k^2 - k_x^2 - k_y^2} \;,\; \Im k_z \leq 0. \tag{4.14}$$

In summary, within a uniform, isotropic, and linear region (it does not have to be unbounded), we can obtain the field distribution at any range location z from that at a reference location ($z = 0$) using

$$\mathbf{E}(x, y, z) = \mathcal{F}_{2\mathrm{D}}^{-1}\left\{\mathcal{F}_{2\mathrm{D}}\left\{\mathbf{E}(x, y, 0)\right\} e^{\mp\mathrm{i}k_z z}\right\}. \tag{4.15}$$

Here, $\mathcal{F}_{2\mathrm{D}}$ denotes 2D FT and $\mathcal{F}_{2\mathrm{D}}^{-1}$ denotes inverse 2D FT. The minus and plus signs in the exponent are taken in the cases $z > 0$ and $z < 0$, respectively. Notice that k_z is a function of k_x and k_y through Eq. (4.14).

Eq. (4.15) must be applied individually to the terms $\mathbf{E}_{i,\mathrm{Rx}}^{\mathrm{inc}}(\mathbf{r}', \omega_m)$ and $\mathbf{E}_{i,\mathrm{Tx}}^{\mathrm{inc}}(\mathbf{r}', \omega_m)$ in the kernel [see Eq. (4.3)] in order to "translate" it to the desired range position. This means that we need to separate these terms. More specifically, we need to separate their phasor values. However, we do not need to separate them as vectors because

$$\mathbf{E}_{i,\mathrm{Rx}}^{\mathrm{inc}}(\mathbf{r}', \omega_m) \cdot \mathbf{E}_{i,\mathrm{Tx}}^{\mathrm{inc}}(\mathbf{r}', \omega_m) = (\hat{\mathbf{p}}_{\mathrm{Rx}} \cdot \hat{\mathbf{p}}_{\mathrm{Tx}}) E_{i,\mathrm{Rx}}^{\mathrm{inc}}(\mathbf{r}', \omega_m) \cdot E_{i,\mathrm{Tx}}^{\mathrm{inc}}(\mathbf{r}', \omega_m), \tag{4.16}$$

where $\hat{\mathbf{p}}_{\mathrm{Rx}} \cdot \hat{\mathbf{p}}_{\mathrm{Tx}}$ is the dot product of their polarization vectors, which we assume does not change significantly with z'.

Estimation of the individual incident field is possible if the reflection PSFs are measured, because in reflection measurements $\mathbf{E}_{i,\mathrm{Rx}}^{(m)\mathrm{inc}} = \mathbf{E}_{i,\mathrm{Tx}}^{(m)\mathrm{inc}}$. Let the i-th experiment correspond to the measurement of the $S_{11}^{(m)}$ parameter at the position (x_i, y_i, z_1), where

[4] The minus sign in the propagator $\exp(-\mathrm{i}k_z z)$ for a wave propagating along $+z$, where $\Re k_z \geq 0$, follows from our definition of the field phasor as $\Re\{F \exp(\mathrm{i}\omega t)\} = f(t)$; see Eq. (1.18).

$z = z_1$ is the plane scanned by antenna 1. The respective PSF is then $S_0^{(m)}(x_i, y_i, z_1) \equiv S_{\text{co},11}^{(m)}(x_i, y_i, z_1; 0, 0, 0)$. Then,

$$E_1^{(m)\text{inc}}(x_i, y_i, z_1; 0, 0, 0) \approx \pm\sqrt{\mathcal{K}^{(m)}(x_i, y_i, z_1; 0, 0, 0)}, \tag{4.17}$$

where the first set of coordinates (x_i, y_i, z_1) shows the position of antenna 1 and the second set of coordinates $(0, 0, 0)$ shows that the field is estimated at the origin. As usual, the subscript in E_1 indicates that antenna 1 transmits. As per Eq. (4.3),

$$\mathcal{K}^{(m)}(x_i, y_i, z_1; 0, 0, 0) = \frac{S_{\text{co},11}^{(m)}(x_i, y_i, z_1; 0, 0, 0)}{\kappa_{11}^{(m)} \Delta\varepsilon_{\text{co}} \Omega_{\text{co}}}. \tag{4.18}$$

The method presented by Eq. (4.17) and Eq. (4.18) does not yield the field polarization. However, experimental methods for the full vectorial characterization of the incident field of antennas exist; see, e.g., [115]. Another limitation arises from the sign arbitrariness in Eq. (4.17). In imaging, we are interested in the relative phase distribution of the field; therefore, the choice of sign is not important as long as it is consistent across all responses. What is more critical here is the arbitrariness in the square root of a complex quantity; namely, $\sqrt{m e^{\text{i}\phi}} = \sqrt{m} e^{\text{i}(0.5\phi \pm n\pi)}$, where n is an integer. Taking the principal square root of the kernel as a function of (x_i, y_i) may result in a $\pm\pi$ phase error. To avoid such errors, the PSFs of all reflection coefficients acquired with the CO, e.g., $S_{\text{co},11}^{(m)}(x_i, y_i, z_1; 0, 0, 0)$, must be unwrapped as functions of (x_i, y_i).

In conclusion, under the assumption that V_{O} is uniform, one measurement of the system PSF with a small scatterer at its center is sufficient to estimate the kernel of the respective data equation. Such a measurement is performed in less than an hour even with slow mechanical scans. It takes less than a minute with electronically switched arrays. Thus, the experimental methods of acquiring the data-equation kernel are usually faster than simulations.

4.3 Qualitative Imaging with Sensitivity Maps

Sensitivity-based qualitative imaging was first proposed in [94, 95] as a detection method. Its spatial resolution and robustness to noise were later studied in [96]. It employs frequency-domain responses, and it can be configured to work with complex-valued responses or with magnitude and phase responses separately, depending on which are available. In far-zone reflection (back-scattered) measurements, it is the phase that carries most of the information about the target, and so its availability is critically important. In the extreme near-zone measurements, back-scattered or forward-scattered, magnitude-only data may suffice. A far-zone scenario where magnitude-only data may be sufficient occurs in the imaging of lossy objects where the signal attenuation is the decisive factor shaping the signal. This section describes how the method works with complex responses such as the scattering parameters since this is the most general case.

4.3.1 Fréchet Derivative of the Data Residual

Let us now consider the simplest objective (or cost) function used in reconstruction:

$$F(\boldsymbol{p}) = \|\boldsymbol{R}_{\text{data}}(\boldsymbol{p})\|, \tag{4.19}$$

which is based only on the data residual,

$$\boldsymbol{R}_{\text{data}}(\boldsymbol{p}) = \boldsymbol{R}(\boldsymbol{p}) - \bar{\boldsymbol{R}}. \tag{4.20}$$

Here, $\boldsymbol{p}$ is a vectorized form of the object's permittivity distribution $\varepsilon_{\text{O}}(\mathbf{r}')$, $\boldsymbol{R}(\boldsymbol{p})$ is the vector of the *modeled* responses, and $\bar{\boldsymbol{R}}$ is the vector of all respective *measured* responses. $\boldsymbol{R}(\boldsymbol{p})$ is obtained through some EM model that maps the object's permittivity distribution to a set of responses; hence, the name *modeled*. $F(\boldsymbol{p})$ is a measure of the similarity between $\bar{\boldsymbol{R}}$ and $\boldsymbol{R}(\boldsymbol{p})$. In the context of the discussion that follows, $\bar{\boldsymbol{R}}$ and $\boldsymbol{R}(\boldsymbol{p})$ are the total responses, i.e., they include both the incident and the scattered responses. In most general terms, every reconstruction method aims at finding a solution p^* that minimizes the cost function $F(\boldsymbol{p})$.

Next, a quadratic (squared ℓ_2) norm is adopted in the form

$$F(\boldsymbol{p}) = 0.5\|\boldsymbol{R}(\boldsymbol{p}) - \bar{\boldsymbol{R}}\|_2^2. \tag{4.21}$$

The responses are frequency-dependent, and, in principle, the vectors $\boldsymbol{R}$ and $\bar{\boldsymbol{R}}$ can incorporate all responses at all frequencies. In practice, however, the strength of the responses at different frequencies can be very different (an order of magnitude or more) due to the frequency-dependent propagation loss. With equal illuminating power, it is the higher-frequency signals that are weaker. Combining responses of substantially different strengths in a single vector leads to a substantially decreased impact of the weak responses on the final image. Yet, it is the weaker high-frequency signals that improve the spatial resolution of the image. This is why, the objective function in Eq. (4.21) is first defined at each (m-th) frequency, so that

$$F^{(m)}(\boldsymbol{p}) = 0.5\|\boldsymbol{R}^{(m)}(\boldsymbol{p}) - \bar{\boldsymbol{R}}^{(m)}\|_2^2, \; m = 1, \ldots, N_\omega. \tag{4.22}$$

Subsequent normalization of the result at each frequency ensures a strength-independent contribution to the final result.

Eq. (4.22) is explicitly written in terms of the permittivity distribution as

$$F^{(m)}\left[\varepsilon(\mathbf{r}')\right] = 0.5\sum_{i=1}^{N_R}\left(\Delta R_i^{(m)} \cdot \Delta R_i^{(m)*}\right)_{[\varepsilon(\mathbf{r}')]}, \; \mathbf{r}' \in V_{\text{O}}, \; m = 1, \ldots, N_\omega, \tag{4.23}$$

where $\Delta R_i^{(m)}\left[\varepsilon(\mathbf{r}')\right] = R_i^{(m)}\left[\varepsilon(\mathbf{r}')\right] - \bar{R}_i^{(m)}$. Its Fréchet derivative consists of the derivatives with respect to the permittivity value at each voxel.

Let the real part of the permittivity of the n-th voxel be $\varepsilon_n' \equiv \Re\varepsilon_{\text{O}}(\mathbf{r}_n')$. Here, $n = 1, \ldots, N_p$, where N_p is the number of voxels comprising V_{O}. Then, the derivative of

$F^{(m)}$ with respect to ε'_n can be expressed as

$$\frac{dF^{(m)}}{d\varepsilon'_n} = 0.5\sum_{i=1}^{N_R}\left[\Delta R_i^{(m)}\frac{d\left(R_i^{(m)}\right)^*}{d\varepsilon'_n} + \left(\frac{dR_i^{(m)}}{d\varepsilon'_n}\right)\left(\Delta R_i^{(m)}\right)^*\right]. \quad (4.24)$$

We recall from Eq. (3.32) in Section 3.2.2 that $\frac{dR_i^{(m)}}{d\varepsilon'_n} = \frac{dR_i^{(m)}}{d\varepsilon_n}$. Also, using Cauchy–Riemann equations, it can be shown that

$$\frac{d\left(R_i^{(m)}\right)^*}{d\varepsilon'_n} = \left(\frac{dR_i^{(m)}}{d\varepsilon'_n}\right)^* = \left(\frac{dR_i^{(m)}}{d\varepsilon_n}\right)^*. \quad (4.25)$$

Thus, Eq. (4.24) reduces to

$$\frac{dF^{(m)}}{d\varepsilon'_n} = \sum_{i=1}^{N_R}\Re\left\{\Delta R_i^{(m)}\cdot\left(\frac{dR_i^{(m)}}{d\varepsilon_n}\right)^*\right\}. \quad (4.26)$$

The Fréchet derivative with respect to the imaginary part ε''_n of the n-th voxel permittivity is found in an analogous manner by making use of the relationship $\frac{dR_i^{(m)}}{d\varepsilon''_n} = -\mathrm{i}\frac{dR_i^{(m)}}{d\varepsilon_n}$ as per Eq. (3.33). The result is

$$\frac{dF^{(m)}}{d\varepsilon''_n} = -\sum_{i=1}^{N_R}\Im\left\{\Delta R_i^{(m)}\cdot\left(\frac{dR_i^{(m)}}{d\varepsilon_n}\right)^*\right\}. \quad (4.27)$$

4.3.2 Reconstruction Formula of the Sensitivity-Map Method

Eq. (4.26) and Eq. (4.27) are at the heart of the imaging procedure of the qualitative sensitivity-map imaging with complex-valued responses. In it, the *modeled* responses are assigned as those of the scatterer-free background, i.e., $R_i^{(m)} \equiv B_i^{(m)}$. Assume that the measurements in the scatterer-free background have been performed and the respective baseline responses $B_i^{(m)}$ are available.[5] Then, the difference between the baseline response and that of the object under test (OUT), $\Delta R_i^{(m)} = B_i^{(m)} - \bar{R}_i^{(m)} = -\bar{S}_i^{(m)}$, is an estimate (with a minus sign) of the scattered portion of the OUT response.

The response sensitivities $dR_i^{(m)}/d\varepsilon_n \equiv dB_i^{(m)}/d\varepsilon_n$ ($n = 1, \ldots, N_p$) are those of the baseline responses. As such, they are a part of the system calibration and are independent of the OUT. Their computation is very fast since it is done through field-based sensitivity formulas such as the one for the S-parameters; see Eq. (3.31). These formulas require the dot-product distribution $\mathbf{E}^{\mathrm{inc}}_{i,\mathrm{Rx}}(\mathbf{r}', \omega_m)\cdot\mathbf{E}^{\mathrm{inc}}_{i,\mathrm{Tx}}(\mathbf{r}', \omega_m)$ in V_{O}, which is also

[5] The baseline measurements are a part of the system calibration.

the kernel of the data equation Eq. (4.2). This kernel is determined through simulations or experiments performed in the background as described in Section 4.2.

The 3D *complex* Fréchet derivative,

$$D^{(m)}(\mathbf{r}'_n) = \sum_{i=1}^{N_R} \left\{ \underbrace{\left(B_i^{(m)} - \bar{R}_i^{(m)}\right)}_{-\bar{S}_i^{(m)}} \cdot \left(\frac{dB_i^{(m)}}{d\varepsilon_n}\right)^* \right\}, \tag{4.28}$$

is a function of the position $\mathbf{r}'_n$ of the n-th voxel, the permittivity of which is a system variable. This function is referred to as the complex *Jacobian map* or *sensitivity map* [95]. A plot of its real part yields the real-permittivity sensitivity map, $dF^{(m)}/d\varepsilon'_n$, whereas the plot of its imaginary part yields the imaginary-permittivity sensitivity map, $dF^{(m)}/d\varepsilon''_n$, with a minus sign.

The sensitivity map obtained with a single response ($N_R = 1$) cannot identify the scattering objects because, after normalization, it is simply a plot of the normalized kernel of the data equation. However, with a multitude of responses, the response derivatives at points where contrast exists add coherently, whereas at points where there is no contrast they add incoherently. The resultant map gains high intensity at points where scattering occurs. Clearly, the larger the number of available responses, the better the performance of the algorithm. The detection accuracy has been shown to improve by adding more Tx and Rx antennas [95]. As expected, the detection errors decrease quickly when a small set of responses ($N_R < 10$) is augmented by yet another response. Eventually, saturation is reached so that adding Tx/Rx points does not improve the image quality significantly.

To understand this principle of coherent addition, consider the expansion of the i-th response $\bar{R}_i^{(m)}$ in Taylor's series about its respective baseline response as

$$\bar{R}_i^{(m)} = B_i^{(m)} + \frac{\partial B_i^{(m)}}{\partial \varepsilon_1}\Delta\varepsilon_1 + \cdots + \frac{\partial B_i^{(m)}}{\partial \varepsilon_{N_p}}\Delta\varepsilon_{N_p} + O^2 \ , i = 1, \ldots, N_R. \tag{4.29}$$

This leads to an approximation of $\bar{S}_i^{(m)}$ in the form

$$\bar{S}_i^{(m)} \approx \frac{\partial B_i^{(m)}}{\partial \varepsilon_1}\Delta\varepsilon_1 + \cdots + \frac{\partial B_i^{(m)}}{\partial \varepsilon_{N_p}}\Delta\varepsilon_{N_p} \ , i = 1, \ldots, N_R. \tag{4.30}$$

Let us multiply both sides of Eq. (4.30) by $\left[\partial B_i^{(m)}/(\partial\varepsilon_n)\right]^*$; do this for all i, and sum over all $i = 1, \ldots, N_R$. The result is

$$\begin{aligned}\sum_{i=1}^{N_R} \bar{S}_i^{(m)} \left[\frac{\partial B_i^{(m)}}{\partial \varepsilon_n}\right]^* \approx{}& \Delta\varepsilon_1 \sum_{i=1}^{N_R} \frac{\partial B_i^{(m)}}{\partial \varepsilon_1}\left[\frac{\partial B_i^{(m)}}{\partial \varepsilon_n}\right]^* + \cdots + \\ & \Delta\varepsilon_n \sum_{i=1}^{N_R} \frac{\partial B_i^{(m)}}{\partial \varepsilon_n}\left[\frac{\partial B_i^{(m)}}{\partial \varepsilon_n}\right]^* + \cdots + \Delta\varepsilon_{N_p} \sum_{i=1}^{N_R} \frac{\partial B_{N_R}^{(m)}}{\partial \varepsilon_{N_p}}\left[\frac{\partial B_{N_R}^{(m)}}{\partial \varepsilon_n}\right]^*.\end{aligned} \tag{4.31}$$

This result (with a minus sign) is exactly the value of the sensitivity map $D^{(m)}$ at the voxel $\mathbf{r}'_n$; see Eq. (4.28). We observe that the coefficient multiplying $\Delta\varepsilon_n$ consists of all positive-real terms, which add without cancellation (coherent addition). On the contrary, the coefficients for all other voxels contain complex-valued terms, which add incoherently. With a large number of responses, Eq. (4.31) produces a sensitivity-map value largely proportional to the contrast $\Delta\varepsilon_n$—the larger the contrast at the n-th voxel, the higher the map intensity. Bearing in mind that the response sensitivities $\partial B_i^{(m)}/\partial\varepsilon_n$ depend on the background-field dot-product distribution alone, we expect that if we provide the same illuminating conditions for all voxels in the OUT, the sensitivity map will be a good representation of the relative distribution of the contrast across the voxels.

4.3.3 Imaging Procedure of the Sensitivity-Map Method

The 3D sensitivity maps obtained at all N_ω frequencies can be combined to obtain an overall image of the OUT. The more frequency samples are available and the wider the frequency band, the better the image quality. Frequency diversity is important in this method because it relies on coherent superposition. It is as important as the spatial diversity, which is brought about by the increase in the number of Tx and Rx locations. As mentioned before, the frequency-dependent sensitivity maps are normalized before they are combined into a final map. One approach is to normalize with respect to the maximum magnitude, where the final sensitivity map is obtained as

$$D(\mathbf{r}'_n) = \frac{1}{N_\omega}\sum_{m=1}^{N_\omega}\frac{D^{(m)}(\mathbf{r}'_n)}{\eta^{(m)}},\ n = 1,\ldots,N_p, \tag{4.32}$$

with $\eta^{(m)}$ being the maximum magnitude value in the m-th frequency sensitivity map:

$$\eta^{(m)} = \max\left\{\left|D^{(m)}(\mathbf{r}'_n)\right|\right\}. \tag{4.33}$$

In [97], similar normalization is used but $\eta^{(m)}$ is determined during the system calibration from the m-th sensitivity map of the CO (the calibration object consisting of a known small scatterer embedded in the center of the background):

$$\eta^{(m)} = \max\left\{\left|D_{\text{co}}^{(m)}(\mathbf{r}'_n)\right|\right\}. \tag{4.34}$$

Such normalization is necessary for a quantitative image. This is discussed shortly. Least-square normalization is also possible [95], where

$$\eta^{(m)} = \frac{1}{N_p}\sqrt{\sum_{n=1}^{N_p}\left|D^{(m)}(\mathbf{r}'_n)\right|^2}. \tag{4.35}$$

All three normalization approaches result in very similar qualitative images because they all normalize the magnitudes of the complex values in the sensitivity maps but do not affect their angles.

The imaging procedure of the sensitivity-map method can be summarized as follows.

1. *Single-Frequency Map Generation*: The sensitivity maps $D^{(m)}(\mathbf{r}'_n)$ $(m = 1, \ldots, N_\omega)$ are computed at each frequency using Eq. (4.28) wherein the response derivatives are computed with Eq. (3.31).
2. *Single-Frequency Map Normalization*: All sensitivity maps are normalized with the chosen coefficient $\eta^{(m)}$ using either one of Eqs. (4.33)–(4.35).
3. *Frequency Map Summation*: The normalized single-frequency sensitivity maps are summed to produce a combined 3D complex map $D(\mathbf{r}'_n)$; see Eq. (4.32).
4. *Image Formatting*: The absolute value of the combined map is taken and may be further normalized to unity magnitude as

$$\bar{D}(\mathbf{r}'_n) = \frac{|D(\mathbf{r}'_n)|}{|D(\mathbf{r}'_n)|_{\max}}. \tag{4.36}$$

This is plotted to produce an image. To improve the dynamic range of the image, the result can be plotted on a logarithmic scale. Since the logarithm cannot be applied to zero, it is recommended to use the formula

$$L(\mathbf{r}'_n) = a \log_{10}\left[\alpha + \bar{D}(\mathbf{r}'_n)\right] \tag{4.37}$$

where $0 < \alpha \leq 1$ and a is often taken as $a = 1$, but it can be adjusted for the best view in the chosen color scheme.

To gain insight into the real and the imaginary part of the contrast, the real and the imaginary parts of the combined map can be plotted separately.

The qualitative sensitivity-map imaging is very fast because it does not involve the inversion of a linear system of equation. It is, however, limited to weak scattering because it assumes a linearized forward model, as exemplified by the first-order Taylor expansion in Eq. (4.30).

4.3.4 Analytical Examples Illustrating the Sensitivity-Map Method

Let us first consider a simple 1D imaging example that illustrates the implementation of the method and highlights the importance of the frequency bandwidth of the acquired responses.

Example 1: Two antennas, at $z = -D/2$ (**A1**) and $z = D/2$ (**A2**) form a 2-port network together with the imaged domain $-d/2 \leq z' \leq d/2$, where $d < D$; see Fig. 4.2. The scattering parameters S_{11} and S_{21} are measured in the frequency band $f_0 - 0.5B \leq f \leq f_0 + 0.5B$. **A1** launches a plane wave toward a very thin dielectric slab positioned at z_s, the thickness of which satisfies $\Delta z_s << \lambda_0$, where $\lambda_0 = \frac{c}{f_0}$ is the wavelength in the background medium at the central frequency f_0. The slab's relative contrast is $\Delta\varepsilon_{r,s}$. Obtain analytically the 1D image produced by the sensitivity-map method, assuming continuous sampling in frequency and using: (a) S_{11} only and (b) S_{21} only.

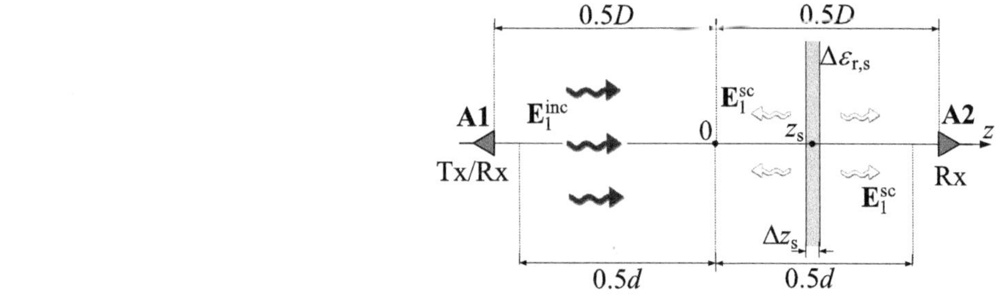

Figure 4.2 Illustration of the setup in Example 1, where a plane wave generated by antenna **A1** illuminates a very thin dielectric slab (thickness is Δz_s) from the left. The slab is assumed infinite in the lateral directions, which renders the problem dependent on z only. The scattered waves are received by **A1** on the left and **A2** on the right. The two antennas form a 2-port MW network, the S-parameters of which are acquired in a finite frequency bandwidth.

Solution: First, we obtain synthetic data (the scattered portion of the S-parameters): $\bar{S}_{11}^{sc}(f)$ and $\bar{S}_{21}^{sc}(f)$. We need these to generate the sensitivity map using Eq. (4.28). For that, we use the S-parameter data equation Eq. (3.58), which we write in the 1D case as

$$\bar{S}_{ij}^{sc}(\omega) = \frac{-i\omega\epsilon_0}{2a_i a_j} \cdot S \cdot \int_{-0.5d}^{0.5d} \Delta\varepsilon_r(z') E_i^{inc}(z') E_j^{inc}(z') dz', \; i, j = 1, 2, \tag{4.38}$$

where S is the cross-section through which the plane waves travel. Also, $\Delta\varepsilon_r = \Delta\varepsilon/\epsilon_0$ is the relative permittivity contrast. Note that we have employed Born's linear approximation. Each of the incident fields can be represented in terms of the power waves exciting their respective ports as[6]

$$E_1^{inc}(\omega, z') = a_1 \sqrt{\frac{\eta_0}{S}} e^{-ik_0(\omega)\cdot(z'+0.5D)} \tag{4.39}$$

$$E_2^{inc}(\omega, z') = a_2 \sqrt{\frac{\eta_0}{S}} e^{ik_0(\omega)\cdot(z'-0.5D)}, \tag{4.40}$$

where η_0 and $k_0(\omega) = \omega\epsilon_0\eta_0$ are the intrinsic impedance and the wavenumber of the background, respectively. Also, $\omega = 2\pi f$ is the radian frequency. Substituting these in Eq. (4.38) and taking into account that the slab is very thin electrically yields

$$\bar{S}_{11}^{sc}(\omega) = -\frac{ik_0(\omega)}{2} \Delta\varepsilon_{r,s} \Delta z_s e^{-i2k_0(\omega)\cdot(z_s+0.5D)} \tag{4.41}$$

$$\bar{S}_{21}^{sc}(\omega) = -\frac{ik_0(\omega)}{2} \Delta\varepsilon_{r,s} \Delta z_s e^{-ik_0(\omega)\cdot D}. \tag{4.42}$$

[6] The power carried by a traveling plane wave of RMS value E through a cross-section S is $P = \frac{E^2}{\eta_0} \cdot S$. On the other hand, $\sqrt{P} = |a|$, where a is the power wave. Therefore, $E = |a| \sqrt{\frac{\eta_0}{S}}$.

Second, we obtain the Fréchet derivatives of the baseline S-parameters, for which we use Eq. (3.31), i.e.,

$$\frac{dB_{ij}}{d\varepsilon_{\mathrm{r}}(z')} = \frac{-\mathrm{i}\omega\epsilon_0}{2a_i a_j}\mathbf{E}_j(z') \cdot \mathbf{E}_i(z')S\Delta z', \tag{4.43}$$

where $\Delta\Omega = S\Delta z'$ is the voxel's volume at z'. Eq. (4.39) and Eq. (4.40) are substituted in Eq. (4.43) to obtain the B_{11} and B_{21} derivatives as

$$\frac{dB_{11}(\omega)}{d\varepsilon_{\mathrm{r}}(z')} = -\frac{\mathrm{i}k_0(\omega)}{2}e^{-\mathrm{i}2k_0(\omega)\cdot(z'+0.5D)}\Delta z', \tag{4.44}$$

$$\frac{dB_{21}(\omega)}{d\varepsilon_{\mathrm{r}}(z')} = -\frac{\mathrm{i}k_0(\omega)}{2}e^{-\mathrm{i}k_0(\omega)D}\Delta z'. \tag{4.45}$$

The next step is to obtain the complex sensitivity maps as a function of the frequency using Eq. (4.28), i.e.,

$$D_{ij}(z', \omega) = -\bar{S}_{ij}^{\mathrm{sc}}(\omega) \cdot \left[\frac{dB_{ij}(\omega)}{d\varepsilon_{\mathrm{r}}(z')}\right]^*. \tag{4.46}$$

The resultant expressions are

$$D_{11}(z', \omega) = -\frac{[k_0(\omega)]^2 \Delta z_{\mathrm{s}} \Delta z' \Delta\varepsilon_{\mathrm{r,s}}}{4} \cos\left[2k_0(\omega) \cdot (z' - z_{\mathrm{s}})\right], \tag{4.47}$$

and

$$D_{21}(z', \omega) = -\frac{[k_0(\omega)]^2 \Delta z_{\mathrm{s}} \Delta z' \Delta\varepsilon_{\mathrm{r,s}}}{4}. \tag{4.48}$$

Note the proportionality between these complex maps and the relative contrast, i.e., $\Re D_{ij} \sim \Re\Delta\varepsilon_{\mathrm{r,s}}$ and $\Im D_{ij} \sim \Im\Delta\varepsilon_{\mathrm{r,s}}$, $i, j = 1, 2$.

We can now obtain the normalized real-permittivity maps:

$$\Re\bar{D}_{11}(z', \omega) = -\cos\left[2(\omega/c) \cdot (z' - z_{\mathrm{s}})\right], \tag{4.49}$$

and

$$\Re\bar{D}_{21}(z', \omega) = -1, \tag{4.50}$$

where $c = \omega/k_0$ is the speed of light. It is apparent at this point that these single-response, single-frequency sensitivity maps do not indicate the position z_{s} of the scattering object. However, the maps are available in the whole frequency band from $f_0 - 0.5B$ to $f_0 + 0.5B$. In its discrete-frequency implementation, the algorithm (see Step 3, *Frequency Map Summation*) sums all normalized single-frequency sensitivity maps. Here, we assume infinite sampling in frequency and that summation translates into integration. Integrating $\Re\bar{D}_{11}$ over all frequencies and normalizing for unity magnitude yield

$$\bar{D}_{11}(z') = \int_{f_0-0.5B}^{f_0+0.5B} \bar{D}_{11}(z', f)df = \cos\left[\frac{2\pi(z' - z_{\mathrm{s}})}{\lambda_0}\right]\left\{\frac{\sin\left[2\pi(z' - z_{\mathrm{s}})B/c\right]}{[2\pi(z' - z_{\mathrm{s}})B/c]}\right\}. \tag{4.51}$$

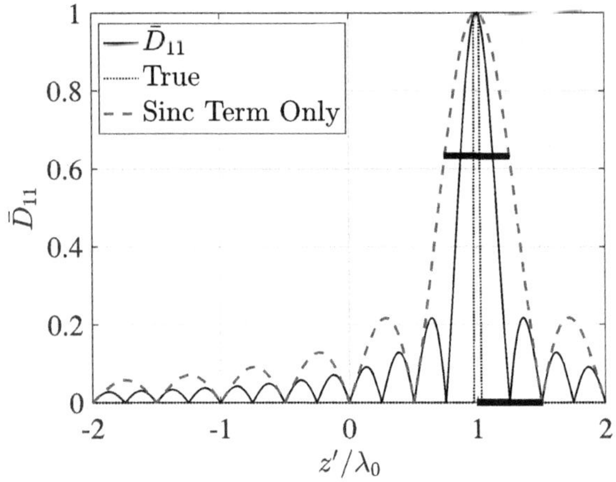

Figure 4.3 Plot of the 1D "image" produced by Eq. (4.51) when $z_s/\lambda_0 = 1$ and $B = f_0$. The $\bar{D}_{11}$ result is shown with a solid line, whereas the *sinc* term alone is shown with a dotted-line. The thick horizontal bars indicate the spatial resolution as determined from the *sinc* function.

Fig. 4.3 shows a plot of the 1D "image" produced by Eq. (4.51) when $z_s/\lambda_0 = 1$ and for a bandwidth $B = f_0$. The dominant behavior is determined by the *sinc* function in the braces (shown in Fig. 4.3 by a dash line), since for small values of $|z' - z_s|$, the cosine function is approximately equal to 1. We observe that the image (solid line in Fig. 4.3) does not reproduce exactly the very narrow box-shaped true distribution of the contrast (sketched in Fig. 4.3 by a dotted line). The contrast distribution approximated by the *sinc* function (dotted line in Fig. 4.3) has a width, which is fundamentally determined by the frequency bandwidth B.

The width of the *sinc* function from its maximum to its first null, $\delta z' = |z'_{\text{null}} - z_s|$, is referred to as the image *spatial resolution limit*. Since the *sinc* term attains its first null when its argument equals π, $\delta z'$ is obtained as

$$\delta z' = \frac{c}{2B}. \tag{4.52}$$

It is worth pointing out that $\delta z'$ is also the width of the *sinc* function at about -4-dB level. This is illustrated in Fig. 4.3 with a thick solid line.

Eq. (4.52) is an important result, which is fundamental in radar imaging. It shows that the image depicts any small target (smaller than $\delta z'$) in the same manner regardless of its actual size. The image maximum does coincide with the target's position, but its width is limited by the available bandwidth B. Greater bandwidth means better (i.e., smaller) resolution limit and therefore an increased ability to distinguish smaller details. When the distance between two scatterers drops below the resolution limit, their respective maxima tend to merge in the image.

We now turn to the sensitivity map associated with the transmission S-parameter; see Eq. (4.50). This map has no dependence whatsoever on the target's position z_s; it is simply a flat line. Integrating over frequency produces $\bar{D}_{21}(z', \omega) = -B$, which does not improve image quality. This is also an important result. It shows the complete loss of imaging capability, or in other words, the resolution limit has become infinite. In Chapter 6, we discuss in more detail the reasons for such a loss of resolution. Here, it suffices to point out that transmission coefficients do not provide resolution along the line of sight between the two antennas. However, they provide good cross-range (or lateral) resolution as we see in the next example.

Example 2: Two antennas aligned along each other's boresight acquire S_{11} and S_{21} as a function of the sampling position x. The first antenna (**A1**) scans the line at $z = -D/2$, whereas the second one (**A2**) scans the line at $z = D/2$; see Fig. 4.4. An electrically small weak scatterer is positioned at x_s in the $z = 0$ plane. The 2D imaged domain (in the xz plane) is assumed to be sufficiently far from the antennas, so that their incident fields can be approximated locally as plane waves. Obtain analytically the line of the 2D sensitivity map $\Re\bar{D}_{11}^{(m)}(x', z')$, which passes through the scatterer, i.e., $\Re\bar{D}_{11}^{(m)}(x', 0)$, in terms of the wavelength $\lambda^{(m)}$ at the m-th frequency. Assume infinite sampling along x that renders S_{11} and S_{21} continuous functions of x. Repeat for $\Re\bar{D}_{21}^{(m)}(x', 0)$.

Solution: The sensitivity map $D_{11}^{(m)}(x')$ at $z' = 0$ is obtained by applying Eq. (4.28) to all reflection responses acquired by **A1**, $\bar{S}_{11}(x, z = -D/2)$. Since $\bar{S}_{11}$ is assumed to be a continuous function of x, the sum is replaced by an integral over x:

$$D_{11}^{(m)}(x') = \int_{-X_{\max}}^{X_{\max}} \left\{ -\bar{S}_{11}^{m,\mathrm{sc}}(x, x_s) \cdot \left[\frac{dB_{11}^{(m)}(x)}{d\varepsilon_r(x')} \right]^* \right\} dx, \quad (4.53)$$

where $X_{\max}$ is half the extent of the scanned aperture. Since the qualitative maps are normalized, we ignore all constants and work with the representative functions only.

Assuming that both the incident and the scattered waves are locally plane,

$$\bar{S}_{11}^{m,\mathrm{sc}}(x, x_s) \sim e^{-\mathrm{i}2k^{(m)}r_1(x,x_s)}, \quad (4.54)$$

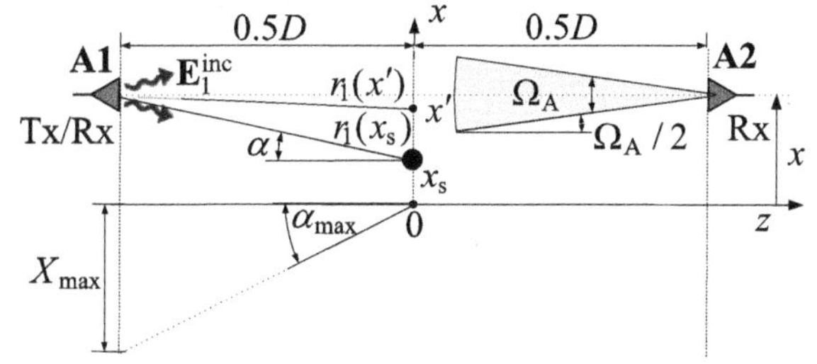

Figure 4.4 Illustration of the setup in Example 2, where two antennas aligned along each other's boresight acquire S_{11} and S_{21} as a function of the sampling position x. A small scatterer is positioned at x_s in the $z = 0$ plane.

where $r_1(x, x_s) = \sqrt{(x_s - x)^2 + (D/2)^2}$ is the distance from **A1** to the scatterer. Observe that all amplitude variations are ignored, and only the phase delays associated with the wave traversing the distance r_1 twice are taken into account. The above result is proportional to the squared phasor of the approximated incident field at x_s produced by **A1** at x. This is in accordance with the S-parameter data equation Eq. (3.58) under the linear Born approximation.

The derivative $dB_{11}^{(m)}(x)/d\varepsilon_r(x')$, on the other hand, is proportional to the squared phasor of the incident field at x' produced by **A1** at x:

$$\frac{dB_{11}^{(m)}(x)}{d\varepsilon_r(x')} \sim e^{-i2k^{(m)} r_1(x,x')}, \tag{4.55}$$

where $r_1(x, x') = \sqrt{(x' - x)^2 + (D/2)^2}$ is the distance from **A1** to the voxel at $(x', 0)$. Therefore,

$$D_{11}^{(m)}(x') \sim \int_{-X_{\max}}^{X_{\max}} e^{-i2k^{(m)} r_1(x,x_s)} \cdot e^{i2k^{(m)} r_1(x,x')} dx. \tag{4.56}$$

Since the object is far from the antennas, we assume that $|x_s - x'| << D/2$. This allows for expressing $r_1(x, x')$ through $r_1(x, x_s)$ as

$$r_1(x, x') = r_1(x, x_s) - (x' - x_s) \sin\alpha, \tag{4.57}$$

where

$$\sin\alpha = \frac{x - x_s}{\sqrt{(x - x_s)^2 + (D/2)^2}}. \tag{4.58}$$

The angle α is depicted in Fig. 4.4. The expression in Eq. (4.56) is now simplified to

$$D_{11}^{(m)}(x') \sim \int_{-X_{\max}}^{X_{\max}} e^{-i2k^{(m)}(x' - x_s)\sin\alpha} dx. \tag{4.59}$$

Changing variables from x to $\sin\alpha$ solves the integral in Eq. (4.59) to produce the normalized sensitivity map

$$\bar{D}_{11}^{(m)}(x') = \Re\bar{D}_{11}^{(m)}(x') = \frac{\sin\left[2k^{(m)}(x' - x_s)\sin\alpha_{\max}\right]}{2k^{(m)}(x' - x_s)\sin\alpha_{\max}}, \tag{4.60}$$

where $\alpha_{\max}$ is the largest angle supported by the acquisition plane with respect to the center of the imaged region; see Fig. 4.4. In deriving Eq. (4.60), it has been assumed that $-\alpha_{\max} \leq \alpha \leq \alpha_{\max}$, which, strictly speaking, is true only for scatterers close to the origin. As such, the result is representative of a best-case scenario and can be used to illustrate the fundamental limitations on the spatial resolution in the lateral direction, referred to as *cross-range resolution*. $\bar{D}_{11}^{(m)}(x')$ is clearly a *sinc* function, from which we can deduce the cross-range resolution limit as

$$\delta x' = \frac{\lambda^{(m)}}{4\sin\alpha_{\max}}. \tag{4.61}$$

Similarly to the case of the range resolution [see Eq. (4.52)], Eq. (4.61) indicates that there is a limit to how small a detail can be and still be faithfully reproduced in the image. The factor with the greatest influence is the wavelength–the shorter it is, the better the cross-range resolution. The size of the aperture matters as well. The case of $\alpha_{\max} \to 90^\circ$ yields the best result but is, of course, practically unachievable because it implies an infinite scanning plane. Moreover, one needs to take into account the antenna beamwidth. This is illustrated in Fig. 4.4 on the side of **A2**. When the antenna beamwidth Ω_A is smaller than $2\alpha_{\max}$ and the cross-range displacement between the antenna and the scattering point $|x - x_s|$ is sufficiently large, the antenna is incapable of receiving the scattered signal. The effective "aperture angle" is then reduced to $\Omega_A/2$. Thus, $\alpha_{\max}$ in Eq. (4.61) is either one-half of the angle supported by the aperture or one-half of the antenna beamwidth, whichever is less. This is why low-gain antennas (but with low back-lobe radiation) are preferred in imaging.

The reader is encouraged to repeat the procedure for the sensitivity map $\bar{D}_{21}^{(m)}(x')$ and to show that it is the same as $\bar{D}_{11}^{(m)}(x')$.

4.3.5 Simulation-Based Example Illustrating the Sensitivity-Map Method

The next example employs simulations with the full-wave EM solver FEKO [122] to emulate the data acquisition with planar raster scanning of two planes on both sides of the inspected object. The distance between the planes is $d = 5$ cm. As shown in Fig. 4.5, the antennas are aligned along each other's boresight while moving together to scan their respective planes. Both, reflection (S_{11}, S_{22}) and transmission (S_{21}) coefficients are collected at 14 equally spaced frequency points ($N_\omega = 14$) from 3 GHz to 16 GHz. The scattered portion of these responses is obtained by subtracting the respective S-parameters obtained from the simulation of the scatterer-free setup (the baseline S-parameters). Since the medium between the antennas is uniform, the baseline simulation does not involve scanning; it simply acquires the S-parameters of the two aligned dipoles in air when the distance between them is $d = 5$ cm.

To accelerate the simulations, simple dipole antennas are used. Dipoles are not wideband as far as impedance match and radiation pattern are concerned; however, in

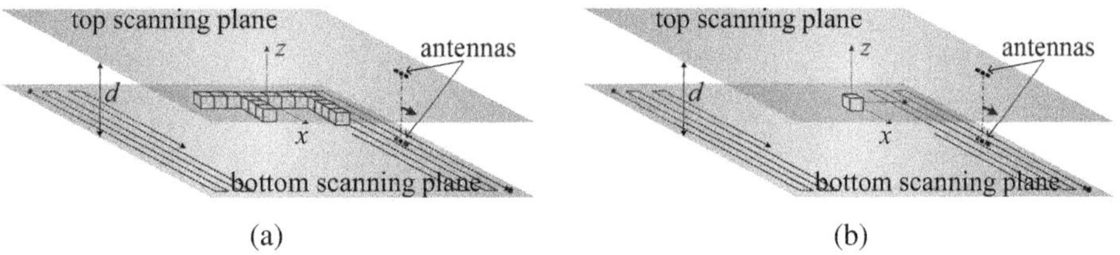

Figure 4.5 Simulation setup in the imaging example for an F-shaped dielectric object with planar scanning. Reflection and transmission coefficients are collected at the planar surfaces on both sides of the inspected object when this object is: (a) an F-shaped OUT, and (b) a voxel-sized cube at the center of the imaged volume that serves as a calibration object (CO).

simulations, we have the freedom to change their physical length at each frequency so that it is always half-wavelength. This ensures a good impedance match and the same illumination pattern across all the frequencies.

The scanned aperture is 20 × 20 cm with a sampling step of 1 cm along both x and y, i.e., the number of samples is $N_x \times N_y = 21 \times 21$. *Nyquist's sampling criterion* requires at least two samples per wavelength, i.e., the spatial sampling step across the aperture Δ_s must satisfy $\Delta_s \leq \lambda/2$, which at 16 GHz (in vacuum) translates into $\Delta_s \leq 9.4$ mm. Thus, the sampling step of 10 mm used here satisfies the Nyquist limit at all frequencies but 16 GHz. In Chapter 6, the impact of the chosen spatial sampling rate is discussed in more detail. Here, we only note that Nyquist's sampling criterion is a guideline that originates from the assumption of far-zone measurements and a worst-case scenario of oblique angles of arrival of the scattered signals. A sampling step only slightly larger than the Nyquist limit is likely to produce useful information. However, a sampling step that exceeds the Nyquist limit by more than 50% is likely to render the high-frequency content of the measurement useless.

The OUT is an F-shaped dielectric bar in the $z = 0$ plane; see Fig. 4.5-a. It has a relative permittivity $\varepsilon_{\mathrm{O,r}} = 1.2 - \mathrm{i}0$, whereas the background medium is air, $\varepsilon_{\mathrm{b,r}} = 1 - \mathrm{i}0$. The relative contrast, $\Delta\varepsilon_{\mathrm{r}} = 0.2 - \mathrm{i}0$, is sufficiently small to ensure successful reconstruction with a linearized scattering model. The F-shaped OUT is constructed of 16 voxel-sized cubicles, each being $1 \times 1 \times 1$ cm^3.

The PSFs are acquired in simulations that emulate a calibration step. As described in Section 4.2, the experimental acquisition of the PSFs is done by placing an electrically small known scattering object in the center of the setup. Here, this is a voxel-sized cube of relative permittivity $\varepsilon_{\mathrm{co,r}} = 1.1 - \mathrm{i}0$, i.e., $\Delta\varepsilon_{\mathrm{co,r}} = 0.1 - \mathrm{i}0$; see Fig. 4.5-b. The same subtraction of the baseline S-parameters is necessary to obtain the scattered portion of the PSFs, which is needed in the inversion.

We could also use an analytical spherical-wave assumption for the PSFs, namely, $\sim \exp[-\mathrm{i}2k_{\mathrm{b}}r(x,y)]/r^2(x,y)$, where $r(x,y)$ is the distance from the center of the setup to the Tx/Rx antenna position (x,y). Since the two antennas at both acquisition planes have common (x,y) coordinates at each measurement, this PSF approximation holds for both the reflection and the transmission PSF. This would save calibration time. Our measurements, however, are done close to or inside the near zone of the antenna, depending on the frequency. Moreover, the object is very close to the scanned planes. We can then expect the above approximation to be fairly inaccurate.

Fig. 4.6 shows the plots of the magnitudes and the phases of the simulated S_{11} and S_{21} PSFs at 4 GHz. We observe that the PSF magnitude distributions are wider in the x direction compared to the y direction. This indicates worse resolution in the x direction as compared to that along y, which is due to the dipole's radiation pattern being narrower in the xz plane than the yz plane. As Fig. 4.5 shows, the dipoles' axes are along x. Thus, the antennas are *omnidirectional* in the yz plane,[7] which implies infinite beamwidth. At

[7] An antenna is omnidirectional when its radiation pattern in one of the principal planes is a circle, i.e., the radiation intensity is the same in all angular direction in this plane.

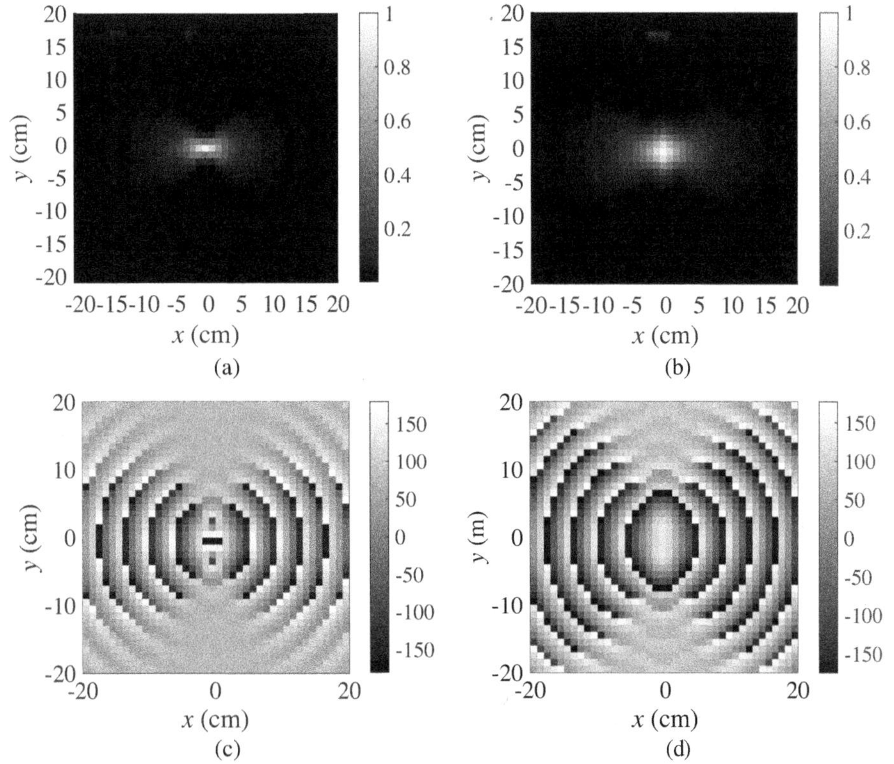

Figure 4.6 Plots of the PSFs at 4 GHz in the simulation example of the reconstruction of an F-shaped dielectric object: (a) $|S_{11}|$ PSF, (b) $|S_{21}|$ PSF. (c) $\angle S_{11}$ PSF, and (d) $\angle S_{21}$ PSF. (Courtesy of Denys Shumakov, Ph.D. student at McMaster University, Canada.)

the same time, the half-wavelength dipole has a beamwidth in the xz plane of about 72°, which limits the effective aperture angle in Eq. (4.61).

Another important observation to make is that the S_{11} PSF is not the same as the S_{21} PSF, which indicates that indeed using the analytical approximation $\exp[-i2k_b r(x,y)]/r^2(x,y)$ would be inadequate.

Using the reconstruction formula Eq. (4.28) and the procedure outlined in Section 4.3.3, the F-shaped object has been reconstructed at three planes: $z = -1, 0$, and $+1$ cm. The image obtained from the absolute value of the qualitative complex sensitivity map (using all frequencies) is shown in Fig. 4.7-a. For comparison, the single-frequency images obtained at 4 GHz and at 16 GHz are also shown in Fig. 4.7-b and Fig. 4.7-c, respectively. We observe that the cross-range resolution improves as the frequency increases. This is in agreement with Eq. (4.61), which shows that a shorter wavelength would result in a lower resolution limit. However, the various artifacts in the single-frequency sensitivity maps tend to cancel once they are added up, leading to an improved image in the final multifrequency image in Fig. 4.7-a. The arms of the F-shape that are parallel to the x axis appear clear whereas the bar along y is blurred along x. This is the result of the poorer resolution along x.

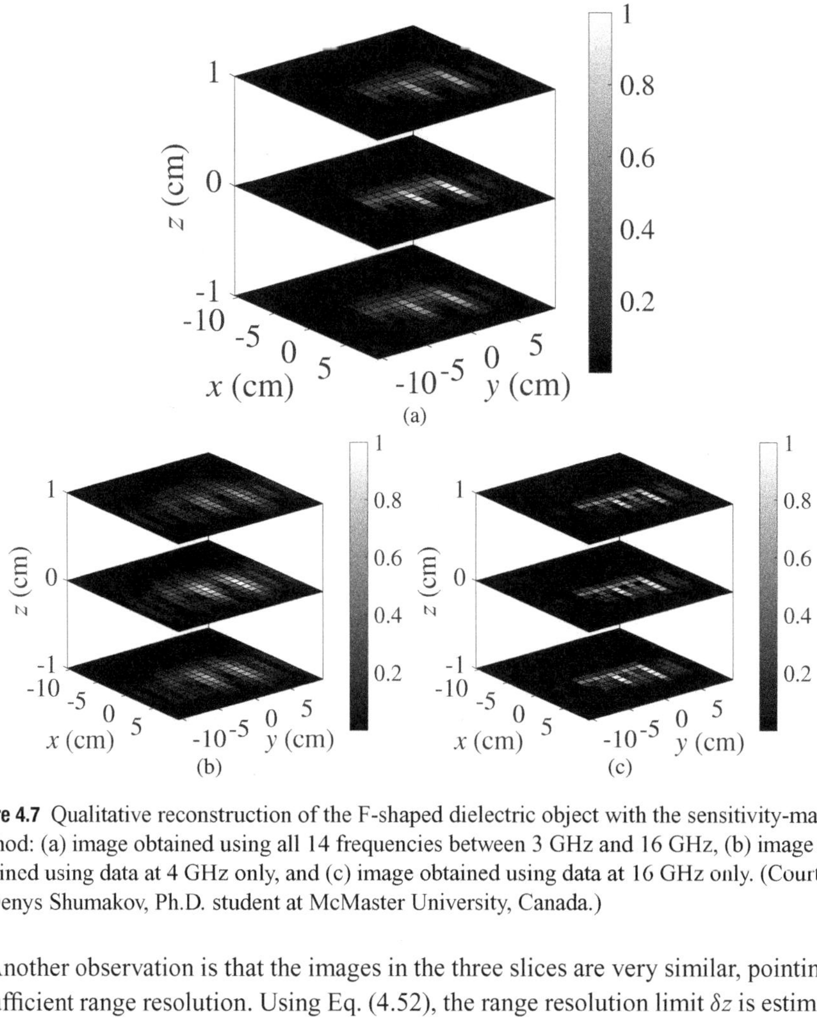

Figure 4.7 Qualitative reconstruction of the F-shaped dielectric object with the sensitivity-map method: (a) image obtained using all 14 frequencies between 3 GHz and 16 GHz, (b) image obtained using data at 4 GHz only, and (c) image obtained using data at 16 GHz only. (Courtesy of Denys Shumakov, Ph.D. student at McMaster University, Canada.)

Another observation is that the images in the three slices are very similar, pointing to insufficient range resolution. Using Eq. (4.52), the range resolution limit δz is estimated as 11.5 mm. The F-shaped object has a thickness of 10 mm with the plane $z = 0$ intersecting it in the middle. At the same time, the image slices are 10 mm apart, which is below δz. Moreover, the F-shaped object is sufficiently thick to have a common surface, with the voxel layers centered at $z = -10$ mm and $z = 10$ mm. Thus, the appearance of the F-shaped object in these slices (with slightly reduced intensity) is to be expected. The time required to generate these images is within milliseconds.

In Section 4.4, a method is presented that extends the sensitivity-map method to a quantitative approach, which not only estimates the actual permittivity values but also improves the image quality in terms of resolution.

4.3.6 Sensitivity-Map Inversion with Convolution

The reconstruction formula Eq. (4.28) is general in the sense that it accommodates responses collected on any arbitrary set of observation points. However, more often, the data are collected on surfaces of canonical shapes (plane, cylinder, etc.) at regular sampling intervals. Let us consider the case of planar acquisition such as the one illustrated in Fig. 4.1. The baseline and OUT responses are then functions of $(x, y) \in S_{\text{Rx}}$, where S_{Rx} is the scanned plane at $z_{\text{Rx}} = const$. Assuming an infinite sampling rate along x and y, Eq. (4.28) can be written as

$$D^{(m)}(x', y', z') = \iint_{S_{\text{Rx}}} \left[-\bar{S}^{(m)}(x, y, z_{\text{Rx}})\right] \cdot \left[\frac{dB^{(m)}(x, y, z_{\text{Rx}})}{d\varepsilon(x', y', z')}\right]^* dxdy, \tag{4.62}$$

where $D^{(m)}(x', y', z')$ is the 2D slice of the complex sensitivity map at the range location z'. If the medium is invariant along x and y,

$$\frac{dB^{(m)}(x, y, z_{\text{Rx}})}{d\varepsilon(x', y', z')} \sim \mathcal{K}^{(m)}(x, y, z_{\text{Rx}}; x', y', z') \sim S_0^{(m)}(x - x', y - y', z_{\text{Rx}}; z'), \tag{4.63}$$

where $S_0^{(m)}(x, y, z_{\text{Rx}}; z')$ is the PSF for a point scatterer in the plane z'; see Eq. (4.7). Therefore,

$$D^{(m)}(x', y', z') \sim -\iint_{S_{\text{Rx}}} \left[\bar{S}^{(m)}(x, y, z_{\text{Rx}})\right] \cdot \left[S_0^{(m)}(x - x', y - y', z_{\text{Rx}}; z')\right]^* dxdy. \tag{4.64}$$

The double integral is the convolution of $\bar{S}^{(m)}(x, y, z_{\text{Rx}})$ and $\left[S_0^{(m)}(-x, -y, z_{\text{Rx}}; z')\right]^*$, or equivalently, it is the cross-correlation of $\bar{S}^{(m)}(x, y, z_{\text{Rx}})$ and $S_0^{(m)}(x, y, z_{\text{Rx}}; z')$. Using the properties of the FT,[8] the sensitivity map can be expressed as

$$D^{(m)}(x', y', z') \sim -\mathcal{F}_{2\text{D}}^{-1}\left\{\mathcal{F}_{2\text{D}}\left\{\bar{S}^{(m)}(x, y, z_{\text{Rx}})\right\} \cdot \left[\mathcal{F}_{2\text{D}}\left\{S_0^{(m)}(x, y, z_{\text{Rx}}; z')\right\}\right]^*\right\}. \tag{4.65}$$

This is the convolution-based reconstruction formula of the sensitivity-map imaging method with planar scanning.

It is instructive to consider the particular case of far-zone measurements of the reflection coefficient, which arises in imaging with monostatic radar. In this case, the PSF can be estimated as

$$S_0^{(m)}(x, y, z_{\text{Rx}}; z') \approx e^{-\text{i}2k^{(m)}\sqrt{x^2+y^2+(z_{\text{Rx}}-z')^2}}, \tag{4.66}$$

where $k^{(m)}$ is the background wavenumber (possibly complex). This PSF can be viewed as a wave emanating from $(x' = 0, y' = 0, z')$ and propagating in a fictitious medium of wavenumber $2k^{(m)}$. It can be represented as a superposition of plane waves in the form

$$S_0^{(m)}(x, y, z_{\text{Rx}}; z') \approx \iint_{k_x\, k_y} e^{-\text{i}k_x x} e^{-\text{i}k_y y} e^{-\text{i}k_z^{(m)}(z_{\text{Rx}}-z')} dk_x dk_y, \tag{4.67}$$

[8] If $g(x) = f^*(-x)$, then in Fourier space, $G(k_x) = F^*(k_x)$.

where $k_z^{(m)} = \pm\sqrt{[2k^{(m)}]^2 - k_x^2 - k_y^2}$ and is subject to the constraints in Eq. (4.13). The reader is encouraged to show that for $z' = z_{\text{Rx}}$,

$$\mathcal{F}_{2\text{D}}\left\{S_0^{(m)}(x, y, z_{\text{Rx}} = z')\right\} = 1. \tag{4.68}$$

It follows from Eq. (4.68) and Eq. (4.9) that

$$\mathcal{F}_{2\text{D}}\left\{S_0^{(m)}(x, y, z_{\text{Rx}}; z')\right\} \approx e^{-\mathrm{i}k_z^{(m)}(z_{\text{Rx}}-z')}. \tag{4.69}$$

Substituting Eq. (4.69) into the sensitivity-map imaging formula, Eq. (4.65), yields

$$D^{(m)}(x', y', z') \sim -\mathcal{F}_{2\text{D}}^{-1}\left\{\mathcal{F}_{2\text{D}}\left\{\bar{S}^{(m)}(x, y, z_{\text{Rx}})\right\} \cdot e^{\mathrm{i}k_z^{(m)}(z_{\text{Rx}}-z')}\right\}. \tag{4.70}$$

This is the imaging formula with back-scattered far-zone data.

That the sensitivity map in Eq. (4.70) is a true qualitative representation of the contrast distribution in a cross-range slice can be easily shown using the data equation Eq. (3.58). It is worth making this derivation because it follows a standard approach to devising direct reconstruction formulas for imaging with far-zone data. The data equation for the reflection S-parameter due to a 2D target in the plane z' is written as

$$\bar{S}^{(m)}(x, y, z_{\text{Rx}}) \sim \iint_{x'y'} \Delta\varepsilon(x', y', z')\left[E_1^{\text{inc}}(x', y', z'; x, y, z_{\text{Rx}})\right]^2 dx'dy'. \tag{4.71}$$

The far-zone incident field is

$$E_1^{\text{inc}}(x', y', z'; x, y, z_{\text{Rx}}) \sim e^{-\mathrm{i}2k^{(m)}\sqrt{(x-x')^2+(y-y')^2+(z_{\text{Rx}}-z')^2}} = S_0^{(m)}(x-x', y-y', z_{\text{Rx}}; z'), \tag{4.72}$$

where $S_0^{(m)}(x, y, z_{\text{Rx}}; z')$ is the PSF defined in Eq. (4.66). Therefore, $\bar{S}^{(m)}(x, y, z_{\text{Rx}})$ in Eq. (4.71) is the convolution of the contrast $\Delta\varepsilon(x', y', z')$ and $S_0^{(m)}(x, y, z_{\text{Rx}}; z')$:

$$\bar{S}^{(m)}(x, y, z_{\text{Rx}}) \sim \mathcal{F}_{2\text{D}}^{-1}\left\{\mathcal{F}_{2\text{D}}\left\{\Delta\varepsilon(x', y', z')\right\} \cdot \mathcal{F}_{2\text{D}}\left\{S_0^{(m)}(x, y, z_{\text{Rx}}; z')\right\}\right\}. \tag{4.73}$$

Using Eq. (4.69), the above is written as

$$\mathcal{F}_{2\text{D}}\left\{\bar{S}^{(m)}(x, y, z_{\text{Rx}})\right\} \sim \mathcal{F}_{2\text{D}}\left\{\Delta\varepsilon(x', y', z')\right\} \cdot e^{-\mathrm{i}k_z^{(m)}(z_{\text{Rx}}-z')}. \tag{4.74}$$

Therefore,

$$\Delta\varepsilon(x', y', z') \sim \mathcal{F}_{2\text{D}}^{-1}\left\{\mathcal{F}_{2\text{D}}\left\{\bar{S}^{(m)}(x, y, z_{\text{Rx}})\right\} \cdot e^{\mathrm{i}k_z^{(m)}(z_{\text{Rx}}-z')}\right\}. \tag{4.75}$$

Comparing Eq. (4.75) to Eq. (4.70), we see that the sensitivity map $D^{(m)}(x', y', z')$ relates to the permittivity contrast as

$$D^{(m)}(x', y', z') \sim -\Delta\varepsilon(x', y', z'). \tag{4.76}$$

In summary, when the data are collected on surfaces of canonical shapes (planes, cylinders, spheres) at regular sampling intervals, the reconstruction formula of sensitivity-map imaging can be cast in the form of a convolution integral, which is solved very quickly using fast Fourier transform (FFT) and inverse FFT algorithms.

4.4 Quantitative Imaging with Scattered-Power Maps

Quantitative imaging with scattered-power maps builds on the qualitative method described in the previous section. The quantitative method was first reported in [97] along with a detailed study of its limitation and robustness to noise.

The method relies on quantitatively accurate PSFs, i.e., the PSFs must not only represent the normalized magnitude and phase distribution of the response to a point scatterer (this is sufficient in qualitative sensitivity-map imaging) but it also has to scale properly, in a complex sense, with the complex contrast of the point scatterer. Quantitatively accurate PSFs can be acquired via calibration measurements using the experimental method described in Section 4.2. The closer the background medium in these measurements is to the averaged electrical properties of the OUT, the better the quantitative output will be since the method remains rooted in a linearized model of scattering.

In far-zone measurement scenarios, "normalized" PSF approximations such as the plane-wave and the isotropic-wave approximations can also be used. For instance, in reflection measurements, the *plane-wave approximation* appears in the form of $e^{-i2\mathbf{k}\cdot\mathbf{r}}$ whereas the *isotropic-wave approximation* is $\frac{\exp(-i2kr)}{r^2}$, where $\mathbf{r}$ is the position vector of the scattering point relative to the Tx/Rx point and $r = |\mathbf{r}|$. These approximations can be improved if the antenna radiation pattern is known; see Section 2.14.2. Such approximations eliminate the need to perform calibration measurements and therefore save time. In fact, they are the only option when the imaged scene is either inaccessible for placements of a small scattering object (e.g., underground imaging) or completely unknown so that we cannot emulate it in the lab before the imaging system is deployed in the field. We immediately observe that analytical approximations of the PSF, although acceptable as normalized far-field distributions, lack the proper scaling, which depends on: (i) the power and the phase of the actual excitation, (ii) the size and the radiation characteristics of the Rx antenna, and (iii) the volume and the contrast of an actual point-like scatterer. This is why the imaged scene must have at least one point (even better a small region), the permittivity of which we know, e.g., the permittivity of the background, which we expect to occupy the fringes of the image. We can use this knowledge to at least approximately scale the result across the whole image.

4.4.1 Derivation of the Quantitative System of Linear Equation

Let us first state the data equation for the scattered portion of the ξ-th type of response acquired at $\mathbf{r}_i$ at the m-th frequency using the linearized model,

$$\bar{S}_{\xi}^{(m)}(\mathbf{r}_i) \approx \kappa_{\xi} \iiint_{V_{\mathrm{O}}} \Delta\varepsilon_{\mathrm{r}}(\mathbf{r}') \left[\mathbf{E}^{\mathrm{inc}}_{\xi,\mathrm{Rx}}(\mathbf{r}', \mathbf{r}_{\mathrm{Rx}}) \cdot \mathbf{E}^{\mathrm{inc}}_{\xi,\mathrm{Tx}}(\mathbf{r}', \mathbf{r}_{\mathrm{Tx}})\right] dv', \tag{4.77}$$

where κ_{ξ} is a known constant that depends on the type of response, e.g., if the ξ-th response is the S_{jk} parameter, $\kappa_{\xi} = -i\omega\epsilon_0/(2a_j a_k)$. The kernel in the square brackets is obtained from the system PSF $S_{0,\xi}^{(m)}(\mathbf{r}_i, \mathbf{r}')$ using the proportionality relationship in

Eq. (4.3), namely,

$$\kappa_\xi \left[\mathbf{E}^{\text{inc}}_{\xi,\text{Rx}}(\mathbf{r}',\mathbf{r}_{\text{Rx}}) \cdot \mathbf{E}^{\text{inc}}_{\xi,\text{Tx}}(\mathbf{r}',\mathbf{r}_{\text{Tx}})\right] = \frac{S^{(m)}_{0,\xi}(\mathbf{r}_i,\mathbf{r}')}{\delta\varepsilon_{\text{r}}\Omega_{\text{co}}}, \tag{4.78}$$

where $\mathbf{r}'$ is the position of the point-like scatterer, $\delta\varepsilon_{\text{r}}$ is its known relative permittivity contrast ($\delta\varepsilon_{\text{r}} = (\varepsilon_{\text{co}} - \varepsilon_{\text{b}})/\epsilon_0$), and Ω_{co} is its volume. As explained in Section 4.2, the PSFs are measured with an electrically small scatterer of known volume and permittivity positioned at the center of V_{O}, $\mathbf{r}' = 0$. These are the respective calibration-object (CO) responses; see Eq. (4.5) and Eq. (4.6). If the acquisition is invariant to translations along a given coordinate axis, this measurement is sufficient to obtain the CO responses for any other point-scatterer position $\mathbf{r}'$ shifted along this axis.

Next, we define the ratio contrast as the OUT contrast distribution normalized by the contrast of the point-like scatterer used to measure the PSFs:

$$\chi(\mathbf{r}') = \frac{\Delta\varepsilon_{\text{r}}(\mathbf{r}')}{\delta\varepsilon_{\text{r}}}. \tag{4.79}$$

Eq. (4.77) can now be written as

$$\bar{S}^{(m)}_{\xi}(\mathbf{r}_i) \approx \frac{1}{\Omega_{\text{co}}} \iiint_{V_{\text{O}}} \chi(\mathbf{r}') S^{(m)}_{0,\xi}(\mathbf{r}_i,\mathbf{r}')dv', \tag{4.80}$$

This is the linearized model relating the measured data to the system PSFs.

Next, we use Eq. (4.80) as the data term in the sensitivity-map reconstruction formula Eq. (4.28), which we write as

$$D^{(m)}(\mathbf{r}') = -\sum_{\xi=1}^{N_T}\sum_{i=1}^{N_S} \bar{S}^{(m)}_{\xi}(\mathbf{r}_i) \cdot \left[\frac{dB^{(m)}_{\xi}(\mathbf{r}_i)}{d\varepsilon(\mathbf{r}')}\right]^*, \tag{4.81}$$

where N_T indicates the number of response types acquired at each observation location $\mathbf{r}_i$ and N_S indicates the number of such locations (the number of spatial samples). Thus, the total number of responses acquired at each frequency is $N_R = N_T \times N_S$. Here, the baseline response derivative can be estimated as the finite-difference ratio

$$\frac{dB^{(m)}_{\xi}(\mathbf{r}_i)}{d\varepsilon(\mathbf{r}')} \approx \frac{S^{(m)}_{0,\xi}(\mathbf{r}_i,\mathbf{r}')}{\delta\varepsilon_{\text{r}}}, \tag{4.82}$$

since the PSF $S^{(m)}_{0,\xi}(\mathbf{r}_i,\mathbf{r}')$ is exactly the change in the baseline response due to a contrast $\delta\varepsilon_{\text{r}}$ in the voxel at $\mathbf{r}'$. The substitution of Eq. (4.82) into Eq. (4.81) yields

$$D^{(m)}(\mathbf{r}') = -\frac{1}{\delta\varepsilon^*_{\text{r}}}\sum_{\xi=1}^{N_T}\sum_{i=1}^{N_S} \bar{S}^{(m)}_{\xi}(\mathbf{r}_i) \cdot \left[S^{(m)}_{0,\xi}(\mathbf{r}_i,\mathbf{r}')\right]^*. \tag{4.83}$$

This is the OUT sensitivity map. In [97], the distribution $M^{(m)}(\mathbf{r}') = -\delta\varepsilon^*_{\text{r}} D^{(m)}(\mathbf{r}')$,

$$M^{(m)}(\mathbf{r}') = \sum_{\xi=1}^{N_T}\sum_{i=1}^{N_S} \bar{S}^{(m)}_{\xi}(\mathbf{r}_i) \cdot \left[S^{(m)}_{0,\xi}(\mathbf{r}_i,\mathbf{r}')\right]^*, \tag{4.84}$$

is termed the *scattered-power map* (or simply *power map*) since it represents a measure of the scattered power emanating from the scattering center at $\mathbf{r}'$ in the OUT compared to the scattered power obtained when a point scatterer is located at $\mathbf{r}'$ in the CO.

Substituting Eq. (4.80) into Eq. (4.83) yields

$$D^{(m)}(\mathbf{r}') = -\frac{1}{\Omega_{\text{co}}\delta\varepsilon_{\text{r}}^*}\sum_{\xi=1}^{N_T}\sum_{i=1}^{N_S}\left[\iiint_{V_\text{O}}\chi(\mathbf{r}'')S_{0,\xi}^{(m)}(\mathbf{r}_i,\mathbf{r}'')dv''\right]\cdot\left[S_{0,\xi}^{(m)}(\mathbf{r}_i,\mathbf{r}')\right]^*. \quad (4.85)$$

The order of summation and integration is exchanged to obtain

$$D^{(m)}(\mathbf{r}') = -\frac{1}{\Omega_{\text{co}}\delta\varepsilon_{\text{r}}^*}\iiint_{V_\text{O}}\left\{\chi(\mathbf{r}'')\sum_{\xi=1}^{N_T}\sum_{i=1}^{N_S}\left[S_{0,\xi}^{(m)}(\mathbf{r}_i,\mathbf{r}'')\right]\cdot\left[S_{0,\xi}^{(m)}(\mathbf{r}_i,\mathbf{r}')\right]^*\right\}dv''. \quad (4.86)$$

If Eq. (4.86) is applied to the CO data obtained with the point-like scatterer at $\mathbf{r}'_n$, it produces the respective CO sensitivity map $D^{(m)}_{\text{co@}\mathbf{r}'_n}(\mathbf{r}')$. Note that this map is not a δ-function, i.e., in general, it contains nonzero values at $\mathbf{r}' \neq \mathbf{r}'_n$. When the point-like scatterer is at $\mathbf{r}'_n$, the CO ratio contrast is

$$\chi_{\text{co}}(\mathbf{r}') = \begin{cases} 1, & \text{if } \mathbf{r}' \in \Omega_{\text{co}} \\ 0, & \text{elsewhere.} \end{cases} \quad (4.87)$$

Also, it is assumed that Ω_{co} is sufficiently small to ensure that the PSF $S_{0,\xi}^{(m)}(\mathbf{r}_i,\mathbf{r}'')$ remains relatively constant for $\forall\mathbf{r}'' \in \Omega_{\text{co}}$. Thus, $D^{(m)}_{\text{co@}\mathbf{r}'_n}(\mathbf{r}')$ (along with the respective power map $M^{(m)}_{\text{co@}\mathbf{r}'_n}(\mathbf{r}')$) is obtained from Eq. (4.86) as

$$D^{(m)}_{\text{co@}\mathbf{r}'_n}(\mathbf{r}') = -\frac{1}{\delta\varepsilon_{\text{r}}^*}\sum_{\xi=1}^{N_T}\sum_{i=1}^{N_S}\left[S_{0,\xi}^{(m)}(\mathbf{r}_i,\mathbf{r}'_n)\right]\cdot\left[S_{0,\xi}^{(m)}(\mathbf{r}_i,\mathbf{r}')\right]^* = -\frac{1}{\delta\varepsilon_{\text{r}}^*}M^{(m)}_{\text{co@}\mathbf{r}'_n}(\mathbf{r}'). \quad (4.88)$$

Comparing Eq. (4.88) with Eq. (4.86), we see that the power map $M^{(m)}(\mathbf{r}')$ for any OUT can be expressed in terms of the CO power maps $M^{(m)}_{\text{co@}\mathbf{r}''}(\mathbf{r}')$, where $\mathbf{r}'' \in V_\text{O}$ indicates the position of the point scatterer:

$$M^{(m)}(\mathbf{r}') = \frac{1}{\Omega_{\text{co}}}\iiint_{V_\text{O}}\chi(\mathbf{r}'')\cdot M^{(m)}_{\text{co@}\mathbf{r}''}(\mathbf{r}')dv'',\; m = 1,\ldots,N_\omega. \quad (4.89)$$

This result can incorporate a wide-band frequency sweep easily by applying the summation step Eq. (4.32) over all single-frequency sensitivity maps of the OUT and the CO, which results in

$$M(\mathbf{r}') = \frac{1}{\Omega_{\text{co}}}\iiint_{V_\text{O}}\chi(\mathbf{r}'')\cdot M_{\text{co@}\mathbf{r}''}(\mathbf{r}')dv''. \quad (4.90)$$

Note, however, that the CO and the OUT power maps at each frequency must be normalized to a common factor $\eta^{(m)}$—this would correspond to dividing both sides of

Eq. (4.89) by $\eta^{(m)}$ before summing over $m = 1, \ldots, N_\omega$. This is why, $\eta^{(m)}$ is first computed from the m-th frequency CO power map and is then used to normalize both the CO and the OUT power maps at ω_m.

Eq. (4.90) is the basis of the quantitative inversion with power maps. Upon discretization of V_O into voxels of volume $\Delta\Omega$, it yields a square linear system of equations of size N_p, where N_p is the number of voxels:

$$\boldsymbol{A}\boldsymbol{x} = \boldsymbol{b}. \tag{4.91}$$

Here, the elements of the vector $\boldsymbol{x}$ contain the unknown ratio contrast at each voxel:

$$x_n = \frac{\Delta\Omega}{\Omega_{\text{co}}} \cdot \chi(\mathbf{r}'_n),\ n = 1, \ldots, N_p. \tag{4.92}$$

The vector $\boldsymbol{b}$ contains the values of the OUT sensitivity map:

$$\boldsymbol{b} = \left[M(\mathbf{r}'_1) \ \cdots \ M(\mathbf{r}'_{N_p}) \right]^T. \tag{4.93}$$

The n-the column of the system matrix is composed of the values of the CO sensitivity map that corresponds to a pointlike scatterer at $\mathbf{r}'_n$, so that

$$\boldsymbol{A} = \begin{bmatrix} M_{\text{co}@\mathbf{r}'_1}(\mathbf{r}'_1) & \cdots & M_{\text{co}@\mathbf{r}'_{N_p}}(\mathbf{r}'_1) \\ \vdots & \ddots & \vdots \\ M_{\text{co}@\mathbf{r}'_1}(\mathbf{r}'_{N_p}) & \cdots & M_{\text{co}@\mathbf{r}'_{N_p}}(\mathbf{r}'_{N_p}) \end{bmatrix}. \tag{4.94}$$

Once the solution $\boldsymbol{x}$ is found, the 3D permittivity distribution of the OUT is recovered as

$$\varepsilon_{\text{r}}(\mathbf{r}'_n) = \varepsilon_{\text{r,b}}(\mathbf{r}'_n) + \left(\frac{\delta\varepsilon_{\text{r}}\Omega_{\text{co}}}{\Delta\Omega} \right) \cdot x_n,\ n = 1, \ldots, N_p. \tag{4.95}$$

The method of solving Eq. (4.91) is critical for the speed and the fidelity of the quantitative reconstruction. Since the size of the system N_p is large (on the order of 10^3 to 10^5), direct inversion or *LU* decomposition is not practical. Iterative linear solvers are preferable. These solvers converge fast with a good initial guess. Such an initial guess is provided by [97],

$$x_n^{(0)} = \frac{M(\mathbf{r}'_n)}{M_{\text{co}@\mathbf{r}'_n}(\mathbf{r}'_n)},\ n = 1, \ldots, N_p, \tag{4.96}$$

which is, in essence, a solution of Eq. (4.91) where all offdiagonal elements of $\boldsymbol{A}$ have been ignored.

There is a good physical argument underlying this choice. A CO map $M_{\text{co}@\mathbf{r}'_n}(\mathbf{r}'_k)$, $k = 1, \ldots, N_p$, is nothing but the scaled complex sensitivity map of the point scatterer, when this scatterer is located at $\mathbf{r}'_n$. Ideally, this map should behave as a 3D δ-function with a peak at $\mathbf{r}'_n$. Of course, this is never the case because that behavior would imply infinitesimal spatial resolution. However, as discussed in Section 4.3, the spatial resolution does improve at higher frequencies and with wider bandwidths (which also implies higher frequencies). Thus, at high frequencies, the system matrix $\boldsymbol{A}$ is diagonally dominant and well-conditioned. It is in such cases, too, that the initial guess in Eq. (4.96) is good, leading to a fast iterative solution.

In the extreme opposite case, where the whole imaged volume is below the resolution limits (i.e., when very low frequencies are employed), all elements of the CO power map tend to be similar, $\boldsymbol{A}$ is ill-conditioned, and the initial guess provided by Eq. (4.96) is going to be far from the true solution. In such cases, the iterative solution is also problematic, in the sense that it takes longer to converge, it may converge to a spurious local minimum, or it may even diverge.

Finally, it is desirable to have an iterative solver that can incorporate physical constraints on the permittivity distribution of the OUT. Typical physical constraints would require that

$$\Re\varepsilon_r(\mathbf{r}') \geq 1, \quad \Im\varepsilon_r(\mathbf{r}') \leq 0. \tag{4.97}$$

Two important comments should be made regarding the limitations of the quantitative power-map method. First, it is based on a linear relationship between the OUT response and the system PSFs; see Eq. (4.80). Therefore, it cannot take into account nonlinear effects such as multiple scattering, coupling between scattering targets, and, ultimately, the nonlinear relationship between the total internal field and the contrast. Thus, it remains limited to weak scattering. Its usefulness in general problems of strong (nonlinear) scattering is twofold: (i) it can provide a good initial guess for a nonlinear reconstruction procedure; and (ii) it can serve as a linear inversion module within nonlinear reconstruction procedures such as the Born iterative method (BIM) and the distorted Born iterative method (DBIM) [31, 46].

Second, an astute reader may observe that we could have solved Eq. (4.80) directly for the unknown $\chi(\mathbf{r}')$ by discretizing the volume integral as we did with Eq. (4.90). Note that Eq. (4.80) can also be used in a multi-frequency scheme with the same normalize-and-sum strategy used to obtain Eq. (4.90) from its single-frequency counterpart. This would require the solution of a rectangular system of equations $\boldsymbol{B}\boldsymbol{x} = \boldsymbol{d}$, which is underdetermined ($N_R < N_p$). This rectangular system is closely related to Eq. (4.91). This solution could be performed using, for example, the singular-value decomposition (SVD) method, which naturally accommodates regularization through the truncated SVD (TSVD) approach; see, for example, [46]. The various algorithms for SVD of a matrix scale as $O(\alpha mn^2 + \beta m^2 n + \gamma n^3)$, where m and n denote the number of rows and columns, respectively, and α, β, and γ are constants [123]. Thus, the TSVD may not be suitable for problems of large size. Other methods exist, typically iterative in nature, that can handle very large underdetermined rectangular systems of equations. In [124], a pseudo-inverse strategy in combination with a Gauss–Seidel iterative update is reported that can solve a sparse system of equations of size $15 \cdot 10^6 \times 100 \cdot 10^6$ in less than 0.5 s using several graphic-processing units (GPUs)!

It can be shown that $\boldsymbol{B}^{\dagger}\boldsymbol{B}\boldsymbol{x} = \boldsymbol{B}^{\dagger}\boldsymbol{d}$ is equivalent to the system $\boldsymbol{A}\boldsymbol{x} = \boldsymbol{B}$ in Eq. (4.91). Here, $\boldsymbol{B}^{\dagger}$ is the Hermitian (or transposed and conjugated) of $\boldsymbol{B}$. It is well known that if the data $\boldsymbol{d}$ are not entirely in the range of the forward operator represented

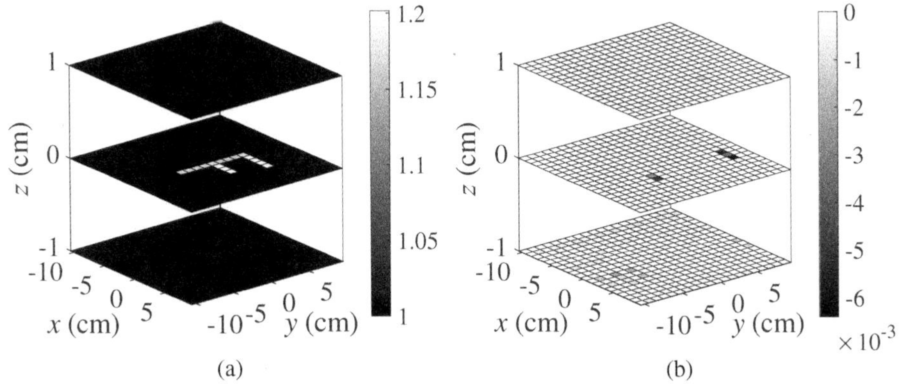

Figure 4.8 Quantitative reconstruction of the F-shaped dielectric object with the power-map method: (a) $\Re\varepsilon_r$, and (b) $\Im\varepsilon_r$. (Courtesy of Denys Shumakov.)

by $\boldsymbol{B}$, the direct solution of $\boldsymbol{Bx} = \boldsymbol{d}$ may not exist. In contrast, under the same conditions, $\boldsymbol{B}^\dagger\boldsymbol{Bx} = \boldsymbol{B}^\dagger\boldsymbol{d}$ may have a solution, which minimizes the least-square error of $\boldsymbol{Bx} = \boldsymbol{d}$; see, for example, [26, 46]. This is also a strong argument in favor of obtaining the PSFs experimentally, which ensures that the system matrix $\boldsymbol{A}$ provides the physically correct functional space spanning all possible responses acquired with the particular imaging setup.

In addition, constructing the power maps has two advantages. The power maps are generated practically instantaneously, which provides images in real time. In addition, a good initial guess of the OUT contrast can be obtained once the power maps are generated; see Eq. (4.96). This is critical in dealing with nonuniqueness and in accelerating the iterative solution to Eq. (4.91).

4.4.2 Examples Illustrating the Scattered-Power Map Method

We first revisit the simulation-based example from Section 4.3.5, where the OUT is an F-shaped dielectric bar of relative permittivity, $\varepsilon_r = 1.2 + i0$, immersed in a background of relative permittivity, $\varepsilon_{r,b} = 1.0 + i0$. The small scatterer in the CO, which is measured to obtain the PSFs, is $\varepsilon_{r,co} = 1.1 + i0$. The power maps are already available for both the CO and the F-shaped OUT from our qualitative solution in Section 4.3.5 since they are simply the multi frequency sensitivity maps before they were normalized for plotting: see, for example, the normalized qualitative F-shaped object shown in Fig. 4.7-a. The linear system of equations Eq. (4.91) is solved using the MATLAB function *lsqlin* [79] and imposing the constraints of Eq. (4.97).

Fig. 4.8 shows the 3D quantitative result. The shape and localization quality of the reconstruction are dramatically improved in comparison with the qualitative map in Fig. 4.7-a. In particular, the range resolution is much improved despite the fact that the separation between the image slices is below the estimated range resolution limit. The

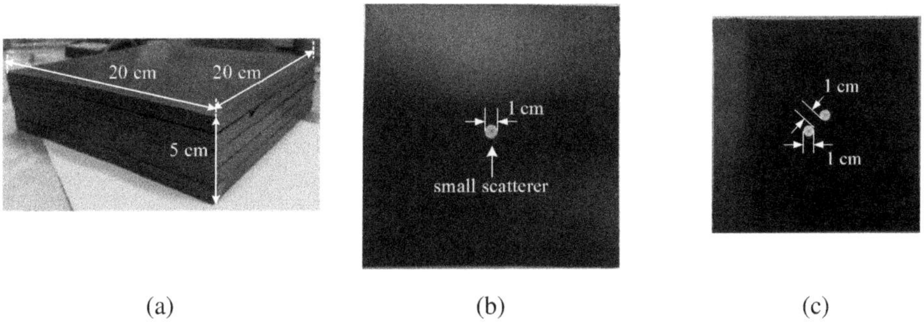

Figure 4.9 Materials used in the example of imaging dielectric inclusions in a lossy dielectric medium: (a) the stack of five absorbing sheets that serves as the reference object (RO) and that is used to acquire the baseline S-parameters; (b) the middle (3rd) absorbing sheet of the calibration object (CO) that hosts a small dielectric cylinder of diameter 1 cm, height 1 cm and relative permittivity $\varepsilon_{\mathrm{r,co}} = 15 - \mathrm{i}0.003$; (c) the middle (3rd) absorbing sheet of the object under test (OUT) that hosts two small dielectric cylinders both of the same permittivity as the dielectric cylinder in the CO and separated by an edge-to-edge distance of 1 cm.

quantitative result is also excellent. To evaluate the fidelity of the reconstruction of a known object, it is customary to use the *relative root-mean-square error* (RRMSE):

$$\mathrm{RRMSE} = \sqrt{\frac{1}{N_p}\sum_{n=1}^{N_p}\frac{\left|\varepsilon_{\mathrm{r}}(\mathbf{r}'_n) - \bar{\varepsilon}_{\mathrm{r}}(\mathbf{r}'_n)\right|^2}{\left|\bar{\varepsilon}_{\mathrm{r}}(\mathbf{r}'_n)\right|^2}}, \tag{4.98}$$

where $\bar{\varepsilon}_{\mathrm{r}}(\mathbf{r}'_n)$ is the true distribution of the complex permittivity. The result in Fig. 4.8 has an RRMSE of about 0.01.

In the next example, we use experimental data collected with planar raster scanning (along x and y) where transmission S-parameters are acquired in the frequency band from 3 GHz to 9 GHz with 86 equally spaced frequency samples. The scanning is performed with a mechanical 2D linear positioner. The Rx side consists of a nine-element switched antenna array, where the elements are broadband bow-tie antennas [118]. Five of them are polarized along the x axis, and four of them are polarized along y. The Tx antenna is a TEM horn [125] polarized along x. Both the Rx and the Tx antennas are designed to operate in direct contact with a dielectric medium of relative permittivity ε_{r} between 8 and 12 in the frequency band between 3 GHz and 9 GHz. The design ensures a good impedance match of return loss better than 6 dB.

Here, the background medium is composed of five stacked absorber sheets; see Fig. 4.9-a.[9] The real part of the relative permittivity of the absorbing sheets is $\varepsilon'_{\mathrm{r,b}} \approx 10$ in the whole frequency band, whereas the dielectric loss tangent $\tan\delta_{\mathrm{d,b}}$ varies almost linearly from 1.0 at 3 GHz down to 0.5 at 9 GHz. Each sheet has dimensions $20 \times 20 \times 1$ cm. Thus the overall thickness of the measured object is 5 cm, which introduces significant attenuation in the signals that varies from about 40 dB at 3 GHz to

[9] Emerson & Cuming Microwave Products Inc. Randolph, MA.

more than 60 dB at 9 GHz. This necessitates the use of a power amplifier on the transmitting end to preserve an acceptable signal-to-noise ratio (SNR), which (as an average over all nine Rx elements) varies from about 10 dB at 3 GHz to 2 dB at 9 GHz. The antennas do not come in direct contact with the surface of the object to prevent friction; they scan at a distance of about 2 mm from the surface on both the Rx and the Tx sides. The scanned area is 15 cm $\times$ 15 cm. The setup that consists of the measurement hardware and the five stacked absorbing sheets is the reference object (RO), which provides the baseline S-parameters.

The PSFs are acquired both via simulations and experimentally. The simulated PSFs do not result in meaningful images, although every effort has been made to represent faithfully the imaging setup in the simulation. It is worth noting that the antennas have fairly complex geometries; they are electrically quite large (especially the receiving array) and contain fine shape details. The discretization mesh has been refined for a mesh convergence error below 2%, which in turn leads to a simulation time longer than a day. The failure of the reconstruction with the simulated PSFs points toward significant modeling errors. However, the reconstruction is successful with the measured PSFs, which are acquired with a single mechanical scan. These are measurements of the CO, which is identical to the RO except for a small cylindrical inclusion (of relative permittivity $\varepsilon_{\mathrm{r,co}} = 15 - \mathrm{i}0.003$) in the center of the stacked absorber sheets; see Fig. 4.9-b.

Fig. 4.9-c shows the middle (third) layer of the OUT, which contains two cylindrical inclusions of the same relative permittivity as the inclusion in the CO. These are spaced so that the edge-to-edge distance between the two is 1 cm.

This imaging task is far more challenging than the simulation-based example with the F-shape due to the noise and uncertainties in the data. The SNR level, which varies between 2 dB and 10 dB implies poor signal quality but this is the typical scenario when measuring lossy objects such as those in tissue imaging, nondestructive testing or underground surveillance.

The CO and the OUT data are obtained by subtracting the baseline (RO) S-parameters from the CO and the OUT S-parameters, respectively. The CO and the OUT power maps are computed using Eq. (4.84) at each frequency. The multi frequency maps are then calculated using the normalize-and-sum formula Eq. (4.32). Then, the system matrix $\boldsymbol{A}$ in Eq. (4.94) is constructed from the CO power maps, and the vector $\boldsymbol{b}$ in Eq. (4.93) is constructed from the values of the OUT power map. The solution of the linear system in Eq. (4.91) in conjunction with Eq. (4.95) provides the quantitative estimate of the complex permittivity distribution in the OUT. The result is shown in Fig. 4.10. The estimated permittivity distribution of the OUT is very good for both the real and the imaginary parts.

4.4.3 Scattered-Power Map Imaging with Convolution

As in the case of the qualitative imaging with sensitivity maps (see Section 4.3.6), the quantitative forward model presented by Eq. (4.90) can be inverted very efficiently if the

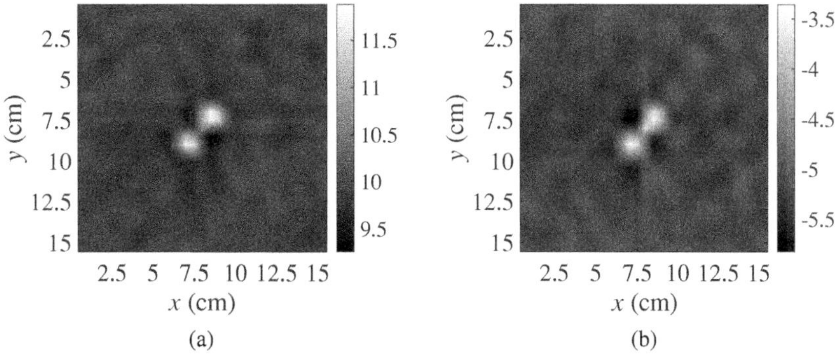

Figure 4.10 Quantitative result for the relative complex permittivity obtained using the scattered-power map method in the imaging of the two dielectric inclusions in a lossy medium using experimental data: (a) real part and (b) imaginary part. (Courtesy of Denys Shumakov.)

imaging system is invariant to translations along some or all of the axes of the coordinate system.

To illustrate this concept, we revisit the planar-scanning scenario where the baseline, the CO, and the OUT responses are functions of $(x, y) \in S_{\rm Rx}$, where $S_{\rm Rx}$ is the scanned plane at $z_{\rm Rx} = const.$ If the baseline responses remain the same as the antennas scan along the x and y axes, the imaging system is translationally invariant along x and y. This requires that the background's electrical properties are independent of x and y. In practice, small variations are inevitable even if the system is carefully designed to minimize positioning errors and reflections from enclosures. However, these variations are suppressed once the baseline responses are subtracted from the CO and the OUT responses to obtain their respective scattered components. As a result, in such imaging systems, the CO power maps $M_{{\rm co}@(x'_n,y'_n,z'_n)}(x', y', z')$ are simply translated versions of the CO power maps $M_{{\rm co}@(0,0,z'_n)}(x', y', z')$ obtained with the point scatterer at the origin of the $z'_n = const.$ plane:

$$M_{{\rm co}@(x'_n,y'_n,z'_n)}(x', y', z') = M_{{\rm co}@(0,0,z'_n)}(x' - x'_n, y' - y'_n, z'). \qquad (4.99)$$

Therefore, Eq. (4.90) can be written as

$$M(x', y', z') = \frac{1}{\Omega_{\rm co}} \iiint\limits_{x''y''z''} \chi(x'', y'', z'') \cdot M_{{\rm co}@(0,0,z'')}(x' - x'', y' - y'', z')dv''. \quad (4.100)$$

This is clearly a 2D convolution in (x', y'), and it can be written as

$$\tilde{M}(k_x, k_y, z') = \frac{\Delta x \Delta y}{\Omega_{\rm co}} \int\limits_{z''} \tilde{\chi}(k_x, k_y, z'') \cdot \tilde{M}_{{\rm co}@(0,0,z'')}(k_x, k_y, z')dz'', \qquad (4.101)$$

where the *tilde* denotes the 2D FT of a function of (x, y) and Δx and Δy are the lateral dimensions of the voxel size. Representing the integral along z'' as a sum leads to

$$\tilde{M}(k_x, k_y, z_p) = \frac{\Delta\Omega}{\Omega_{\rm co}} \sum_{q=1}^{N_z} \tilde{\chi}(k_x, k_y, z_q) \cdot \tilde{M}_{{\rm co}@(0,0,z_q)}(k_x, k_y, z_p), \; p = 1, \ldots, N_z, \quad (4.102)$$

where $\Delta\Omega = \Delta x \Delta y \Delta z$ is the voxel's volume, Δz is the sampling step along z, and N_z is the number of image slices along z. We can now solve a small square ($N_z \times N_z$) linear system of equations at each point in Fourier space $\boldsymbol{\kappa} = (k_x, k_y)$:

$$\boldsymbol{M}_{\boldsymbol{\kappa}} \boldsymbol{x}_{\boldsymbol{\kappa}} = \boldsymbol{m}_{\boldsymbol{\kappa}}, \tag{4.103}$$

where

$$\boldsymbol{x}_{\boldsymbol{\kappa}} = \frac{\Omega}{\Omega_{\text{co}}} \left[\tilde{\chi}(\boldsymbol{\kappa}, z_1) \cdots \tilde{\chi}(\boldsymbol{\kappa}, z_{N_z}) \right]^T,$$

$$\boldsymbol{m}_{\boldsymbol{\kappa}} = \left[\tilde{M}(\boldsymbol{\kappa}, z_1) \cdots \tilde{M}(\boldsymbol{\kappa}, z_{N_z}) \right]^T,$$

$$\boldsymbol{M}_{\boldsymbol{\kappa}} = \begin{bmatrix} \tilde{M}_{\text{co}@(0,0,z_1)}(\boldsymbol{\kappa}, z_1) & \cdots & \tilde{M}_{\text{co}@(0,0,z_{N_z})}(\boldsymbol{\kappa}, z_1) \\ \tilde{M}_{\text{co}@(0,0,z_1)}(\boldsymbol{\kappa}, z_2) & \cdots & \tilde{M}_{\text{co}@(0,0,z_{N_z})}(\boldsymbol{\kappa}, z_2) \\ \vdots & \ddots & \vdots \\ \tilde{M}_{\text{co}@(0,0,z_1)}(\boldsymbol{\kappa}, z_{N_z}) & \cdots & \tilde{M}_{\text{co}@(0,0,z_{N_z})}(\boldsymbol{\kappa}, z_{N_z}) \end{bmatrix}.$$

Once we have $\tilde{\chi}(\boldsymbol{\kappa}, z_p)$ for $\forall \boldsymbol{\kappa}$ and for all $p = 1, \ldots, N_z$, the 3D ratio contrast is recovered through 2D inverse Fourier transform slice by slice:

$$\chi(x', y', z'_p) = \mathcal{F}_{2\text{D}}^{-1} \left\{ \tilde{\chi}(k_x, k_y, z'_p) \right\}, \ z'_p = z_1, \ldots, z_{N_z}. \tag{4.104}$$

Finally, the relative complex permittivity of the OUT is calculated as

$$\varepsilon_{\text{r}}(x', y', z'_p) = \varepsilon_{\text{b,r}}(x', y', z'_p) + \delta\varepsilon_{\text{r}} \cdot \chi(x', y', z'_p). \tag{4.105}$$

The solution described above is significantly faster than the one described in Section 4.4.1. The reason is that the direct inversion in 3D space requires the solution of the very large system of equations in Eq. (4.91), where the system matrix $\boldsymbol{A}$ is of size $N_p \times N_p$. Here, $N_p = N_x N_y N_z$ is the total number of voxels in the imaged volume. In contrast, the convolution-based approach solves $N_x N_y$ small linear systems of equations, each of them having a system matrix of size $N_z \times N_z$; see Eq. (4.103). This is because when the 2D FFT is applied to a function of x and y, which is sampled evenly at N_x and N_y points, respectively, the resultant function of k_x and k_y has the same number of samples in k-space. Thus, the convolution-based approach needs to solve $N_x N_y$ systems of the kind in Eq. (4.103). It is faster to solve $N_x N_y$ systems of size $N_z \times N_z$ than to solve one system of size $N_x N_y N_z \times N_x N_y N_z$, especially because usually $N_z < N_\xi$, $\xi = x, y$. Depending on the size of the problem, the computation with the convolution-based approach can be two to three orders of magnitude faster. Of course, the price to be paid for this speed is that the method is applicable only to imaging systems that are invariant to translations in two of the three coordinate axes.

4.5 Linear Inversion with Pulsed Radar: Synthetic Focusing

The time-domain imaging systems are essentially radars that illuminate the OUT and collect the scattered pulsed signals as temporal waveforms. Just like the frequency-sweep S-parameter measurements, the goal is to collect the signals over as many spatial points as possible, which is achieved either by scanning the antennas over the acquisition surfaces or by RF switching. The acquisition technique where the radar collects data over a line or a surface, one spatial sample at a time, is referred to as *synthetic aperture radar* (SAR). The SAR data acquisition is no different from the frequency-sweep data acquisition over surfaces, except that its signals are functions of time.

Most of the pulsed radar systems are *ultra-wideband* radio systems. The term *ultra-wideband* (UWB) has a specific meaning, which is strictly adhered to in the engineering literature. A radio system is UWB if it has all of the following features:

- It occupies a part of the spectrum ranging from 100 MHz to about 10 GHz.
- It has a large fractional bandwidth b_{f}.
- It has a sufficiently large absolute bandwidth B, which must exceed a set value B_{min} (according to the various national and international conventions, B_{min} may be as low as 50 MHz and as high as 500 MHz).

The fractional bandwidth is defined as

$$b_{\mathrm{f}} = \frac{B}{f_{\mathrm{c}}} = 2\left(\frac{f_{\mathrm{max}} - f_{\mathrm{min}}}{f_{\mathrm{max}} + f_{\mathrm{min}}}\right), \tag{4.106}$$

where $f_{\mathrm{c}} = 0.5(f_{\mathrm{max}} + f_{\mathrm{min}})$ is the center frequency, whereas f_{min} and f_{max} are the minimum and maximum frequencies, respectively. The minimum and maximum frequencies correspond to a threshold value (typically at -10 dB) in the spectral power density of the UWB system. The fractional bandwidths of UWB systems exceed 0.25.

UWB radar is a rapidly growing technology, which has already found numerous applications and which holds enormous potential for further applications in detection and imaging [126, 127]. Examples of current applications include the following:

- Nondestructive testing in civil engineering: detection of cracks, bubbles, rust, defects in rebar (reinforcing steel bar in concrete), disaggregation of construction material, detection of foreign bodies, and detection of infrastructure behind the wall and under the floor (studs, wiring, pipes).
- Ground-penetrating radar (GPR) [128, 129, 130]: detection of objects close to the surface such as landmines, valuable metallic objects, pipes and cables, and tunnels.
- Search and rescue: detection of people buried under rubble or snow.
- Automotive radar: collision avoidance.
- Geology and archeology: buried objects and remnants of roads and buildings.

- Security and law enforcement: through-the-wall imaging [131].
- Medical engineering: imaging, life-sign monitoring [126, 132–138].

A comprehensive description of modern UWB radar electronics, the signal waveforms, and the signal-processing methods in various applications can be found in [126, 127, 139]. Here, the goal is to introduce the reader to the basic principles of image generation from time-domain signals.

4.5.1 The Linearized Time-Domain Forward Model

Let us revisit the scalar time-domain model of scattering in Eq. (1.258) and restate it in terms of the (total) field generated by the Tx antenna, u_{Tx}; the impulse field response in the background when the Rx antenna operates in a transmitting mode, $h_{\text{Rx}}^{\text{inc}}$; and the contrast κ:

$$s(\mathbf{r}_{\text{Rx}}, t; \mathbf{r}_{\text{Tx}}) = \iiint_{V_{\text{O}}} \kappa(\mathbf{r}') \left[h_{\text{Rx}}^{\text{inc}} * \frac{\partial^2 u_{\text{Tx}}}{\partial t'^2} \right]_{(\mathbf{r}_{\text{Rx}}, t; \mathbf{r}')} dv'. \tag{4.107}$$

Here, $s(\mathbf{r}_{\text{Rx}}, t; \mathbf{r}_{\text{Tx}})$ is the scattered portion of the response waveform received at $\mathbf{r}_{\text{Rx}}$ when the Tx antenna is at $\mathbf{r}_{\text{Tx}}$, and $*$ denotes convolution. The contrast is expressed through the speed of light in the background υ_{b} and that in the scatterer, υ_{s} as $\kappa(\mathbf{r}') = \upsilon_{\text{sc}}^{-2}(\mathbf{r}') - \upsilon_{\text{b}}^{-2}$ [see Eq. (1.204)].

Under the assumption of weak scattering, $u_{\text{Tx}} \approx u_{\text{Tx}}^{\text{inc}}$. As per Eq. (1.252), u_{Tx} is the convolution of the background field impulse response at $\mathbf{r}'$ and the excitation waveform $w_{\text{Tx}}(t)$:

$$u_{\text{Tx}}(\mathbf{r}', t'; \mathbf{r}_{\text{Tx}}) \approx u_{\text{Tx}}^{\text{inc}}(\mathbf{r}', t'; \mathbf{r}_{\text{Tx}}) = h_{\text{Tx}}^{\text{inc}}(\mathbf{r}', t; \mathbf{r}_{\text{Tx}}) * w_{\text{Tx}}(t). \tag{4.108}$$

Using the properties of the convolution, Eq. (4.107) is written as

$$s(\mathbf{r}_{\text{Rx}}, t; \mathbf{r}_{\text{Tx}}) \approx \iiint_{V_{\text{O}}} \kappa(\mathbf{r}') \left[h_{\text{Rx}}^{\text{inc}} * h_{\text{Tx}}^{\text{inc}} * w_{\text{Tx}}'' \right]_{(\mathbf{r}_{\text{Rx}}, t; \mathbf{r}_{\text{Tx}}; \mathbf{r}')} dv', \tag{4.109}$$

where the double prime in w'' denotes a second derivative in time.

The term in the square brackets in Eq. (4.109),

$$\mathcal{K}(\mathbf{r}_{\text{Rx}}, t; \mathbf{r}_{\text{Tx}}; \mathbf{r}') = h_{\text{Rx}}^{\text{inc}}(\mathbf{r}', t'; \mathbf{r}_{\text{Rx}}) * h_{\text{Tx}}^{\text{inc}}(\mathbf{r}', t'; \mathbf{r}_{\text{Tx}}) * w_{\text{Tx}}''(t'), \tag{4.110}$$

is the *time-domain resolvent kernel* of the linearized scattering equation. It is independent of the OUT, and we can assume that it is known, at least approximately.

Let $s_0(\mathbf{r}_{\text{Rx}}, t; \mathbf{r}_{\text{Tx}})$ be the PSF obtained with a very small scatterer of contrast κ_{co} and volume Ω_{co} at the origin of V_{O}, $\mathbf{r}' = 0$. As per Eq. (4.109), this PSF can be approximated as

$$s_0(\mathbf{r}_{\text{Rx}}, t; \mathbf{r}_{\text{Tx}}) \approx \kappa_{\text{co}} \left[h_{\text{Rx}}^{\text{inc}} * h_{\text{Tx}}^{\text{inc}} * w_{\text{Tx}}'' \right]_{(\mathbf{r}_{\text{Rx}}, t; \mathbf{r}_{\text{Tx}}; 0)} \Omega_{\text{co}}. \tag{4.111}$$

Here, we have adopted the notation CO (in the subscript of κ_{co}) from the previous chapter, where the term *calibration object* (CO) was introduced to denote an object where

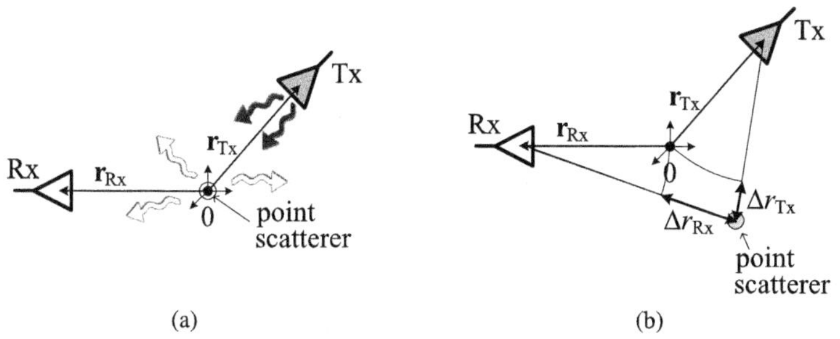

Figure 4.11 Illustration of measurements with a small (point) scatterer: (a) PSF acquisition with the scatterer at the origin; and (b) acquisition when the point scatterer is off center and the entailing changes in the signal path lengths are Δr_{Tx} and Δr_{Rx}.

an electrically very small (point) scatterer is embedded in the otherwise scatterer-free setup. The terminology originates from the fact that such a setup is used to acquire the system's point-spread functions (PSFs) as part of its calibration. Fig. 4.11-a shows a sketch of a setup where the PSF is acquired with a small scatterer at the origin and where the Rx and Tx antennas are positioned at $\mathbf{r}_{\mathrm{Rx}}$ and $\mathbf{r}_{\mathrm{Tx}}$, respectively.

From the PSF in Eq. (4.111), we can obtain the kernel associated with the particular Rx and Tx antennas when $\mathbf{r}' = 0$:

$$\mathcal{K}_0(\mathbf{r}_{\mathrm{Rx}}, t; \mathbf{r}_{\mathrm{Tx}}) \equiv \mathcal{K}(\mathbf{r}_{\mathrm{Rx}}, t; \mathbf{r}_{\mathrm{Tx}}; \mathbf{r}' = 0) = \frac{s_0(\mathbf{r}_{\mathrm{Rx}}, t; \mathbf{r}_{\mathrm{Tx}})}{\kappa_{\mathrm{co}}\Omega_{\mathrm{co}}}. \tag{4.112}$$

The proportionality holds between the respective FTs as well:

$$\tilde{\mathcal{K}}_0(\mathbf{r}_{\mathrm{Rx}}, \omega; \mathbf{r}_{\mathrm{Tx}}) = \frac{\tilde{S}_0(\mathbf{r}_{\mathrm{Rx}}, \omega; \mathbf{r}_{\mathrm{Tx}})}{\kappa_{\mathrm{co}}\Omega_{\mathrm{co}}}. \tag{4.113}$$

To perform reconstruction based on the forward model in Eq. (4.109), we need the kernel as a function of $\mathbf{r}' \in V_{\mathrm{O}}$. This can be found easily using our knowledge of $\tilde{\mathcal{K}}_0(\mathbf{r}_{\mathrm{Rx}}, \omega; \mathbf{r}_{\mathrm{Tx}})$ if the medium is uniform.

Fig. 4.11-b illustrates the change of the point-scatterer location in the measurement with the same Rx and Tx antennas used to acquire the PSF. This change entails changes in the signal paths from the Tx antenna to the scatterer and from the scatterer to the Rx antenna:

$$\begin{aligned} \Delta r_{\mathrm{Tx}} &= |\mathbf{r}' - \mathbf{r}_{\mathrm{Tx}}| - |\mathbf{r}_{\mathrm{Tx}}|, \\ \Delta r_{\mathrm{Rx}} &= |\mathbf{r}' - \mathbf{r}_{\mathrm{Rx}}| - |\mathbf{r}_{\mathrm{Rx}}|. \end{aligned} \tag{4.114}$$

The kernel obtained from this point-scatterer response, $\tilde{\mathcal{K}}(\mathbf{r}_{\mathrm{Rx}}, \omega; \mathbf{r}_{\mathrm{Tx}}; \mathbf{r}')$, can be approximated as being a time-shifted version of $\tilde{\mathcal{K}}_0(\mathbf{r}_{\mathrm{Rx}}, \omega; \mathbf{r}_{\mathrm{Tx}})$. In the frequency domain, this is expressed as

$$\tilde{\mathcal{K}}(\mathbf{r}_{\mathrm{Rx}}, \omega; \mathbf{r}_{\mathrm{Tx}}; \mathbf{r}') \approx \tilde{\mathcal{K}}_0(\mathbf{r}_{\mathrm{Rx}}, \omega; \mathbf{r}_{\mathrm{Tx}}) e^{-\mathrm{i}k(\Delta r_{\mathrm{Tx}} + \Delta r_{\mathrm{Rx}})}, \tag{4.115}$$

where $k_b = \omega/v_b$ and v_b are the background wavenumber and phase velocity, respectively. Note that here we have neglected the effect on the field magnitude. The corresponding time-domain relationship is

$$\mathcal{K}(\mathbf{r}_{Rx}, t; \mathbf{r}_{Tx}; \mathbf{r}') \approx \mathcal{K}_0(\mathbf{r}_{Rx}, t - \Delta t(\mathbf{r}'); \mathbf{r}_{Tx}), \ \Delta t = \frac{\Delta r_{Tx}(\mathbf{r}') + \Delta r_{Rx}(\mathbf{r}')}{v_b}, \quad (4.116)$$

i.e., the kernel for a point-scatterer at $\mathbf{r}'$ is a time-shifted version of that for a point-scatterer at the origin.

4.5.2 Imaging through Cross-Correlation

In the previous section, it was discussed that in a uniform unbounded medium we can estimate the kernel of the forward model from the response to a point scatterer at the assumed center of the imaged volume. This simple time-shifting (or migration) strategy is central to all radar-based imaging methods. In the following, this strategy is applied to a multitude of responses acquired through multiple Tx and Rx locations.

Assume that we have a set of N_R time-domain responses, $s_i(t)$, $i = 1, \ldots, N_R$. We can express each of them through the forward model in Eq. (4.109) using the respective kernel $\mathcal{K}_i(t, \mathbf{r}') \equiv \mathcal{K}(\mathbf{r}_{Rx,i}, t; \mathbf{r}_{Tx,i}; \mathbf{r}')$. The discretization of the volume integral into N_p voxels (each of volume $\Delta\Omega$) leads to

$$s_i(t) = \sum_{n=1}^{N_p} \kappa(\mathbf{r}'_n)\mathcal{K}_i(t, \mathbf{r}'_n)\Delta\Omega, \ i = 1, \ldots, N_R. \quad (4.117)$$

To obtain an estimate of the contrast strength at the voxel at $\mathbf{r}'_p$, we take the cross-correlation (denoted with $\otimes$) of each response with its respective kernel at $\mathbf{r}'_p$ and we sum over all cross-correlations. The result is

$$X(t, \mathbf{r}'_p) = \sum_{i=1}^{N_R} \mathcal{K}_i(t, \mathbf{r}'_p) \otimes s_i(t) = \sum_{n=1}^{N_p} \kappa(\mathbf{r}'_n) \sum_{i=1}^{N_R} \left[\mathcal{K}_i(t, \mathbf{r}'_p) \otimes \mathcal{K}_i(t, \mathbf{r}'_n)\right]. \quad (4.118)$$

With a sufficiently large number of responses, $X(t, \mathbf{r}'_p)$ is dominated by the auto-correlation term

$$\kappa(\mathbf{r}'_p) \sum_{i=1}^{N_R} \left[\mathcal{K}_i(t, \mathbf{r}'_p) \otimes \mathcal{K}_i(t, \mathbf{r}'_p)\right], \quad (4.119)$$

because the remaining terms are not only smaller but they also add incoherently. Therefore, the energy of $X(t, \mathbf{r}'_p)$,

$$I(\mathbf{r}'_p) = \int_t X^2(t, \mathbf{r}'_p)dt, \quad (4.120)$$

is a measure of $|\kappa(\mathbf{r}'_p)|$, and it determines the intensity of the image at $\mathbf{r}'_p$. This is the reconstruction principle underlying UWB radar imaging. The result is qualitative, but the processing is very fast, allowing for real-time imaging.

4.5.3 Steering Filters

Each cross-correlation term, $\mathcal{K}_i(t, \mathbf{r}'_p) \otimes s_i(t)$, in Eq. (4.118) represents a signal-processing operation on the i-th response, which is described mathematically by [11]

$$X_i(t, \mathbf{r}') = \int_{-\infty}^{\infty} \mathcal{K}_i(\tau, \mathbf{r}') s_i(\tau + t) d\tau. \tag{4.121}$$

Here, we have dropped the subscript in $\mathbf{r}'_p$. Note that all responses are real-valued functions of time. In the frequency domain, the correlation appears in the form

$$\tilde{X}_i(\omega, \mathbf{r}') = \tilde{\mathcal{K}}_i^*(\omega, \mathbf{r}') \cdot \tilde{S}_i(\omega). \tag{4.122}$$

Next we derive $\mathcal{K}_i(\tau, \mathbf{r}')$ and its FT. These functions, known as *steering filters*, are the cornerstone in the formation of the image using time-domain data.

In view of Eq. (4.115), the kernel for the i-th response at any position $\mathbf{r}' \in V_O$ is expressed as

$$\tilde{\mathcal{K}}_i(\omega, \mathbf{r}') \approx \tilde{\mathcal{K}}_{0,i}(\omega) e^{-ik_b\left[\Delta r_{\text{Tx},i}(\mathbf{r}') + \Delta r_{\text{Rx},i}(\mathbf{r}')\right]}, \tag{4.123}$$

where $\tilde{\mathcal{K}}_{0,i}(\omega)$ is the kernel obtained from the i-th PSF (point scatterer at the origin) and

$$\begin{aligned} \Delta r_{\text{Tx},i}(\mathbf{r}') &= |\mathbf{r}' - \mathbf{r}_{\text{Tx},i}| - |\mathbf{r}_{\text{Tx},i}|, \\ \Delta r_{\text{Rx},i}(\mathbf{r}') &= |\mathbf{r}' - \mathbf{r}_{\text{Rx},i}| - |\mathbf{r}_{\text{Rx},i}|, \end{aligned} \tag{4.124}$$

in accordance with Eq. (4.114). Here, $\mathbf{r}_{\text{Tx},i}$ and $\mathbf{r}_{\text{Rx},i}$ are the Tx and Rx positions in the i-th experiment, respectively. Thus, the FTs of the cross-correlation coefficients in Eq. (4.118) appear in the form

$$\begin{aligned} \mathcal{F}\left\{\mathcal{K}_i(t, \mathbf{r}'_p) \otimes \mathcal{K}_i(t, \mathbf{r}'_n)\right\} &= \tilde{\mathcal{K}}_i^*(\omega, \mathbf{r}'_p) \cdot \tilde{\mathcal{K}}_i(\omega, \mathbf{r}'_n) = \\ & \left|\tilde{\mathcal{K}}_{0,i}(\omega)\right|^2 e^{ik_b\left[\Delta r_{\text{Tx},i}(\mathbf{r}'_p) + \Delta r_{\text{Rx},i}(\mathbf{r}'_p)\right]} \cdot e^{-ik_b\left[\Delta r_{\text{Tx},i}(\mathbf{r}'_n) + \Delta r_{\text{Rx},i}(\mathbf{r}'_n)\right]}, \quad i = 1, \ldots, N_R. \end{aligned} \tag{4.125}$$

With measurements in the far zone, we can assume that $\left|\tilde{\mathcal{K}}_{0,i}(\omega)\right|$ is more or less the same for all responses and it can be factored out of Eq. (4.118) for the purposes of qualitative imaging. This leads to the further simplification of the FTs of the cross-correlation coefficients, namely,

$$\mathcal{F}\left\{\mathcal{K}_i(t, \mathbf{r}'_p) \otimes \mathcal{K}_i(t, \mathbf{r}'_n)\right\} \sim e^{ik_b\left[\Delta r_{\text{Tx},i}(\mathbf{r}'_p) - \Delta r_{\text{Tx},i}(\mathbf{r}'_n) + \Delta r_{\text{Rx},i}(\mathbf{r}'_p) - \Delta r_{\text{Rx},i}(\mathbf{r}'_n)\right]}, \quad i = 1, \ldots, N_R. \tag{4.126}$$

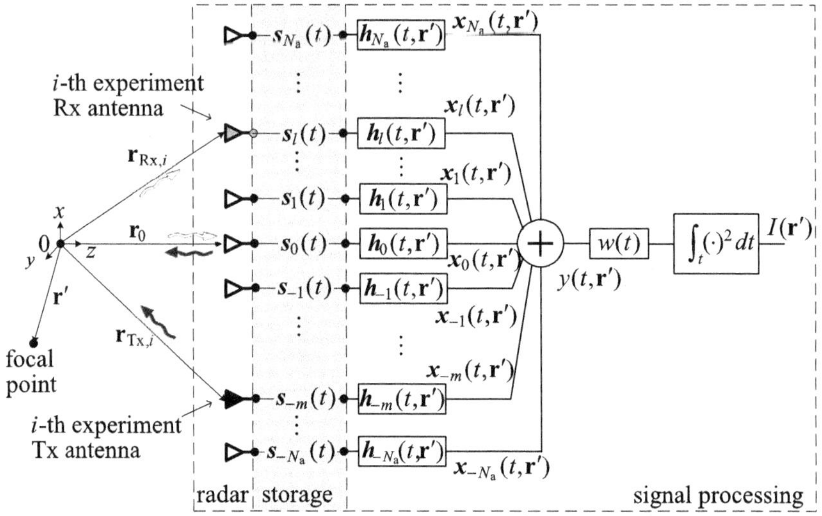

Figure 4.12 Illustration of a generic multistatic UWB imaging system. Each (i-th) response corresponds to a pair of Tx/Rx antennas. Typically, while one antenna transmits, all others receive. Thus, one measurement produces $2N_a + 1$ responses. The monostatic measurement with antenna **A0** can be used to set a phase reference. An Rx antenna feeds the signal into a bank of steering filters, each corresponding to a Tx antenna. To produce the image intensity at $\mathbf{r}'$, all steering filters in the array are tuned to $\mathbf{r}'$. This is repeated for all voxels in the imaged volume.

Note that the FT of the auto-correlation coefficient ($\mathbf{r}'_n \equiv \mathbf{r}'_p$) now equals 1, whereas the FTs of the cross-correlation coefficients are complex exponentials of unity magnitude that vary with $\mathbf{r}'_n$. This explains the coherent addition of the auto-correlation terms and the incoherent addition of the cross-corelation terms in Eq. (4.118).

We now return to the FTs of the kernels $\mathcal{K}_i(t, \mathbf{r}')$, $i = 1, \ldots, N_R$, used to process the signals through Eq. (4.121). To be able to sum over all X_i [see Eq. (4.121], the kernels must have a common phase reference. This is not the case in Eq. (4.123), where $\Delta r_{\mathrm{Tx},i}$ and $\Delta r_{\mathrm{Rx},i}$ are computed relative to the Tx and Rx positions in the i-th experiment ($|\mathbf{r}_{\mathrm{Tx},i}|$ and $|\mathbf{r}_{\mathrm{Rx},i}|$) [see Eq. (4.124)]. The choice of a phase reference is illustrated through Fig. 4.12, which shows an array of $2N_a + 1$ elements, with **A0** being the antenna at its center. It also shows the i-th experiment with the involved Tx and Rx antennas and their positions $\mathbf{r}_{\mathrm{Tx},i}$ and $\mathbf{r}_{\mathrm{Rx},i}$.

Consider the monostatic measurement with the antenna **A0** at the array center and a point scatterer at the origin; see Fig. 4.12. In this measurement, $\mathbf{r}_{\mathrm{Rx}} = \mathbf{r}_{\mathrm{Tx}} = \mathbf{r}_0$. The kernel $\tilde{\mathcal{K}}_0(\mathbf{r}_0, \omega; \mathbf{r}_0)$ obtained from this PSF response can provide a reference. The kernel corresponding to the i-th experiment with the point scatterer at the origin can be obtained as a delayed version of $\tilde{\mathcal{K}}_0(\mathbf{r}_0, \omega; \mathbf{r}_0)$:

$$\tilde{\mathcal{K}}_{0,i}(\omega) \equiv \tilde{\mathcal{K}}_0(\mathbf{r}_{\mathrm{Rx},i}, \omega; \mathbf{r}_{\mathrm{Tx},i}) \approx \tilde{\mathcal{K}}_0(\mathbf{r}_0, \omega; \mathbf{r}_0) e^{-ik_b(r_{\mathrm{Tx},i} + r_{\mathrm{Rx},i} - 2r_0)}. \tag{4.127}$$

Here, $r_{\mathrm{Tx},i} = |\mathbf{r}_{\mathrm{Tx},i}|$, $r_{\mathrm{Rx},i} = |\mathbf{r}_{\mathrm{Rx},i}|$, $r_0 = |\mathbf{r}_0|$, $(r_{\mathrm{Tx},i} + r_{\mathrm{Rx},i})$ is the signal path in the i-th PSF measurement, and $2r_0$ is the signal path associated with $\tilde{\mathcal{K}}_0(\mathbf{r}_0, \omega; \mathbf{r}_0)$. Substituting

Eq. (4.127) into Eq. (4.123) yields

$$\tilde{\mathcal{K}}_i(\omega; \mathbf{r}') \approx \tilde{\mathcal{K}}_0(\mathbf{r}_0, \omega; \mathbf{r}_0) \cdot e^{\mathrm{i}2k_\mathrm{b}r_0} \cdot e^{-\mathrm{i}k_\mathrm{b}\left[|\mathbf{r}'-\mathbf{r}_{\mathrm{Tx},i}|+|\mathbf{r}'-\mathbf{r}_{\mathrm{Rx},i}|\right]}. \quad (4.128)$$

Since $\tilde{\mathcal{K}}_0(\mathbf{r}_0, \omega; \mathbf{r}_0)$ is common in all N_R kernels, it can be factored out. Thus, we arrive at an expression for $\tilde{\mathcal{K}}_i(\omega, \mathbf{r}')$ in the form

$$\tilde{\mathcal{K}}_i(\omega; \mathbf{r}') \approx e^{\mathrm{i}2k_\mathrm{b}r_0} \cdot e^{-\mathrm{i}k_\mathrm{b}\left[|\mathbf{r}'-\mathbf{r}_{\mathrm{Tx},i}|+|\mathbf{r}'-\mathbf{r}_{\mathrm{Rx},i}|\right]}. \quad (4.129)$$

The spectral function,

$$\tilde{H}_i(\omega; \mathbf{r}') \equiv \tilde{\mathcal{K}}_i^*(\omega; \mathbf{r}') \approx e^{-\mathrm{i}2k_\mathrm{b}r_0} \cdot e^{\mathrm{i}k_\mathrm{b}\left[|\mathbf{r}'-\mathbf{r}_{\mathrm{Tx},i}|+|\mathbf{r}'-\mathbf{r}_{\mathrm{Rx},i}|\right]}, \quad (4.130)$$

can be viewed as a filtering transfer function applied to the received signal $\tilde{S}_i(\omega)$ [see Eq. (4.122)]. Explicitly, the use of this filter in Eq. (4.122) yields the "filtered" response in the frequency domain as

$$\tilde{X}_i(\omega, \mathbf{r}') = \left\{e^{\mathrm{i}k_\mathrm{b}\left[|\mathbf{r}'-\mathbf{r}_{\mathrm{Tx},i}|+|\mathbf{r}'-\mathbf{r}_{\mathrm{Rx},i}|-2r_0\right]}\right\} \cdot \tilde{S}_i(\omega). \quad (4.131)$$

The impulse response corresponding to Eq. (4.130) is

$$h_i(t; \mathbf{r}') = \delta\left(t - t_0 + \frac{|\mathbf{r}' - \mathbf{r}_{\mathrm{Tx},i}| + |\mathbf{r}' - \mathbf{r}_{\mathrm{Rx},i}|}{v_\mathrm{b}}\right), \quad (4.132)$$

where $t_0 = 2r_0/v_\mathrm{b}$ is some reference time. When convolved with the i-th signal waveform [see Eq. (4.121)], h_i shifts it to the left on the temporal axis by

$$\Delta t_i = \frac{|\mathbf{r}' - \mathbf{r}_{\mathrm{Tx},i}| + |\mathbf{r}' - \mathbf{r}_{\mathrm{Rx},i}|}{v_\mathrm{b}} - t_0\,. \quad (4.133)$$

Explicitly, the use of the impulse response of Eq. (4.132) in Eq. (4.121) yields the "filtered" response in the time domain as

$$X_i(t, \mathbf{r}') = \int_{-\infty}^{+\infty} h_i(-\tau; \mathbf{r}') s_i(\tau + t) d\tau = s_i(t + \Delta t_i). \quad (4.134)$$

Note that $h_i(t; \mathbf{r}')$ corresponds to the conjugated kernel as per Eq. (4.130). Therefore, it must be time-reversed on substitution in Eq. (4.121), leading to the result shown in Eq. (4.134). It is straightforward to show that Eq. (4.134) is in fact the convolution of the filter $h_i(t; \mathbf{r}')$ and the signal $s_i(t)$.

The temporal shift Δt_i aims at aligning the possible scattering emanating at $\mathbf{r}'$ and contained in the i-th response with that in all other responses. The value of t_0, which is common for all responses, does not need to be related to the center of the array. Its main purpose is to ensure that after the temporal migration is applied to the signal it does not become acausal, i.e., it remains within positive time. Such causality would be fully ensured if $t_0 = T_{\max}$, where $T_{\max}$ is the temporal length of the recorded radar responses.

The function $h_i(\omega; \mathbf{r}')$ in Eq. (4.132) is referred to as the i-th response *steering filter* since it effectively "steers" the array focus toward the imaged point $\mathbf{r}'$. This synthetic process of steering is similar to the use of the phase shifters to steer the beam of an antenna array in the desired direction.

Note that the steering filter in Eq. (4.132) is often referred to as the *migration* steering filter. It is the simplest and also the most basic steering filter since the various more sophisticated filters are derived from it. Often the migration steering filter is improved by adding an amplitude factor $a(\mathbf{r}_{\text{Rx}}, \mathbf{r}_{\text{Tx}}, \mathbf{r}')$, which takes into account the spreading loss of the scattered and/or transmitted wave, the radiation patterns of the Tx and Rx antennas, and possible attenuation in the propagation medium; see, for example [126]. Since this factor is not a function of time, it is referred to as the *gain factor*. More sophisticated filters also take into account frequency-dispersion effects as well as reflections from planar interfaces, such as those arising in the imaging in a layered background medium, e.g., the interface between the ground surface and air in GPR.

Finally, actual measurements of the radar PSF [see Eq. (4.111) and Eq. (4.112)] can provide high-fidelity system-specific steering filters. Just like in the case of frequency-sweep systems (see Section 4.2), the PSF of a radar imaging system can be measured with a very small scattering object at the center of the envisioned imaged volume [140]. It consists of all N_R radar responses in this measurement scenario: $s_0(\mathbf{r}_{\text{Rx},i}, t; \mathbf{r}_{\text{Tx},i})$, $i = 1, \ldots, N_R$. The transfer function of the steering filter in Eq. (4.130) can now be represented accurately as

$$\tilde{H}_i(\omega; \mathbf{r}') \equiv \tilde{K}_i^*(\omega; \mathbf{r}') = \tilde{S}_0^*(\mathbf{r}_{\text{Rx},i}, \omega; \mathbf{r}_{\text{Tx},i}) \cdot e^{-\mathrm{i}2k_b r_0} \cdot e^{\mathrm{i}k_b\left[|\mathbf{r}'-\mathbf{r}_{\text{Tx},i}|+|\mathbf{r}'-\mathbf{r}_{\text{Rx},i}|\right]}, \quad (4.135)$$

where $\tilde{S}_0(\mathbf{r}_{\text{Rx},i}, \omega; \mathbf{r}_{\text{Tx},i})$ is the FT of $s_0(\mathbf{r}_{\text{Rx},i}, t; \mathbf{r}_{\text{Tx},i})$. In Eq. (4.135), we have factored out the multiplicative term $(\kappa_{\text{co}}\Omega_{\text{co}})^{-1}$, which is common to all responses; see Eq. (4.113). Further, we can set the phase reference so that $2k_b r_0 \equiv \omega T_{\max}$. This sets r_0 as the distance to the farthest point from which a back-scattered signal can be received. With this setting, the steering filter in Eq. (4.135) becomes

$$\tilde{H}_i(\omega; \mathbf{r}') = \tilde{S}_0^-(\mathbf{r}_{\text{Rx},i}, \omega; \mathbf{r}_{\text{Tx},i}) \cdot e^{\mathrm{i}k_b\left[|\mathbf{r}'-\mathbf{r}_{\text{Tx},i}|+|\mathbf{r}'-\mathbf{r}_{\text{Rx},i}|\right]}, \quad (4.136)$$

where

$$\tilde{S}_0^-(\mathbf{r}_{\text{Rx},i}, \omega; \mathbf{r}_{\text{Tx},i}) = \tilde{S}_0^*(\mathbf{r}_{\text{Rx},i}, \omega; \mathbf{r}_{\text{Tx},i}) \cdot e^{-\mathrm{i}\omega T_{\max}} \quad (4.137)$$

is exactly the FT of the acausal (or time-reversed) counterpart of the PSF $s_0(\mathbf{r}_{\text{Rx},i}, t; \mathbf{r}_{\text{Tx},i})$ [see Eq. (1.187)].

Applying the steering filter in Eq. (4.136) to the i-th signal,

$$\tilde{X}_i(\omega, \mathbf{r}') = \left\{\tilde{S}_0^-(\mathbf{r}_{\text{Rx},i}, \omega; \mathbf{r}_{\text{Tx},i}) \cdot e^{\mathrm{i}k_b\left[|\mathbf{r}'-\mathbf{r}_{\text{Tx},i}|+|\mathbf{r}'-\mathbf{r}_{\text{Rx},i}|\right]}\right\} \cdot \tilde{S}_i(\omega) \quad (4.138)$$

is nothing but the signal's correlation with the i-th PSF adjusted for a scattering center at $\mathbf{r}'$. The impulse response corresponding to the filter's transfer function in Eq. (4.136) is explicitly stated as

$$h_i(t;\mathbf{r}') = s_0(\mathbf{r}_{\mathrm{Rx},i}, T_{\max} - t; \mathbf{r}_{\mathrm{Tx},i}) * \delta\left(t + \frac{|\mathbf{r}' - \mathbf{r}_{\mathrm{Tx},i}| + |\mathbf{r}' - \mathbf{r}_{\mathrm{Rx},i}|}{v_{\mathrm{b}}}\right) =$$
$$s_0\left[\mathbf{r}_{\mathrm{Rx},i}, T_{\max} - \left(t + \frac{|\mathbf{r}' - \mathbf{r}_{\mathrm{Tx},i}| + |\mathbf{r}' - \mathbf{r}_{\mathrm{Rx},i}|}{v_{\mathrm{b}}}\right); \mathbf{r}_{\mathrm{Tx},i}\right]. \quad (4.139)$$

4.5.4 Image Generation by Migration: Delay and Sum

The image generation by migration is the simplest and the most basic approach to image generation with pulsed signals. It is also known as *delay-and-sum beam forming*. The various algorithms exploiting this principle are widely known in the MW literature as delay-and-sum (DAS) algorithms. The DAS reconstruction is prevalent in GPR. In modified forms it has been also employed in medical-imaging research on alternative methods for early-stage breast cancer detection; see, for example, [126, 132, 133, 136, 137, 138].

Fig. 4.12 shows the signal-flow schematic of the formation of the image using the steering filters and the DAS strategy. In a typical radar measurement, while one antenna transmits, all others receive. For example, if all $2N_{\mathrm{a}}+1$ antennas are used to transmit and receive, then at the end of the experiment each antenna would have produced a set of $2N_{\mathrm{a}}+1$ responses (one response for each transmission):

$$\boldsymbol{s}_l(t) = \left[s_{l,-N_{\mathrm{a}}}(t) \ \cdots \ s_{l,N_{\mathrm{a}}}(t)\right]^T, \ l = -N_{\mathrm{a}}, \ldots, N_{\mathrm{a}}. \quad (4.140)$$

An Rx antenna has a bank of steering filters, each corresponding to a transmission from a particular antenna:

$$\boldsymbol{h}_l(t;\mathbf{r}') = \left[h_{l,-N_{\mathrm{a}}}(t;\mathbf{r}') \ \cdots \ h_{l,N_{\mathrm{a}}}(t;\mathbf{r}')\right]^T, \ l = -N_{\mathrm{a}}, \ldots, N_{\mathrm{a}}. \quad (4.141)$$

Each signal $s_{l,m}(t)$ is convolved with its respective steering filter $h_{l,m}(t,\mathbf{r}')$ to produce the output

$$x_{l,m}(t,\mathbf{r}') = s_{l,m}(t) * h_{l,m}(t;\mathbf{r}'), \ l,m = -N_{\mathrm{a}}, \ldots, N_{\mathrm{a}}. \quad (4.142)$$

Thus, each (l-th) channel in Fig. 4.12 produces a set of functions of time and $\mathbf{r}'$:

$$\boldsymbol{x}_l(t,\mathbf{r}') = \left[x_{l,-N_{\mathrm{a}}}(t;\mathbf{r}') \ \cdots \ x_{l,N_{\mathrm{a}}}(t;\mathbf{r}')\right]^T, l = -N_{\mathrm{a}}, \ldots, N_{\mathrm{a}}. \quad (4.143)$$

Summation is carried out over the elements of each channel's output $\boldsymbol{x}_l(t)$, $l = -N_{\mathrm{a}}, \ldots, N_{\mathrm{a}}$, and then over all channels to produce the overall output:

$$y(t,\mathbf{r}') = \sum_{l=-N_{\mathrm{a}}}^{N_{\mathrm{a}}} \sum_{m=-N_{\mathrm{a}}}^{N_{\mathrm{a}}} x_{l,m}(t,\mathbf{r}'). \quad (4.144)$$

The result is usually "trimmed" with a windowing function $w(t)$ before producing the intensity of the voxel at $\mathbf{r}'$, $I(\mathbf{r}') = \int_t y^2(t,\mathbf{r}')dt$. The windowing function, which is centered at the reference time t_0, suppresses possible parasitic signals that may be due

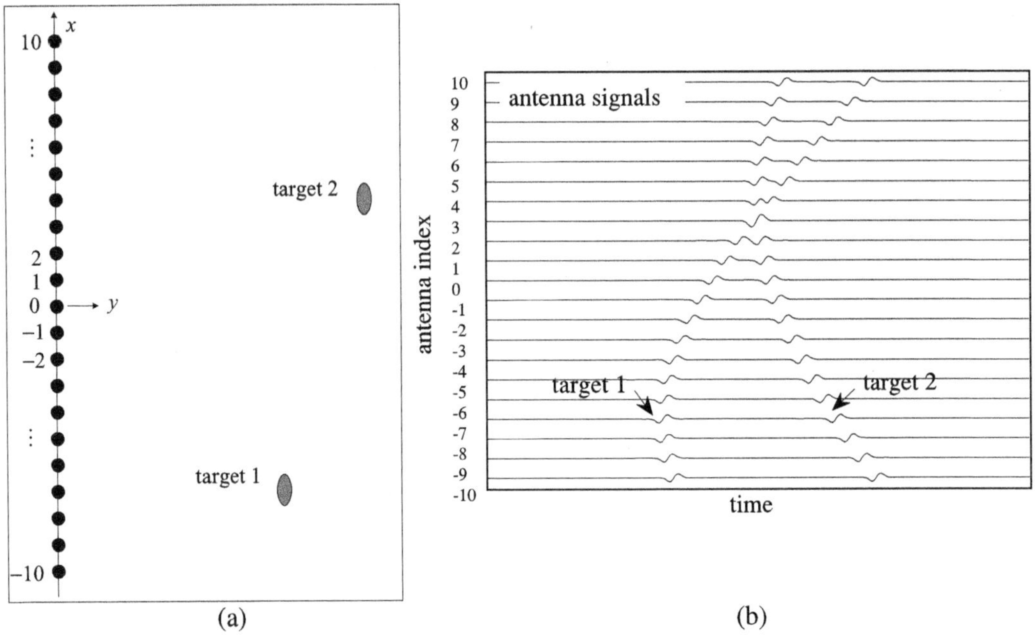

Figure 4.13 A synthetic example of a linear array of 21 isotropic radiators evenly distributed along the x axis with a separation of 10 mm. (a) A monostastic acquisition is performed to image the 2D scene in the xy plane where two small targets are present: target 1 is at $\mathbf{r}'_1 = (-70, 80)$ mm and target 2 is at $\mathbf{r}'_2 = (40, 130)$ mm. (b) The acquired back-scattered responses are shown, each containing the two time-delayed pulse returns from the targets. The amplitude changes due to the spatial spreading loss are ignored.

to early-time coupling (leakage) between the Tx and Rx antennas as well as late-time reflections from enclosures, the "ringing" due to the limited bandwidth of the antennas, insufficient cancellation after the filtered responses are added and so on. Note that insufficient cancellation is more pronounced with fewer responses. Thus, the windowing function can improve significantly the dynamic range of the image. Its temporal width can be determined from a calibration measurement with a large flat metallic screen positioned in the far zone of an antenna element; see, for example [126]. The back-scattered signal from such a measurement is a representation of the field waveform in the far zone, and its truncation at a certain threshold can determine the proper temporal extent of $w(t)$.

Fig. 4.13-a shows a simple 2D arrangement where a monostatic array of isotropic radiators lies on the x axis. The separation between the array elements is 10 mm. An UWB pulse with spectral content between 1.0 GHz and 9.0 GHz is emitted by each radiator sequentially, and the back-scattered time-domain response is received. Two small targets are present in the xy plane where target 1 is at $\mathbf{r}'_1 = (-70, 80)$ mm whereas target 2 is at $\mathbf{r}'_2 = (40, 130)$ mm. Fig. 4.13-b shows the back-scattered responses recorded by the 21 array elements. We observe that each response contains two time-delayed pulses

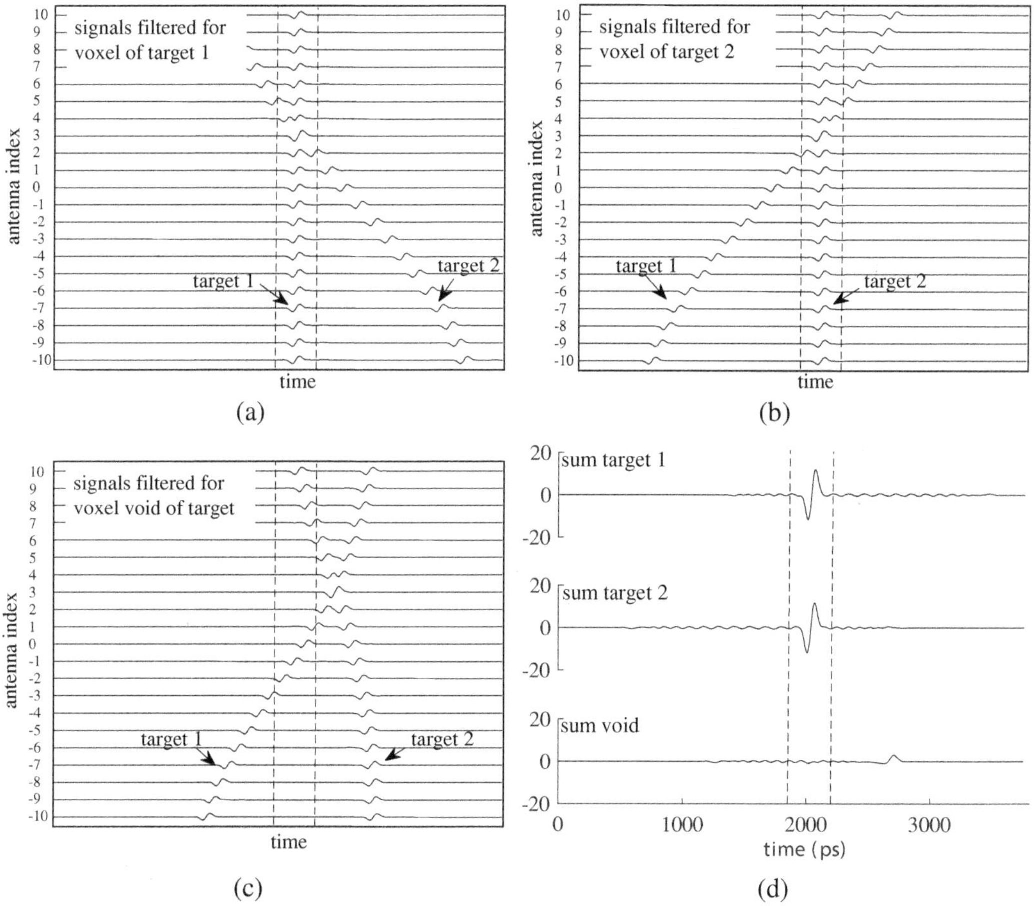

Figure 4.14 Filtered responses in the synthetic example with a linear array and two small targets, where the steering filters are tuned to the position of: (a) target 1, (b) target 2, and (c) the voxel void of scatterers. (d) The result of summing the 21 responses filtered for the three voxel positions: targets 1 and 2 and the void voxel.

due to the incident field bouncing off each target. The illustration ignores the change in the signal amplitude due to the spreading loss.

Fig. 4.14-a shows the filtered (or migrated, or back-propagated) responses with the steering filters being tuned to the position $\mathbf{r}'_1$ of target 1. We observe that the target 1 back-scattered pulses align in all responses, which will lead to coherent superposition upon summation. Fig. 4.14-b shows the migrated responses with the steering filters being tuned to the position $\mathbf{r}'_2$ of target 2. Now it is the pulses arriving from target 2 that are aligned in time. Fig. 4.14-c shows the migrated responses with the steering filters being tuned to the point at $\mathbf{r}'_{\text{void}} = (0, 100)$ mm, which is void of scatterers. We observe that there is no particular alignment in the back-scattered pulses, and therefore, adding all responses will result in an incoherent superposition. Finally, Fig. 4.14-d shows the

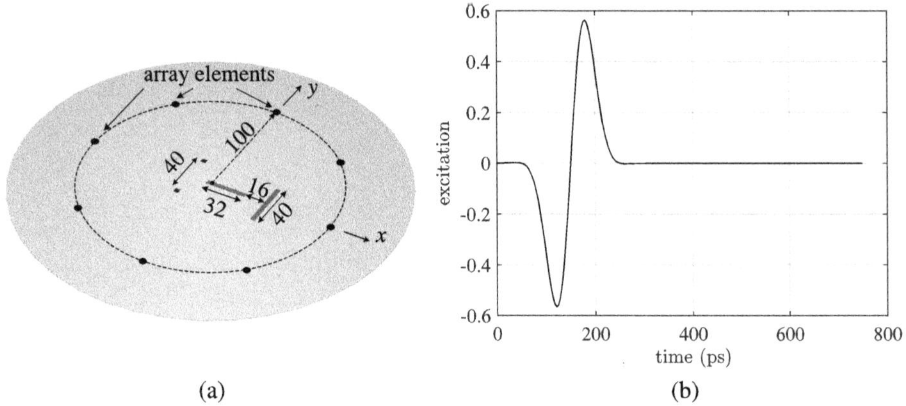

Figure 4.15 The simulation setup for the example of DAS imaging using a multistatic circular array consisting of eight equispaced elements. The array radius is 100 mm. (a) Layout of the array placement together with the imaged object, which consists of two small square dielectric objects, 4 mm on a side, with the distance between them being 40 mm; a dielectric bar oriented along x of length 32 mm; and a dielectric bar oriented along y of length 40 mm. All dimensions in the plot are in mm. (b) The waveform of the excitation pulse.

result of summing the filtered responses for the three voxel positions: those for targets 1 and 2 and the one corresponding to a void voxel.

It is clear that the energy values at the voxels of targets 1 and 2 are significantly larger than the energy value for the void voxel even if windowing is not applied. Here, $I(\mathbf{r}_1') = I(\mathbf{r}_2') \approx 2988$, whereas $I(\mathbf{r}_{\text{void}}') \approx 171$. However, the windowing reduces the energy associated with the void voxel substantially, whereas it has a negligible effect on the energy associated with the voxels that have scatterers. Thus, it increases the ratio of the maximum image value to the image noise, i.e., it improves the dynamic range of the image. In this example, after windowing, $I_w(\mathbf{r}_1') = I_w(\mathbf{r}_2') \approx 2927$ whereas $I_w(\mathbf{r}_{\text{void}}') \approx 14$. The temporal extent of the windowing function is indicated by two vertical dotted lines in all four plots in Fig. 4.14.

4.5.5 Simulation Example

The DAS reconstruction algorithm with the migration filters is tested in a simple example, where the data have been acquired with the time-domain EM simulator MEFiSTo-3D [10]. To accelerate the data acquisition, the simulator is used in its 2D TM$_z$ mode [21], where the field is completely described by three components only: E_z, H_x, and H_y. The limitations associated with the assumption of a 2D TM$_z$ field were discussed in Section 2.13.

We employ a circular array of eight equispaced point-like sources with the array radius being 100 mm; see Fig. 4.15-a. The medium is vacuum ($\varepsilon_{\text{r,b}} = 1$), and the simulation employs a reflection-free boundary condition in order to emulate an open medium. The sources are coincident with eight field probes that record the E_z field waveforms.

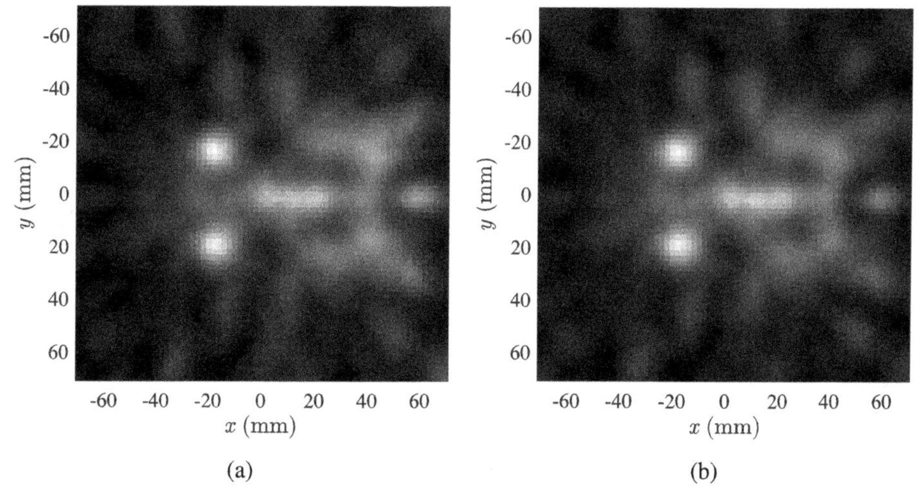

Figure 4.16 Qualitative images in the circular-array example produced by the simple migration filters in a DAS algorithm: (a) migration filters as defined in Eq. (4.132), and (b) migration filters as defined in Eq. (4.132) augmented by the gain factor, which takes into account the spreading loss.

Both the sources and the probes have an extent of one pixel, i.e., one mesh element in the simulator's 2D grid, which is 2×2 mm. In each simulation, one source point is excited while all probes receive. This produces eight responses simultaneously. Eight such simulations are performed, exciting the source points sequentially. In total, 64 responses are obtained. The time required by each simulation is on the order of 10 seconds, which is quite acceptable and allows for the exploration of various OUT arrangements. This example can be re-created with any available time-domain simulation engine, including FDTD simulators, some of which are available as freeware; see, e.g., *openEMS* [141].

As shown in Fig. 4.15-a, the imaged object consists of four components: two small square dielectric objects, 4 mm on a side, with the distance between them being 40 mm; a dielectric bar oriented along x of length 32 mm; and a dielectric bar oriented along y of length 40 mm. The relative permittivity of the two small squares is 50, that of the dielectric bar along x is 25, and that of the dielectric bar along y is 20.

Fig. 4.15-b shows the excitation pulse, the bandwidth of which lies between approximately 1.25 GHz and 8.75 GHz (at 3 dB level).

Fig. 4.16-a shows the image produced by the DAS algorithm using the simple migration filters in Eq. (4.132). Fig. 4.16-b shows the image produced by the migration filters augmented by the gain factor $\sqrt{r_i(\mathbf{r}')r_j(\mathbf{r}')}$, where r_i is the distance between the imaged pixel (at $\mathbf{r}'$) and the i-th probe, and r_j is the distance between the imaged pixel and the j-th source ($i, j = 1, \ldots, 8$). Note that here the free-space spreading loss is assumed to be of the form $\sim 1/\sqrt{\rho}$ as dictated by the fundamental solution of the 2D wave equation; see Section 1.11.

We observe that the two small objects are detected very well, although they appear as round spots in the image rather than as squares. This is expected because their size is less than $\lambda_{\min}/8$, where $\lambda_{\min}$ is the minimum wavelength corresponding to the maximum frequency (8.25 GHz). The dielectric bar oriented along x is also well defined. On the other hand, the dielectric bar oriented along y is rather distorted due to the presence of nearby artifacts. These artifacts are caused by strong coupling and multiple scattering, which occurs between the two sizable dielectric bars. Both of them are close to $\lambda_{\min}$ and are also close to each other, which causes strong multiple scattering and resonance effects not only in the space between them but also in their own dielectric volumes. Such scattering is a strongly nonlinear function of the local permittivity values. The migration filters in Eq. (4.132) assume a very simple time-delay model of propagation, where the wave travels only to and from a pixel but does not bounce between pixels. They do not include effects such as standing waves or resonances. This is the reason why there are many strong artifacts in the region where the two dielectric bars come close to each other. The gain factor does suppress somewhat the artifacts, but in general, it cannot compensate for the multiple scattering and the resonance effects.

In conclusion, the DAS reconstruction algorithms are based on a linearized model of scattering and are thus limited to weak scattering as defined by the constraints of Born's linear approximation. Their greatest advantages are speed and simplicity. However, for strongly heterogeneous objects, they can only serve to provide an approximate initial guess, which must be further refined using nonlinear reconstruction approaches.

4.6 Brief Overview of Other Linear Reconstruction Methods for Pulsed Radar

There are a few reconstruction methods that are linear in nature yet provide better qualitative results than the simple DAS method. The improvement is due to a better forward model of the scattering process. It was already mentioned that the steering filters can be constructed from the time-domain PSF of the radar system measured with a small scatterer at the origin $\mathbf{r}' = 0$ of the imaged volume; see Eq. (4.139). Such measurement-based steering filters improve the algorithm performance significantly.

The method that uses the measured PSF as a steering filter is slower than the DAS reconstruction with a simple signal migration since, in addition to the signal shift and sum, a convolution of each response with its PSF is performed. Nonetheless, it is sufficiently fast to perform in real time even with hundreds of responses. The disadvantage is that a calibration measurement is needed. Moreover, in order to perform only one calibration measurement with the small scatterer at the origin, the measurement setup must be invariant to translations. Only under such conditions can the point-scatterer responses for any position $\mathbf{r}' \neq 0$ be obtained from the PSF. Also, it is clear that such PSF-based reconstruction still cannot take into account multiple scattering and mutual coupling effects. Just like the DAS method, it views the OUT as a collection of point scatterers, the individual contributions of which superimpose to form the overall OUT response.

4.6.1 Radar Imaging through 3D Deconvolution

Another method, which exploits 3D deconvolution, performs even faster than the DAS reconstruction; see, for example, [128, chapter 7], [140]. It employs the measured system PSF. The deconvolution is 3D because it operates in two spatial dimensions (over which the PSF is measured) and in time. It is limited to acquisition systems that are invariant to translations in two of the coordinate axes.

As explained in Section 4.5.2 and Section 4.5.3, the steering filters can be viewed as back-propagated (or migrated) versions of the system PSF, whereas their convolutions with the responses (see the filters $\boldsymbol{h}_i$, $i = -N_a, \ldots, N_a$ in Fig. 4.12) are nothing but cross-correlation operations that estimate signal-to-PSF similarity. With the system PSF being invariant to translations along any given axis, the summation over the filtered responses collected over this axis translates into a spatial convolution. This hints of the possibility of finding a solution through a spatio-temporal deconvolution.

Next, the 3D deconvolution concept is illustrated through the case of planar scanning where monostatic data are acquired over a plane $z_A = const$. We have to revisit Eq. (4.118), where we specify that

$$\mathcal{K}_i(t, \mathbf{r}') \equiv \mathcal{K}(\mathbf{r}_{\mathrm{Rx},i}, t; \mathbf{r}_{\mathrm{Tx},i}; \mathbf{r}'). \tag{4.145}$$

In the monostatic scenario,

$$\mathbf{r}_{\mathrm{Rx},i} = \mathbf{r}_{\mathrm{Tx},i} = (x_i, y_i, z_A). \tag{4.146}$$

Assume that the sampling grid formed by (x_i, y_i) on the z_A plane is dense enough to satisfy Nyquist's sampling criterion. This allows for dropping the index i and for treating analytically all functions of the spatial coordinates (x, y).

Further, assume we have acquired the system PSF $s_0(x, y, t)$ [see Eq. (4.112)], and therefore, the kernel $\mathcal{K}(x, y, t; \mathbf{r}')$ is available to us. Here, $\mathbf{r}' = (x', y', z')$ indicates the position of a point scatterer, whereas (x, y) is the position at the z_A plane where the PSF is acquired. Let us introduce the shorthand notation $\mathcal{K}_{z'}(t; x, y) \equiv \mathcal{K}(x, y, t; 0, 0, z')$. With reference to Eq. (4.118), replace the sum over all N_R responses with an integral over the acquisition plane. The cross-correlation function of the imaged point is now obtained as

$$X(t; x', y', z') = \iint\limits_{x\,y} \mathcal{K}_{z'}(t; x - x', y - y') \otimes s(t; x, y) dxdy = \iiint\limits_{x''\,y''\,z''} \kappa(x'', y'', z'')\cdot$$
$$\left\{ \iint\limits_{x\,y} \left[\mathcal{K}_{z'}(t; x - x', y - y') \otimes \mathcal{K}_{z''}(t; x - x'', y - y'')\right] dxdy \right\} dx''dy''dz''. \tag{4.147}$$

Notice that the first double integral over x and y is the 2D spatial cross-correlation of $\mathcal{K}_{z'}(t, x, y)$ and $s(t; x, y)$ in addition to their temporal cross-correlation. Similarly, the double integral within the braces is the spatial cross-correlation of the kernels at positions z' and z'' evaluated at $(x' - x'', y' - y'')$. Thus, Eq. (4.147) can be written as

$$X(t; x', y', z') = \mathcal{K}_{z'}(t; x, y) \otimes_{3D} s(t; x, y) = \iiint_{x''\, y''\, z''} \kappa(x'', y'', z'') \cdot \left[\mathcal{K}_{z'}(t; x' - x'', y' - y'') \otimes_{3D} \mathcal{K}_{z''}(t; x' - x'', y' - y'')\right] dx''dy''dz'', \quad (4.148)$$

where $\otimes_{3D}$ denotes 3D cross-correlation.

We next make an important observation that the range locations z' and z'' are related to the temporal variable t. An approximation of this relationship utilizes the linear expression

$$\mathcal{K}_{z''}(t; x, y) = \mathcal{K}_{z'}(t - t''; x, y), \ t'' = \frac{2(z'' - z')}{v_b}. \quad (4.149)$$

This is a simple statement of the fact that a shift of the point scatterer along range entails a temporal response delay due to a longer traveling path to and from the scatterer. It is approximate because it ignores the lateral displacement between the antenna and the point target. Thus, the integration variable z'' in Eq. (4.148) can be replaced by t''. Moreover, the kernels that depend on z' and z'' already include the cross-correlation in these variables through their temporal variables. Thus, Eq. (4.148) can be stated as

$$\mathcal{K}_{z'}(t; x, y) \otimes_{3D} s(t; x, y) = \kappa(x, y, t) *_{3D} \left[\mathcal{K}_{z'}(x, y, t) \otimes_{3D} \mathcal{K}_{z'}(x, y, t)\right], \quad (4.150)$$

where $*_{3D}$ denotes 3D convolution in the space of x, y, and t. In terms of the 3D FTs of the involved quantities, Eq. (4.150) is written as

$$\tilde{\mathcal{K}}^*_{z'}(\omega; k_x, k_y)\tilde{S}(\omega; k_x, k_y) = \tilde{\kappa}(\omega; k_x, k_y)\tilde{\mathcal{K}}_{z'}(\omega; k_x, k_y)\tilde{\mathcal{K}}^*_{z'}(\omega; k_x, k_y). \quad (4.151)$$

Eq. (4.151) is the basis of the 3D radar imaging using deconvolution. Often, the kernel is available only at $z' = 0$ since the PSF is measured only with the target at $(x', y', z') = (0, 0, 0)$. In this case, $\mathcal{K}_{z'}(t; x, y)$ can be obtained from $\mathcal{K}_0(t; x, y)$ via the time-delay relationship in Eq. (4.149). If the kernel is available from measurements with the point scatterer at several range locations, the image is improved as Eq. (4.151) can be applied with all of these kernels and the so-obtained results for $\tilde{\kappa}(\omega; k_x, k_y; z')$ can be properly post-processed to produce a 3D image.

Eq. (4.151) suggests that the FT of the OUT contrast function can be computed as

$$\tilde{\kappa}(\omega; k_x, k_y) = \frac{\tilde{\mathcal{K}}^*_{z'}(\omega; k_x, k_y)\tilde{S}(\omega; k_x, k_y)}{\tilde{\mathcal{K}}_{z'}(\omega; k_x, k_y)\tilde{\mathcal{K}}^*_{z'}(\omega; k_x, k_y)}. \quad (4.152)$$

Note that this formula is applied at each (k_x, k_y) pair and at each frequency ω, i.e., if the involved quantities are represented numerically by 3D matrices, these matrices are multiplied/divided element by element. This is a fast computation. We should also note that when multiple kernels are available (at different z'), Eq. (4.152) can be applied with each kernel, and a system of equations can be formed at each (k_x, k_y) pair using Eq. (4.151) instead of solving a single equation as in Eq. (4.152).

Eq. (4.152) is usually not used directly because the solution may be unstable, especially if the signal-to-noise ratio (SNR) is low. A regularized version is given by [140]

$$\tilde{\kappa}(\omega; k_x, k_y) = \frac{\tilde{\mathcal{K}}_{z'}^*(\omega; k_x, k_y)\tilde{S}(\omega; k_x, k_y)}{\tilde{\mathcal{K}}_{z'}(\omega; k_x, k_y)\tilde{\mathcal{K}}_{z'}^*(\omega; k_x, k_y) + \alpha(\omega)}, \tag{4.153}$$

where α is a regularization parameter. Details on the choice of α and its impact on the image can be found in [128, chapter 7], [126, 140]. Often α is taken as the inverse of the SNR (linear ratio) of the measurement system. The reconstruction formula Eq. (4.153) is known as *Wiener deconvolution* or *Wiener inverse filter* [128, chapter 7], [126, 140].

The final step of the deconvolution-based reconstruction is to obtain the 3D image in real space from $\tilde{\kappa}(\omega; k_x, k_y)$. There are various ways to perform this task; see, for example, [128]. If $\mathcal{K}_{z'}(t; x, y)$ is available from measurements at multiple range locations, an obvious possibility is to use the 2D inverse Fourier transform with respect to (k_x, k_y) to obtain a 2D image of the contrast at each z' plane as

$$\kappa(\omega, x', y'; z') = \mathcal{F}_{2D}^{-1}\left\{\tilde{\kappa}_{z'}(\omega; k_x, k_y)\right\}, \tag{4.154}$$

where $\tilde{\kappa}_{z'}(\omega; k_x, k_y)$ is obtained as in Eq. (4.153) with $\tilde{\mathcal{K}}_{z'}(\omega; k_x, k_y)$. Notice that the so-obtained contrast is a function of the frequency; therefore, this approach can handle significant dispersion in the OUT constitutive parameters.

If the contrast dispersion can be ignored, the UWB nature of the signals can be exploited to obtain a 3D image with a single kernel and in a single post-processing step. For that, the function $\tilde{\kappa}(\omega; k_x, k_y)$ is mapped onto the 3D Fourier space defined by (k_x, k_y, k_z) using the eigenvalue equation

$$k_z^2 = \left(\frac{2\omega}{v_b}\right)^2 - k_x^2 - k_y^2. \tag{4.155}$$

Note the factor 2 in the term $k_b = 2\omega/v_b$, which accounts for the fact that in a monostatic scenario the signals track a path twice. Since the function is likely sampled uniformly along ω as a result of a 3D FFT processing, the mapping of the ω axis onto the k_z axis through the nonlinear relation in Eq. (4.155) also requires interpolation and resampling. Then, a 3D image can be obtained using

$$\kappa(x', y', z') = \mathcal{F}_{3D}^{-1}\left\{\tilde{\kappa}(k_x, k_y, k_z)\right\}. \tag{4.156}$$

4.6.2 Radar Imaging through Time Reversal

The mathematical foundations of the imaging methods based on time reversal were introduced in Section 1.6.4. There, two types of methods were mentioned that employ time reversal: the *experimental time reversal* and the *computational time reversal*, using a terminology from [26].

Experimental time reversal originated in acoustics with the work of Fink *et al.*; see, for example, the excellent reviews in [142–145]. It does not aim at generating an image of a scatterer. The goal is to focus the wave-field energy onto a scatterer, the location of which is not known beforehand. The greatest advantage of the method is that it works in a heterogeneous medium and no prior knowledge is necessary about the background and/or the scattering object. Its success in practical acoustic systems has been enabled by the availability of compact piezoelectric transducers that are linear and capable of nearly instantaneous measurement of the pressure waveforms. The typical emitted pulse is at a central frequency of 3.5 MHz modulated by a pulse lasting in the range of several tens of nanoseconds [143]. These pulses can be sampled and digitized with modern analog-to-digital converters (ADC). Moreover, the wavelength of the acoustic waves at 3.5 MHz is about a half-millimeter in water. It stays in this range for most of the human tissue types. This translates into a spatial resolution limit that is a fraction of a millimeter. Thus, time-reversal focusing was first developed for biomedical applications such as ultrasound imaging and lithotripsy (destruction of stones in the kidney). Systems for the nondestructive testing of metallurgical samples have also been developed.

The most important component of a time-reversal (TR) experiment is the time-reversal cavity (TRC) or the time-reversal mirror (TRM). The TRC refers to a dense network of transceivers on a surface completely surrounding the volume of interest, whereas the TRM is deployed on a planar surface on one side of the volume. In both cases, the purpose is to record the scattered field waveforms at the surface and to re-transmit them *simultaneously* in a time-reversed fashion from the same locations. If the temporal length of the recorded back-scattered signals is $T_{\max}$, then $T_{\max}$ is also the length of retransmitted excitation waveforms. Explicitly, if the i-th transceiver records a signal $s_i(t)$, $t \in [0, T_{\max}]$, then it transmits the signal $s_i^{\dagger}(t) = s_i(T_{\max} - t)$ during the back-propagation experiment.

As explained in Section 1.6.4, the so-generated field (the back-propagated field) converges toward its sources, i.e., the scattering centers. In effect, it forms an interference pattern that is maximized at the locations of the scattering objects at $t \approx T_{\max}$.

A complete TR experiment consists of several steps. In the first step, a plane or isotropic wave illuminates the unknown object whose scattering responses are recorded by the TRM (or the TRC). In the second step, the TRM time-reverses the response waveforms and injects them back into the object. Now, the illuminating field is significantly focused onto the scattering centers in the object. Since the scattering is stronger if the illuminating field at the scatterer's location is stronger, in this second experiment, the scattering from the targets is greatly enhanced compared to the original experiment, where the incident field illuminated the whole volume more or less uniformly. Moreover, compared to the first set of scattering responses, the second set exhibits greater differences between the contributions of stronger targets than between the weaker ones. This is because the back-propagated field attains stronger values at the locations of stronger scatterers, thereby enhancing their contribution over that of the weaker ones.

The process is repeated several times. Each time, the scattering from the strongest scatterer is enhanced compared to that of the weaker ones. This iterative succession of TR experiments is expected to converge since the retransmitted field will eventually attain only one maximum at the location of the strongest scatterer, effectively "ignoring" the presence of the weaker scatterers. The utility of such "iterative time reversal" is explained by Fink as follows [142]: "After the first time-reversed illumination, the weakest target will be illuminated more weakly and will reflect a wavefront much fainter than the one coming from the strongest target. After some iterations, the process will converge and produce a wavefront focused on the most reflective target. This process converges if the target separation is sufficient to avoid the illumination of one target by the real acoustic image of the other target" In other words, if multiple scattering and mutual coupling between scatterers are negligible, one can obtain a set of scattering responses at the TRM, which is entirely due to one target only, the strongest one.

Once the response set due to the strongest target is identified, the iterative time-reversal sequence can be repeated. This time, in the very first time-reversal step, the responses of the strongest target are removed, leaving those of the second strongest target to be extracted.

The reader should be aware that despite the apparent simplicity of the concept of experimental time reversal, it poses numerous technical challenges. The challenges are far greater with microwaves due to the need for much higher time-sampling rates. This complicates the construction of the TRM electronics and creates challenges associated with the antenna design and the channel multiplexing. Nonetheless, numerous systems have been reported [146–151], which indicates that experimental time reversal with microwaves is an active area of research.

We next turn to computational time-reversal imaging. It aims at reconstructing an image from a single scattering experiment with the object under test (OUT) by employing knowledge (usually approximate) of the background medium. A detailed treatment of the computational time-reversal methods based on the signal-processing techniques DORT[10] [152, 153] and MUSIC[11] [154, 155, 156] is available in [26] along with an extensive list of references. Note that both the DORT and the MUSIC time-reversal methods operate in the frequency domain. Both employ the singular-value decomposition of the *multistatic data matrix* $\boldsymbol{K}$, which, in the context of S-parameter measurements, is exactly the S-matrix acquired with the OUT. It can be shown that the resulting singular vectors correspond to the scattering responses produced by the individual scattering centers in the OUT, whereas the singular values are representative of the contrast at these centers [26, 156].

The final step in computational time-reversal imaging relies on the availability of Green's function of the background. In essence, this step generates an image by computing where the time-reversed signals (i.e., the singular vectors) would focus upon injection into the background medium; hence, the name *computational time-reversal imaging*. This step is a numerical implementation of the back-propagation integral equation Eq. (1.121) when applied to frequency-domain signals. The explicit back-propagation integral equation in the frequency domain is given in Eq. (1.201). The reader is reminded that experimental time reversal achieves wave-field focusing onto the scatterers (one by one), but it does not generate an image. In contrast, computational time reversal generates an image, but it needs Green's function of the background to do so. It is important

[10] DORT is derived from *décomposition de l'opérateur de retournement temporel*, meaning *decomposition of the time-reversal operator*.

[11] MUSIC is derived from *multiple signal classification*.

to note that, unlike experimental time reversal, the DORT and MUSIC reconstruction methods are *not* restricted to loss-free media since the "computational" Green function can accommodate losses. Neither are these computational methods restricted to a homogeneous background medium (similarly to experimental time reversal). However, if Green's function of the background is not measured (e.g., through the PSF measurements described in Section 4.2), then simplifying assumptions are necessary. In contrast, experimental time reversal achieves effective "physical" wave focusing onto the targets without "knowing" what the background is.

Computational time-reversal imaging can be carried out using EM simulators directly. In essence, this approach emulates a TRM experiment. The data are the time-domain scattered-field responses measured by an array of probes on a surface surrounding the OUT; therefore, time-domain simulators are used such as those based on the FDTD and the TLM methods. Just like in the TRM experiments, the simulation employs excitation points coincident with the probes in the actual measurement. The excitation waveforms are the time-reversed responses measured with the OUT, and they are injected simultaneously. If the temporal extent of the recorded responses is $[0, T_{\max}]$, the length of the simulation that back-propagates these responses is also $T_{\max}$. At the end of the simulation, the interference pattern formed by the injected time-reversed waves should indicate the locations of the scattering centers. These locations correspond to the maxima in the interference-pattern magnitude distribution. A large number of publications have reported successful image reconstruction using this simulation-based approach to time-reversal imaging with microwave signals; see, for example, [157–163]. The major limitations of these methods are associated with: (i) the modeling errors of the simulation (i.e., the inability to account for all influencing factors in the actual experiment) and (ii) an inability to account for multiple scattering effects and the mutual coupling between closely spaced scattering centers. Additionally, the EM simulations of 3D problems are time consuming and real-time performance is usually not feasible.

To illustrate the efficacy of the simulation-based time-reversal imaging, we revisit the example in Section 4.5.5; see Fig. 4.15. The total number of recorded responses in this example is 64 corresponding to the eight experiments, where the point sources in the eight-element circular array are excited one by one. In each experiment, all eight probes (which are coincident with the point sources) record the scattered waveforms. Each waveform contains 15 001 time samples. Just like in the DAS reconstruction example, two simulations are carried out: (i) the background (scatterer-free) scenario, which yields the incident-field responses, and (ii) the OUT scenario, which yields the total-field responses. The incident-field responses are then subtracted from the total ones to produce the scattered-field responses. These 64 scattered responses are then time-reversed. The eight time-reversed responses ($s_{ij}^{\dagger}(t)$, $j = 1, \ldots, 8$) corresponding to the i-th receiver are summed to produce the excitation of the i-th transmitter in the back-propagation simulation. This is repeated for all eight probes ($i = 1, \ldots, 8$) to obtain the excitations for all transmitters in the back-propagation simulation. All eight transmitters are excited simultaneously in the back-propagation simulation, which takes exactly 15,001 steps in time. It is reiterated that the simulation runs forward in time just like in the experimental time reversal. It is carried out with the same time-domain simulator

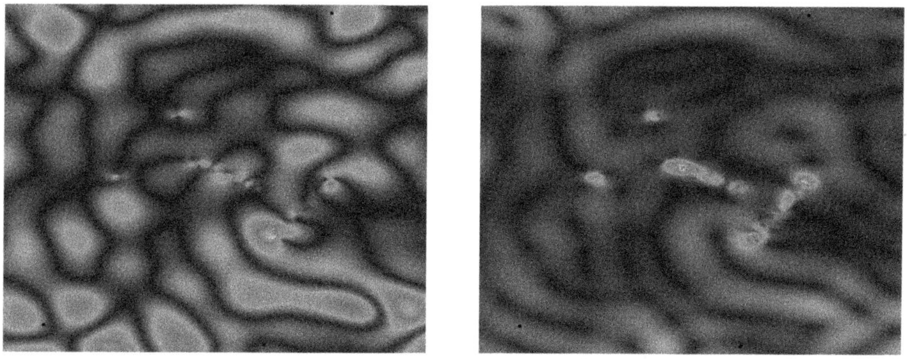

(a) (b)

Figure 4.17 Qualitative images in the circular-array example produced by a simulation-based time reversal. The images are obtained by plotting the field-magnitude distribution in the final step of the back-propagation simulation: (a) **E** field magnitude and (b) **H** field magnitude.

MEFiSTo-3D Pro [10] that was used to emulate the background and OUT measurements. Therefore, there are no modeling errors or noise. The reader is also reminded that this example is carried out in a background of vacuum, i.e., there are no losses. The final image is obtained by simply plotting the magnitude of the field distribution at the final step of the back-propagation simulation.

Fig. 4.17-a shows the image obtained by plotting the **E** field magnitude distribution. Significant artifacts are observed; however, the scattering objects are clearly visible. In fact, a comparison with the images obtained through the DAS reconstruction (see Fig. 4.16) reveals sharper focusing, i.e., better spatial resolution. Interestingly, the image obtained by plotting the **H** field magnitude distribution (see Fig. 4.17-b) contains far fewer artifacts, and it shows clearly the superior resolution achieved with the time-reversal method. Note also that the success of the time-reversal reconstruction depends critically on the ability to capture the responses along the entire boundary enclosing the OUT. Since we sample the scattered responses at only eight locations, image artifacts are expected, even if there are no modeling errors.

5 Linear Inversion in Fourier Space

In this chapter, methods for image reconstruction, both qualitative and quantitative, are described that solve a linearized model of scattering in Fourier space. They are referred to as *spectral-domain* reconstruction methods. In all of these methods, the data sets, which are originally functions of the observation position in real space (e.g., x and y), are subjected to Fourier transformation. The reconstruction of the object's contrast function is then performed in Fourier space, e.g., with respect to k_x and k_y (the Fourier variables corresponding to x and y). This necessitates a final step of an inverse Fourier transform to return the result in real space. This category of reconstruction methods includes MW holography, diffraction tomography, and a great variety of techniques used to produce imagery from synthetic aperture radar (SAR) measurements.

The spectral-domain reconstruction methods are well studied. Their greatest advantage is that they are fast, allowing for image generation in real time. Their disadvantage is that, like other linear inversion methods, they are limited to problems where multiple-scattering and mutual-coupling effects in the object under test (OUT) can be neglected. They are the workhorse of the real-time MW and millimeter-wave imaging systems with various applications in concealed weapon detection [62, 63, 164, 165, 166], nondestructive testing [167–173], medical-imaging research [174–179], antenna measurements [180, 181], and many other areas.

The subject of reconstruction in Fourier space is extensive, and it would be impossible to present it in its entirety here. The goal here is to introduce its basic principles and to give the reader an appreciation for the mathematical beauty of the methodology and its power in real-life applications. MW holography is chosen as the method through which the reader is introduced to the subject. There are several reasons for this choice. First, modern MW holography is akin to SAR reconstruction; in fact, it was developed as an extension of SAR to 3D imaging with data acquired on planes. Thus, it can serve as a sound basis for further studies of the various SAR imaging methods. Second, when applied in a cylindrical coordinate system, MW holography can also be viewed as an extension of diffraction tomography to 3D imaging. Third, MW holography is arguably the most widely used real-time reconstruction method applied to data in the MW and millimeter-wave frequency ranges.

In the following discussion, we consider the most common scenario where the responses are the S-parameters of the MW network formed by the imaged object and the acquisition setup, including the transmitting (Tx) and receiving (Rx) antennas.

5.1 Brief Historical Notes on Microwave Holography

Optical holography originated with the work of Gabor [182, 183] in the late 1940s. At the time, Gabor aimed to improve the quality of the images obtained by electron microscopes, but he demonstrated the feasibility of his method with light waves. He was the first to develop the basic principle of capturing the magnitude and phase of a wave as an interference pattern formed by this wave and a known reference wave. He also developed the mathematical foundations for the image reconstruction from such intereference patterns. A review can be found in [14].

Gabor's experimental setup was improved by Leith and Upatnieks [184, 185] in the early 1960s to achieve images of striking clarity and depth. The critical improvement was the introduction of the off-axis reference beam [186, Introduction by Leith]. Due to the availability of lasers and the work of Leith and Upatnieks, optical holography gained wide attention and led to a new generation of 3D imaging systems.

Soon after (in the late 1960s), optical holography was expanded to other wave fields, most notably acoustics [187–189]. There is a fundamental difference between the goals of the original optical holography and those of acoustical holography. The former aims at reproducing faithfully the optical perception of a 3D object from its recording, i.e., when the object is no longer available for viewing. In contrast, acoustic holography, and indeed acoustic imaging in general, aims at producing an image of an object that is optically obscured. The acoustic field penetrates many media that are impenetrable to light, e.g., human tissue and fluids, metals, and ceramics. Nonetheless, the principles of producing an image are the same in both methods.

MW holography was developed in the 1960s at about the same time as acoustical holography [190–194]. For reviews of the state of the-art at the time, the reader is referred to [195–198]. Similarly to acoustical holography, MW holography aims at the image reconstruction of targets obscured by barriers, which are optically impenetrable but are transparent to microwaves.

In its early stage, MW holography closely resembled the optical method for data acquisition (also referred to as the *recording step*) and image reconstruction (the *reconstruction step*). Just like in the optical holography method of Leith and Upatnieks [184], the recording step aimed at obtaining the *hologram* of the object.[1] The hologram is essentially a planar record of the intensity of an interference pattern obtained in a single-frequency measurement with two coherent waves. In an optical hologram, the record is imprinted on a photographic plate (the recording medium). In MW holography, this could be the intensity pattern acquired through the scanning of an antenna over the acquisition plane, the antenna being connected to a simple diode detector (a receiver that measures only the signal intensity) [196].

The holographic recording step is illustrated in Fig. 5.1. Two coherent waves, the *illuminating beam* and the *reference beam*, are obtained from a MW monochromatic

[1] The name *hologram* is due to Gabor [183]. It is derived from the Greek word *holos* (for *all* or *whole*), indicating the fact that the photograph contains the complete information (amplitude and phase) of the object's scattered wave. Moreover, it allows for the reconstruction of a 3D image of the object.

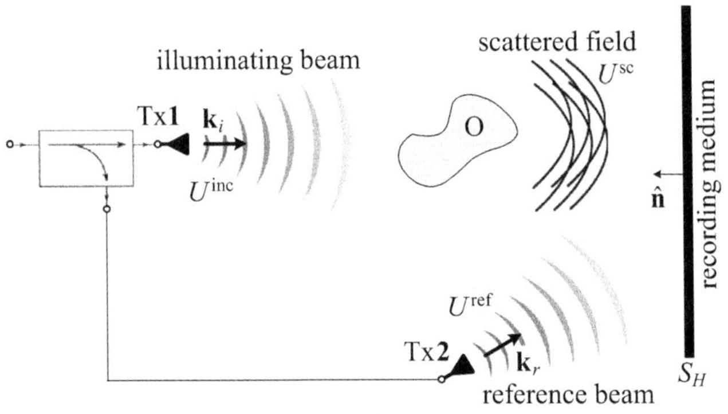

(a)

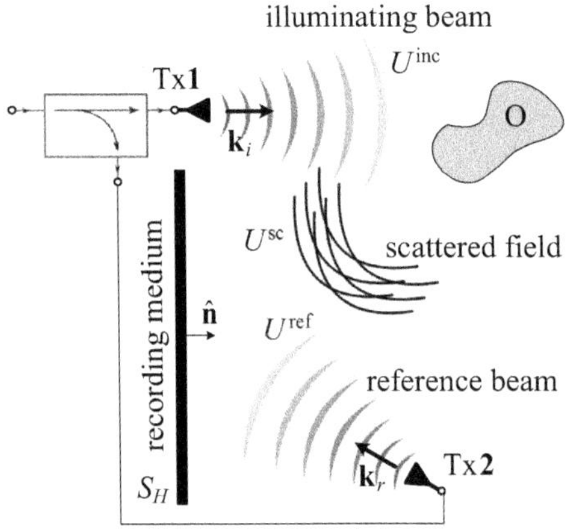

(b)

Figure 5.1 Illustration of the recording step in MW holography that closely resembles the hologram generation in optical holography with an off-axis reference beam. Two coherent waves are obtained from a MW monochromatic generator by splitting the generator's output with a directional coupler or a power splitter. The illuminating beam U^{inc} causes scattering from the imaged object (O), U^{sc}, that interferes with the reference wave U^{ref}. The interference pattern is recorded on the "hologram" plane S_H. The interference pattern contains information about both the magnitude and the phase of U^{sc}. (a) The hologram plane S_H is positioned to capture the forward-scattered wave U^{sc}, which allows for reconstructing the object's *transmittance function*. (b) S_H is positioned to capture the back-scattered U^{sc}, which allows for reconstructing the object's *reflectivity function*.

generator by splitting the generator's output with a directional coupler or a power splitter. The illuminating beam U^{inc} causes scattering from the imaged object (O), U^{sc}, that interferes with the reference wave U^{ref}. The reference wave must *not* interact with the object (O). The interference pattern is recorded on the "hologram" plane S_H. Fig. 5.1-a illustrates the case in which the hologram plane S_H is positioned to capture the forward-scattered U^{sc}, which allows for reconstructing the object's *transmittance function*. To block the incident wave diffracting around the object, absorbing screens may be employed [196]. Alternatively, diffused illumination directly backing the object has also been used [198]. Fig. 5.1-b shows a case where S_H is positioned to capture the back-scattered wave U^{sc}, which allows for reconstructing the object's *reflectivity function*. In general, S_H can be positioned at any desired angle with respect to the illumination as long as it captures the useful portion of U^{sc} without being exposed directly to the illuminating beam.

The recording medium at S_H records the intensity distribution

$$I_H = \left(U^{sc} + U^{ref}\right)\left(U^{sc} + U^{ref}\right)^* = |U^{sc}|^2 + |U^{ref}|^2 + (U^{sc})^* U^{ref} + U^{sc}(U^{ref})^*, \tag{5.1}$$

where * denotes conjugation. It is clear from Eq. (5.1) that the hologram contains the amplitude and the phase information of U^{sc} in the third and the fourth terms. If the reference wave U^{ref} is known, then U^{sc} can be extracted from each one of these terms. This is the task of the reconstruction step of holography.

With modern MW instruments (e.g., vector network analyzers (VNAs)), we can easily obtain both the amplitude and the phase of U^{sc}. Thus, the recording step is simple to implement. Yet, intensity MW detectors are cheaper and often smaller in size than VNAs. This is why, to this day, some MW holographic approaches employ intensity measurement; see, e.g., [199].

What is important to realize is that holography exploits both the magnitude and the phase information imprinted on the hologram to generate 3D images and to enhance resolution compared to the 2D images obtained from amplitude measurements only. In fact, in imaging instrumentation, holography is defined as an interferometric technique for recording the amplitude and the phase of a wave [174].

In the reconstruction step of optical holography, the processed photographic hologram is illuminated by the coherent *reconstruction beam*, which is usually (but not necessarily) of the same frequency as the reference beam U^{ref}. In this step, the object is absent and the illuminating beam is off, but the same geometrical arrangement of the hologram and the reference beam is used as that at the time of the hologram recording. Upon passing through the hologram, the reconstruction beam acquires the phase and amplitude modulation of the original wave, the intensity pattern of which created the hologram in the first place. The wave emerging from the illuminated hologram is mathematically described by

$$U_I = U^{ref} I_H = U^{ref}|U^{ref}|^2 + U^{ref}|U^{sc}|^2 + (U^{sc})^*(U^{ref})^2 + U^{sc}|U^{ref}|^2. \tag{5.2}$$

The first term is the uniform reference wave, which does not carry any information about the object. The second term can be made negligibly small compared to the others by making the reference beam much stronger than the scattered wave. The last term is clearly proportional to the original scattered wave produced by the object during the recording step. With the off-axis reference beam proposed by Leith and Upatnieks, this wave can be focused in a direction different from that corresponding to the third term. An observer intercepting this wave will view the 3D image of the recorded object. Note that in the original arrangement of Gabor, there was no reference beam, or equivalently, one could think of a system where the illuminating and reference beams were coincident.

The reconstruction step in the original MW holography systems also closely resembled that in optics. The MW hologram was scaled down to a size that fit on a photographic hologram and was subsequently processed using laser light; see, e.g., [196, 200].

Today, the tasks performed by the optics are emulated by computer algorithms. As these algorithms developed, their resemblance to existing methods for signal processing in SAR became evident. Leith makes the following observation in [195]: "There exists in the field of communication science a variety of techniques that resemble holography to various degrees, both in concept and in their mathematical formulation. Further, when these processes are carried out with the aid of coherent optical systems, the resemblance to holography becomes striking indeed. Yet these processes have developed quite independently of holography and in no way depend upon principles originating in holography." He proceeds to exemplify this resemblance in four well-known radar techniques: the SAR, chirp radar, rotating-target-imaging system, and beam forming, which he refers to as "quasi-holographic." It is these quasi-holographic techniques that form the basis of modern MW holography.

To this day, the terms *SAR imaging* and *MW holography* are used to refer to reconstruction approaches that are mathematically very similar. In both, exploiting the phase information is at the heart of the image generation. The difference is in the dimensionality of the problem. SAR works with wideband radar data acquired on a line. The location of the imaged surface is known.[2] Thus SAR methods reconstruct a 2D image of the target reflectivity on this known surface. In contrast, MW holography works with wideband data acquired on a surface, which enables the reconstruction of a 3D image; see, for example, [62, 63]. In effect, it merges 2D single-frequency holography reconstruction with wideband SAR reconstruction.

Next, a general method for 3D image generation is described that uses the principles of MW holography. This method was first described in [118], assuming a data equation for the **E** field. Here, it is cast in the framework of the S-parameter data equation Eq. (3.58). The method of Amineh *et al.* can be viewed as a generalization of the method of Sheen *et al.* [62, 63] that allows for employing the experimentally derived kernel of the data equation; see Eq. (4.7). Such a generalization is important in the case of near-field imaging.

5.2 3D Holographic Inversion with Wideband Planar Data Acquisition

In the following, planar data acquisition is assumed as shown in Fig. 5.2. The data are in the form of S-parameters acquired at the planes $z_1 = -d/2$ and/or $z_2 = d/2$ on both sides of the OUT. The theory accommodates both back-scattered signals acquired with single-sided measurements and forward-scattered signals where the Tx and Rx antennas reside on two parallel planes with the object in between. The S-parameters are functions of x and y as well as the frequency ω.

[2] SAR techniques are used to image the earth's surface by an antenna mounted on an aircraft.

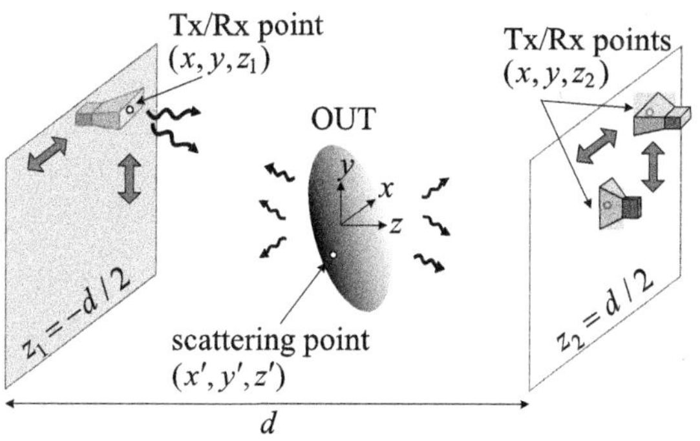

Figure 5.2 An illustration of the setup of planar acquisition. The S-parameters are functions of x and y and, in general, can be measured on two mutually parallel planes on opposite sides of the OUT. If the OUT is accessible from one side only, then all Tx and Rx positions reside on one plane only. Mechanical scanning or electronic switching can be used to achieve coverage of the whole acquisition plane along x and y (see large double-headed arrows).

As per Eq. (3.58), the scattered portions of the S-parameter responses are expressed as

$$S_\xi(x,y;\omega) = \kappa_\xi(\omega) \iiint_{x'y'z'} \Delta\varepsilon(x',y',z') \left[\mathbf{E}^{\text{inc}}_{\xi,\text{Rx}} \cdot \mathbf{E}_{\xi,\text{Tx}}\right]_{(x',y',z';x,y;\omega)} dx'dy'dz', \quad \xi = 1,\ldots,N_T, \tag{5.3}$$

where the subscript ξ denotes the measurement associated with the ξ-th pair of Tx/Rx antennas. In the case of a reflection measurement, one antenna may be both transmitting and receiving. What is important is that, while scanning, the Tx and Rx antennas in the ξ-th measurement are in a fixed configuration with respect to each other. Thus, their position $\mathbf{r}_\xi$ is fully described by the coordinates (x, y).[3] The coefficient $\kappa_\xi(\omega) = -\mathrm{i}\omega/(2a_{\xi,\text{Rx}}a_{\xi,\text{Tx}})$ depends on the root-power incident waves at the Tx and Rx ports, $a_{\xi,\text{Tx}}$ and $a_{\xi,\text{Rx}}$, respectively. $\mathbf{E}^{\text{inc}}_{\xi,\text{Rx}}(x',y',z';x,y;\omega)$ is the incident-field distribution when the ξ-th Rx antenna operates in a transmitting mode and it is positioned at (x, y). The reader is reminded that the incident field is by definition the field obtained in the absence of an OUT. $\mathbf{E}_{\xi,\text{Tx}}(x',y',z';x,y;\omega)$ is the total field in the OUT due to the Tx antenna, when this antenna's position is at $\mathbf{r}_\xi = (x, y)$. The contrast $\Delta\varepsilon(\mathbf{r}') = \varepsilon_\text{O}(\mathbf{r}') - \varepsilon_\text{b}(\mathbf{r}')$, $\mathbf{r}' = (x',y',z')$, is unknown and is the subject of reconstruction. Here, ε_O is the OUT permittivity and ε_b is that of the background (assumed known).

[3] The position $\mathbf{r}_\xi = (x, y)$ associated with the ξ-th measurement does not necessarily reflect the lateral positions of the Tx and Rx antennas. For example, the scanned Tx antenna position may be $\mathbf{r}_{\xi\text{Tx}} \equiv \mathbf{r}_\xi = (x, y)$, but the Rx antenna may be deliberately offset with respect to the Tx antenna along x, i.e., $\mathbf{r}_{\xi\text{Rx}} = (x - \Delta x, y)$. The offset Δx is preserved during the scan; thus, $\mathbf{r}_\xi$ alone is a sufficient indicator of the observation position.

In the case of Born's linearized model of scattering, Eq. (5.3) is written as

$$S_\xi(x, y; \omega) = \kappa_\xi(\omega) \iiint\limits_{x'y'z'} \Delta\varepsilon(x', y', z') \left[\mathbf{E}^{\text{inc}}_{\xi,\text{Rx}} \cdot \mathbf{E}^{\text{inc}}_{\xi,\text{Tx}}\right]_{(x',y',z';x,y;\omega)} dx'dy'dz'. \tag{5.4}$$

We recognize the field dot product in Eq. (5.4) as the resolvent kernel that was first introduced in Section 4.2:

$$\mathcal{K}_\xi(x', y', z'; x, y; \omega) = \left[\mathbf{E}^{\text{inc}}_{\xi,\text{Rx}} \cdot \mathbf{E}^{\text{inc}}_{\xi,\text{Tx}}\right]_{(x',y',z';x,y;\omega)}. \tag{5.5}$$

In a homogeneous or layered medium, where the layers are in xy planes, the resolvent kernel for any position $\mathbf{r}_\xi = (x, y)$ of the Tx/Rx pair can be obtained from the reference kernel,

$$\mathcal{K}_{0\xi}(x', y', z'; \omega) = \mathcal{K}_\xi(x', y', z'; 0, 0; \omega), \tag{5.6}$$

using a simple coordinate translation:

$$\mathcal{K}_\xi(x', y', z'; x, y; \omega) = \mathcal{K}_{0\xi}(x' - x, y' - y, z'; \omega). \tag{5.7}$$

Substituting Eq. (5.7) in Eq. (5.4) yields

$$S_\xi(x, y; \omega) = \kappa_\xi(\omega) \iiint\limits_{x'y'z'} \Delta\varepsilon(x', y', z')\mathcal{K}_{0\xi}(x' - x, y' - y, z'; \omega)dx'dy'dz'. \tag{5.8}$$

To cast Eq. (5.8) in the form of a 2D convolution in x' and y', we utilize the relationship Eq. (4.7) between the kernel and the respective point-spread function (PSF) $S_{0\xi}(x, y; \omega)$, i.e.,

$$\mathcal{K}_{0\xi}(x' - x, y' - y, z'; \omega) \approx \frac{S_{0\xi}(x - x', y - y'; z'; \omega)}{\kappa_\xi(\omega)\Delta\varepsilon_{\text{co}}\Omega_{\text{co}}}. \tag{5.9}$$

Here, $\Delta\varepsilon_{\text{co}} = \varepsilon_{\text{co}} - \varepsilon_{\text{b}}$ is the permittivity contrast of the small scatterer at $(x', y', z') = (0, 0, z')$ with which the PSF $S_{0\xi}(x, y; z'; \omega)$ is acquired, whereas Ω_{co} is its volume. Substituting Eq. (5.9) in Eq. (5.8) leads to

$$S_\xi(x, y; \omega) = \frac{1}{\Delta\varepsilon_{\text{co}}\Omega_{\text{co}}} \iiint\limits_{x'y'z'} \Delta\varepsilon(x', y', z')S_{0\xi}(x - x', y - y'; z'; \omega)dx'dy'dz'. \tag{5.10}$$

In Eq. (5.10), the integration over x' and y' is a 2D convolution of $\Delta\varepsilon$ and $S_{0\xi}$. Thus, taking the Fourier transform (FT) of Eq. (5.10) with respect to x and y produces

$$\tilde{S}_\xi(k_x, k_y; \omega) = \frac{1}{\Delta\varepsilon_{\text{co}}\Omega_{\text{co}}} \int\limits_{z'} \tilde{F}(k_x, k_y, z')\tilde{S}_{0\xi}(k_x, k_y; z'; \omega)dz', \tag{5.11}$$

where $\tilde{S}_\xi(k_x, k_y; \omega) = \mathcal{F}_{2\text{D}}\left\{S_\xi(x, y; \omega)\right\}$, $\tilde{S}_{0\xi}(k_x, k_y; z'; \omega) = \mathcal{F}_{2\text{D}}\left\{S_{0\xi}(x, y; z'; \omega)\right\}$ and $\tilde{F}(k_x, k_y, z')$ is the 2D FT of the contrast in the z' plane:

$$\tilde{F}(k_x, k_y, z') = \mathcal{F}_{2\text{D}}\left\{\Delta\varepsilon(x', y', z')\right\}. \tag{5.12}$$

In the following, we exploit the fact that the signal dependences on the variables ω and z' in Eq. (5.11) are related. For example, in a uniform background, the PSF of a reflection coefficient for a point scatterer at z' can be expressed as

$$\tilde{S}_{0\xi}(k_x, k_y; z'; \omega) = \tilde{S}_{0\xi}(k_x, k_y; 0; \omega) \cdot e^{-i2k_z z'}, \tag{5.13}$$

where $\tilde{S}_{0\xi}(k_x, k_y; 0; \omega)$ is the PSF with the point scatterer at $z' = 0$. This follows from the angular spectrum representation Eq. (4.11). Here, as per the eigenvalue relation in Eq. (4.10),

$$k_z^2 = \omega^2 \mu_b \varepsilon_b - k_x^2 - k_y^2. \tag{5.14}$$

For propagating field modes, where $k_z > 0$, Eq. (5.13) implies a linear relationship between z' and time t, namely, $z' = v_z(t - t_0)/2$, where t_0 is some reference time. Here, $v_z = \omega/k_z$ is the phase-velocity component in the z direction. Thus, the FT with respect to time (i.e., the signal dependence on ω), is also representative of its dependence on the position of the scattering center z'. In general, and especially in the case of a lossy medium and the presence of evanescent field modes, the relationship between z' and time is not that of a simple proportionality. This is why it is advantageous to avoid such assumptions and to discretize Eq. (5.11) directly as

$$\tilde{S}_\xi^{(m)}(k_x, k_y) = \frac{1}{\Delta\varepsilon_{co}\Omega_{co}} \sum_{n=1}^{N_z} \Delta z_n \tilde{F}(k_x, k_y, z'_n) \tilde{S}_{0\xi}^{(m)}(k_x, k_y; z'_n),$$
$$m = 1, \ldots, N_\omega, \xi = 1, \ldots, N_T, \tag{5.15}$$

where Δz_n is the n-th discretization interval in the z (range) direction, N_z is the number of imaged range locations z'_n, and the subscript $^{(m)}$ indicates the m-th frequency sample. This is the discrete forward model used by the holographic reconstruction.

Eq. (5.15) can be used to construct a relatively small linear system of equations at each point $\boldsymbol{\kappa} = (k_x, k_y)$ in Fourier space, which relates the unknown contrast to the available data. The data vector $\boldsymbol{d}(\boldsymbol{\kappa})$ is a vectorized arrangement of all N_T responses at all N_ω frequencies, i.e.,

$$\boldsymbol{d}(\boldsymbol{\kappa}) = \left[\boldsymbol{d}_1^T(\boldsymbol{\kappa}) \cdots \boldsymbol{d}_{N_T}^T(\boldsymbol{\kappa})\right]^T_{N_T N_\omega \times 1}, \tag{5.16}$$

where

$$\boldsymbol{d}_\xi(\boldsymbol{\kappa}) = \left[\tilde{S}_\xi^{(1)}(\boldsymbol{\kappa}) \cdots \tilde{S}_\xi^{(N_\omega)}(\boldsymbol{\kappa})\right]^T, \xi = 1, \ldots, N_T. \tag{5.17}$$

Thus, the size of $\boldsymbol{d}(\boldsymbol{\kappa})$ is $N_D = N_T N_\omega$, where N_D is referred to as the number of data points. The vector,

$$\boldsymbol{f}(\boldsymbol{\kappa}) = \frac{1}{\Delta\varepsilon_{co}\Omega_{co}} \left[\Delta z_1 \tilde{F}(\boldsymbol{\kappa}; z'_1) \cdots \Delta z_{N_z} \tilde{F}(\boldsymbol{\kappa}, z'_{N_z})\right]^T, \tag{5.18}$$

contains the values of the unknown contrast at the point $\boldsymbol{\kappa}$ in Fourier space for all N_z range positions. Finally, the system matrix $\boldsymbol{A}(\boldsymbol{\kappa})$ is constructed from the PSFs as

$$\boldsymbol{A}(\boldsymbol{\kappa}) = \begin{bmatrix} \boldsymbol{A}_1(\boldsymbol{\kappa}) \\ \vdots \\ \boldsymbol{A}_{N_T}(\boldsymbol{\kappa}) \end{bmatrix}_{N_D \times N_z}, \tag{5.19}$$

where

$$\boldsymbol{A}_\xi(\boldsymbol{\kappa}) = \begin{bmatrix} \tilde{S}_{0\xi}^{(1)}(\boldsymbol{\kappa}; z_1') & \cdots & \tilde{S}_{0\xi}^{(1)}(\boldsymbol{\kappa}; z_{N_z}') \\ \vdots & \ddots & \vdots \\ \tilde{S}_{0\xi}^{(N_\omega)}(\boldsymbol{\kappa}; z_1') & \cdots & \tilde{S}_{0\xi}^{(N_\omega)}(\boldsymbol{\kappa}; z_{N_z}')(\boldsymbol{\kappa}) \end{bmatrix}_{N_\omega \times N_z}, \xi = 1, \ldots, N_T. \tag{5.20}$$

The inversion is performed by solving a system,

$$\boldsymbol{A}(\boldsymbol{\kappa}) \cdot \boldsymbol{f}(\boldsymbol{\kappa}) = \boldsymbol{d}(\boldsymbol{\kappa}), \tag{5.21}$$

at each $\boldsymbol{\kappa} = (k_x, k_y)$. Once all of these systems are solved for $f(k_x, k_y; z_n')$, $n = 1, \ldots, N_z$, the FT of the object's contrast $\tilde{F}(k_x, k_y, z_n')$ is recovered using Eq. (5.18), i.e.,

$$\tilde{F}(k_x, k_y, z_n') = \frac{\Delta\varepsilon_{\mathrm{co}}\Omega_{\mathrm{co}}}{\Delta z_n} f(k_x, k_y; z_n'). \tag{5.22}$$

Finally, the inverse FT is applied to $\tilde{F}(k_x, k_y, z_n')$ to reconstruct a 2D slice of the contrast $\Delta\varepsilon(x', y', z')$ at each z_n' plane:

$$\Delta\varepsilon(x', y', z_n') = \mathcal{F}_{\mathrm{2D}}^{-1}\left\{\tilde{F}(k_x, k_y, z_n')\right\}, n = 1, \ldots, N_z. \tag{5.23}$$

Note that often the coefficient $(\Delta\varepsilon_{\mathrm{co}}\Omega_{\mathrm{co}})^{-1}$ in the forward models Eq. (5.10) and Eq. (5.15) is not known accurately or is not known at all. The latter case arises when the PSFs are represented via analytical approximations such as spherical or locally plane waves. In such cases, this coefficient is set equal to unity. The resulting reconstructed contrast has only a qualitative value, and the image is generated by simply plotting the normalized contrast distribution, namely,

$$I(x', y', z_n') = \frac{|\Delta\varepsilon(x', y', z_n')|}{|\Delta\varepsilon|_{\max}}, n = 1, \ldots, N_z, \tag{5.24}$$

where $|\Delta\varepsilon|_{\max}$ is the maximum absolute value of the contrast across all voxels of the imaged volume.

5.3 Spatial Sampling Rate and Its Impact on the Computational Requirements

The choice of the frequencies ω_m ($m = 1, \ldots, N_\omega$) is critical to the spatial resolution of images. It has already been demonstrated through some simple examples in

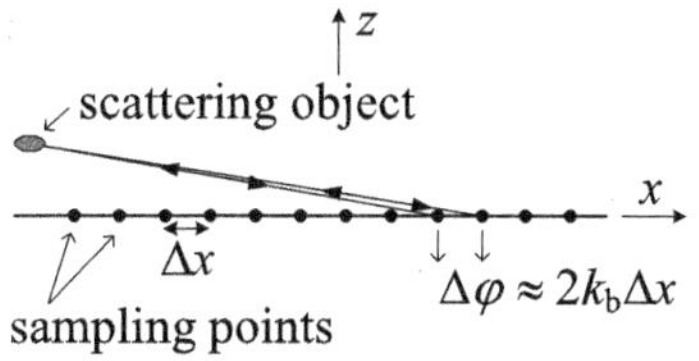

Figure 5.3 An illustration of a scenario where the phase difference between the signals received at two neighboring sampling points approaches the maximum value of $2k_b\Delta x$. Here, Δx is the sampling step along x and k_b is the wavenumber. This scenario arises in a monostatic (reflection) measurement, where the lateral position x' of the scattering center is much larger than its range position z'. This causes near-grazing angles of arrival of the back-scattered waves.

Section 4.3.4 how the frequency bandwidth B affects the range resolution $\delta z'$ [see Eq. (4.52)] and how the wavelength $\lambda^{(m)}$ affects the cross-range resolution $\delta x'$ [see Eq. (4.61)]. More rigorous derivations are provided in Chapter 6. In general, scattering experiments at higher frequencies (shorter wavelengths) and wider bandwidths ensure better resolution. However, there are various factors that limit the maximum frequency and the available bandwidth, such as the penetration ability of the radiation and the bandwidth of the antennas and the electronics.

Let us assume for now that the maximum frequency and the bandwidth of the illuminating radiation are already determined in accordance with the required resolution performance and the physical and technical constraints. The next step in the imaging system design is to make sure that we make the best use of the chosen frequencies. To be able to make use of all frequency components of the scattered signals, the spatial sampling rate of the scan must satisfy the Nyquist criterion, i.e., the phase shift from one sample to the next must be less than π at any given frequency. As per our convention (see Fig. 5.2), the lateral (or cross-range) positions are x and y . The respective cross-range sampling steps Δx and Δy must be chosen so that

$$k_\zeta \Delta\zeta < \pi, \zeta \equiv x, y. \tag{5.25}$$

Therefore, $\Delta\zeta$ is limited by the maximum k_ζ value, which is

$$k_\zeta^{\max} \approx 2k_b = \frac{4\pi}{\lambda_b}, \zeta \equiv x, y. \tag{5.26}$$

Here, k_b is the background's wavenumber and λ_b is the wavelength. The value of $k_\zeta^{\max} \approx 2k_b$ in Eq. (5.26) comes from an estimation of the maximum possible phase shift that occurs in a reflection-coefficient measurement where the scattered wave arrives at an extremely oblique (grazing) angle. This scenario is illustrated in Fig. 5.3. At such angles, the wave vector $\mathbf{k}_b = k_x\hat{\mathbf{x}} + k_y\hat{\mathbf{y}} + k_z\hat{\mathbf{z}}$ can be deemed to be almost parallel to the acquisition surface, so that either k_x or k_y could approach $k_b = |\mathbf{k}_b|$. Since the wave travels to and from the target, the cumulative difference in the signal path associated with two neighboring acquisition points is approximately $2\Delta x$, which is the reason for the factor of 2 in $k_\zeta^{\max} \approx 2k_b$. In forward-scattering measurements, the cumulative

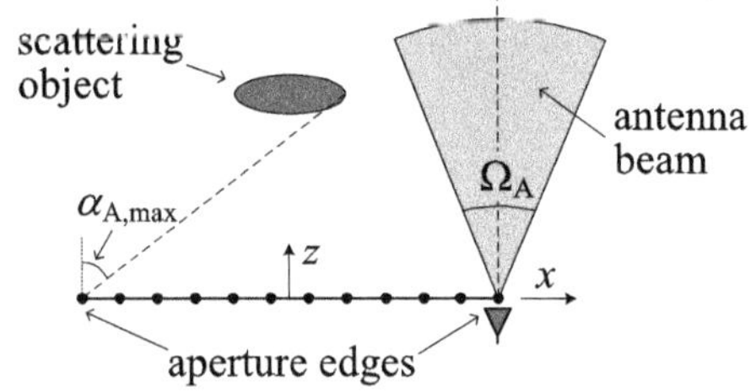

Figure 5.4 Illustration of limited target views due to the limited aperture size and the antenna beamwidth. The finite aperture size does not allow for capturing back-scattered waves that arrive at angles of incidence greater than $\alpha_{A,max}$ even if the Rx antenna is isotropic. If the Rx antenna has a beamwidth $\Omega_A < \alpha_{A,max}$, the view of the target is further limited because, when the antenna scans close to the aperture edge, the target is out of its beam.

difference in the signal path cannot exceed $2\Delta x$ either; thus, the same upper bound for k_ζ^{max} holds.

Finally, the substitution of Eq. (5.26) in Eq. (5.25) leads to the recommended choice of the spatial sampling step as

$$\Delta\zeta \leq \Delta\zeta_{max} = \frac{\lambda_{min}}{4}, \zeta \equiv x, y, \tag{5.27}$$

where λ_{min} is the shortest wavelength in the background. As a word of caution, this analysis is based on the assumption that the back-scattered field is composed of traveling plane waves. If the antennas and the scattering object are in each other's extreme near zone, evanescent waves are present, which may exhibit fast spatial variations that could increase k_ζ^{max} beyond $2k_b$. In fact, as shown in Chapter 6, such increase in the "k-space bandwidth" of the data improves the spatial resolution beyond what is known as the *diffraction limit*.[4] Thus, in measurements involving evanescent fields, a smaller sampling step is recommended in order to capture fast field variations in space.

On the other hand, in far-zone measurements, a reception at grazing angles is often not achievable due to the finite size of the aperture and the limited beamwidth of the receiving antennas. In this case, Eq. (5.26) should be augmented as

$$k_\zeta^{max} \approx 2k_b \sin\alpha_{max} = \frac{4\pi \sin\alpha_{max}}{\lambda_b}, \zeta \equiv x, y, \tag{5.28}$$

where $\alpha_{max} \in [0, 90^\circ]$ is the maximum angle at which the OUT can be viewed. The limitations on α_{max} are illustrated in Fig. 5.4. It is clear from this illustration that α_{max} is limited either by the maximum angle subtended by the acquisition aperture $\alpha_{A,max}$

[4] The term *diffraction limit* is also known as *Abbe's diffraction limit* in recognition of the German physicist Ernst Abbe. Abbe formulated and illustrated with experiments in microscopy the resolution limitations that arise from the finite optical wavelength [201]. For a discussion of Abbe's theory, the reader is referred to [14, chapter 8].

or by half of the antenna beamwidth $\Omega_A/2$, whichever is smaller, i.e., $\alpha_{max} = \min(\alpha_{A,max}, \Omega_A/2)$. With Eq. (5.28), the spatial sampling recommendation in Eq. (5.27) is relaxed:

$$\Delta\zeta_{max} = \frac{\lambda_{min}}{4\sin\alpha_{max}}, \; \zeta \equiv x, y. \tag{5.29}$$

This explains why many practical systems are designed with a sampling step close to $\lambda_{min}/2$, yet they make full use of the available frequencies and achieve good image resolution.

The reconstruction approach described in Section 5.2 operates on the 2D FTs of the responses, $\tilde{S}_\xi^{(m)}(k_x, k_y)$, $\xi = 1, \ldots, N_T$, $m = 1, \ldots, N_\omega$; see the data vector defined through Eq. (5.16) and Eq. (5.17). Thus, the first step that the reconstruction algorithm takes is to perform 2D FFT on the $N_D = N_T N_\omega$ data sets, $S_\xi^{(m)}(x, y)$. Unless zero padding is employed, the number of samples along x and y, N_x and N_y, is the same as the number of samples along k_x and k_y, respectively. If the spatial sampling intervals are Δx and Δy, then the 2D FFT produces an output on a uniform (k_x, k_y) grid, the extent of which is defined by the limits $\pm K_x$ and $\pm K_y$ such that

$$K_x = \frac{\pi}{\Delta x}, \; K_y = \frac{\pi}{\Delta y}. \tag{5.30}$$

This is a direct result of Nyquist's sampling limit. The sampling step along k_x in the 2D Fourier space is then

$$\Delta k_x = \frac{2K_x}{N_x} = \frac{2\pi}{N_x \Delta x} = \frac{2\pi}{D_x}, \tag{5.31}$$

where D_x is the physical extent of the acquisition aperture along x. The sampling step along k_y is determined in the same manner.

It is clear that if the spatial sampling step is chosen as $\Delta\zeta < \Delta\zeta_{max}$ ($\zeta \equiv x, y$), in accordance with Eq. (5.27), then $K_\zeta > 2k_b$. This may be advantageous in near-field imaging since it allows for taking full advantage of evanescent modes, which will result in improved spatial resolution of the image. On the other hand, too large an extent in Fourier space may allow for high-spatial-frequency noise, which will corrupt the imaging result. Thus, the optimal choice of the spatial sampling rate is a compromise between a better spatial resolution and suppression of the high-spatial-frequency noise, which depends on the system SNR. In practice, sampling rates are usually chosen close to the limit in Eq. (5.27). If noise levels are high, the reconstruction through the solution of the linear systems in Eq. (5.21) is ill-posed. This type of ill-posedness manifests itself through high sensitivity of the solution to small variation in the data. In this case, k-space low-pass filtering (e.g., Gaussian or Wiener filtering) improves the result. K-space filtering has an effect similar to that of regularization using, for example, truncated singular-value decomposition (TSVD), Tikhonov methods, and Wiener deconvolution [202, 203]. Note that the reconstruction Eq. (5.19) is solved at each (k_x, k_y) point, and therefore, direct filtering in Fourier space is the most straightforward "regularization" approach.

Last but not least, one has to consider the impact of the chosen sampling on the computational cost of the reconstruction. The holographic method described in Section 5.2 is quick. It can reconstruct the permittivity distribution over a million voxels within seconds. This is typical for the Fourier-space direct reconstruction schemes. However, such speeds are achieved through the careful selection of both the spatial and the frequency sampling. A significant part of the computational effort is performing the 2D FT on the data (all N_T responses at all N_ω frequencies). The typical computational cost of a 2D FFT on a set of $N_x \times N_y$ samples scales as $O(N_x N_y \log_2(N_x N_y))$ [204]. Although this is an efficient numerical process, keeping the number of spatial samples to a minimum is important.

Casting the contrast solution $\tilde{F}(k_x, k_y, z'_n)$ from Fourier space back into real space in order to obtain $\Delta\varepsilon(x', y', z'_n)$, $n = 1, \ldots, N_z$, requires N_z 2D inverse FFTs. The computational cost of an inverse FFT is the same as that of the FFT. Typically, $N_z \ll N_T N_\omega$. This is why the computational cost of this operation is significantly smaller than that of the FFT performed on the data. Note that the lateral voxel size in the reconstructed volume $\Delta x' \times \Delta y'$ is usually chosen to match the sampling steps, i.e., $\Delta\zeta' = \Delta\zeta$, $\zeta \equiv x, y$. This simplifies the implementation significantly.

An important advantage of the holographic approach described in Section 5.2 is that the computational effort in solving the linear systems in Eq. (5.21) is small. This is due to its "divide and conquer" strategy. Instead of solving the problem in its entirety in Fourier space, it solves a small linear system of equations at each point $\boldsymbol{\kappa} = (k_x, k_y)$. It is significantly faster to solve $N_x N_y$ linear systems of size $N_D \times N_z$ (where $N_D = N_T N_\omega$) than to solve one linear system of size $(N_x N_y N_D) \times (N_x N_y N_z)$. The typical values of N_z and N_ω are anywhere between 3 and 20, whereas those of N_x and N_y are on the order of 100 to 1000. In an N-port reciprocal antenna system, $N_T = (N^2 + N)/2$. Thus, with reflection-only measurements, $N_T = 1$, whereas in a 2-port reflection/transmission measurement, $N_T = 3$, namely, S_{11}, S_{22}, and $S_{12} = S_{21}$.

5.4 Frequency Sampling Rate

The frequency sampling interval Δf is determined in a manner similar to the spatial sampling. A change in frequency entails a change in the phase delay of the back-scattered signal since this phase delay can be estimated as $\mathcal{P}(\omega) \approx 2k_b(\omega)R$, where $k_b = \omega\sqrt{\varepsilon_b \mu_b}$ and R is the distance from the antenna to the scattering center. The maximum phase delay occurs at the maximum target range $R = R_{max}$; therefore, Δf determines that range. The difference between the phase delays at two consecutive frequency samples of a back-scattered signal at any given receiver must be smaller than π to avoid *aliasing* in the Fourier-domain solution [204]. In our case, if Δf is not chosen sufficiently small, aliasing means that scattering from targets beyond the maximum range R_{max} will be spuriously moved below that range, thereby "mixing" them with the targets within range. The condition

$$2\Delta k_b(\omega) R_{max} < \pi, \tag{5.32}$$

leads to the requirement

$$\Delta f < \frac{v_b}{4R_{max}}. \tag{5.33}$$

Here, v_b is the phase velocity in the background.

The physical meaning of Eq. (5.33) is readily understood with time-domain (pulsed radar) measurements in mind. We know that the maximum frequency span F_c beyond which aliasing occurs (also known as the *Nyquist critical frequency*) is determined by the temporal sampling step Δt as [204]

$$F_c = \frac{1}{2\Delta t}. \tag{5.34}$$

To avoid aliasing, F_c must be aligned with the maximum frequency F_{max} of the radar signal (typically determined at -10 dB down from the maximum spectral density), i.e., $F_c \approx F_{max}$. When the 1D FFT is applied to a temporal signal containing N_t samples, the samples in the frequency domain are $N_f = N_t$. It follows from Eq. (5.34) that

$$\Delta f = \frac{1}{2T_{max}}, \tag{5.35}$$

where $T_{max} = N_t \Delta t$ is the length of the temporal record. We want T_{max} to be sufficiently long to allow for the back-scattered signals from the most remote targets to arrive at the receiver, i.e., $T_{max} = 2R_{max}/v_b$. Substituting this in Eq. (5.35) results in Eq. (5.33).

In typical near-field imaging scenarios where R_{max} is anywhere between a couple of centimeters to a meter, the frequency sampling intervals are in the range of hundreds of MHz. Frequency oversampling helps to offset the impact of noise to a certain extent, but it also increases the computational toll since a data set at each frequency must be subjected to a 2D FFT. It also increases the number of measurements and, therefore, increases the acquisition time. Thus, the choice of the frequency sampling step Δf is a compromise between range, noise suppression, and time restrictions.

In the reconstruction technique described in Section 5.2, frequency oversampling may increase the condition number of the system matrices $\boldsymbol{A}(\boldsymbol{\kappa})$; see Eq. (5.19) and Eq. (5.20). This happens when the system PSFs are very similar from one frequency to another, resulting in very similar rows in $\boldsymbol{A}_{\xi}(\boldsymbol{\kappa})$. Again, regularization methods such as Tikhonov regularization can help overcome the matrix ill conditioning.

5.5 Construction of the System Matrices: The PSFs

The system matrices in the linear systems underlying the reconstruction described in Section 5.2 are built from the system PSFs; see Eq. (5.19) and Eq. (5.20). How to obtain these matrices was addressed in much detail in Section 4.2. There, it was emphasized that when measurements are performed in the extreme near field of the antennas and especially if the medium is layered rather than homogeneous, it is desirable to measure the PSFs since the analytical approximations may not be adequate. The PSFs are measured with a small (point) scatterer at the center $\mathbf{r}' = (x', y', z') = (0, 0, 0)$ of the imaged volume. However, the system matrices $\boldsymbol{A}(\boldsymbol{\kappa})$ need the PSFs obtained with point scatterers at all discrete range positions given by $(0, 0, z'_n)$, $n = 1, \ldots, N_z$. These can be measured, which is a simple approach, but requires additional calibration measurements.

If the background is loss free, homogeneous, and unbounded, analytical approximations of the PSFs can be employed. For example, with a monostatic scan where the target is in the far zone of the antennas, the PSF can be approximated as

$$S_0(x, y, \omega) \sim e^{-\mathrm{i}2k_\mathrm{b}\sqrt{x^2+y^2+z_A^2}}, \tag{5.36}$$

where z_A is the position of the scanned plane and $R_0 = \sqrt{x^2 + y^2 + z_A^2}$ is the distance from the antenna at (x, y, z_A) to the point target at $(x', y', z') = (0, 0, 0)$. The factor of 2 in the argument of the exponent accounts for the signal path to and from the target. Note that the amplitude decay has been ignored since it has little impact on the *relative* PSF dependence on x and y when the target is far from the scanned aperture. Also, the subscript ξ has been omitted since there is only one type of response ($N_T = 1$), namely, the reflection S-parameter. With the approximation in Eq. (5.36), the PSF for a point scatterer at $z' \neq 0$ is readily obtained as

$$S_0(x, y; z'; \omega) \sim e^{-\mathrm{i}2k_\mathrm{b}\sqrt{x^2+y^2+(z'-z_A)^2}}. \tag{5.37}$$

The 2D Fourier transform of the PSFs in Eq. (5.37) can be computed to construct the system matrix in Eq. (5.20). It should be pointed out that these PSFs are not properly scaled with respect to the actual excitation they include neither the amplitude decay due to the free-space loss factor nor the directional properties of the Tx antenna. Nor do they include the free-space loss of the scattered wave. Thus, the obtained solution for the contrast is only qualitative.

5.6 Matrix-Free 3D Holography with Reflection Coefficient Measurements

It is worth following up on the discussion in the previous section about reconstruction with far-zone reflection-coefficient measurements wherein the PSFs are approximated analytically. It has been noted that the 2D FT of the PSFs in Eq. (5.37) provides the elements of the matrices Eq. (5.20) used to solve for the contrast through the systems of equations in Eq. (5.21). A little bit more analytical work shows that a solution can be obtained without solving the systems of equations in Eq. (5.21) by performing a 3D inverse FFT instead. The computational cost is comparable, but the solution is mathematically elegant. This solution has been proposed by Sheen *et al.* in a millimeter-wave system for concealed weapon detection; see, for example, [62, 63].

The PSF in Eq. (5.37) can be decomposed into an infinite superposition of plane waves as [62]

$$e^{-i2k_b\sqrt{x^2+y^2+(z'-z_A)^2}} = \iint e^{-ik_x x - ik_y y - ik_z(z'-z_A)} dk_x dk_y, \tag{5.38}$$

where k_x, k_y, and k_z relate to the background wavenumber k_b through the eigenvalue equation

$$k_x^2 + k_y^2 + k_z^2 = (2k_b)^2. \tag{5.39}$$

It is emphasized that the medium is assumed to be uniform and loss free. In addition, with far-zone measurements, only propagating modes are present; therefore, both k_x and k_y are limited to values in the interval $[-2k_b, 2k_b]$. Thus, k_z is real and determined from

$$k_z = \sqrt{4k_b^2 - k_x^2 - k_y^2} = \sqrt{4(\omega/\upsilon_b)^2 - k_x^2 - k_y^2}. \tag{5.40}$$

Here, υ_b is the phase velocity in the background. Changing variables in Eq. (5.38) as $\kappa_x = -k_x$ and $\kappa_y = -k_y$ allows for expressing it as

$$e^{-i2k_b\sqrt{x^2+y^2+(z'-z_A)^2}} = \iint e^{i\kappa_x x + i\kappa_y y - ik_z(z'-z_A)} d\kappa_x d\kappa_y. \tag{5.41}$$

The right side is the inverse FT of $e^{-ik_z(z'-z_A)}$; therefore,

$$\mathcal{F}_{2D}\left\{e^{-i2k_b\sqrt{x^2+y^2+(z'-z_A)^2}}\right\} = e^{-ik_z(z'-z_A)}. \tag{5.42}$$

This is the desired FT of the PSF, which relates the unknown contrast to the data in Fourier space through Eq. (5.11):

$$\tilde{S}_0(k_x, k_y, 0; z'; \omega) = e^{-ik_z(z'-z_A)}. \tag{5.43}$$

Note that ξ ($\xi = 1, \ldots, N_T$) has been omitted because only reflection coefficients are measured, i.e., $N_T = 1$. Substituting Eq. (5.43) into Eq. (5.11) results in

$$\tilde{S}(k_x, k_y; \omega) = \int_{z'} \tilde{F}(k_x, k_y, z') e^{-ik_z(z'-z_A)} dz' = e^{ik_z z_A} \int_{z'} \tilde{F}(k_x, k_y, z') e^{-ik_z z'} dz'. \tag{5.44}$$

The integral on the right is the FT of $\tilde{F}(k_x, k_y, z')$ with respect to the variable z', which corresponds to k_z in Fourier space. It follows that Eq. (5.44) can be written as

$$\tilde{S}(k_x, k_y; \omega) = \tilde{F}(k_x, k_y, k_z) e^{ik_z z_A}, \tag{5.45}$$

where $\tilde{F}(k_x, k_y, k_z)$ is the 3D FT of the target reflectivity $f(x', y', z')$.

Next, note that the variables ω [on the left of Eq. (5.45)] and k_z [on the right of Eq. (5.45)] are related through Eq. (5.40). Then, the reconstruction formula for the reflectivity can be written as [62]

$$f(x', y', z') = \mathcal{F}_{3D}^{-1}\left\{\mathcal{F}_{2D}\left\{S(x, y; \omega)\right\} \cdot e^{-i\sqrt{[2k_b(\omega)]^2 - k_x^2 - k_y^2}\, z_A}\right\}. \tag{5.46}$$

Eq. (5.46), although mathematically succinct, is not directly applicable with a conventional 3D FFT algorithm, which requires uniform sampling. The data are usually sampled uniformly along x, y, and ω. Thus, at each frequency, the data 2D FT $\tilde{S}(k_x, k_y; \omega) = \mathcal{F}_{2D}\{S(x, y; \omega)\}$ are also sampled uniformly in k_x, k_y, and ω but not in k_z because of the nonlinear relationship between ω and k_z; see Eq. (5.40). Thus, the data needs to be interpolated and resampled for uniformly spaced points along k_z.

5.7 Simulation-Based Example

To illustrate the MW holographic reconstruction with planar data acquisition, let us revisit the simulation-based example from Section 4.3.5, where the OUT is an F-shaped dielectric bar of relative permittivity $\varepsilon_r = 1.2 + i0$ immersed in a background of relative permittivity $\varepsilon_{r,b} = 1.0 + i0$. The setup is exactly as described in Section 4.3.5 and so it is not described again. However, it is instructive to comment on several setup parameters that have bearing on our prior discussions concerning the choice of sampling intervals in space and frequency. The S-parameters are simulated at $N_\omega = 14$ frequencies evenly distributed from 3 GHz to 16 GHz, included. This means that the frequency sampling step is $\Delta f = 1$ GHz. According to Eq. (5.33), this sets the maximum range extent of the OUT to about 7.5 cm. At the same time, the imaged volume has a range extent of 5 cm. Thus, Δf is sufficiently small.

Further, Eq. (5.29) dictates a spatial sampling step along x of $\Delta x \leq c/(4 f_{\max} \sin \alpha_{x,\max})$, where c is the speed of the wave in vacuum and $f_{\max} = 16$ GHz. The half-wavelength dipole beamwidth in its elevation plane is about 78°; therefore, $\alpha_{x,\max} \approx 39°$, and $\sin \alpha_{x,\max} \approx 0.63$. Thus, the sampling step along x should be chosen as $\Delta x \leq 7.4$ mm. In the y direction, the angle $\alpha_{y,\max}$ is limited by the finite aperture, where $\sin \alpha_{y,\max} \approx 0.89$. The dipole antenna is omnidirectional in its azimuth plane and has no influence on $\alpha_{y,\max}$. Therefore, the sampling step along y should be $\Delta y \leq 5.3$ mm. Yet, we will use the data acquired with $\Delta x = \Delta y = 10$ mm as was done in the example in Section 4.3.5. This spatial sampling does not allow for the best use of the data acquired at the aperture edges, but should still allow for good reconstruction compared to that with the sensitivity-based method and the quantitative scattered-power method. This is indeed the case as is shown next.

The data are processed with the reconstruction method described in Section 5.2. The PSFs are acquired by emulating the measurements of the same voxel-sized scatterer of 1 cm^3 and relative permittivity $\varepsilon_{co,r} = 1.1 - i0$ as described in Section 4.3.5. These PSFs are then processed with the 2D FFT to obtain the elements of the system matrices in Eqs. (5.19) and (5.20), namely, $\tilde{S}_{0\xi}^{(m)}(k_x, k_y; z'_n)$. Here, $\xi = 1, 2, 3$ denotes the response type (S_{11}, S_{22}, $S_{21} = S_{12}$); $m = 1, \ldots, 14$ is the frequency index; and $n = 1, 2, 3$ is the index of the imaged plane. Three planes are imaged: $z'_1 = -10$ mm, $z'_2 = 0$, and $z'_3 = 10$ mm, where $z'_2 = 0$ intersects the F-shaped bar midway along its vertical extent of 1 cm; see Fig. 4.5-a.

Fig. 5.5-a shows a plot of the normalized magnitude of the reconstructed contrast,

$$I(x', y', z') = \frac{|\Delta\varepsilon(x', y', z')|}{|\Delta\varepsilon|_{\max}}. \quad (5.47)$$

This qualitative image shows better quality in terms of cross-range and range resolution compared to the qualitative sensitivity map in Fig. 4.7-a. This is expected since the sensitivity-map method relies on the coherent addition of cross-correlation terms without solving any linear system of equations. In contrast, the holography method solves a system of equations, which matches the data point by point to the PSF-based scattering model in Fourier space. The holographic inversion takes longer than the sensitivity-based reconstruction, which is practically instantaneous (on the order of milliseconds), but it is still fast. In this example, the holographic reconstruction (implemented in MATLAB [79]), takes less than a second on a personal computer.[5] The rectangular linear systems in Eq. (5.21) are solved using MATLAB's function *pinv*, which implements the Moore-Penrose pseudoinverse of a matrix. These systems are small—the system matrices $\boldsymbol{A}(\boldsymbol{\kappa})$ are of size $N_D \times N_z$, where $N_D = N_T N_\omega = 3 \cdot 14 = 42$ and $N_z = 3$. Thus, their solution is quick. Since the number of spatial samples is $N_x \times N_y = 21 \times 21$, the number of sampling points in real space and in Fourier space is 441. Therefore, there are 441 system of equations to be solved. The computational effort in solving these systems is comparable to that required by the 2D FFTs of the 42 data sets.

The images in Fig. 5.5-b,c show the quantitative output of the reconstruction. This quantitative output is possible due to the availability of the exact system PSFs, for which we know the volume Ω_{co} and the contrast $\Delta\varepsilon_{\text{co}}$ of the small scatterer. Thus, the scaling factor $\Delta\varepsilon_{\text{co}}\Omega_{\text{co}}/\Delta z_n$ in Eq. (5.22) is available. Observe that the real part of the relative permittivity of the F-shaped object is reconstructed well with very few artifacts and insignificant nonphysical values ($\varepsilon_{\text{r}}' < 1$) associated with a small number of background voxels. The reconstruction of the imaginary relative permittivity shows artifacts that are relatively small (two orders of magnitude smaller than the values of ε_{r}'). Nonphysical values, although small, are also present. This illustrates one drawback of the inversion in Fourier space, namely, the difficulties in imposing physical constraints on the contrast solution. When the contrast function is computed through a solution in real space, such as the scattered-power map (SPM) method described in Section 4.4, the constraints $\Re\varepsilon_{\text{r}} \geq 1$ and $\Im\varepsilon_{\text{r}} \leq 0$ can be imposed easily. This is why the SPM quantitative images of the F-shaped bar in Fig. 4.8 do not contain voxels with nonphysical values.

Finally, note that this example involves a scattering object of very small contrast, which ensures the success of the linear reconstruction. The presence of strong scatterers of large and complex shapes inevitably leads to poor imaging results because the linearized model of scattering that underlines methods such as the SPM and holography cannot represent nonlinear effects due to the multiple scattering and the mutual coupling between the components of a complex target.

5.8 Measurement-Based Example

The next example employs the planar data acquisition apparatus that was also used in the experimental example presented in Section 4.4.2. This time, the imaged object is a teddy

[5] CPU: 2.4 GHz Intel Core i5, RAM: 8 GB/1600 MHz/DDR3, MATLAB ver. 2014b.

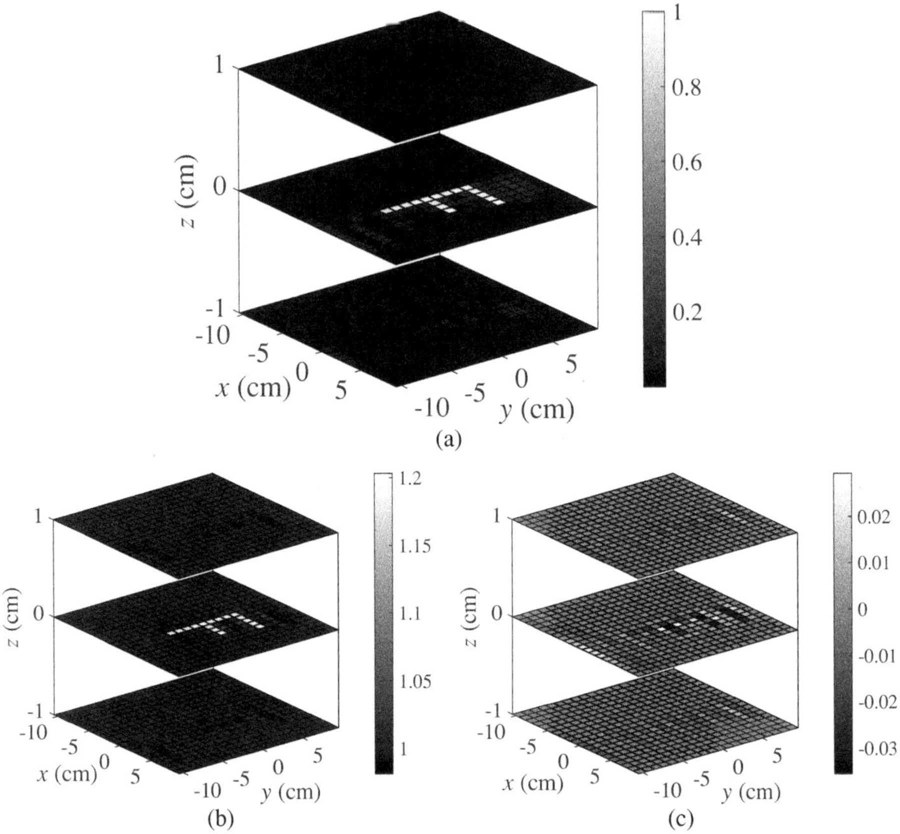

Figure 5.5 Reconstruction of the F-shaped dielectric object (see Section 4.3.5) with the holography method described in Section 5.2: (a) qualitative image of the normalized contrast magnitude, (b) quantitative estimate of the real part of the OUT relative permittivity, $\Re\varepsilon_r$, and (c) quantitative estimate of the imaginary part of the OUT relative permittivity, $\Im\varepsilon_r$. The actual relative permittivity of the F-shaped bar is $\varepsilon_r = 1.2 + i0$, whereas that of the background is $\varepsilon_r = 1.0 + i0$. (Courtesy of Daniel Tajik, M.A.Sc. student at McMaster University, Canada.)

bear plush toy and the background is air. Fig. 5.6 shows the toy taped to the platform of the scanner. The vertical size of the toy, measured from the back of the head to the tip of the nose, is approximately 10 cm. Laid flat on the positioning platform, the toy occupies an area of approximately 20 cm in width (across the arms) and 23 cm in length. The scanned area is 29 cm by 29 cm, and the sampling step along x and y is 5 mm. At each sampling position, the S_{21} parameter is acquired from 8 GHz to 12 GHz, with 41 evenly spaced frequency points. The S_{21} parameter is measured at the ports of two antennas aligned along each other's boresight. The antennas are open-ended WR90 waveguides. The distance between them is approximately 16.5 cm. The antennas operate in air. The scanning platform carries the OUT along x and y while the antennas are fixed at the lateral center of the scanner. The vertical position of the platform (along z) is adjustable and fixed at a distance of about 1.5 cm from the aperture of the bottom antenna.

Figure 5.6 Photo of the teddy bear plush toy afixed to the scanner's platform. The top antenna (a WR90 open-end waveguide) is visible in the top left corner of the photo.

First, the S_{21} PSF is acquired experimentally in a calibration object (CO) measurement where a small scatterer is scanned between the two antennas in a plane that corresponds approximately to the range location of the toy's body. The small scatterer is a dielectric cylinder of relative permittivity $\varepsilon_{r,co} \approx 12 - i0$. Its diameter is 0.5 cm and its height is 1 cm. This PSF is used to construct the system matrices in Eq. (5.21).

The teddy bear toy is scanned twice. In the first measurement, the toy is scanned as is, whereas in the second experiment two items are placed inside. The first item is a cross-shaped dielectric object of relative permittivity 12, fabricated from Eccostock HiK material [205]. The length of the bars making up the cross is 3 cm. The cross's thickness is 1 cm. It is embedded in the center of the torso of the teddy bear. The second item is placed in the right hand of the teddy bear. It is L-shaped and consists of two perpendicular arms, each being 2 cm long. It is 1 cm thick as well. It is cut from an absorber sheet [205], the relative permittivity of which is approximately 10. Its loss tangent is about 0.5.

Fig. 5.7 shows the image slices in a plane intersecting the toy midway. Fig. 5.7-a shows the normalized magnitude of the contrast distribution in the OUT that consists of the teddy bear toy without any items placed inside, whereas Fig. 5.7-b shows the OUT with the teddy bear that has two hidden items inside. In Fig. 5.7-a, the brightest objects correspond to the toy's eyes and nose, which are made of plastic and apparently feature higher dielectric contrast than the fabric and the synthetic stuffing of the toy's body. In Fig. 5.7-b, the high-permittivity hidden items are the brightest objects although the shape of the toy is still clearly visible. The shape of the embedded items is not well resolved, which is expected since the holography reconstruction is diffraction limited.

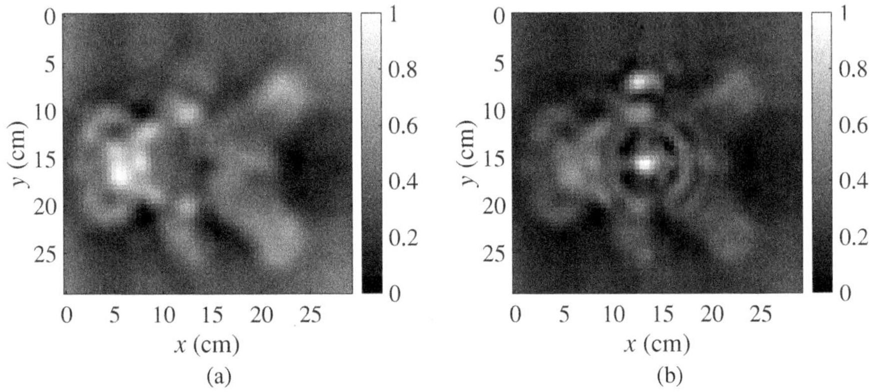

Figure 5.7 Qualitative images obtained through the experiments with the teddy bear plush toy: (a) the toy measured without any items inside; and (b) the toy measured with two items hidden inside: a cross-shaped item of $\varepsilon_r \approx 12 - i0$ in the middle of the torso and an L-shaped item of $\varepsilon_r \approx 10 - i5$ in the right arm. (Courtesy of Daniel Tajik.)

The wavelength at 12 GHz is about 25 mm in air, whereas the shape details of the two embedded items are on the order of a cubic centimeter.

5.9 Further Reading on Fourier-Space Reconstruction

What has been presented so far opens only a small doorway into the extensive subject of MW and millimeter-wave imaging employing Fourier transforms. The goal has been to introduce the reader to the terminology and the basic concepts, thereby facilitating the understanding of the great variety of more advanced algorithms and sophisticated applications in real-life imaging systems. It is impossible to list here even a fraction of the published papers, which appear in the engineering, physics, geophysics, remote-sensing, and medical-imaging literature. This section tries to systematically group the various approaches in categories and to provide some representative references.

5.9.1 Far-Field Holography

Sheen *et al.* at the Pacific Northwest National Laboratory have developed planar and cylindrical systems for the detection of on-body concealed weapons that utilize 3D holography reconstruction [62, 63] on wideband signals centered at about 30 GHz, 35 GHz, 100 GHz, and 350 GHz. 3D reconstruction of a full-body scan is achieved in less than 1 s. The team has also developed various imaging systems in the low *giga-hertz* range for applications in GPR and the imaging of vehicles.

This team proposed the reconstruction formula in Eq. (5.46) and then used it to process the reflected signals of a monostatic measurement on a plane [62]. Note that this formula relies on a far-zone approximation of the kernel of the scattering integral, which is approximated as e^{-i2kr}. Here, r is the distance from the Tx/Rx antenna to the imaged voxel. They have also developed a monostatic apparatus for cylindrical acquisition [63], the data of which are processed in Fourier space as well. This time, the 2D Fourier space is defined over the variables (k_ϕ, k_z) corresponding to the Tx/Rx position (ϕ, z).

A very similar reconstruction method for cylindrical data has been reported by Detlefsen *et al.* [164].

MW holography for data acquired on spherical surfaces has also been developed [206, 169] mostly for the nondestructive evaluation of antennas and radomes, as well as for the localization of reflector surface distortions and of defective radiating elements of antenna arrays.

Wang *et al.* have proposed a 3D MW holographic approach based on reflection measurements with a single transmitter and multiple receivers. The measurements are at a single frequency, but depth information is achieved by physically changing the distance between the plane of the Rx antenna array and the imaged object [207]. An interesting investigation is carried out regarding the configuration of the Rx array in [208], wherein spiral, random, and regularly spaced array configurations are compared. The authors conclude that the spiral and the random antenna arrays deliver clearer and more accurate holographic images.

5.9.2 Indirect Holography

The team of Smith *et al.* proposed what they called *indirect MW holography*, which closely resembles the methods of optical holography and those of the early MW holography experiments [181, 173]. More specifically, the resemblance is in the *recording step* since intensity (or power) measurements are carried out on the acquisition plane. The authors refer to the method as "indirect" since the amplitude and phase are recovered from the intensity measurements. This is in contrast to what they call a "direct" measurement of the amplitude and phase, e.g., using a VNA.

To form the intensity pattern, the system uses a guided reference wave instead of a reference beam transmitted by an antenna. The reference wave is coherent with the illuminating wave because it is derived by diverting part of the transmitter's power using a directional coupler. As shown in Fig. 5.1, in the early MW holography setups, this diverted power is fed to the antenna that transmits the reference wave. In indirect holography, this power is fed into a MW network that emulates the behavior of an off-axis reference plane wave. The critical component in this network is a phase shifter, which changes the wave's phase synchronously with the location of the recording probe in a linear fashion. In fact, the use of phase shifters and waveguides to emulate a reference beam was suggested by Tricoles and Farhat in [196]. They pointed out that this approach has the important advantage of generating a reference wave free of multipath and scattering from the environment.

Let the angle of incidence of the reference plane wave toward the acquisition plane S_H be θ_r. This is the angle between the vectors $\hat{\mathbf{n}}$ and $\mathbf{k}_r$ in Fig. 5.1. With an off-axis reference illumination, $\theta_r \neq 0$. Thus, the wave vector $\mathbf{k}_r$ of the reference wave has a component parallel to S_H, namely, $\mathbf{k}_{r||} = k_x\hat{\mathbf{x}} + k_y\hat{\mathbf{y}}$, where $\hat{\mathbf{x}}$ and $\hat{\mathbf{y}}$ define the 2D Cartesian coordinate system on S_H. Let the sampling steps along x and y be denoted as Δx

and Δy, respectively. Then, the phase shift between the reference point at $x = y = 0$ and a grid point at $(i\Delta x, j\Delta y)$, $i, j \in \mathbb{Z}$, is $\Delta\Phi = [k_x(i\Delta x) + k_y(j\Delta y)]^{1/2}$. This is the phase shift that is synthesized by the phase shifter in the MW network producing the reference wave. This reference wave is added to the signal received by the antenna probe at S_H using a hybrid tee. It is the power of this total wave that is measured by the power meter in the receiver. This measurement is represented by the classical hologram expression in Eq. (5.1).

The reconstruction step in this approach is performed computationally. The output of the power meter as a function of the probe's position, $I_H(x, y)$, is Fourier transformed and filtered in Fourier space to extract the magnitude and phase information of the scattered field. Once the complex scattered field is available, the usual holographic inversion can be performed to estimate the object's reflectivity function.

The indirect holography technique was first applied to the problem of determining antenna radiation patterns and reconstructing complex near-field distributions; see, e.g., [181]. The method has also been applied to the imaging of metallic objects [173], dielectric objects [209], and breast-tissue phantoms [210].[6]

5.9.3 Diffraction Tomography

Diffraction tomography (DT) emerged in the fields of acoustics and microwave imaging as an alternative to the straight-ray tomography developed for X-ray imaging. Its principles are closely related to those of holographic reconstruction and can be traced back to the classical paper of Wolf [211]. DT began as a generalization of the conventional straight-ray tomography in the late 1970s and the early 1980s with applications in acoustics [212–216], optics (using a scalar models) [217, 218], and microwaves [219, 220, 221, 222, 223] also employing scalar models. This subject is well studied and described in many books along with extensive bibliographies; see, for example, [224, 26, 225, 46].

Unlike X-rays, acoustic and EM waves do not travel along straight paths, the main reason being that the scattering objects making up the OUT are comparable or smaller than the wavelength in most imaging applications. Therefore, the inversion method of X-ray tomography, which is based on a straight-path propagation model, cannot be applied. It is this straight-path propagation model that allows for employing the *Fourier Slice Theorem*, which is central to X-ray computed tomography (CT) [224, 225]. A 3D version of the Fourier Slice Theorem has also been developed [226] and applied to generate 3D images from radar measurements.

However, if the interaction of the object and the field is modeled with the *scalar*-wave equation and if this interaction is weak (i.e., the linear Born or Rytov approximations hold), then reconstruction approaches similar to CT can be derived. The basic mathematical principle of these approaches is the *Fourier Diffraction Theorem* [224]. It can

[6] The term *tissue phantom* refers to artificial objects, the electrical properties of which mimick those of a tissue sample or an organ. Tissue phantoms are widely used during the development of an imaging method or a system because they allow for extensive experimentation without the need to involve humans.

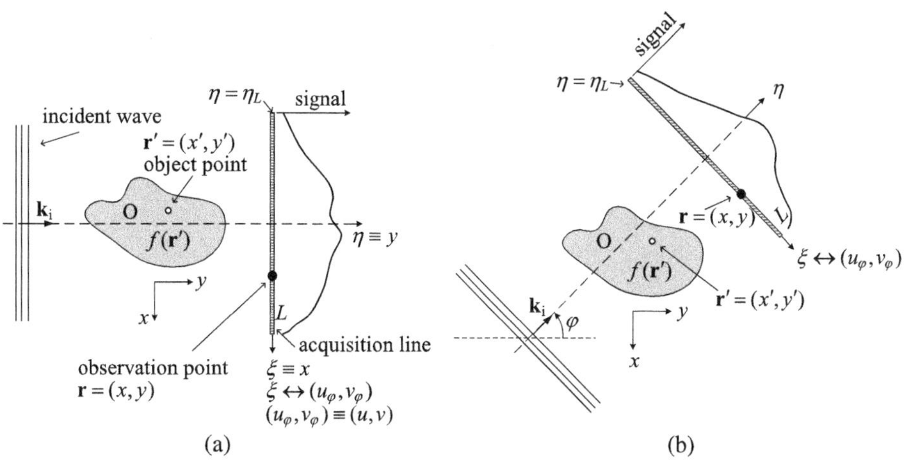

Figure 5.8 Illustration of the setup of transmission DT: (a) an experiment where the coordinate system of the acquisition setup (ξ, η) is aligned with the reference one (x, y), (b) an experiment where the acquisition setup is at an angle $\varphi = 45^\circ$ with respect to the reference coordinate system. The object is fixed in the reference coordinate system.

be shown that the Fourier Diffraction Theorem reduces to the Fourier Slice Theorem when the wavelength tends to zero [224]. Next, the derivation of the Fourier Diffraction Theorem is given following steps similar to a proof found in [224].

The Fourier Diffraction Theorem employs the linearized scalar Born model of scattering. It also assumes a plane-wave illumination as illustrated in Fig. 5.8. Fig. 5.8-a shows the setup of transmission tomography in one of the several DT experiments needed to generate an image. Note that the acquisition line L is perpendicular to the wave vector $\mathbf{k}_i$ of the illumination. In Fig. 5.8-a, the object (O) is illuminated from the left by the plane wave, the wave vector of which is aligned with the y axis of the reference coordinate system (x, y), $\mathbf{k}_i = k_b\hat{\mathbf{y}}$. Here, k_b is the wavenumber of the background, which is assumed homogeneous. In this case, L is aligned with the x axis. The object is in a fixed position with respect to the reference coordinate system, whereas the acquisition setup (the plane-wave source together with the probes on the acquisition line L) rotates to take measurements at different viewing angles $\varphi \in [0, 360^\circ]$. A coordinate system (ξ, η) can be associated with the acquisition setup. At each DT experiment, this coordinate system rotates to assume a position at an angle φ with respect to the reference coordinate system (x, y). Fig. 5.8-a shows the DT experiment where $\varphi = 0$, whereas in Fig. 5.8-b $\varphi = 45^\circ$.

For simplicity, let us consider the 2D DT, which is the most common implementation.[7] It generates a 2D slice image of a 3D object (indicated by O in Fig. 5.8) in the plane defined by the acquisition line L and the wave vector $\mathbf{k}_i$ of the incident wave. It is this application in imaging, that gave tomography its name, which is derived from

[7] Diffraction tomography can be extended to 3D problems with the plane wave illumination arriving at angles spanning all elevation and azimuth directions [26].

the Greek *tomos* for *piece* or *part*. The data collected on L in MW imaging are usually complex (magnitude and phase).

Let us revisit the scattering equation Eq. (5.4) and adapt it to a 2D scenario in the (x, y) plane. Two assumptions are now made. First, we assume that the illuminating incident field $\mathbf{E}_{\text{Tx}}^{\text{inc}}$ has the same polarization as $\mathbf{E}_{\text{Rx}}^{\text{inc}}$ (the field that would be generated by the Rx antenna if it were to operate as a transmitter). Second, we assume that the scattering does not alter the field polarization. Then, a 2D scalar form of Eq. (5.4) can be written as

$$S(\mathbf{r}) = \iint_{\mathbf{r}'} f(\mathbf{r}') \left[U_{\text{Rx}}^{\text{inc}} U_{\text{Tx}}^{\text{inc}} \right]_{(\mathbf{r}', \mathbf{r})} d\mathbf{r}', \tag{5.48}$$

where $\mathbf{r} = (x, y) \in L$ is the 2D position vector of the observation, $\mathbf{r}' = (x', y') \in \text{O}$ is the position inside the object, $f(\mathbf{r}')$ is the qualitative estimate of the object's contrast function, and $S(\mathbf{r})$ is the recorded signal. All these positions are defined in the reference coordinate system, in which the object is stationary.

The plane-wave illumination in two dimensions is represented as

$$U_{\text{Tx}}^{\text{inc}}(\mathbf{r}') = e^{-\mathrm{i}\mathbf{k}_{\mathrm{i}} \cdot \mathbf{r}'}, \tag{5.49}$$

where $\mathbf{k}_{\mathrm{i}} = \hat{\mathbf{x}} k_x + \hat{\mathbf{y}} k_y$. The plane wave satisfies the eigenvalue relation $k_x^2 + k_y^2 = k_{\mathrm{b}}^2$, where k_{b} is assumed to be real-valued. Notice that the incident field in Eq. (5.49) does not depend on $\mathbf{r}$ since it is a plane wave, which is fully characterized by the direction of its wave vector $\mathbf{k}_{\mathrm{i}}$. It is this wave vector that changes its direction in the sequence of DT experiments to provide a full 360° view of the object.

The assumption of a plane-wave illumination is a distinguishing feature of DT. In contrast, the far-field holographic approaches such as those in [63, 164] assume an incident field in the form of a locally plane spherical wave $\sim e^{-\mathrm{i}k_{\mathrm{b}}r}$; the methods presented in [64, 65, 66] employ various forms of improvements on this assumption such as simulated incident fields or a spherical wave augmented by the antenna far-field radiation pattern. The near-field holographic approach in [118] exploits the convolution of the contrast function with the measured system PSF and does not need to make an assumption about the incident wave.

Further, DT adopts the 2D fundamental solution to the scalar Helmholtz equation, $\frac{\mathrm{i}}{4} H_0^{(2)}(k_{\mathrm{b}}\rho)$, $\rho = |\mathbf{r}' - \mathbf{r}|$ (see Table 1.3), as that representing $U_{\text{Rx}}^{\text{inc}}$. We remember that $U_{\text{Rx}}^{\text{inc}}$ is in effect Green's function in the data equation of scattering, which explains this choice. It also makes it clear that, from a physical standpoint, the DT model assumes that the Rx antennas behave like 2D point probes, which in turn would behave like 2D point sources in a Tx mode.

In comparison, the above-mentioned far-field holographic approaches use isotropic spherical waves to approximate not only the illuminating field but also Green's function.

Now the 2D DT model of scattering is written as

$$S(\mathbf{r}) = \frac{\mathrm{i}}{4} \iint_{\mathbf{r}'} f(\mathbf{r}') \cdot e^{-\mathrm{i}\mathbf{k}_{\mathrm{i}} \cdot \mathbf{r}'} \cdot H_0^{(2)}(k_{\mathrm{b}}\rho) d\mathbf{r}', \tag{5.50}$$

where $\rho = |\mathbf{r}' - \mathbf{r}|$. The next step in the DT model development, is the plane-wave expansion of Hankel's function [3, chapter 6], [120, chapter 2]

$$H_0(k_b|\mathbf{r}' - \mathbf{r}|) = \frac{1}{\pi}\int_{-\infty}^{\infty} \frac{1}{k'_y} e^{-i[k'_x(x'-x)+k'_y(y'-y)]} dk'_x, \tag{5.51}$$

where $k'^2_y = k^2_b - k'^2_x$. Here, k'_x and k'_y can be thought of as being the components of the (complex) wave vector of a uniform plane wave. It is complex because, in general, k'_y may assume imaginary values if $|k'_x| > k_b$. Such imaginary values imply evanescent waves, which decay rapidly and are usually negligible beyond distances larger than 10 wavelengths from the point source [224].

Eq. (5.51) is a superposition of plane waves that yields Hankel's function such that it represents a wave launched at the receiver's point (x, y) toward the scattering point (x', y'). There is an ambiguity in the eigenvalue formula $k'^2_y = k^2_b - k'^2_x$, from which k'_y is determined for a given k'_x. Let us initially consider the scenario in Fig. 5.8-a, where $y > y'$, for $\forall y'$. To represent waves traveling toward the object, we choose the solutions with k'_y such that

$$k'_y = -\sqrt{k^2_b - k'^2_x}. \tag{5.52}$$

The plane-wave expansion in Eq. (5.51) is critical in inverting the relationship between the signal $S(\mathbf{r})$ and the contrast $f(\mathbf{r}')$.

Eq. (5.50) is written for an observation point on L (i.e., $y = y_L = \eta_L$) and Eq. (5.51) is substituted to yield

$$S(x, y_L) = \frac{i}{4\pi}\iint_{\mathbf{r}'} \left\{ f(\mathbf{r}') \cdot e^{-i\mathbf{k}_i \cdot \mathbf{r}'} \cdot \int_{-\infty}^{\infty} \frac{1}{k'_y} e^{-i[k'_x(x'-x)+k'_y(y'-y)]} dk'_x \right\} d\mathbf{r}'. \tag{5.53}$$

With the plane-wave illumination in Fig. 5.8-a, we have $e^{-i\mathbf{k}_i \cdot \mathbf{r}'} = e^{-ik_b y'}$. Then,

$$S(x, y_L) = \frac{i}{4\pi}\int_{-\infty}^{\infty} dk'_x \iint_{x'y'} \frac{f(x', y')}{k'_y} e^{-i[k'_x(x'-x)+k'_y(y'-y_L)]} \cdot e^{-ik_b y'} dx' dy'. \tag{5.54}$$

We rewrite Eq. (5.54) so that it becomes clear that it contains the 2D FT of the object function $f(x', y')$:

$$S(x, y_L) = \frac{i}{4\pi}\int_{-\infty}^{\infty} dk'_x \frac{e^{i(k'_x x + k'_y y_L)}}{k'_y} \underbrace{\iint_{x'y'} f(x', y') e^{-i[k'_x x' + (k'_y + k_b) y']} dx' dy'}_{\tilde{F}(k'_x, k'_y + k_b)}, \tag{5.55}$$

where $\tilde{F}(k'_x, k'_y)$ is the 2D FT of the object's contrast function. Then,

$$S(x, y_L) = \frac{i}{4\pi}\int_{-\infty}^{\infty} \frac{e^{i(k'_x x + k'_y y_L)}}{k'_y} \tilde{F}(k'_x, k'_y + k_b) dk'_x. \tag{5.56}$$

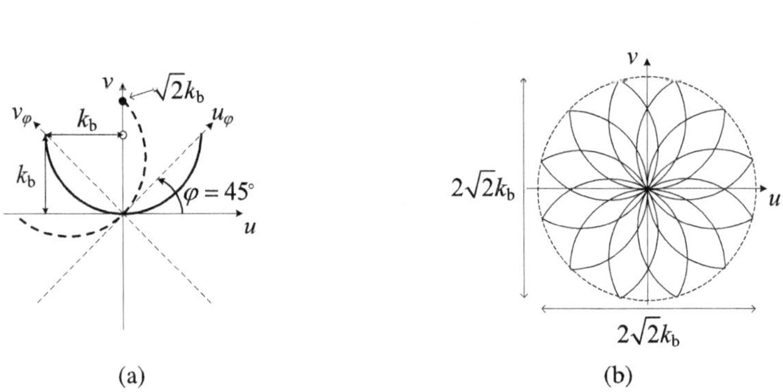

Figure 5.9 Illustration of the semicircular arcs where the 2D object's FT $\tilde{F}(u, v)$ becomes available as a result of mapping the 1D FT of the signal $\tilde{S}(k_x)$ using the equation of the Fourier Diffraction Theorem Eq. (5.59): (a) the arcs obtained from the measurements at $\phi = 0^\circ$ (solid line) and $\phi = 45^\circ$ (dash line) and (b) multiple arcs obtained from the measurements at angles from 0° to 330° with a step of $\Delta\varphi = 30^\circ$.

Taking the FT of both sides of Eq. (5.56) with respect to x leads to

$$\tilde{S}(k_x, y_L) = \frac{\mathrm{i}}{4\pi} \int_{-\infty}^{\infty} dk'_x \frac{e^{\mathrm{i}k'_y y_L}}{k'_y} \tilde{F}(k'_x, k'_y + k_\mathrm{b}) \int_{-\infty}^{\infty} e^{\mathrm{i}k'_x x} e^{-\mathrm{i}k_x x} dx, \tag{5.57}$$

where we notice that we can make use of the improper integral

$$\int_{-\infty}^{\infty} e^{\mathrm{i}(k'_x - k_x)x} dx = 2\pi\,\delta(k'_x - k_x). \tag{5.58}$$

Substituting Eq. (5.58) into Eq. (5.57) leads to

$$\tilde{S}(k_x, y_L) = -\frac{\mathrm{i}}{2} \frac{1}{\sqrt{k_\mathrm{b}^2 - k_x^2}} e^{-\mathrm{i}\sqrt{k_\mathrm{b}^2 - k_x^2} \cdot y_L} \tilde{F}\left(k_x, k_\mathrm{b} - \sqrt{k_\mathrm{b}^2 - k_x^2}\right), \; |k_x| \leq k_\mathrm{b}, \tag{5.59}$$

where we have made the substitution for k_y using Eq. (5.52). We have also imposed the requirement that $|k_x| \leq k_\mathrm{b}$ in order to be able to work with a real-valued coordinate k_y in Fourier space.

Eq. (5.59) states the Fourier Diffraction Theorem. It shows that the 1D FT of the signal collected at L relates to the values of the 2D FT of the contrast function on a line in the 2D Fourier space. Let the axes spanning this 2D Fourier space be u and v as shown in Fig. 5.9-a. The points belonging to this line have the coordinates $u = k_x$ and $v = k_\mathrm{b} - \sqrt{k_\mathrm{b}^2 - k_x^2}$. These coordinates satisfy the line equation

$$u^2 + (v - k_\mathrm{b})^2 = k_\mathrm{b}^2. \tag{5.60}$$

It follows that the line on which the contrast's FT $\tilde{F}$ becomes available is a circular arc centered on the v axis a distance k_b away from the origin; see Fig. 5.9-a. In fact, this is a semicircular arc since $|v| \leq k_\mathrm{b}$.

If we now repeat the experiment with the whole setup at an angle of φ with respect to the reference coordinate system (see Fig. 5.8-b), and obtain the 1D FT of the data $\tilde{S}(\xi)$, $\tilde{F}$ becomes available on a semicircular arc, which is rotated at an angle φ. Fig. 5.9-a shows the arc for data obtained at $\varphi = 45^\circ$ with a dotted line. The process can be repeated for as many angles $\varphi \in [0, 360^\circ]$ as is practical to fill in the Fourier space (u, v) with data points for $\tilde{F}(u, v)$. Fig. 5.9-b shows all the arcs for data that are obtained with measurements at angles from 0° to 330° with a step of $\Delta\varphi = 30^\circ$. It is clear that all these data points lie within a circle of radius $\sqrt{2}k_\text{b}$. The so-obtained data can be interpolated on a regular 2D grid, which is convenient for the 2D inverse FT that produces the 2D image as a function of (x, y). Also, we should note that denser sampling can be achieved using wideband frequency data since the radius of the projection circle in Fourier space is $\sqrt{2}k_\text{b}$; see Fig. 5.9-b. The higher the frequency, the larger the radius of the semicircular arcs; therefore, the Fourier-space span of the data is also larger.

The DT imaging procedure described here employs 1D FTs on the data collected at the receiver line L at all angles φ, followed by interpolation in Fourier space and an inverse Fourier transform to produce the image in real (x, y) space. An alternative but conceptually similar method is offered by the *filtered back-propagation* reconstruction [214] where there is no need for interpolation in Fourier space. The method is described in detail in [26, chapter 8] as well as in [51, chapter 6].

DT is a relatively simple real-time reconstruction method. However, there are a few practical issues that one has to be aware of. First, as mentioned earlier, DT is rooted in a scalar model of scattering, and therefore, it views the inspected object as a collection of isotropic point scatterers that do not alter the polarization of the illuminating field. This limitation is intrinsic to most of the other direct inversion methods and, in fact, to all real-space and Fourier-space methods considered in this book.

Second, DT makes the assumption that the incident field is a plane wave. Plane wave illumination of the whole OUT is not easy to implement at MW frequencies, especially when the OUT is at a distance of several wavelengths or less from the antennas. The field distribution in the reactive near-field region and in the radiating near-field (Fresnel) region of MW antennas is very complicated with rapidly changing magnitude, phase, and polarization; see, for example, [15]. Thus, the plane-wave illumination is practically achievable only in the far zone of the illuminating sources. Even then careful design is required to avoid amplitude and lateral phase variations in the whole volume where OUTs are placed. This problem is well known to antenna engineers who design anechoic chambers for antenna testing. Achieving a sufficiently large *quiet zone* in the chamber, where the illuminating field resembles closely a plane wave, is a difficult design task. The need to place the OUT in the far zone of the MW source is a serious limitation, especially when the OUT exhibits substantial loss. This is the case in tissue imaging, which became the first application of interest in the early research on DT with microwaves [220, 221, 222, 227]. Another such application of DT is the imaging of buried targets [228, 229]. This is why, microwave DT usually reconstructs first the induced source of scattering,

$$S_\text{ind}(\mathbf{r}') = f(\mathbf{r}')U_\text{Tx}^\text{inc}(\mathbf{r}', \mathbf{r}), \tag{5.61}$$

in the forward model of Eq. (5.48). The reader is encouraged to reexamine the derivation of the Fourier Diffraction Theorem in Eq. (5.59), wherein the incident field is made part of the unknown distribution $f(\mathbf{r}')$ instead of assuming that $U_{\text{Tx}}^{\text{inc}} = e^{-k_b y'}$. The resulting form of the theorem written for the 2D FT of the induced source is

$$\tilde{S}(k_x, y_L) = -\frac{\mathrm{i}}{2}\frac{1}{\sqrt{k_b^2 - k_x^2}} e^{-\mathrm{i}\sqrt{k_b^2-k_x^2}\cdot y_L}\tilde{S}_{\text{ind}}\left(k_x, -\sqrt{k_b^2 - k_x^2}\right), \ |k_x| \leq k_b. \qquad (5.62)$$

Once the induced source is found within the OUT, in principle, the contrast can be recovered if we know $U_{\text{Tx}}^{\text{inc}}(\mathbf{r}', \mathbf{r})$. In summary, the assumption of plane-wave illumination is not strictly necessary in DT and can be easily avoided, thereby avoiding complications in the hardware realization.

The need to know $U_{\text{Tx}}^{\text{inc}}(\mathbf{r}', \mathbf{r})$ brings us to the third important limitation. DT, just like most linear reconstruction approaches, employs the linear Born approximation, in which the total internal field is replaced by the incident one, $U_{\text{Tx}}^{\text{inc}}(\mathbf{r}', \mathbf{r})$. This limits the method to weak scattering; see [48] for a detailed study of the limitation of DT within the linear Born and Rytov approximations. It is this limitation, along with a resolution constrained by Abbe's diffraction limit, that impeded progress toward the biomedical applications of MW DT. Arguably, this impediment has spurred a whole new trend in MW imaging toward nonlinear and optimization-based reconstruction approaches tailored to the specifics of MW radiation [46] and its interaction with living tissue.

Fourth, 2D DT assumes that Green's function is the fundamental solution to the 2D Helmholtz equation (Hankel's function $H_0^{(2)}(k_b\rho)$). This implies several settings that must be observed in the hardware implementation: (a) the medium must be uniform and unbounded (reflection-free) in the imaged plane; (b) the geometry of the whole setup, the OUT included, must comply with a 2D EM wave mode; and (c) the Rx antennas must closely resemble point-like omnidirectional probes. With regard to item (b), as discussed in Section 2.13, the 2D TM$_z$ mode allows for the reduction of Maxwell's equation to a single 2D scalar-wave equation in terms of E_z.[8] Here, z is the direction orthogonal to the imaged plane. Such a mode requires that the vertical E_z field component must satisfy Neumann's boundary condition $\frac{\partial E_z}{\partial z} = 0$ in the imaged plane. This is a condition of vertical symmetry, which can be easily observed in the acquisition hardware by carefully choosing the antennas and the enclosures. However, it can hardly be expected that the OUT observes such a symmetry. Asymmetry of the OUT about the plane being imaged causes 3D effects that the 2D scalar-wave equation cannot represent. This led Semenov *et al.* to reach the following conclusion in their study of MW tomography for biomedical imaging [230]: "We concluded that usage of the 2-D model of diffraction for a 3-D object 'slice' reconstruction is an inappropriate method for microwave imaging. An object should be reconstructed as a 3-D structure and only then visualised as a 2-D sliced structure." The challenges of MW diffraction tomography in biomedical imaging along

[8] The 2D TE$_z$ mode is also described by a single 2D scalar-wave equation but in terms of H_z. However, realizing a TE$_z$ mode of operation is not practical.

with the emerging nonlinear reconstruction approaches (at that time) are described in a review article by Bolomey and Pichot [231].

The discussion on MW tomography would not be complete without mentioning an interesting development named *chirp-pulse microwave computed tomography* (CP-MCT). The first CP-MCT system was developed by Miyakawa [232, 233] in 1993 for the purpose of imaging the temperature distribution inside biological tissues. Later, the method was developed for applications in tissue imaging and imaging for breast-cancer detection [234–237]. In this method, a Tx antenna transmits a chirp pulse,[9] the frequency of which changes from 1 GHz to 2 GHz or from 2 GHz to 3 GHz within a sweep time that can be tuned to assume values from 20 ms to 200 ms. On a diametrically opposite side (relative to the imaged object), the Rx antenna receives the signal. The received signal is mixed with a reference signal that is obtained by simply diverting part of the signal generated by the chirp transmitter. The output of the mixer yields a "beat" frequency, which is a measure of the signal delay. Chirp radar has one remarkable property—with proper filtering of the beat frequencies, only the signals arriving along the shortest signal path can be selected. It is the attenuation of these signals that is being measured, similarly to the X-ray attenuation measured by CT systems. By filtering out all signal components due to multipath and diffraction, the CP-MCT system emulates a straight-path propagation. This allows use of the conventional computed tomography (CT) reconstruction based on the Fourier Slice Theorem, namely, the filtered back-projection algorithm [51], to map the attenuation coefficient.

[9] A chirp pulse is a pulse, the frequency of which changes with time, typically in a linear fashion. The chirp pulse is characterized by its minimum and maximum frequencies, ω_1 and ω_2, as well as its sweep time T_S, i.e., the time taken to increase the frequency from its minimum to its maximum value. A linear-chirp pulse is mathematically described as $s(t) \sim \sin(\omega_1 t + 0.5Kt^2)$, where $K = (\omega_2 - \omega_1)/T_S$.

6 Performance Metrics in Imaging

There are several basic metrics used to evaluate the performance of a MW imaging system. They share similar definitions and methods of evaluation with the optical, acoustic, magnetic-resonance, or X-ray imaging systems. These metrics can be divided into two groups: (i) those that measure the quality of the raw data produced by the acquisition hardware and (ii) those that measure the quality of the final images.

The choice of metrics used to evaluate the quality of the final images is dependent on whether the images are qualitative or quantitative. The accuracy of the quantitative images is usually evaluated with the relative root-mean-square error (RRMSE), which we already used in some examples; see Eq. (4.98). This approach requires an experiment with an object, the permittivity distribution of which is known exactly. The evaluation of the accuracy of a qualitative image is somewhat subjective because it depends on how the image is displayed and interpreted. This is why, in medical imaging, various metrics have been introduced such as the signal-to-noise ratio of an ideal observer (SNRI), the image contrast resolution, the spectrum of the noise equivalent quanta (NEQ), the SNR of the decision statistics, the detective quantum efficiency, and others; see, e.g., [238, 239, 240]. They are studied in relation to the clinical specificity and sensitivity of an imaging method.[1] These metrics are not common in MW imaging and they are beyond the scope of this text.

Here, we focus on the first group of metrics since they are independent of the reconstruction approach used to process the data. The most common metrics related to the performance of the hardware are: (i) the best achievable *spatial resolution* δ_ξ, where ξ indicates the direction in which the resolution is assessed, (ii) the *data signal-to-noise ratio* SNR_d, (iii) the data dynamic range D_d, and (iv) the *physical contrast sensitivity*.

We start with the spatial resolution, which dictates the choice of the frequency and the bandwidth of the system. Moreover, it plays a role in the design of the acquisition surfaces and the mutual placement of the transmitting (Tx) and the receiving (Rx) antennas. It should be noted that the spatial resolution of an image is also dependent on the specifics of the reconstruction algorithm and whether evanescent-field information is available in the acquired data. Here, we focus on the case with far-zone measurements

[1] The clinical sensitivity of a detection method refers to how frequently it reports that an abnormality exists in the instances where one actually exists [241]. The clinical specificity refers to how frequently the detection correctly reports a normal finding when no abnormality exists [241].

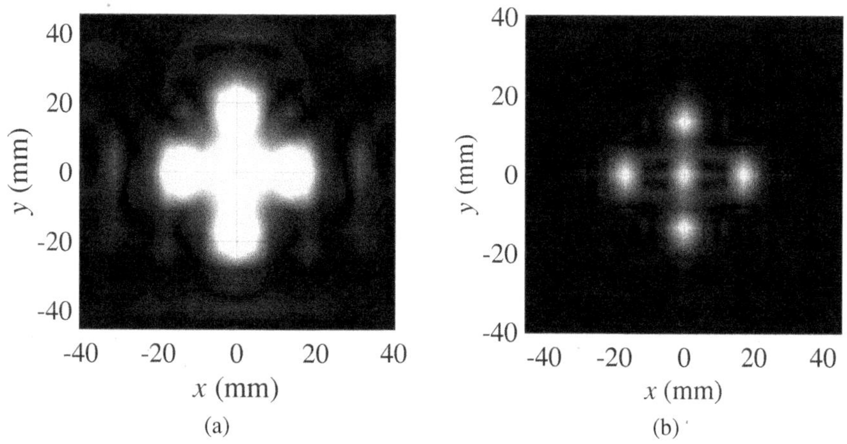

Figure 6.1 Two images of the same object that consists of five spheres in a cross configuration. The image in (a) exhibits poorer spatial resolution compared to the one in (b). (Reprinted with permission from [242].)

where evanescent field components are negligible. This case is much simpler and amenable to analytical treatment than the case of near-field measurements.

6.1 Fundamental Spatial Resolution Limits with Far-Zone Measurements

The spatial resolution δ_ξ ($\xi = x, y, z$) gives the smallest shape detail of the object under test (OUT) along ξ that is faithfully reproduced in the image. The reader is familiar with the concept of a *pixel*[2] in a 2D digital image, which is the smallest building element of the image. The color and intensity are constant within a pixel. Thus, any variations of the color and intensity in an area within the actual object that is smaller than that represented by the pixel cannot be reproduced by the image. The size of the pixel sets the spatial resolution limit of a digital image. Similarly, δ_x, δ_y, and δ_z set the size of the smallest 3D element in the image of the OUT within which the contrast is relatively constant, regardless of the actual contrast variations in the OUT.

The resolution of an image determines the ability to discern one object from another. Fig. 6.1-a and Fig. 6.1-b show the images of five spheres (each of diameter 7.6 mm) lying in a common plane and located at the center and the edges of a cross configuration. The center-to-center distance between the central sphere and an edge sphere is 15 mm. The image in Fig. 6.1-b has better spatial resolution compared to that in Fig. 6.1-a, allowing the viewer to clearly discern the five spheres.

There is a systematic approach to the evaluation of the spatial resolution limits. A point-like scattering object, referred to as the *scattering probe*, is imaged. The width of the image point-spread function (IPSF) in any given direction is a measure of the

[2] The word *pixel* is derived from *picture element*.

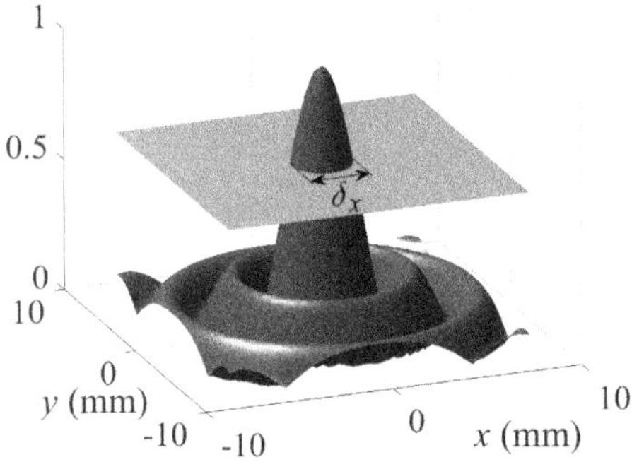

Figure 6.2 An illustration of the image point-spread function (IPSF) in a 2D image of a target. The width of the IPSF along x, δ_x, is a measure of the resolution limit along x. The width of the IPSF is usually taken at -4 dB, which corresponds to about 0.631 from the maximum. This level is shown by a plane cut.

resolution limit in this direction. Fig. 6.2 shows a typical 2D IPSF of a scattering probe. A cut at the -4 dB level is shown, which corresponds to about 0.631 from the maximum. This level is commonly chosen because it corresponds to the width of the *sinc* function from its maximum to its first null; see Fig. 4.3.

In a measurement-based study of the resolution limits of a system, the scattering probe must be significantly smaller than the background wavelength λ_b (less than $\lambda_b/8$), so that its shape has a negligible effect on the scattered field it produces. Yet, it needs to be sufficiently large to be registered by the acquisition system. A smaller size of the probe can be compensated by a larger contrast to achieve stronger scattering. In practical terms, the probe's scattering signal must rise above the noise and uncertainty of the measurement system. One should also be aware of the polarization dependence of the radar cross-section of the probe.

The case of far-zone measurements in a uniform background is amenable to analytical studies of the spatial resolution limits. These limits are an important guideline in the imaging system design. In the following analysis, we can use the S-parameter data equation Eq. (3.58), which is written for the case of weak scattering as

$$S(\mathbf{r}, \omega) = \kappa \iiint_{V'} \Delta\varepsilon(\mathbf{r}')\mathbf{E}_{\mathrm{Tx}}^{\mathrm{inc}}(\mathbf{r}', \mathbf{r}_{\mathrm{Tx}}; \omega) \cdot \mathbf{E}_{\mathrm{Rx}}^{\mathrm{inc}}(\mathbf{r}', \mathbf{r}; \omega) dv'. \tag{6.1}$$

Here, S denotes a scattering response measured at the position $\mathbf{r}$, κ is a known constant as per Eq. (3.58), $\Delta\varepsilon$ is the contrast that is to be reconstructed, ω is the frequency, and $\mathbf{r}_{\mathrm{Tx}}$ is the position of the Tx antenna. The incident field vectors $\mathbf{E}_{\mathrm{Tx}}^{\mathrm{inc}}$ and $\mathbf{E}_{\mathrm{Rx}}^{\mathrm{inc}}$ are due to the Tx and the Rx antennas, respectively, when the latter is assumed to operate in a transmitting mode. The reader is reminded that $\mathbf{E}_{\mathrm{Tx}}^{\mathrm{inc}}$ approximates the total field due to the Tx antenna under the linear Born approximation, whereas $\mathbf{E}_{\mathrm{Rx}}^{\mathrm{inc}}$ is Green's vector function for the particular setup; see Sections 2.11 and 3.3.

Let us now assume that the Tx and Rx antennas are in a fixed configuration during the scan, i.e., $\mathbf{r}_{\mathrm{Tx}} = \mathbf{r} + \Delta\mathbf{r}_{\mathrm{Tx}}$ with $\Delta\mathbf{r}_{\mathrm{Tx}} = (\Delta x_{\mathrm{Tx}}, \Delta y_{\mathrm{Tx}}, \Delta z_{\mathrm{Tx}})$ being a constant vector. This is a common scenario in measurements since either the whole apparatus is scanned

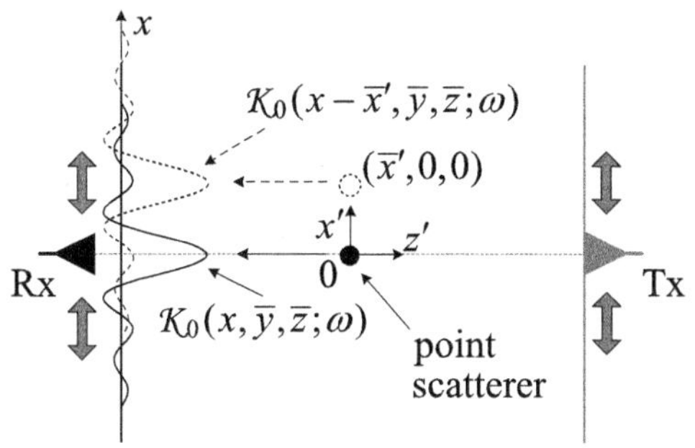

Figure 6.3 An illustration of the coordinate translation of the resolvent kernel described by Eq. (6.4). The kernel $\mathcal{K}(\bar{x}', 0, 0; x, y, z; \omega)$, shown with the dotted line, can be obtained from $\mathcal{K}_0(x, y, z; \omega)$, shown with the solid line, by translating the latter along the x axis by $\bar{x}'$. These kernels are proportional to the responses recorded with the point scatterer at $(\bar{x}', 0, 0)$ and at $(0, 0, 0)$, respectively. In particular, $\mathcal{K}_0(x, y, z; \omega)$ is the data PSF, i.e., the response to a point scatterer at the origin $(0, 0, 0)$. The mutual position of the Tx and Rx antennas is fixed as the scan is performed. In this illustration, the antennas are on two opposite sides of the scanned region aligned along each other's boresight. The double-head arrows show that the antennas scan along the line $(x, \bar{y}, \bar{z})$ to produce the responses as a function of x.

relative to a fixed object or the object is scanned relative to a fixed apparatus. In this case, $\mathbf{r}$ is sufficient to describe the positions of both the Rx and Tx antennas.

The kernel of the integral equation Eq. (6.1) can now be defined as

$$\mathcal{K}(\mathbf{r}', \mathbf{r}; \omega) = \mathbf{E}_{\text{Tx}}^{\text{inc}}(\mathbf{r}', \mathbf{r} + \Delta\mathbf{r}_{\text{Tx}}; \omega) \cdot \mathbf{E}_{\text{Rx}}^{\text{inc}}(\mathbf{r}', \mathbf{r}; \omega). \tag{6.2}$$

Since the background is uniform, the kernel in Eq. (6.2) is invariant to translations along x, y, and z of the whole setup (Tx/Rx antennas and the imaged object). Let the origin or the coordinate system coincide with $\mathbf{r}' = 0 \in V'$ and let the kernel at this position (as a function of the position $\mathbf{r}$ of the Tx/Rx pair) be

$$\mathcal{K}_0(\mathbf{r}; \omega) \equiv \mathcal{K}(\mathbf{r}' = 0, \mathbf{r}; \omega). \tag{6.3}$$

Then,

$$\mathcal{K}(\mathbf{r}', \mathbf{r}; \omega) = \mathcal{K}_0(\mathbf{r} - \mathbf{r}'; \omega). \tag{6.4}$$

Note that $\mathcal{K}_0(\mathbf{r}; \omega)$ in Eq. (6.3) is nothing but the response observed when a point scatterer resides at $\mathbf{r}' = 0$ (less a constant factor), i.e., it is proportional to the system point-spread function (PSF). This was already discussed in Section 4.2; see Eq. (4.3). Similarly, the kernel $\mathcal{K}(\mathbf{r}', \mathbf{r}; \omega)$ is proportional to the response observed when a point scatterer is at $\mathbf{r}'$. The coordinate translation described by Eq. (6.4) is illustrated in Fig. 6.3 for a shift of the point scatterer along the x axis.

The substitution of Eq. (6.4) in Eq. (6.1) leads to

$$S(x, y, z; \omega) = \kappa \iiint_{x'y'z'} \Delta\varepsilon(x', y', z')\mathcal{K}_0(x - x', y - y', z - z'; \omega) dx'dy'dz'. \tag{6.5}$$

Hereafter, we drop the constant κ since it is inconsequential for the resolution limits. Eq. (6.5) is in the form of a 3D convolution. Thus, in 3D Fourier space, Eq. (6.5) can be written as

$$\tilde{S}(k_x, k_y, k_z; \omega) = \tilde{F}(k_x, k_y, k_z)\tilde{\mathcal{K}}_0(k_x, k_y, k_z; \omega), \tag{6.6}$$

where $\tilde{F}$ is the 3D FT of the contrast $\Delta\varepsilon$.

Since the data are not available in all space, the representation in Eq. (6.6) appears to be of little value. In particular, if the data are acquired on a plane $z = \bar{z}$, then the FT with respect to z is not feasible. However, recall that if the wave field-here this is the scattering response $S(x, y, \bar{z}; \omega)$—satisfies the Helmholtz equation in a uniform open medium, its dependence on z is inherently related to its dependence on the frequency ω through Eq. (4.9), i.e.,

$$\tilde{S}(k_x, k_y, z; \omega) = \tilde{S}(k_x, k_y, \bar{z}; \omega)e^{\mp k_z(z-\bar{z})}, \tag{6.7}$$

where k_z satisfies the eigenvalue equation

$$k_z^2 = \left(\frac{\omega}{\upsilon_b}\right)^2 - k_x^2 - k_y^2. \tag{6.8}$$

Above, we assumed that the background is loss free and its wavenumber $k_b = \omega/\upsilon_b$ is real. Here, υ_b is the speed of light in the background. Further, we make the assumption that the wave field represented by the response $S(x, y, \bar{z}; \omega)$ does not contain evanescent modes (i.e., $k_z^2 \geq 0$) and therefore, k_z is real,

$$k_z = \pm\sqrt{\left(\frac{\omega}{\upsilon_b}\right)^2 - k_x^2 - k_y^2}, k_z \in \mathbb{R}. \tag{6.9}$$

A real k_z is necessary in order to employ analysis in Fourier space. If $\bar{z} > 0$ (remember that the origin has been assumed to lie within the volume V' of the scattering object), then $k_z \geq 0$, and vice versa. This is because the scattered signal propagates along $+z$ toward the acquisition plane; therefore, the wave vector of each partial plane wave is expected to be at an acute angle with respect to the z axis.

Eq. (6.9) shows a dependence $k_z(k_x, k_y, \omega)$, which maps the 3D Fourier space (k_x, k_y, ω) onto the 3D Fourier space (k_x, k_y, k_z). Setting $z = \bar{z}$ in Eq. (6.7), and with Eq. (6.9) in mind, we write

$$\tilde{S}(k_x, k_y, k_z; \bar{z}) = \tilde{S}\left(k_x, k_y, \bar{z}; \omega(k_x, k_y, k_z)\right). \tag{6.10}$$

The dependence of ω on the Fourier-space coordinates k_x, k_y, and k_z follows from Eq. (6.8) and is explicitly stated as

$$\omega = \upsilon_b\sqrt{k_x^2 + k_y^2 + k_z^2}. \tag{6.11}$$

Note that we have chosen the plus-sign solution of Eq. (6.8). The requirement that ω is positive is a physical constraint. Eq. (6.11) shows that a plane $\omega = const.$ in the 3D Fourier space (k_x, k_y, ω) maps onto a hemisphere in the space of (k_x, k_y, k_z); this hemisphere is centered at the origin and has a radius of ω/υ_b. The mapping is onto a hemisphere and not onto a sphere because k_z is restricted to either positive or negative values as explained after Eq. (6.9). Thus, having dense frequency sampling in a wide frequency bandwidth of the data $S(x, y, \bar{z}; \omega)$ and the system PSF $\mathcal{K}_0(x, y, \bar{z}; \omega)$ provides Fourier data points that fill in a hemisphere in the space spanned by (k_x, k_y, k_z).

We can now return to Eq. (6.6) and restate it as

$$\tilde{S}\left(k_x, k_y, k_z(k_x, k_y, \omega)\right) = \tilde{F}(k_x, k_y, k_z)\tilde{\mathcal{K}}_0\left(k_x, k_y, k_z(k_x, k_y, \omega)\right), \quad (6.12)$$

which emphasizes that the dependence on the third axis (k_z) in Fourier space is contained in the frequency sweep of the response $S(x, y, \bar{z}; \omega)$ and the system PSF $\mathcal{K}_0(x, y, \bar{z}; \omega)$. The 3D Fourier transforms $\tilde{S}$ and $\tilde{\mathcal{K}}_0$ are obtained by first applying 2D Fourier transformation to $S(x, y, \bar{z}; \omega)$ and $\mathcal{K}_0(x, y, \bar{z}; \omega)$, respectively, which produces $\tilde{S}(k_x, k_y, \omega)$ and $\tilde{\mathcal{K}}_0(k_x, k_y, \omega)$. This is followed by the mapping onto the space of (k_x, k_y, k_z) using Eq. (6.9). The location of the observation plane $\bar{z}$ has been omitted in Eq. (6.12) for brevity, but planar data acquisition is assumed in the following discussion.

Eq. (6.12) implies that the qualitative estimate of the contrast,

$$\check{f}(x, y, z) = \mathcal{F}_{3D}^{-1}\left\{\tilde{F}(k_x, k_y, k_z)\right\}, \quad (6.13)$$

can be, in principle, obtained as

$$\check{f}(x, y, z) = \mathcal{F}_{3D}^{-1}\left\{\frac{\tilde{S}(k_x, k_y, k_z)}{\tilde{\mathcal{K}}_0(k_x, k_y, k_z)}\right\}. \quad (6.14)$$

This is a formal reconstruction formula that is rarely used directly because the data as well as the PSF always suffer from incompleteness, noise, and uncertainties. But it can be seen as a "maximum likelihood solution" [65], which is in a form suitable for resolution studies.

Let us now consider a point scatterer at (x_0, y_0, z_0), the contrast of which is described by a 3D δ-function:

$$f_\delta(x', y', z') = \delta(x' - x_0)\delta(y' - y_0)\delta(z' - z_0). \quad (6.15)$$

Its 3D FT is

$$\tilde{F}(k_x, k_y, k_z) = e^{-ik_x x_0}e^{-ik_y y_0}e^{-ik_z z_0}. \quad (6.16)$$

According to Eq. (6.12), the data acquired with the point scatterer appear in Fourier space in the form

$$\tilde{S}(k_x, k_y, k_z) = e^{-ik_x x_0}e^{-ik_y y_0}e^{-ik_z z_0} \cdot \tilde{\mathcal{K}}_0(k_x, k_y, k_z). \quad (6.17)$$

We now make the following assumptions: (1) the PSF $\mathcal{K}_0$ is known exactly, and (2) both the data and the PSF are free of noise and uncertainties. This leaves only one fundamental practical limitation, namely, the extent of the data and the PSF in Fourier space:

$$-K_\xi \le k_\xi \le K_\xi, K_\xi \ge 0, \xi = x, y, z. \quad (6.18)$$

As discussed in Section 5.3, $|k_x|$ and $|k_y|$ are limited to a maximum value of $2k_b$ with far-zone measurements; see Fig. 5.3. In practice, this maximum value is rarely achievable because it requires an infinite acquisition aperture and isotropic antennas.

Here, we consider the best possible scenario of $K_\xi = 2k_b^{max}$ ($\xi = x, y$), where $k_b^{max} = \omega_{max}/v_b$ and ω_{max} is the maximum frequency in the radiation.

Regarding the limits of k_z, as discussed after Eq. (6.9), k_z is positive if the acquisition plane is to the right of the scatterer on the z axis, and it is negative if the acquisition plane is on the left. With two acquisition planes on both sides of the scatterer (see Fig. 6.3), both positive and negative k_z spectral values are available. Note that with a single-sided measurement, k_z is limited to positive-only or negative-only values. The choice of sign in this case has no effect on the final result.

The maximum value of $|k_z|$ is determined by the maximum frequency of the radiation, i.e., $K_z = 2k_b^{max}$. This follows from the far-zone representation of the PSF as $\mathcal{K}_0(x, y, \bar{z}; \omega) \sim e^{-i2k_b r}$, which arises in reflection-coefficient measurements; see Section 5.6. Here, $r = \sqrt{x^2 + y^2 + \bar{z}^2}$. The factor of 2 in the exponent's argument is due to the wave traversing the distance r twice as it propagates to and from the point scatterer. Similarly to Eq. (5.38), the plane-wave spectrum of such a PSF is given by

$$e^{-i2k_b\sqrt{x^2+y^2+\bar{z}^2}} = \iint e^{-i(k_x x + k_y y + k_z \bar{z})} dk_x dk_y, \tag{6.19}$$

where the "effective" wavenumber is now $k = 2k_b$ and k_z obeys

$$k_z = \pm\sqrt{(2k_b)^2 - k_x^2 - k_y^2}. \tag{6.20}$$

The maximum limit $K_z = 2k_b^{max}$ results when $k_x = k_y = 0$ at the highest frequency. Note that Eq. (6.20) also dictates that with single-sided measurements the minimum value of $|k_z|$ is zero, which occurs when $k_x^2 + k_y^2 = k_b^2$ at any frequency. The physical meaning of the case with $k_z = 0$ is in the existence of waves in the PSF plane-wave spectrum that approach the acquisition aperture at extreme oblique incidence.

We can now apply the reconstruction formula Eq. (6.14) to the case of the point scatterer defined in Eq. (6.15). With the assumption that the PSF $\mathcal{K}_0$ is known exactly, the substitution of Eq. (6.17) into Eq. (6.14) along with the limits of the Fourier variables leads to

$$\breve{f}_\delta(x, y, z) \sim \int_{-2k_b^{max}}^{2k_b^{max}} \int_{-2k_b^{max}}^{2k_b^{max}} \int_{-2k_b^{max}}^{2k_b^{max}} e^{-ik_x x_0} e^{-ik_y y_0} e^{-ik_z z_0} \cdot e^{ik_x x} e^{ik_y y} e^{ik_z z} dk_x dk_y dk_z. \tag{6.21}$$

The solution of this integral is

$$\breve{f}_\delta(x, y, z) \sim \frac{\sin\left[2k_b^{max}(x - x_0)\right]}{\left[2k_b^{max}(x - x_0)\right]} \cdot \frac{\sin\left[2k_b^{max}(y - y_0)\right]}{\left[2k_b^{max}(y - y_0)\right]} \cdot \frac{\sin\left[2k_b^{max}(z - z_0)\right]}{\left[2k_b^{max}(z - z_0)\right]}. \tag{6.22}$$

It is now clear that the estimated contrast distribution is a 3D *sinc* function, which is different from the actual δ-distribution in Eq. (6.15). The reason for this difference lies in the limited extent of the data in 3D Fourier space.

Eq. (6.22) allows for determining the resolution limits δ_x, δ_y, and δ_z as the distance from the *sinc* function's major maximum at (x_0, y_0, z_0) to its first null along the respective axis; see Fig. 4.3. The first null of $\frac{\sin x}{x}$ occurs at $x = \pi$. The resolution limits are thus obtained from

$$2k_b^{max}\delta_\xi = \pi, \xi = x, y, z, \tag{6.23}$$

which can be stated in terms of the shortest wavelength $\lambda_b^{min} = 2\pi/k_b^{max}$ as

$$\delta_\xi = \frac{\lambda_b^{min}}{4}, \xi = x, y, z. \tag{6.24}$$

Eq. (6.24) states the *fundamental resolution limit*, which is the best possible image resolution achievable with far-zone measurements.

In practice, the spectral limits K_ξ ($\xi = x, y$) in Eq. (6.18) are less than $2k_b$ due to the limited size of the aperture and the limited beamwidth of the Tx and Rx antennas. A more realistic estimate of the cross-range spectral limits is given by Eq. (5.28), viz.

$$K_\xi = 2k_b \sin\alpha_{max}, \xi = x, y, \tag{6.25}$$

where α_{max} is either the maximum angle subtended by the acquisition aperture relative to the center of the imaged volume or half the antenna beamwidth, whichever is less. Using the integration limits specified by Eq. (6.25) in Eq. (6.21) leads to the cross-range resolution limits given by

$$\delta_\xi = \frac{\lambda_b^{min}}{4\sin\alpha_{max}}, \xi = x, y. \tag{6.26}$$

These are worse than the fundamental limit in Eq. (6.24).

The fundamental range resolution limit δ_z is also compromised by practical limitations. For example, with single-sided measurements, the theoretical lower limit of the integral over k_z in Eq. (6.21) equals zero, which results in a range resolution limit of

$$\delta_z = \frac{\lambda_b^{min}}{2}. \tag{6.27}$$

Monostatic SAR is an example of such measurements. Moreover, in SAR, the distance to the target D is much larger than the spotlight of the radar beam on the ground. As a result, k_x and k_y are much smaller than k_b. Thus, the lower limit for k_z in Eq. (6.21) is actually $k_z^{min} \approx 2k_b^{min}$, not zero. The resulting range resolution limit is now

$$\delta_z = \frac{v_b}{2(f_{max} - f_{min})} = \frac{v_b}{2B}, \tag{6.28}$$

where B is the bandwidth of the radiation in *hertz*.

The fundamental resolution limits obtained in Eq. (6.24) hold for the case of far-zone measurements, and they are referred to as the *radar diffraction limits*. They show that the wavelength of the radiation has a critical impact on the ability to resolve shape details in the reconstructed images. The impact of these limits was first observed in optics (in the operation of microscopes and telescopes) and they were theoretically derived by Abbe [201] in 1873. Since Abbe considered the resolution limit when imaging an optical source (not an optical scatterer), his result for the diffraction limit is stated as $\lambda/2$ as opposed to the radar diffraction limit, which is $\lambda/4$. To this day, the wavelength dependence of the image spatial resolution is referred to as *Abbe's diffraction limit*.

What can be expected with measurements in the near zone, i.e., when the OUT is in the reactive near zone or the Fresnel zone of the antennas? Unfortunately, the internal field distributions and, therefore, the system PSF are complicated and antenna-specific. This is why closed-form solutions are not available. However, it is well known that the near-zone field contains quasi-static and evanescent components, which vary rapidly in space. This in turn could lead to more rapid variation of the received signal as a function of x and y. Thus the Fourier spectrum of the signal is expected to broaden beyond $2k_{\mathrm{b}}^{\max}$. This can potentially result in a spatial resolution better than Abbe's diffraction limit.

6.2 General Bistatic Formula for Spatial Resolution Limits

In the previous section, the radar diffraction limit was obtained with the assumption that the Tx/Rx antenna pair performs a 3D sweep over x and y (the lateral or cross-range coordinates) as well the frequency ω. We observed that the resolution limits are smaller (i.e., better) when the spectral extent of the data is larger. In the cross-range resolution in particular, the importance of the signal acquisition at the aperture edges became apparent because it is these signals that feature the largest cross-range spatial frequencies k_x and k_y. These spatial frequencies can be viewed as the projections of the wave vector of a plane wave onto the acquisition plane, which is also the plane in which the cross-range resolution is estimated. Therefore, there is a link between the directions of the wave vectors of the incident and scattered waves and the direction along which the resolution limit is estimated.

This link is made apparent in a method proposed by Walterschield *et al.* [243], which evaluates the resolution limit in any given direction for any given arrangement of a Tx and Rx antenna pair (bistatic arrangement). The value of the method lies in its ability to evaluate the resolution performance of any geometric configuration of the antenna placement and the acquisition surfaces or lines. The fundamental range and cross-range spatial resolution limits derived in the previous section give the best possible performance with measurements taken on a plane wherein the data sets are functions of x, y, and ω. In contrast, the approach described next does not assume spatial scanning, i.e., it holds at a single (arbitrary) geometrical arrangement of the Tx antenna, the Rx antenna, and a point scatterer. It assumes only the availability of broadband information within the bandwidth B.

6.2.1 Resolving a Point Source

The starting premise is that the radial spatial resolution limit $\delta_r^\star$ when imaging a point source a distance R away from a receiver is

$$\delta_r^\star = \frac{v_{\mathrm{b}}}{B}, \tag{6.29}$$

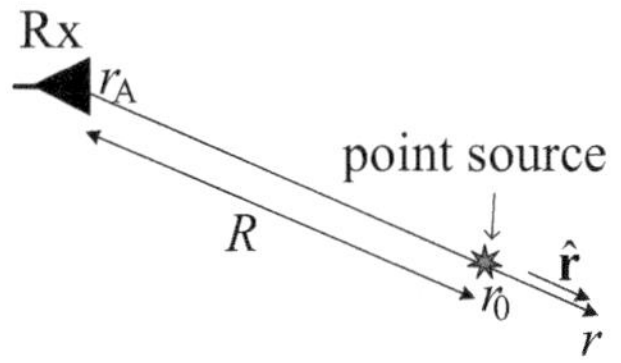

Figure 6.4 Illustration showing an antenna receiving a signal from a point source a distance R away. The antenna does not scan a surface; it is fixed in space. The axis r passes through the antenna and the point source. The positions of the antenna and the point source on r are given by r_A and r_0, respectively. The only "sweep" performed by the receiver is that in frequency ω, which is sufficient to localize the source along r with resolution given by Eq. (6.29).

provided R is sufficiently large to place the source in the antenna far zone. The radial resolution is the one along the line connecting the source and the antenna. Eq. (6.29) reminds us of the fundamental range-resolution formula in Eq. (6.28) and, indeed, it is derived in a very similar manner.

With reference to Fig. 6.4, let us define the radial axis r so that it passes through the point source at r_0 and the Rx antenna at r_A. The far-zone signal picked up by the antenna can be approximated as

$$S(r_A, \omega) \sim e^{-ik_b|r_A - r_0|}. \tag{6.30}$$

At the same time, the scalar forward model of this radiation problem for a frequency-independent source $\mathcal{S}(\mathbf{r}')$ (a δ-source in time) in the far zone can be written as

$$S(\mathbf{r}, \omega) = \iiint_{V'} \mathcal{S}(\mathbf{r}') e^{-ik_b|\mathbf{r}' - \mathbf{r}|} dv', \tag{6.31}$$

where $e^{-ik_b|\mathbf{r}' - \mathbf{r}|}$ is a far-zone approximation of Green's function. For a source the extent of which is very small in the lateral directions compared to $R = |\mathbf{r}' - \mathbf{r}|$, Eq. (6.31) reduces to the 1D integral

$$S(\dot{r}_A, \omega) \sim \int \mathcal{S}(r') e^{-ik_b|r' - r_A|} dr', \tag{6.32}$$

where we have used $\mathbf{r} = r_A \hat{\mathbf{r}}$ and $\mathbf{r}' = r' \hat{\mathbf{r}}$.

Let us assume that the source $\mathcal{S}(r')$ is nonzero only for $r' > r_A$. The result for the resolution limit does not change if we work with the assumption that $\mathcal{S}(r')$ is nonzero only for $r' < r_A$. With $\mathcal{S}(r') \neq 0$ only for $r' > r_A$, Eq. (6.32) is written as

$$S(r_A, \omega) \sim e^{ik_b r_A} \int_{-\infty}^{\infty} \mathcal{S}(r') e^{-ik_b r'} dr'. \tag{6.33}$$

We recognize the above integral as the 1D FT of the source distribution, which leads to the simple forward model in Fourier space:

$$S(r_A, \omega) e^{-ik_b r_A} \sim \tilde{\mathcal{S}}(k_b). \tag{6.34}$$

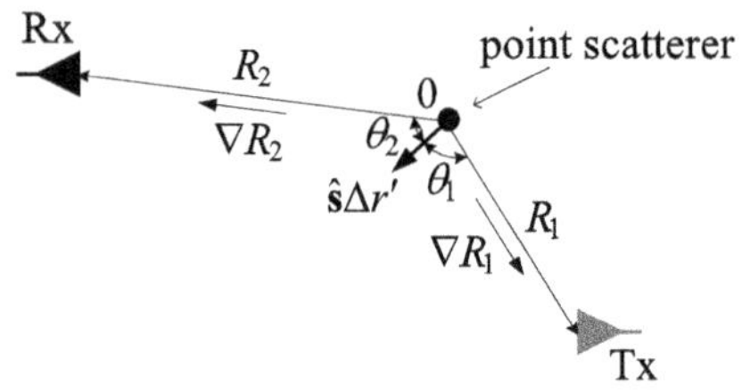

Figure 6.5 A bistatic Tx/Rx antenna arrangement with a point scatterer at the origin. The distances from the scatterer to the Tx and Rx antennas are denoted as R_1 and R_2, respectively. An incremental change in the position of the scatterer is described by $\hat{\mathbf{s}}\Delta r'$ with $\hat{\mathbf{s}}$ being a unit vector. The receiver's ability to register this change depends on the angles θ_1 and θ_2 between $\hat{\mathbf{s}}$ and ∇R_1 and ∇R_2, respectively.

Note that $k_b = \omega/v_b$, which allows mapping $S(r_A, \omega)$ into $S(r_A, k_b)$ easily. The source reconstruction formula is then

$$\mathcal{S}(r') \sim \mathcal{F}_{1D}^{-1}\left\{S(r_A, k_b)e^{-ik_b r_A}\right\}. \tag{6.35}$$

Let us now apply Eq. (6.35) with the signal received from the point source; see Eq. (6.30). To be consistent with our previous assumption that the source position satisfies $r' > r_A$, Eq. (6.30) is written as

$$S(r_A, \omega) \sim e^{-ik_b(r_0 - r_A)}. \tag{6.36}$$

Substituting Eq. (6.36) into Eq. (6.35) yields the following estimate of the point source:

$$\breve{\mathcal{S}}_\delta(r') = \mathcal{F}_{1D}^{-1}\left\{e^{-ik_b r_0}\right\}. \tag{6.37}$$

The practical limitations on the frequency bandwidth set the limits of the inverse FT integral:

$$\breve{\mathcal{S}}_\delta(r') = \int_{k_b^{\min}}^{k_b^{\max}} e^{ik_b(r'-r_0)}dk_b \sim \frac{\sin\left[0.5(k_b^{\max} - k_b^{\min})(r' - r_0)\right]}{\left[0.5(k_b^{\max} - k_b^{\min})(r' - r_0)\right]}. \tag{6.38}$$

Setting the argument of the *sinc* function equal to π yields exactly the radial resolution limit in Eq. (6.29).

6.2.2 Resolution Formula for Bistatic Measurements

Next we focus on the bistatic radar scenario, which is illustrated in Fig. 6.5. Let the distance from the Tx antenna to the point scatterer at the origin be R_1 and that from the point scatterer to the Rx antenna be R_2. Thus, the total signal path through the scatterer is $R = R_1 + R_2$. Let the position of the scatterer change incrementally by $\delta r'\hat{\mathbf{s}}$, where $\hat{\mathbf{s}}$

is a unit vector. This invokes a small change ΔR in the overall signal path R, which is determined from

$$\Delta R = \delta r' \hat{\mathbf{s}} \cdot \nabla R. \tag{6.39}$$

Here, $\hat{\mathbf{s}} \cdot \nabla R = \hat{\mathbf{s}} \cdot (\nabla R_1 + \nabla R_2)$ is the sum of the directional derivatives of the distances R_1 and R_2 in the direction $\hat{\mathbf{s}}$ of the scatterer's change in position. Note that both gradients, ∇R_1 and ∇R_2, must be taken in a common coordinate system. The natural origin of this coordinate system is the position of the point scatterer.

The change in the distance ΔR traveled by the signal from the Tx antenna through the scatterer to the Rx antenna can be detected by the receiver only if it is larger than its radial resolution limit in Eq. (6.29). Therefore, the minimum detectable ΔR is

$$\Delta R_{\min} = \frac{v_b}{B}. \tag{6.40}$$

Setting ΔR from Eq. (6.39) to $\Delta R = \Delta R_{\min}$ yields [244]

$$\delta r'(\hat{\mathbf{s}}) = \frac{v_b}{B} \frac{1}{|\hat{\mathbf{s}} \cdot \nabla R|}. \tag{6.41}$$

This is the general resolution limit of bistatic radar. In terms of the angles θ_1 and θ_2 between ∇R_1 and ∇R_2 and the direction $\hat{\mathbf{s}}$ along which we want to determine the spatial resolution limit, Eq. (6.41) can be written as

$$\delta r'(\hat{\mathbf{s}}) = \frac{v_b}{B} \frac{1}{|\cos\theta_1 + \cos\theta_2|}. \tag{6.42}$$

The angles θ_1 and θ_2 are illustrated in Fig. 6.5.

Eq. (6.41) helps us choose the acquisition surfaces or lines scanned by the Tx and Rx antennas since it provides insight into the resolution capabilities associated with each acquisition location. Let us consider as an example the monostatic scenario, where $\theta_1 = \theta_2$ regardless of the position of the scatterer relative to the Tx/Rx antenna pair. This scenario is a special case of the bistatic one when the Tx and Rx antennas coincide. Fig. 6.6-a provides an illustration. There are two mutually orthogonal directions along which we want to analyze the resolution limit: the range direction $\hat{\mathbf{s}}_r$ and the cross-range direction $\hat{\mathbf{s}}_\perp$. Fig. 6.6-a shows these directions for two scatterers that are within the antenna beam. If the scatterer is not within the beam, there is no signal at the receiver, and discussing resolution is pointless. The respective resolution limits are the range resolution limit δ_r and the cross-range one $\delta_\perp$. In the case of $\hat{\mathbf{s}}_r$, $\theta_1 = \theta_2 = \pi$. Then δ_r is determined from Eq. (6.42) as

$$\delta_r = \delta r'(\hat{\mathbf{s}}_r) = \frac{v_b}{2B}, \tag{6.43}$$

which is the same as the SAR range resolution limit obtained in Eq. (6.28).

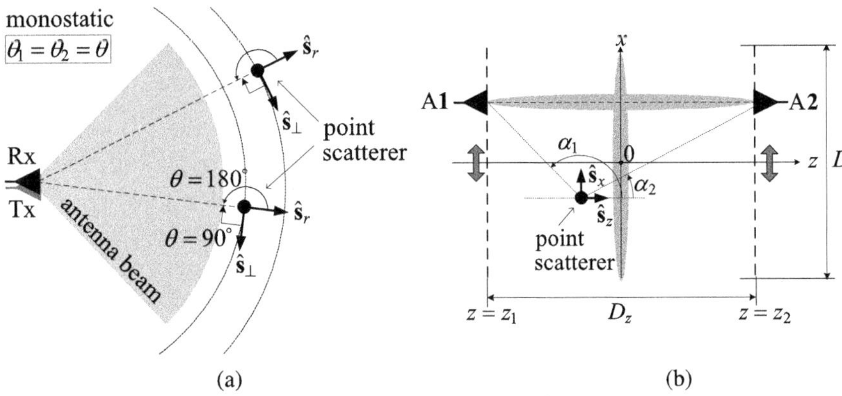

Figure 6.6 (a) Illustration showing a monostatic measurement. At each position of the point scatterer, two orthogonal directions are defined: the radial (or range) direction $\hat{\mathbf{s}}_r$ and the tangential (or cross-range) direction $\hat{\mathbf{s}}_\perp$. The range δ_r and cross-range $\delta_\perp$ resolution limits are determined along $\hat{\mathbf{s}}_r$ and $\hat{\mathbf{s}}_\perp$, respectively. The angles θ_1 and θ_2 in the resolution formula Eq. (6.42) are equal, $\theta_1 = \theta_2 = \theta$. The values they attain with respect to $\hat{\mathbf{s}}_r$ and $\hat{\mathbf{s}}_\perp$ are π and $\pi/2$, respectively. (b) Illustration of a scanning setup where two antennas, A**1** and A**2**, scan two parallel planes on both sides of the imaged volume. The antennas are aligned along each other's boresight. Both of them can transmit and receive. The directions in which the resolution limits are estimated are described by the unit vectors $\hat{\mathbf{s}}_x$ and $\hat{\mathbf{s}}_z$. The shaded areas show regions where the transmission-coefficient data lack range-resolution information.

In the case of $\hat{\mathbf{s}}_\perp$, $\theta_1 = \theta_2 = \pi/2$. Therefore, $\delta_\perp \to \infty$, which indicates a complete lack of resolution ability. This example reveals that the frequency sweep in the bandwidth B provides only range (or depth) resolution. To obtain cross-range (or lateral) resolution, the antennas must also "sweep" in the two spatial lateral directions.

Let us now expand the monostatic configuration shown in Fig. 6.6-a to the two-antenna acquisition in Fig. 6.6-b. In this setup, the two antennas, A**1** and A**2**, are aligned along each other's boresight. Both of them can transmit and receive. Thus, we have two monostatic acquisitions, namely, the reflection S-parameters S_{11} and S_{22}. We also have two bistatic acquisitions of the two transmission coefficients S_{21} and S_{12}. But if the system is reciprocal, then $S_{21} = S_{12}$.

Further, the two antennas scan together the two mutually parallel planes that are orthogonal to their boresight. Let the antenna boresight be along z, whereas the lateral axes are x and y. Fig. 6.6-b shows a cut of the arrangement in the xz plane. A**1** scans in the plane $z = z_1$ and A**2** scans at $z = z_2$. The goal in this example is to investigate the resolution limits along range δ_z and along cross-range δ_x. The respective unit vectors, $\hat{\mathbf{s}}_z$ and $\hat{\mathbf{s}}_x$, are shown in Fig. 6.6-b for an arbitrary position of the point scatterer. The analysis for δ_y is analogous to that for δ_x.

It is convenient to associate two angles, α_1 and α_2, with the position of a point scatterer, (x', z'), which can be anywhere in the region limited by $z' \in [z_1, z_2]$; see Fig. 6.6-b. These are the angles between the lines connecting the scatterer with the antennas and the z axis.

Using Eq. (6.41) or Eq. (6.42), the range resolution limits associated with S_{11}, S_{22}, and S_{21} are determined as

$$\delta_{z,S_{ii}} = \frac{v_b}{2B|\cos\alpha_i|}, i = 1, 2, \tag{6.44}$$

$$\delta_{z,S_{21}} = \frac{v_b}{B|\cos\alpha_1 + \cos\alpha_2|}. \tag{6.45}$$

We observe that the best range resolution of $\delta_{z,S_{ii}} = \frac{v_b}{2B}$, $i = 1, 2$, with the reflection coefficients is obtained when the point scatterer lies along the antennas' boresight, i.e., when $\alpha_1 = \pi$ and $\alpha_2 = 0$. This result is consistent with Eq. (6.28), which was obtained with the Fourier analysis of the data acquired on a plane. On the other hand, in this same case, $\delta_{z,S_{21}} \to \infty$, indicating the complete loss of range-resolution information in the transmission-coefficient data. The region of space around the antennas' boresight, where there is loss of range resolution with transmission-coefficient measurements, is shaded in gray in Fig. 6.6-b. Another case where $\delta_{z,S_{21}} \to \infty$ occurs is the case when the point scatterer is in the plane $z' = 0$ (the plane midway between the two antennas). Then $\alpha_1 = \pi - \alpha_2$. The region of space around the plane $z' = 0$, where range resolution is lost with transmission-coefficient data, is shaded in gray in Fig. 6.6-b. In general, with a finite aperture size $D_x \times D_y$, the range resolution obtained with boresight transmission data is always worse than that obtained with reflection data. The smaller D_x and D_y are compared to the distance D_z between the apertures, the worse the range resolution is with such data.

The loss of range resolution with the transmission data obtained via planar scanning of two boresight-aligned antennas was investigated in [244], where it was proposed to augment such systems with Rx antennas that are not aligned with the boresight of the Tx antenna. Both the Tx and the Rx antenna must have relatively wide beams to enable the reception of scattering at oblique angles. This solution targets systems where reflection-coefficient measurements are not feasible. This is often the case when objects of high loss are imaged, e.g., living tissue, concrete or moist wood. With the same antenna used as a transmitter and a receiver, the back-scattered signals from the target are orders of magnitude weaker compared to the reflections arising at the antenna terminals due to impedance mismatch. This makes the separation of the signal from the inherent antenna reflection difficult, even with a careful system calibration. In such cases, a system entirely based on forward scattering is desirable, but care must be taken to improve the range resolution.

From Eq. (6.41), the cross-range resolution limits associated with S_{11}, S_{22}, and S_{21} are obtained as

$$\delta_{x,S_{ii}} = \frac{v_b}{2B|\sin\alpha_i|}, i = 1, 2, \tag{6.46}$$

$$\delta_{x,S_{21}} = \frac{v_b}{B|\sin\alpha_1 + \sin\alpha_2|}. \tag{6.47}$$

The best cross-range resolution with both reflection and transmission measurements is obtained when $\alpha_{1,2} \to \pm 90^\circ$. This is why the lateral sizes D_x and D_y of the acquisition surface in planar scans are critical in improving the overall cross-range resolution quality of the image. To maintain a wide range of viewing angles it is also important to use wide-beam antennas.

6.3 Physical Assessment of the Imaging System

The physical assessment of the imaging system aims to determine the quality of the data before they are processed to produce an image. The raw-data quality is critical for the final image fidelity, and it can be assessed independently from the reconstruction or the image-processing algorithms. It is desirable to be able to quantify the physical merits of the data-acquisition hardware through common metrics. This allows for comparing data-acquisition systems objectively, helps in designing the system architecture and its components such as the antennas and the receivers. It also facilitates the choice of the appropriate image processing strategy depending on the data quality.

The physical assessment of the imaging systems has been systematically studied in medical imaging. Objective metrics now exist that are unified across the various medical-imaging modalities (X-ray, magnetic resonance imaging, ultrasound, etc.) [238, 239, 240, 245]. The metrics for the physical assessment of MW and millimeter-wave imaging systems exist, but there appears to be no unified system of definitions and assessment methods.

Next, three basic hardware performance measures are presented that closely resemble those in medical imaging. Engineers are familiar with them because they are used to evaluate the quality of radio receivers and to test instrumentation and communication links.

6.3.1 System Dynamic Range D_{sys} and Response Dynamic Range D_{r}

The dynamic range is the ratio of the largest to the smallest values that a certain quantity can assume. This ratio is usually specified in the logarithmic units of *decibels* (dB). In the assessment of radio receivers, the dynamic range represents the range of powers arriving from the antenna that a receiver can process and produce digital/audio/video output of acceptable quality. If the signal is too weak, it is masked by noise. Every receiver has a noise floor referred to as the *receiver sensitivity*. When the input signal power drops below this floor, the receiver is unlikely to recover any information from it. The input power is also limited from above, usually because of the nonlinearities in the mixers and the amplifiers, which saturate. The nonlinearities distort the output signals, leading to unacceptable quality.

The largest possible dynamic range of a MW imaging system is that of the RF sensor or the measurement instrument. The sensor consists of the antenna and a radio receiver. In most cases, the antenna is a linear device, the dynamic range of which is so large that it can be assumed infinite. Therefore, the dynamic range of the RF sensor is that of the RF receiver.[3] This dynamic range is determined by the ratio of the largest to the smallest signal power that the receiver can measure:

$$D_{Rx} = \frac{P_{max}}{P_{min}}. \tag{6.48}$$

P_{max} is usually limited by the input compression levels of the low-noise amplifiers and the mixers in the radio front end [84, 246] whereas P_{min} is the receiver sensitivity. It is common to express both P_{max} and P_{min} in dB.[4] Then the receiver dynamic range in dB is given by

$$D_{Rx,dB} = P_{max,dB} - P_{min,dB}. \tag{6.49}$$

Since vector network analyzers (VNAs) are often used to set up MW imaging systems, they can serve as an example here. The maximum operating level of most modern VNAs is about +10 dBm at the test port [84].[5] The noise floor of the VNA receivers is in the range of −150 dBm to −90 dBm depending on the *resolution bandwidth* (RBW) and the number of averaged samples [84, 246].[6,7] Since the VNAs are test instruments, they allow for different RBW and data-averaging options, which result in different noise-floor levels. More specifically, lowering the RBW and increasing the number of averaged data reduces the thermal-noise floor of the measurement but it also increases the measurement time.

The RBW of the VNA measurement is set by the receiver bandwidth B. This is the bandwidth of the intermediate-frequency (IF) filter. Decreasing B reduces the noise floor because it filters out thermal noise that is outside the bandwidth of the subsequent digital filter. The thermal noise-floor power N_T is expressed in terms B with the well-known relation

$$N_T = kTB \text{ (W)}, \tag{6.50}$$

where $k \approx 1.38 \times 10^{-23}$ (J/K) is Boltzmann's constant, T is the temperature in degrees Kelvin, and B (Hz) is the bandwidth. As an example, let us calculate N_T in dB at room temperature $T = 290°$K (about 17°C) in a bandwidth $B = 1$ Hz:

$$N_{T,dB}(1 \text{ Hz}) \approx 10\log_{10}\left(1.38 \cdot 10^{-23} \cdot 290 \cdot 1\right) \approx -203.98 \text{ (dB)}. \tag{6.51}$$

[3] The antenna contributes to the system noise, but it is insignificant compared to other noise sources such as mixers, amplifiers, and measurement uncertainties.

[4] The value in dB of the power level P measured in *watts* is given by $P_{dB} = 10\log_{10} P$.

[5] The value in dBm of the power level P_{mW} in *milliwatts* is given by $P_{dBm} = 10\log_{10} P_{mW} = P_{dB} + 30$.

[6] The RBW is the smallest frequency bandwidth in *hertz* that can be resolved in a frequency sweep. It is analogous to the concept of spatial resolution in imaging. An RBW of 1 kHz means that if the signal's spectrum varies within any given "frequency bin" that is 1-kHz wide, this variation is not going to register.

[7] Modern VNAs allow for averaging of the frequency-swept S-paramemeters. The VNA continuously cycles through the measurement of the S-parameters. One cycle would typically take a fraction of a second, allowing for a number (N_a) of data sets to be collected and averaged. The averaging is done in complex space and at each frequency. The SNR improves by 3 dB for every factor-of-two increase in N_a, but it also increases the measurement time by a factor of two [246].

If $B = 1$ kHz, then we simply add 30 to the result in Eq. (6.51) to obtain $N_{\mathrm{T,dB}}(1\ \mathrm{kHz}) = -173.98$ dB. It is clear that reducing the RBW reduces thermal noise. Unfortunately, this also means increasing the measurement time; see [246] for concrete examples.

Moreover, the thermal noise power is not the only contributor to P_{min}. There is also electronic noise, which adds to the overall noise that appears at the radio output. The amount of this noise, when referenced to the input of the receiver, is conveniently expressed in terms of the receiver *noise figure* $\mathrm{NF_{dB}}$.[8] The NF is a known parameter for a given receiver, i.e., it can be measured [247]. Thus, the receiver sensitivity can be expressed as

$$P_{\mathrm{min,dB}} = N_{\mathrm{T,dB}} + \mathrm{NF_{dB}}. \tag{6.52}$$

Substituting the above in Eq. (6.49) yields

$$D_{\mathrm{Rx,dB}} = P_{\mathrm{max,dB}} - N_{\mathrm{T,dB}} - \mathrm{NF_{dB}}. \tag{6.53}$$

This is the receiver dynamic range. As an example, let us calculate the receiver dynamic range in the case of $P_{\mathrm{max,dBm}} = 10$ (or -20 dB), $N_{\mathrm{T,dB}} = -173.98$ (the result obtained at $T = 290^{\circ}$K and $B = 1$ kHz), and $\mathrm{NF_{dB}} = 13$. The result is $D_{\mathrm{Rx,dB}} = 140.98$ dB. This is a good value for receivers in the MW bands.

The actual dynamic range of a MW measurement system $D_{\mathrm{sys,dB}}$ is lower than that of the receiver. The decrease is due to the fact that the maximum power $\bar{P}_{\mathrm{max,dB}}$ that reaches the receiver is smaller than $P_{\mathrm{max,dB}}$, i.e., $\bar{P}_{\mathrm{max,dB}} < P_{\mathrm{max,dB}}$. To ensure safe operation of the equipment, the transmitted power needs to be chosen so that $P_{\mathrm{max,dB}}$ is never exceeded at the receiver input regardless of the object that is being measured. In MW imaging experiments, there are a number of factors that limit the amount of MW power that can be transmitted, the most important of which are the various regulations regarding EM interference and RF radiation safety. The latter regulations are especially important in biomedical measurements involving humans.

Further, the thermal and electronic noise of the receiver are not the only contributors to the stochastic noise of the overall system. Due to environmental factors such as temperature and humidity, the performance of components such as cables, connectors, directional couplers, and mixers can change over time considerably. This is referred to as *component drift* [84]. This is why calibrating the VNA is recommended before each measurement. Nonetheless, the short-term behavior of the component drift is unpredictable. Another factor is the noise of the transmitter, both in amplitude and phase. In summary, the system dynamic range $D_{\mathrm{sys,dB}}$ is significantly smaller than that of the receiver because of: (i) maximum available power at the receiver input that is smaller

[8] The noise figure is a measure of the deterioration of the system SNR at the output as compared to the SNR at its input

$$\mathrm{NF_{dB}} = 10\log_{10}\left(\frac{\mathrm{SNR_{in}}}{\mathrm{SNR_{ou}}}\right).$$

than the maximum power the receiver can handle and (ii) increased noise and uncertainties due to factors external to the receiver. $D_{\text{sys,dB}}$ can be expressed as

$$D_{\text{sys,dB}} = \bar{P}_{\text{max,dB}} - N_{\text{sys,dB}} < D_{\text{Rx,dB}}, \tag{6.54}$$

where $N_{\text{sys,dB}}$ represents the overall system noise floor that includes thermal noise, electronic noise, and uncertainties of stochastic nature that are external to the receiver. The noise floor of the VNAs can be easily measured and displayed as a function of frequency. To accomplish this, the instrument is first calibrated with the respective system interconnects (cables or transmission lines, connectors, and adapters). Then all ports (to which the Tx and the Rx antennas are to be connected in an actual experiment) are loaded with precision 50-Ω terminations. This ensures that there is no signal at any of the ports. The trace for a desired S-parameter is then displayed to show directly the noise signal as a function of frequency. Analogous procedure, can be used for custom-built measurement systems.

There are additional factors that need to be considered in an imaging system. There are a great deal of uncertainties arising in imaging experiments that do not exist in the conventional MW device measurements with VNAs. In device measurements, the S-parameters are measured at a small number of observation locations, the ports. Typical MW devices have 1 to 4 ports. This allows for a rigidly fixed setup that can be carefully calibrated. The noise floor of such a "fixed" system can indeed be estimated using Eq. (6.54) or can be measured as described in the previous paragraph.

In contrast, imaging experiments require spatial scanning where we have hundreds, even thousands, of observation locations. No VNA or RF switch can accommodate so many ports. Thus, mechanical scanning or customized MW electronics are employed in the form of large electronically switched arrays of sensors. Mechanical scanning suffers from vibrations and positioning errors that may surpass the noise floor of a "fixed" measurement system by orders of magnitude. Electronically switched sensor arrays suffer from interelement coupling, variations in performance from one array element to another, and uncertainties associated with the complex RF switching circuitry.

For these reasons, the noise and uncertainties of a MW imaging system are evaluated experimentally. This is accomplished via repetitive measurements of the same object. It is desirable for the object to be uniform so that the signal strength does not vary significantly from one position to another. Usually the object is simply the measurement setup in the absence of an OUT. The so-obtained responses are functions of position and frequency (or time). Moreover, there can be both reflection and transmission coefficients involved. Since the average strengths of the reflection coefficients may be significantly different from those of the transmission coefficients, the uncertainties should be evaluated separately. The same applies to the dependence on frequency if it affects the average signal strength. This is often the case for measurements in a lossy environment such as tissue where the attenuation grows rapidly with frequency.

Let $\boldsymbol{r}$ denote a vectorized data set of N responses (e.g., S-parameters) acquired as a function of position in the scatterer-free setup, which we call the reference object (RO). It is desirable to perform several scans over time so that $\boldsymbol{r}$ contains a statistically significant sample that can capture environmental and drift effects in the whole system, the transmitter included. The RO is uniform; thus, ideally, all elements of $\boldsymbol{r}$ should be the same. In practice, both systematic and stochastic factors introduce slight variations. With a careful calibration, the systematic variations can be de-embedded. We assume that $\boldsymbol{r}$ contains the RO responses after all systematic errors have been removed. The stochastic variations cannot be removed, and they define the uncertainty level of the measurement. This level is quantified by the standard deviation

$$\sigma(\boldsymbol{r}) = \sqrt{\frac{1}{N-1}\sum_{i=1}^{N}|r_i - \bar{r}|^2}, \bar{r} = \frac{1}{N}\sum_{i=1}^{N} r_i, \tag{6.55}$$

which defines the noise floor of the responses. Note that the elements of $\boldsymbol{r}$ are complex-valued. Now, the dynamic range of the responses can be defined as

$$D_{\mathrm{r,dB}} = \bar{P}_{\mathrm{max,dB}} - 20\log_{10}\sigma(\boldsymbol{r}). \tag{6.56}$$

A larger data dynamic range is desirable because it allows for capturing both weak and strong scattering signals. It is especially important to reduce $\sigma(\boldsymbol{r})$ since it is the limiting factor in capturing weak scattering and, therefore, is critical to the contrast sensitivity of the system, which is discussed in Section 6.3.3.

6.3.2 Data Signal-to-Noise Ratio ($\mathrm{SNR_d}$)

There are two main methods for establishing the data quality in magnetic resonance imaging (MRI) [248–251] and near-infrared tomography [252]. Both are based on examining the images obtained with the data produced by the particular acquisition hardware. The first method uses two separate images—one image depicts a uniform target (the noise image) and the other depicts a known object. In the framework of MW imaging, the uniform target is to be interpreted as the RO (or the background) that we used in the previous section to define the noise floor of the responses $\sigma(\boldsymbol{r})$; see Eq. (6.55). The image of the uniform target is used to evaluate the image noise level since, ideally, it must show zero contrast. The known object, on the other hand, is to be interpreted as a *calibration object* (CO), which we introduced in Chapter 4. Usually, this is a small scatterer placed at the center of the otherwise uniform volume. The region where the small scatterer is known to reside defines the *region of interest* (ROI) in the image. It is the mean value of the signal in the ROI that is compared to the noise level in the noise image to estimate the data signal-to-noise ratio, $\mathrm{SNR_d}$.

The second method uses only one image, that of the CO. In the CO image, in addition to the ROI , a *feature-free region* (FFR) is defined, where the contrast value is expected to be zero. The method evaluates the noise by taking the standard deviation of the signal

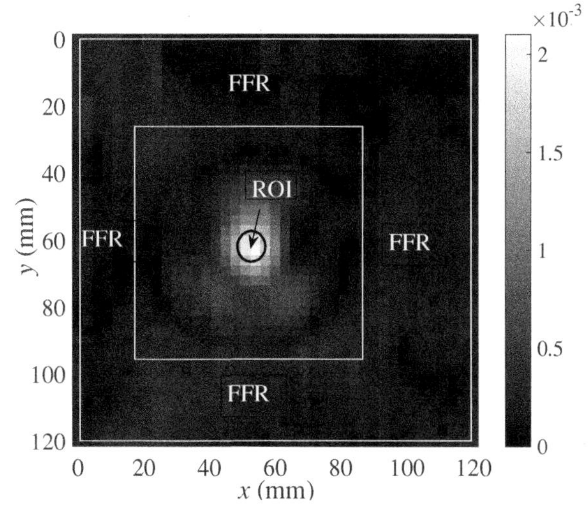

Figure 6.7 2D image of the magnitudes of the calibrated transmission S-parameters obtained in a CO measurement. The ROI is shown with a black circle. This is the known location of the small scatterer. The FFR lies between the two white squares.

in the FFR. The signal strength is the mean value in the ROI. Thus, this method differs from the first one in only evaluating the signal noise from the FFR instead of using a separate RO measurement and the respective noise image.

With some adjustment, the methods just described can be effectively used to evaluate the data quality in MW imaging. The data submitted to the reconstruction algorithms in MW imaging undergo a calibration step that extracts the scattered component of the response from two measurements, the RO (or baseline) measurement and the OUT measurement. The former is carried out during the system calibration whereas the latter is the actual imaging experiment. Let $\boldsymbol{r}^{(m)}$ denote the OUT responses acquired at the m-th frequency arranged in a vector. Each element of $\boldsymbol{r}^{(m)}$ is a complex-valued response at a given observation position on the acquisition surface. The respective baseline responses are represented by the vector $\boldsymbol{b}^{(m)}$. The data submitted to the reconstruction algorithm are obtained via subtraction of the baseline response from the OUT one at each observation position, i.e.,

$$\boldsymbol{d}^{(m)} = \boldsymbol{r}^{(m)} - \boldsymbol{b}^{(m)}. \tag{6.57}$$

Note that in the MW imaging experiments, the baseline responses are, in general, not zero, and neither are they negligible.

To evaluate the data quality, the OUT must be known so that the ROI is well defined. This is where a CO can be used. Fig. 6.7 shows a 2D image of the magnitudes of the calibrated transmission S-parameters obtained in a CO measurement. The calibration is performed as per Eq. (6.57). The ROI is a circular region corresponding to the location of the actual cylindrical small scatterer, the cross-section of which is 10 mm in diameter.

The length of the cylinder along z is 10 mm. The data voxel values within the ROI comprise $\boldsymbol{d}_{\text{ROI}}^{(m)}$. The FFR is the region confined between the two white squares. Notice that the inner boundary of the FFR is significantly detached from the ROI. This ensures that the FFR does not include variations due to the diffraction pattern of the scattering from the small cylinder. For that, the acquisition area must be sufficiently large. Alternatively, two consecutive RO measurements can be taken. The difference between the two can provide the noise image, the whole area of which is feature free. In either case, the data within the FFR comprise $\boldsymbol{d}_{\text{FFR}}^{(m)}$.

The quality of the data that the imaging setup can provide at the m-th frequency is evaluated through the data SNR defined as the power ratio [253]

$$\text{SNR}_{\text{d}}^{(m)} = \left[\frac{\left|\bar{d}_{\text{ROI}}^{(m)}\right|}{\sigma\left(\boldsymbol{d}_{\text{FFR}}^{(m)}\right)}\right]^2, \bar{d}_{\text{ROI}}^{(m)} = \frac{1}{N}\sum_{i=1}^{N} d_{\text{ROI},i}^{(m)}, \tag{6.58}$$

which can also be expressed in dB as

$$\text{SNR}_{\text{d,dB}}^{(m)} = \left|\bar{d}_{\text{ROI}}^{(m)}\right|_{\text{dB}} - 20\log_{10}\sigma\left(\boldsymbol{d}_{\text{FFR}}^{(m)}\right). \tag{6.59}$$

Here, the standard deviation $\sigma(\boldsymbol{d}_{\text{FFR}}^{(m)})$ of the data in the FFR is calculated using Eq. (6.55).

The metric offered by Eq. (6.58) was used in [253] to evaluate experimentally the performance of MW imaging systems for breast-cancer detection. There, the RO was constructed as a uniform medium (a slab) of permittivity values close to the averaged electrical properties of breast tissue, namely, $\epsilon_{\text{r}} \approx 10$ and $\tan\delta_{\text{d}} \approx 0.5$. The small scatterer in the CO, on the other hand, was chosen to have a permittivity value close to the averaged permittivity of breast cancerous tissue, namely $\epsilon_{\text{r}} \approx 50$. Its size of about 1 cm^3 was chosen to match approximately the resolution-cell volume.[9]

Fig. 6.8 shows the data SNR versus the frequency of a MW imaging system that uses planar raster scanning to examine breast-tissue phantoms [254]. The result suggests that the data are of low quality in frequencies greater than 8 GHz. This is due to the rapid growth of signal attenuation in the tissue as the frequency increases. The decrease in the SNR for frequencies below 3 GHz is due to the poor impedance match at the Rx and Tx antennas.

The choice of the contrast $\Delta\varepsilon_{\text{CO}}$ and the size Ω_{CO} of the small scatterer in the CO affects the result produced by Eq. (6.58) since, under the linear Born approximation, the scattered signal depends linearly on the product of the two, $\Delta\varepsilon_{\text{CO}}\Omega_{\text{CO}}$; see, for example, Section 4.2. It was suggested in [253] to use metallic scatterers of volume commensurate with the resolution cell to obtain the best achievable SNR_{d}. Such an SNR estimate would be independent of the choice of $\Delta\varepsilon_{\text{CO}}$.

[9] The resolution cell is a cubicle the sides of which are equal in length to the spatial resolution limits.

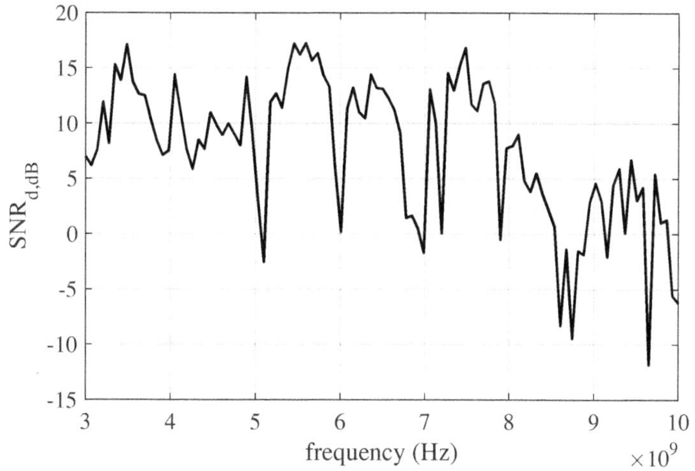

Figure 6.8 Example of the evaluation of the data SNR of a MW imaging system, which uses planar raster scanning to examine tissue phantoms mimicking breast tissue. (Reprinted with permission from [254].)

An SNR metric that would not depend on $\Delta\varepsilon_{\text{CO}}$ and/or Ω_{CO} has to rely only on the RO measurements, which produce the baseline responses $\boldsymbol{b}^{(m)}$. The RO volume is feature free by default and a ROI cannot be defined. In [255], an SNR metric was proposed that is entirely based on $\boldsymbol{b}^{(m)}$:

$$\text{SNR}_{\text{d}}^{(m)} = \left[\frac{|\bar{b}^{(m)}|}{\sigma(\boldsymbol{b}^{(m)})} \right]^2, \bar{b}^{(m)} = \frac{1}{N} \sum_{i=1}^{N} b_i^{(m)}. \tag{6.60}$$

Notice that this SNR metric uses the baseline responses $\boldsymbol{b}^{(m)}$ directly since the CO responses are not available and the calibration step in Eq. (6.57) is inapplicable. In fact, Eq. (6.60) is equivalent to the data dynamic range defined in Eqs. (6.55)–(6.56). Here, too, it is recommended to repeat the RO measurements several times in order to obtain a significant statistical sample of FFR values and to capture component drift and environmental influences. In [255], this metric was used to estimate analytically the measurement uncertainties in pulsed-radar imaging systems for breast-cancer detection. It was applied separately to the magnitude and the phase of the time-domain data that were processed with the inverse FT.

The data SNR in Eq. (6.60) can be viewed as the best-case scenario that Eq. (6.58) could describe. In the framework of Eq. (6.58), the numerator in Eq. (6.60), $|\bar{b}^{(m)}|^2$, would be the strongest possible averaged signal power in the ROI. To understand this, imagine that the CO measurement involves transmission coefficients. If the small scatterer in the CO blocks completely the illuminating field, then most of the CO responses within the ROI will be zero. As a result of the subtraction calibration step Eq. (6.57), the actual data will have the value of the RO (or baseline) responses with a minus sign; thus, $\left|\bar{d}_{\text{ROI}}^{(m)}\right|^2 \approx |\bar{b}^{(m)}|^2$.

On the other hand, the variance of the baseline responses $\sigma^2(\boldsymbol{b}^{(m)})$ in Eq. (6.60) cannot exceed the variance of the FFR calibrated responses $\sigma^2(\boldsymbol{d}^{(m)}_{\text{FFR}})$ in Eq. (6.58). The reader is reminded that $\boldsymbol{d}^{(m)}_{\text{FFR}} = \boldsymbol{r}^{(m)}_{\text{FFR}} - \boldsymbol{b}^{(m)}_{\text{FFR}}$ as per Eq. (6.57). We assume that the variance of the data in the FFR region is the same as that in the whole RO region, i.e., $\sigma^2(\boldsymbol{b}^{(m)}_{\text{FFR}}) = \sigma^2(\boldsymbol{b}^{(m)})$. If the stochastic noise in the two data sets $\boldsymbol{r}^{(m)}_{\text{FFR}}$ and $\boldsymbol{b}^{(m)}_{\text{FFR}}$ is uncorrelated, it is not going to cancel on subtracting one set from the other. In fact, uncorrelated noise adds up, i.e.,

$$\sigma^2\left(\boldsymbol{r}^{(m)}_{\text{FFR}} - \boldsymbol{b}^{(m)}_{\text{FFR}}\right) = \sigma^2\left(\boldsymbol{r}^{(m)}_{\text{FFR}}\right) + \sigma^2\left(\boldsymbol{b}^{(m)}_{\text{FFR}}\right). \tag{6.61}$$

In the FFR, it is expected that $\bar{r}^{(m)} = \bar{b}^{(m)}$. It follows that

$$\sigma^2\left(\boldsymbol{r}^{(m)}_{\text{FFR}}\right) \approx \sigma^2\left(\boldsymbol{b}^{(m)}_{\text{FFR}}\right), \tag{6.62}$$

and

$$\sigma^2\left(\boldsymbol{d}^{(m)}_{\text{FFR}}\right) \approx 2\sigma^2\left(\boldsymbol{b}^{(m)}_{\text{FFR}}\right) \approx 2\sigma^2\left(\boldsymbol{b}^{(m)}\right). \tag{6.63}$$

Thus, it is estimated that the SNR formula in Eq. (6.60) would produce a value that is at least twice as large as the one produced by Eq. (6.58).

6.3.3 Physical Contrast Sensitivity

The physical contrast sensitivity (or simply the contrast sensitivity) aims at quantifying the smallest detectable change in the complex permittivity per unit volume. It is clear that the contrast sensitivity depends on the nominal state of the object. Thus, it is common to evaluate it in the scatterer-free uniform background, i.e., the RO. Thus, the data set relevant to this discussion is that of the baseline responses $\boldsymbol{b}^{(m)}$. In the following, the superscript (m) associated with a frequency index is omitted for brevity, but is implied in all relations.

The term *contrast sensitivity* is closely related to an image metric known as *contrast resolution*, which quantifies the ability to distinguish between differences in intensity in an image. The contrast sensitivity can be viewed as the contrast resolution of the system as computed from the baseline responses.

In [256], the contrast sensitivity in a unit volume is defined for the case of S-parameter measurements as

$$\delta\bar{\epsilon}_{1\text{v}} = \frac{1}{|\varepsilon_{\text{b}}|} \cdot \frac{\sigma(\boldsymbol{b})}{\max\left|\frac{\partial \boldsymbol{b}}{\partial \varepsilon}\right|_{1\text{v}}} \;(\text{m}^3), \tag{6.64}$$

where ε_{b} is the permittivity of the background and $\sigma(\boldsymbol{b})$ is the standard deviation of the baseline signal (i.e., its noise floor); see Eq. (6.55). The response derivatives comprising the vector $\partial\boldsymbol{b}/\partial\varepsilon$ can be analytically expressed using the S-parameter sensitivity formula Eq. (3.11), viz.

$$\frac{dS_{ij}(\mathbf{r})}{d\varepsilon} = \frac{-i\omega}{2a_i a_j} \iiint_{V'} \mathbf{E}_j(\mathbf{r}') \cdot \mathbf{E}_i(\mathbf{r}') dv' \;\left(\text{F}^{-1} \times \text{m}\right), \tag{6.65}$$

where i and j are the Rx and Tx ports, respectively, and V' is the volume of constant permittivity ε. Let $b_k \equiv S_{ij}(\mathbf{r}_k)$ represent the k-th baseline S-parameter acquired at the position $\mathbf{r}_k$. For a contrast object of sufficiently small volume V', Eq. (6.65) allows for expressing the derivative of b_k per unit volume as

$$\left[\frac{db_k}{d\varepsilon(\mathbf{r}')}\right]_{1\mathrm{v}} = \frac{-i\omega}{2a_i a_j}\mathbf{E}_j(\mathbf{r}') \cdot \mathbf{E}_i(\mathbf{r}') \left(\mathrm{F}^{-1} \times \mathrm{m}^{-2}\right), \quad k = 1, \ldots, N, \tag{6.66}$$

where $\mathbf{r}'$ is the position of the voxel at which the derivative is calculated and N is the number of baseline responses comprising $\boldsymbol{b}_{N\times 1}$. From Eq. (6.66), it is clear that the response derivative is dependent on both $\mathbf{r}'$ and $\mathbf{r}_k$. It is the absolute value of the maximum response derivative that is used in the denominator of Eq. (6.64) to estimate the best (the smallest) possible contrast sensitivity. Eq. (6.66) is useful in estimating the system contrast sensitivity via simulations of the background measurements since the field distributions $\mathbf{E}_j(\mathbf{r}')$ and $\mathbf{E}_i(\mathbf{r}')$ are readily available.

The baseline response derivatives can also be obtained experimentally from the measurements of a CO, where an electrically small scatterer of known contrast $\Delta\varepsilon_{\mathrm{CO}}$ and known volume Ω_{CO} is placed at the center of the uniform background. In this case,

$$\left[\frac{db_k}{d\varepsilon(\mathbf{r}')}\right]_{1\mathrm{v}} \approx \frac{r_{\mathrm{CO},k} - b_k}{\Delta\varepsilon_{\mathrm{CO}}\Omega_{\mathrm{CO}}}, \tag{6.67}$$

where $r_{\mathrm{CO},k}$ is the k-th response acquired in the CO measurement and $(r_{\mathrm{CO},k} - b_k)$ estimates the change in the k-th baseline response as a result of the contrast presented by the small scatterer at $\mathbf{r}'$.

7 Looking Forward: Nonlinear Reconstruction

This book has introduced the reader to the principles of EM scattering theory along with the basic linear (or direct) reconstruction approaches of MW and millimeter-wave imaging. It has been emphasized that the linear reconstruction methods suffer from limitations stemming from the linearizing approximations of the forward model of scattering. The linearized models are incapable of taking into account multiple scattering and mutual-coupling effects. Such effects dominate the scattering in complex heterogeneous objects such as living tissue, luggage items, and structural components in civil engineering.

This is why the forefront of MW imaging research is focused on reconstruction approaches that can tackle nonlinear scattering. Yet, research and development in MW nonlinear reconstruction demand familiarity with the basics of scattering theory and linear reconstruction along with some understanding of the nature of the MW signals and measurements. Familiarity with linear reconstruction methods is especially important because these are often employed as modules in nonlinear inversion strategies.

It is the author's hope that this text has aided the novice on a difficult journey through a thick forest of equations in mathematical physics and an overwhelming gallery of clever inversion techniques to the highly technical subject of MW metrology. By no means is the present text exhaustive. Numerous references have been provided throughout and many more exist in the scientific and engineering literature. MW imaging is a vast and dynamic field of research, and one has to keep an eye on new developments.

But most importantly, it is the author's hope that the reader will continue this journey. The next big step is *nonlinear reconstruction*. The monograph of Pastorino [46] dedicates special attention to this subject and contains an extensive list of references. The applications of microwaves in medical imaging almost exclusively employ nonlinear strategies. Overviews of these applications can be found in [93, chapter 7], [257], and [134, 135].

Here is a taste of what lies ahead. The nonlinear reconstruction approaches share two common characteristics. First, they are quantitative, i.e., they are, in principle, capable of recovering the complex permittivity distribution of the object under test (OUT). This is a consequence of their second common characteristic: they do not employ linearizing approximations regarding the internal field distribution in the OUT such as the Born or Rytov approximations that we discussed in Sections 1.13 through 1.18. Thus, the forward model they are based on is nonlinear with respect to the unknown contrast. This is apparent from the data equation Eq. (3.58) for the scattered portion of an S-parameter response, which we can write as

$$S(\mathbf{r}, \omega) = \kappa \iiint_{V'} \Delta\varepsilon(\mathbf{r}')\mathbf{E}_{\mathrm{Tx}}^{\mathrm{tot}}(\mathbf{r}', \mathbf{r}_{\mathrm{Tx}}; \omega) \cdot \mathbf{E}_{\mathrm{Rx}}^{\mathrm{inc}}(\mathbf{r}', \mathbf{r}; \omega) dv', \mathbf{r} \notin V'. \tag{7.1}$$

Here, ω is the frequency, $\mathbf{r}$ is the observation position, $\Delta\varepsilon(\mathbf{r}')$ is the contrast to be reconstructed, $\mathbf{E}_{\mathrm{Tx}}^{\mathrm{tot}}(\mathbf{r}', \mathbf{r}_{\mathrm{Tx}}; \omega)$ is the total internal field due to the Tx antenna, and $\mathbf{E}_{\mathrm{Rx}}^{\mathrm{inc}}(\mathbf{r}', \mathbf{r}; \omega)$ is the incident field that is generated by the Rx antenna if it were to operate in a transmitting mode in the scatterer-free background. The constant $\kappa = \frac{-i\omega}{2a_{\mathrm{Tx}}a_{\mathrm{Rx}}}$ is known since a_{Tx} and a_{Rx} are the root-power waves exciting the ports of the Tx and the Rx antennas, respectively. It is clear that $S(\mathbf{r}, \omega)$ depends on the Tx and Rx field distributions *normalized to the strength of the respective excitation.* Therefore, it is independent of the actual strength of the port excitations.

What is important here is that $\mathbf{E}_{\mathrm{Tx}}^{\mathrm{tot}}$ is an implicit function of $\Delta\varepsilon$ and is thus unknown. This implicit dependence is governed by Maxwell's equations, and in real-life scenarios it is not tractable analytically. Thus, Eq. (7.1) is nonlinear with respect to $\Delta\varepsilon$ where the nonlinearity is rooted in the unknown field quantity $\mathbf{E}_{\mathrm{Tx}}^{\mathrm{tot}}$, the behavior of which is constrained by Maxwell's equations. We are now ready to understand the problem of nonlinear reconstruction as a constrained optimization problem. This problem can be mathematically stated as

$$\begin{aligned} &\Delta\varepsilon^*(\mathbf{r}') = \arg\min_{\Delta\varepsilon} \left\| \bar{\boldsymbol{S}} - \boldsymbol{S}\left(\Delta\varepsilon, \mathbf{E}_{\mathrm{Tx}}^{\mathrm{tot}}(\Delta\varepsilon)\right) \right\| \\ &\text{subject to } \mathcal{M}\left\{\mathbf{E}_{\mathrm{Tx}}^{\mathrm{tot}}(\Delta\varepsilon)\right\} = 0, \end{aligned} \tag{7.2}$$

where $\bar{\boldsymbol{S}}$ is the set of measured scattered responses, $\boldsymbol{S}$ are the responses that the forward EM model would produce for a given $\Delta\varepsilon$, and $\|\cdot\|$ is a suitable norm. For brevity of notation, the dependence of $\bar{\boldsymbol{S}}$ and $\boldsymbol{S}$ on the observation position $\mathbf{r}$ and the frequency ω, has been suppressed but this dependence is implied. $\mathcal{M}$ denotes the linear operator of Maxwell's equations. The constraint,

$$\mathcal{M}\left\{\mathbf{E}_{\mathrm{Tx}}^{\mathrm{tot}}(\Delta\varepsilon)\right\} = 0, \tag{7.3}$$

indicates that $\mathbf{E}_{\mathrm{Tx}}^{\mathrm{tot}}$ must satisfy Maxwell's equations in V' with all associated boundary conditions and medium properties.

In plain terms, Eq. (7.2) states that we aim at finding a contrast distribution $\Delta\varepsilon^*(\mathbf{r}')$, $\mathbf{r}' \in V'$, such that it minimizes the difference between the measured responses and the responses that an EM model would predict with this contrast distribution. The EM model can be a data equation such as Eq. (7.1), but it also can be any other solver, simulations included. The "subject to" line states the constraint, which insists on verifying that not only the data match is achieved but also Maxwell's equations are satisfied. Thus, to recover the contrast $\Delta\varepsilon$, the nonlinear reconstruction must also recover $\mathbf{E}_{\mathrm{Tx}}^{\mathrm{tot}}$.

The nonlinear reconstruction methods aim at solving Eq. (7.2). The good news is that there are a plethora of methods developed by the mathematical science of applied optimization that deal with constrained nonlinear optimization; see, for example [258, 259]. The bad news is that the number of unknowns implied by the contrast distribution

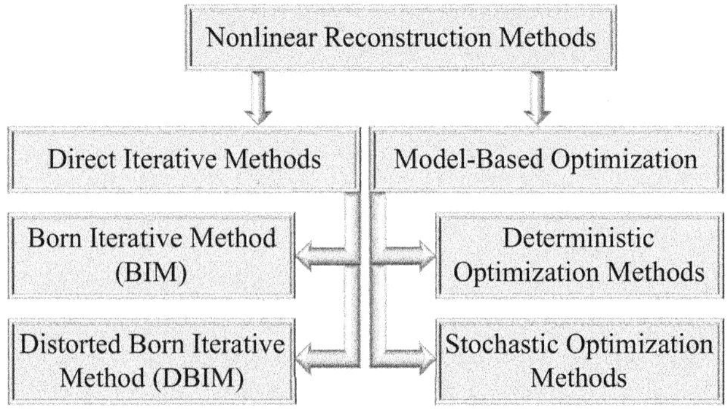

Figure 7.1 Classification of the nonlinear reconstruction methods.

$\Delta\varepsilon(\mathbf{r}')$ is large. It is determined by the number of voxels comprising the imaged volume V' and is on the order of hundreds even thousands. This is an extremely complex optimization problem with too many degrees of freedom. Such problems are prone to spurious solutions that arise when the optimization process is trapped in a local minimum. Another practical issue arises due to the need to employ an EM solver, which ensures that Maxwell's equations are satisfied. Full-wave solvers are notoriously time consuming and various methods to accelerate the computations have been pursued.

Despite all difficulties, the methods of nonlinear reconstruction promise to advance MW imaging toward numerous practical applications. Current research encompasses two classes of methods [260]: (a) the direct iterative methods and (b) the model-based optimization methods. Fig. 7.1 shows a chart of the classes of nonlinear reconstruction methods. The model-based optimization methods form a very large class of methods and are not reviewed in detail here. As described earlier, these methods approach the inversion problem in Eq. (7.2) as a general nonlinear-optimization task where the variables describing $\Delta\varepsilon(\mathbf{r}')$ are tuned iteratively in an EM simulation to minimize the data error. A great variety of nonlinear-optimization algorithms exist, both deterministic and stochastic as shown in Fig. 7.1. The reader is referred to [46] for a detailed review with extensive references describing various applications in MW imaging.

The direct iterative methods deserve a better look in the context of this book because they are the ones using linear reconstruction as modules inside the iterative algorithm. They include the Born iterative method (BIM) and the distorted Born iterative method (DBIM); see, for example, [31, chapter 9]. The distinguishing feature of these methods is that they use the Born series [see Section 1.13 and Eq. (1.306)] to successively improve the accuracy of the contrast estimate and that of the total field. At the core of these methods is a linear reconstruction engine, which may employ any of the techniques discussed in Chapters 4 and 5. Linear inversion is also referred to as direct inversion, hence the name of this class of iterative approaches.

A flowchart of a typical implementation of the BIM is shown in Fig. 7.2. There, $K(\mathbf{r}') = \mu_0\omega^2\Delta\varepsilon(\mathbf{r}')$ is used to shorten the first group of equations. The algorithmic steps are described in mathematical terms by the equations that are right next to the

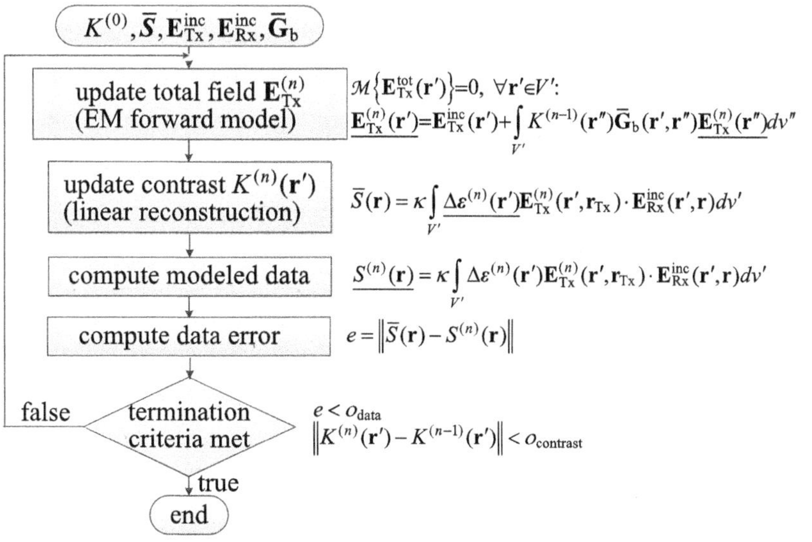

Figure 7.2 Flowchart of the Born iterative method (BIM). The notation $K(\mathbf{r}') = \mu_0\omega^2\Delta\varepsilon(\mathbf{r}')$ is used to shorten the first group of equations. The quantities updated at each step are underlined.

respective block in the flowchart. The quantities being updated in each step are underlined in the equations. The algorithm is initialized by submitting the initial guess for the contrast distribution $K^{(0)}(\mathbf{r}')$, the data set $\bar{S}$ that contains all scattered responses $S(\mathbf{r}, \omega)$, the field distributions $\mathbf{E}_{\text{Tx}}^{\text{inc}}(\mathbf{r}', \mathbf{r}_{\text{Tx}}; \omega)$ and $\mathbf{E}_{\text{Rx}}^{\text{inc}}(\mathbf{r}', \mathbf{r}; \omega)$, and Green's background tensor $\bar{\mathbf{G}}_b$. In the first step, the EM forward model solves Maxwell's equations with the prescribed contrast distribution to determine what the total internal field is, $\mathbf{E}_{\text{Tx}}^{(n)}(\mathbf{r}', \mathbf{r}_{\text{Tx}}; \omega)$, $\mathbf{r}' \in V'$, for all frequencies and Tx positions. The integral equation solving the vector Helmholtz equation Eq. (2.86) is given in Fig. 7.2 as one possible EM forward model, but any solver can be used. This problem is linear with respect to $\mathbf{E}_{\text{Tx}}^{(n)}$. In the second step, a linear reconstruction algorithm updates the contrast. Since the total internal field is available from the previous step, this inversion problem is also linear. The linear reconstruction solution is usually not exact. Thus, the data equation is used in the third step to calculate the actual responses $S^{(n)}(\mathbf{r})$ that the model would predict with the so found contrast. In the fourth step this is used to calculate the data error, i.e., the disagreement between the measured responses and those that the model predicts for the current contrast iterate. The data error is a suitable norm of the difference between the measured and the modeled scattered responses. This is followed by a convergence check based on certain termination criteria. The algorithm terminates if the data error drops below a specified threshold o_{data}. If it is larger than the threshold, the algorithm returns to the first step to update the total internal field. The data error is a mandatory criterion, but additional criteria are also used, e.g., the difference between the previous and the current contrast iterate. This difference is a good indicator of whether the algorithm is converging, diverging, or "wandering" around in a small region of the variable space

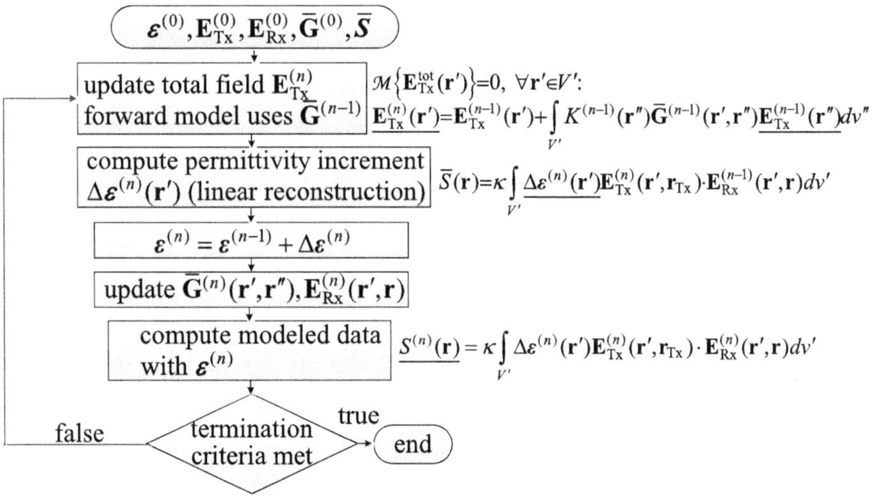

Figure 7.3 Flowchart of the distorted Born iterative method (DBIM). Note that $K^{(n)}(\mathbf{r}') = \mu_0\omega^2\Delta\varepsilon^{(n)}(\mathbf{r}')$. The quantities updated at each step are underlined.

(the space spanned by the contrast values) without making any progress in improving the data error.

To appreciate the strategy of the BIM one has to recognize that it iterates between the state equation (the one that produces the total internal field due to the Tx antenna) and the data equation (the one that estimates the contrast). In doing so, each one of these equations is rendered linear.

The DBIM is another nonlinear reconstruction method that relies on a linear-reconstruction module. Its flowchart is shown in Fig. 7.3. It differs from the BIM by viewing the contrast distribution of each previous iteration as the background medium in the current iteration. In contrast, the background medium in the BIM is fixed and is usually assumed uniform. As a result, in BIM Green's background tensor $\bar{\mathbf{G}}_b$, $\mathbf{E}_{\text{Tx}}^{\text{inc}}$, and $\mathbf{E}_{\text{Rx}}^{\text{inc}}$ are all "fixed." All of these are subject to updating in the DBIM. The advantage of the DBIM over the BIM is that it converges faster [31]. Its disadvantage is that it is sensitive to noise. This sensitivity is counteracted by regularization methods in the linear-reconstruction step [31]. Moreover, the DBIM is numerically intensive because it requires the computation of Green's function and all Tx and Rx field distribution at each iteration.

An important fact must be stated here regarding the equivalence of the DBIM to the optimization-based Newton methods solving Eq. (7.2). In [31], Chew states, "This iterative procedure is like Newton's method in solving a nonlinear integral equation, except that the solution is regularized at every iteration." A rigorous proof that the DBIM is equivalent to the Newton–Kantorovich method is due to Remis and van den Berg [261].

In conclusion, the nonlinear reconstruction methods of MW imaging constitute a vibrant fast-growing field of research with great hopes for advancement in biomedical applications as well as applications in security surveillance and nondestructive testing.

This progress is expected to accelerate due to two main technological advancements: in computing technology and the equally remarkable advent of RF and MW integrated circuits operating well into the 70-GHz range. Sensitive yet cheap software-defined radios and entire radar systems fit on chips the size of which is measured in square millimeters. It is only a matter of time before this technology finds its way into the design of imaging sensor arrays that will be capable of acquiring tens of thousands of low-noise responses within a second. This will translate into larger and higher-quality data sets, and, therefore, better images.

Appendix A Maxwell's Equations

Maxwell's equations are the axioms of electromagnetic (EM) theory—they are not subject to proof and they are the starting point of every EM model. They are based on extensive experimental evidence, and to date, they are our best physical model of the observed EM phenomena. Textbooks and advanced texts explaining the origins of Maxwell's equations and discussing their applications are in abundance. Here, these equations are summarized in their time-domain and frequency-domain forms.

A.1 Maxwell's Equations in the Time Domain

The first Maxwell equation is the law of magnetic induction, i.e., Faraday's law. It is stated in differential form as

$$-\nabla \times \boldsymbol{\mathcal{E}} = \frac{\partial \boldsymbol{\mathcal{B}}}{\partial t} + \boldsymbol{\mathcal{M}}_{\mathrm{i}}, \tag{A.1}$$

while its integral form (as per Stokes' theorem) is

$$-\oint_C \boldsymbol{\mathcal{E}} \cdot d\mathbf{c} = \frac{\partial}{\partial t} \iint_{S_C} \boldsymbol{\mathcal{B}} \cdot d\mathbf{s} + \mathcal{I}_{\mathrm{m}}. \tag{A.2}$$

All quantities above are functions of position and time $(\mathbf{r}, t)$. Also, $\boldsymbol{\mathcal{E}}$ is the electric field vector (also known as the electric intensity vector) measured in *volts per meter* $(\mathrm{V} \cdot \mathrm{m}^{-1})$; $\boldsymbol{\mathcal{B}}$ is the magnetic flux density vector (also known as the magnetic induction vector), measured in *tesla* or *webers per square meter* $(\mathrm{T} \equiv \mathrm{Wb} \cdot \mathrm{m}^{-2})$; $\boldsymbol{\mathcal{M}}_{\mathrm{i}}$ is the (fictitious) magnetic current density[1] measured in $\mathrm{V} \cdot \mathrm{m}^{-2}$; t is time (in *seconds*); and S_C is a surface bounded by the contour C. In Eq. (A.2), $\mathcal{I}_{\mathrm{m}} = \iint_{S_C} \boldsymbol{\mathcal{M}}_{\mathrm{i}} \cdot d\mathbf{s}$ is the (fictitious) magnetic current, measured in *volts*.

[1] There is no physical evidence for the existence of magnetic monopoles, i.e., magnetic charge carriers analogous to the electric charge careers such as electrons. Therefore, the magnetic source quantities such as the magnetic charge density ρ_{m} and the magnetic current density $\boldsymbol{\mathcal{M}}_{\mathrm{i}}$ are considered fictitious. Although fictitious, these EM sources render Maxwell's equations symmetric and prove to be a useful mathematical tool when solving boundary-value problems through an application of the equivalence principle. In effect, they are employed as sources, the action of which is equivalent to that of physically existing EM fields and sources.

In microwave (MW) imaging, we are mostly concerned with the existence and the reconstruction of the physical sources. In this case, the fictitious magnetic sources in Maxwell's equations are set to zero.

The second Maxwell equation is Ampère's law generalized by Maxwell to include the displacement current in addition to the current representing the actual electric-charge transport (i.e., conduction current and/or current sources). It is stated in differential form as

$$\nabla \times \mathcal{H} = \frac{\partial \mathcal{D}}{\partial t} + \mathcal{J}, \tag{A.3}$$

while its integral form is

$$\oint_C \mathcal{H} \cdot d\mathbf{c} = \frac{\partial}{\partial t} \iint_{S_C} \mathcal{D} \cdot d\mathbf{s} + \mathcal{I}_{\rm e}. \tag{A.4}$$

As in Faraday's laws, all quantities in Ampère's laws are functions of position and time, $(\mathbf{r}, t)$. $\mathcal{H}$ is the magnetic field vector (also known as the magnetic intensity vector) measured in *amperes per meter* $(\mathrm{A} \cdot \mathrm{m}^{-1})$; $\mathcal{D}$ is the electric flux density vector (also known as the electric induction vector), measured in *coulombs per square meter* $(\mathrm{C} \cdot \mathrm{m}^{-2})$; $\mathcal{J}$ is the electric current density, measured in $\mathrm{A} \cdot \mathrm{m}^{-2}$; and $\mathcal{I}_{\rm e} = \iint_{S_C} \mathcal{J} \cdot d\mathbf{s}$ is the electric current, measured in *amperes*. The derivative $\frac{\partial \mathcal{D}}{\partial t}$ in Eq. (A.3) is the displacement-current density while $\mathcal{J}$ represents the current density associated with charge transport. The latter is usually written explicitly in terms of two components as $\mathcal{J} = \mathcal{J}_{\rm i} + \mathcal{J}_{\rm c}$, where $\mathcal{J}_{\rm i}$ is the impressed (or source) current density and $\mathcal{J}_{\rm c}$ is the conduction (or ohmic) current density induced in the medium by the field. In the forward EM problem, the former is known while the latter is unknown because it is a function of the unknown field vectors.

In electricity, the concept of *convection* currents is also used to describe current that is *not* due to the EM field, i.e., current describing the charge transport induced by external forces (e.g., mechanical, thermal, or chemical). The convection currents are important in multi-physics forward models since they serve as the link between the EM model (Maxwell's equations) and the other physical models. They are known if Maxwell's equations are solved independently from the other physical models. Then, the convection currents are part of $\mathcal{J}_{\rm i}$. However, when the solution of the involved physical models is mutually dependent (or coupled), the convection currents become unknown in a system of equations where Maxwell's equations are only a subset. They are then included in $\mathcal{J}_{\rm c}$.

Similarly to the current density, the total electric current $\mathcal{I}_{\rm e}$ may also consist of two types of currents: impressed and conduction currents, i.e., $\mathcal{I}_{\rm e} = \mathcal{I}_{\rm i} + \mathcal{I}_{\rm c}$.

The two "curl" Maxwell equations are mutually coupled and they form a system of two vectorial equations. They provide the mathematical model of electromagnetism. However, mathematically, they are not sufficient to obtain a solution because they contain five vectorial unknowns: $\mathcal{E}, \mathcal{D}, \mathcal{H}, \mathcal{B}$, and $\mathcal{J}_{\rm c}$ (the impressed currents $\mathcal{J}_{\rm i}$ are known by definition). They are complemented by the constitutive relations, which, in general, can be stated as [9, 262, 263]:

$$\mathcal{D} = \mathcal{F}_D(\mathcal{E}, \mathcal{B}), \tag{A.5}$$

$$\mathcal{H} = \mathcal{F}_H(\mathcal{E}, \mathcal{B}), \tag{A.6}$$

and

$$\mathcal{J}_\mathrm{c} = \mathcal{F}_J(\mathcal{E}, \mathcal{B}). \tag{A.7}$$

The constitutive operators $\mathcal{F}_D$, $\mathcal{F}_H$, and $\mathcal{F}_J$ can be in general nonlinear tensorial functionals; see, for example, [9].[2]

The simplest case is that of a linear, isotropic, and dispersion-free medium, when

$$\mathcal{D} = \epsilon\mathcal{E}, \;\; \mathcal{B} = \mu\mathcal{H}, \;\; \mathcal{J}_\mathrm{c} = \sigma_\mathrm{e}\mathcal{E}, \tag{A.8}$$

where ϵ (measured in *farads per meter*, $\mathrm{F \cdot m^{-1}}$), μ (measured in *henry per meter*, $\mathrm{H \cdot m^{-1}}$), and σ_e (measured in *siemens per meter*, $\mathrm{S \cdot m^{-1}}$) are the medium's permittivity, permeability and specific conductivity, respectively. These are referred to as the medium constitutive parameters (or simply the medium parameters). In this simple case, the medium parameters are independent of the field strength (the medium is *linear*, implying the linear constitutive equations stated earlier), independent of the field polarization (the medium is *isotropic*, i.e., the parameters are simple scalars as opposed to tensors), and independent of the frequency (the medium is *dispersion-free*). The medium parameters may still be functions of position, in which case the medium is *heterogeneous* as opposed to a *uniform* or *homogeneous* medium, the parameters of which are constant in space.

It is worth noting that the linear constitutive relations in Eq. (A.8) are incapable of modeling magnetic losses that become relevant in AC regimes of operation. This is expected when it comes to hysteresis losses because they are inherently linked to the medium magnetic nonlinearity. However, even if the magnetic medium is linear, AC losses exist. This case is important in practice because any medium, if interacting with sufficiently weak fields, is linear. The losses in a linear magnetic medium are due to the alternating magnetization at the molecular (or domain) level. This loss mechanism is analogous to the dielectric polarization loss. As long as the medium is capable of magnetization (i.e., it contains magnetic dipoles), the alternating field, however weak it may be, will encounter the damping effect of magnetic loss in addition to the dielectric-polarization or conduction losses (if present).

A simple mathematical model that can accommodate AC magnetic loss in a linear dispersion-free medium is provided by the constitutive relation

$$\frac{\partial \mathcal{B}}{\partial t} = \mu \frac{\partial \mathcal{H}}{\partial t} + \sigma_\mathrm{m}\mathcal{H}, \tag{A.9}$$

[2] The constitutive relations can also be stated in a mathematically equivalent form in which $\mathcal{D}$, $\mathcal{B}$, and $\mathcal{J}_\mathrm{c}$ are functions of $\mathcal{E}$ and $\mathcal{H}$ [41, 263].

where $\sigma_{\rm m}$ is an *equivalent* (or *effective*) magnetic conductivity measured in *ohms per meter* ($\Omega \cdot \rm m^{-1}$). Eq. (A.9) can be viewed as an extension of the relation $\mathcal{B} = \mu\mathcal{H}$ in Eq. (A.8) to the case of AC fields. Note that Eq. (A.9) implies that with static fields $\sigma_{\rm m} = 0$, which correctly reflects our experimental observations that magnetic charges and magnetic conduction currents do not exist.

Here, the physical meaning of the electric conductivity $\sigma_{\rm e}$ in the case of time-varying fields is clarified. In this situation, $\sigma_{\rm e}$ incorporates not only the loss effects associated with conduction (i.e., charge transport through the medium), but also the dielectric polarization loss. Thus, it is an effective parameter just like $\sigma_{\rm m}$. Distinguishing between the AC conduction and polarization losses in dielectrics through measurements is difficult. This is why the effective electric conductivity of a given material is the only loss parameter reported (usually as a function of frequency). For example, the effective electric conductivity is commonly used to describe the EM losses in the various biological tissues. On the other hand, in the electronics industry, the losses of a dielectric material are typically described by the so-called dielectric loss tangent. This loss parameter, however, is better suited for the frequency-domain EM model. It is discussed in the next section.

With the introduction of the effective magnetic and electric conductivities, the time-domain Maxwell equations in the case of a linear, isotropic, and dispersion-free medium become symmetric with respect to the electric and magnetic quantities:

$$-\nabla \times \mathcal{E} = \mu\frac{\partial \mathcal{H}}{\partial t} + \sigma_{\rm m}\mathcal{H} + \mathcal{M}_{\rm i}, \tag{A.10}$$

$$\nabla \times \mathcal{H} = \epsilon\frac{\partial \mathcal{E}}{\partial t} + \sigma_{\rm e}\mathcal{E} + \mathcal{J}_{\rm i}. \tag{A.11}$$

This time-domain model is often used in MW imaging since it is common to have a linear, isotropic, and heterogeneous medium. A widely used numerical method to solve Eqs. (A.10) and (A.11) is the finite-difference time-domain (FDTD) method [264]. Applications such as biomedical imaging, underground radar, and nondestructive testing of materials usually deal with relatively weak fields, for which the medium is linear. Also, the medium's anisotropy is often negligible. On the other hand, many materials exhibit significant dispersion in the MW band. For example, materials of high water content have permittivity and electric conductivity, which are both dependent on the frequency in the RF, MW, and millimeter-wave bands. This calls for proper treatment of the constitutive relations in the time-domain analyses, which involves convolution in time.

The other two Maxwell equations are known as the "divergence" equations. The first one, the Gauss law of electricity, states the relation between the electric flux density $\mathcal{D}$ and the electric charge density $\rho_{\rm e}$ (measured in *coulombs per cubic meter*, $\rm C \cdot m^{-3}$):

$$\nabla \cdot \mathcal{D} = \rho_{\rm e}. \tag{A.12}$$

Here, just like in the case of the electric current density, $\rho_{\rm e}$ can be divided into two contributions: the impressed charged density $\rho_{\rm i}$ and the charged density associated with the conduction current ρ_c, i.e., $\rho_\text{e} = \rho_\text{i} + \rho_\text{c}$. The association of each charge quantity to its respective current quantity is through the conservation-of-charge law:

$$\nabla \cdot \mathcal{J} = -\frac{\partial \rho_\text{e}}{\partial t}, \tag{A.13}$$

which can be written for the total electric current and charge or separately for their impressed and conduction components.

As per the Gauss theorem, Eq. (A.12) can be restated as an integral equation in the form

$$\oiint_S \mathcal{D} \cdot d\mathbf{s} = \iiint_{V_S} \rho_\text{e} dv = q_\text{e}, \tag{A.14}$$

where V_S is the volume bounded by the closed surface S and q_e is the electric charge (measured in *coulombs*, C) enclosed by S. Note that while the Gauss law of electricity states the electric charge as the "divergence" source of the EM field, the Ampère law, see Eq. (A.3) and Eq. (A.4), states the electric current as its rotational (or "curl") source.

The Gauss law of magnetism is analogous, and it constitutes the fourth (and last) Maxwell equation. It states the relationship between the magnetic flux density $\mathcal{B}$ and the fictitious magnetic charge density m. Physical reality requires that $m = 0$, in which case we must state Maxwell's fourth equation as

$$\nabla \cdot \mathcal{B} = 0, \tag{A.15}$$

or, in an integral form,

$$\oiint_S \mathcal{B} \cdot d\mathbf{s} = 0. \tag{A.16}$$

The mathematical statement that the Gauss law of magnetism makes is that the magnetic field is a divergence-free or rotational field. A more descriptive way of saying the same thing is that the force lines of the magnetic field are closed. This is equivalent to the physical assertion that magnetic monopoles do not exist. We emphasize that this assertion is axiomatic and derived from extensive observations. Theoretical physics, however, argues in favor of the possibility of the existence of magnetic charges [262], so, we should keep an open mind.

In case fictitious magnetic currents are involved in the analysis (e.g., when the equivalence principle is employed), we are compelled to admit fictitious magnetic charge density m such that it relates to the magnetic current through the continuity law $\nabla \cdot \mathcal{M} = -\frac{\partial m}{\partial t}$. Then, the Gauss law of magnetism becomes fully analogous to that in electricity. In its differential form, for example, it states

$$\nabla \cdot \mathcal{B} = m. \tag{A.17}$$

The two "curl" Maxwell equations together with the constitutive relations form a complete mathematical model of the electrodynamic field in the sense that with the proper boundary conditions they provide a unique solution [1, 3]. One may wonder then about the utility of the "divergence" equations; especially because they follow from the curl equations and the conservation of charge Eq. (A.13). To see how this is true, take the divergence of the two sides of Eq. (A.3) and make use of Eq. (A.13) to arrive at Eq. (A.12). Analogously, taking the divergence of the two sides of Eq. (A.1), where the fictitious source is set to zero ($\mathcal{M}_i = 0$), and assuming zero initial conditions for the magnetic field, we arrive at Eq. (A.15). In other words, as long as the sources observe the conservation of charge (they should, because the conservation of charge is one of the fundamental laws of physics), the divergence equations are not independent from the curl equations.

Nonetheless, the divergence equations are very useful because they provide an explicit link from the field to the electric charge. In fact, they play a critical role in electrostatic and magnetostatic modeling. Note that the curl Maxwell equations Eq. (A.1) and Eq. (A.3) do not provide an explicit mathematical model for the static field where the electric field becomes independent from the magnetic field:

$$\nabla \times \mathcal{E} = 0, \tag{A.18}$$

$$\nabla \times \mathcal{H} = \mathcal{J}. \tag{A.19}$$

Eq. (A.18) is simply a statement that the static electric field is irrotational (or conservative), but by itself it cannot provide a unique solution for the field because it does not involve the charge sources. Such a solution becomes possible only after complementing Eq. (A.18) with the Gauss law of electricity. Eq. (A.19), on the other hand, shows explicitly the relation between the magnetic field and its source (the electric current), but it, too, is insufficient to provide a unique solution by itself. Eq. (A.19) defines the curl of the magnetic field. However, from Helmholtz's theorem, we also need the divergence of the field to obtain a unique solution for the given boundary conditions. This divergence is provided by the Gauss law of magnetism.

The model of the EM field is completed by the equation describing the force it exerts on charges, the Lorentz force. For a point charge q_e, this force is

$$\mathcal{F} = q_e \mathcal{E} + q_e \mathbf{v} \times \mathcal{B}, \tag{A.20}$$

where $\mathbf{v}$ is the charge velocity. Notice that the vectors $\mathcal{E}$ and $\mathcal{B}$ determine the force per unit charge. This is why they are often referred to as the EM force vectors. The force density $\mathcal{F}'$ (measured in $\mathrm{N} \cdot \mathrm{m}^{-3}$) can be conveniently expressed in terms of the charge and current densities as

$$\mathcal{F}' = \rho_e \mathcal{E} + \mathcal{J} \times \mathcal{B}. \tag{A.21}$$

The Lorentz force equation is critically important in multi-physics modeling. It provides the fundamental link between the EM field and any other kind of force field that may need to be modeled (chemical, thermal, acoustic, etc.).

A.2 Maxwell's Equations in the Frequency Domain

Using the field phasors, which are functions of position and frequency, $(\mathbf{r}, \omega)$, the two curl Maxwell equations (in differential form) are stated as

$$\nabla \times \mathbf{H} = \mathrm{i}\omega\mathbf{D} + \mathbf{J}_\mathrm{i}, \tag{A.22}$$

$$-\nabla \times \mathbf{E} = \mathrm{i}\omega\mathbf{B} + \mathbf{M}_\mathrm{i}. \tag{A.23}$$

Here, $\mathbf{E}$ and $\mathbf{H}$ are the vector phasors of the electric and magnetic field intensities, respectively, whereas $\mathbf{D}$ and $\mathbf{B}$ are the vector phasors of the electric and magnetic flux densities, respectively. Also, $\mathbf{J}_\mathrm{i}$ and $\mathbf{M}_\mathrm{i}$ are the phasors of the impressed electric and magnetic current densities, respectively. It must be emphasized that $\mathbf{J}_\mathrm{i}$ describes the known excitation, and it does *not* include the conduction current density $\mathbf{J}_\mathrm{c}$. In fact, in the frequency-domain Maxwell equations, the conduction current density is absorbed by the term $j\omega\mathbf{D}$ as explained shortly.

In the frequency domain, the constitutive relations are commonly stated through the general bi-anisotropic equations (in the Boys-Post form) [263]:

$$\mathbf{D} = \boldsymbol{\varepsilon} \cdot \mathbf{E} + \boldsymbol{\xi} \cdot \mathbf{B} \tag{A.24}$$

$$\mathbf{H} = \boldsymbol{\zeta} \cdot \mathbf{E} + \boldsymbol{\mu}^{-1} \cdot \mathbf{B}, \tag{A.25}$$

where $\boldsymbol{\varepsilon}$, $\boldsymbol{\mu}$, $\boldsymbol{\xi}$, and $\boldsymbol{\zeta}$ are the constitutive dyadics. These are usually given in the form of second-rank Cartesian tensors, i.e., 3×3 matrices relating the triple-component field vectors. As in the time-domain model, the constitutive dyadics may depend on the field strength (nonlinear medium) and on position (heterogeneous medium), and they may differ at each frequency of interest (dispersive medium). Dispersion is handled in a simpler way in the frequency-domain constitutive relations as compared to the time-domain relations since the convolution integrals in time are replaced by simple multiplications.

In MW imaging, most of the imaged domains are well approximated by the assumption of an isotropic linear medium, in which case the constitutive relations simplify to

$$\mathbf{D} = \varepsilon\mathbf{E}, \tag{A.26}$$

$$\mathbf{B} = \mu\mathbf{H}, \tag{A.27}$$

where ε and μ are the medium's permittivity and permeability (scalar complex-valued functions of position and frequency), respectively It is customary to express these complex constitutive parameters through their real and imaginary parts as $\varepsilon = \varepsilon' - \mathrm{i}\varepsilon''$ and $\mu = \mu' - \mathrm{i}\mu''$ where ε'' and μ'' are positive real. It can be shown that the negative imaginary parts of the complex permittivity and permeability are a consequence of the energy loss in the material. Positive imaginary parts imply energy gain in the material and are thus nonphysical. The real parts ε' and μ' tend asymptotically to the static permittivity and permeability ϵ and μ defined through Eq. (A.8) in the low-frequency limit. In this case, they are positive real as well. However, the behavior of the material can be very different at high frequencies where ε' and μ' can attain negative or zero values [262, 12].

As mentioned before, the conduction current phasor $\mathbf{J}_c$ is absorbed by the phasor of the displacement current $i\omega\mathbf{D}$; see Eq. (A.23). Accordingly, the constitutive relation between the conduction current phasor $\mathbf{J}_c$ and the electric field phasor $\mathbf{E}$ (Ohm's law in differential form, $\mathbf{J}_c = \sigma_e \mathbf{E}$) is absorbed by Eq. (A.26). It can be shown that the conductivity σ_e contributes to ε by adding to its imaginary part:

$$\varepsilon = \varepsilon' - i\left(\varepsilon' \tan\delta_d + \frac{\sigma_e}{\omega}\right). \tag{A.28}$$

Here, $\tan\delta_d$ is the dielectric loss tangent describing the polarization loss. As discussed in the previous section, the loss contributions of the conduction effects (represented by σ_e) and those of the polarization effects (represented by $\tan\delta_d$) are difficult to discern through measurements. Thus, the way the imaginary permittivity is expressed in Eq. (A.28) is rarely used. It is expressed either as

$$\varepsilon'' = \varepsilon' \tan\delta_d, \tag{A.29}$$

or as

$$\varepsilon'' = \frac{\sigma_e}{\omega}. \tag{A.30}$$

One can calculate the effective conductivity from the loss tangent and vice versa using

$$\sigma_e = \omega\varepsilon' \tan\delta_d. \tag{A.31}$$

In analogy to Eq. (A.29), the magnetic loss is usually described by the magnetic loss tangent $\tan\delta_m$:

$$\mu'' = \mu' \tan\delta_m. \tag{A.32}$$

Equivalently, one can use the effective magnetic conductivity as

$$\mu'' = \frac{\sigma_m}{\omega}. \tag{A.33}$$

A final note with regard to the complex constitutive parameters concerns their expression in terms of the *relative* permittivity $\varepsilon_r = \frac{\varepsilon}{\epsilon_0}$ and the *relative* permeability $\mu_r = \frac{\mu}{\mu_0}$. Here, ϵ_0 and μ_0 are the permittivity and the permeability of vacuum. The relative permittivity and permeability are the preferred way of describing the EM properties of the various materials and media.

The substitution of the constitutive relations Eq. (A.26) and Eq. (A.27) into Eq. (A.23) and Eq. (A.22) leads to the frequency-domain Maxwell equations in the case of a linear isotropic medium:

$$-\nabla \times \mathbf{E} = i\omega\mu\mathbf{H} + \mathbf{M}_i, \tag{A.34}$$

$$\nabla \times \mathbf{H} = i\omega\varepsilon\mathbf{E} + \mathbf{J}_i. \tag{A.35}$$

These equations, together with the properly defined boundary conditions for the field vectors [1, 55] provide a unique solution to the forward EM problem.

The "divergence" Maxwell equations for the phasors of the electric flux density $\mathbf{D}$ and the magnetic flux density $\mathbf{B}$ follow directly from the time-domain Gauss laws of electricity and magnetism as

$$\nabla \cdot \mathbf{D} = \rho_\mathrm{i} \tag{A.36}$$

$$\nabla \cdot \mathbf{B} = m_\mathrm{i}. \tag{A.37}$$

Note that here the charge densities, ρ_i and m_i, are those of the sources only. As noted before, the divergence equations above follow from the curl equations, Eq. (A.22) and Eq. (A.23), and the conservation-of-charge laws in phasor form, e.g., for the electric charge,

$$\nabla \cdot \mathbf{J}_\mathrm{i} = -\mathrm{i}\omega\rho_\mathrm{i}. \tag{A.38}$$

For the magnetic sources, we have $\nabla \cdot \mathbf{M}_\mathrm{i} = -\mathrm{i}\omega m_\mathrm{i}$.

The integral formulations of Maxwell's equations in the frequency domain are obtained from their differential forms by making use of the Stokes and Gauss integral theorems. They are summarized next.

1. Ampère's law

$$\oint_C \mathbf{H} \cdot d\mathbf{c} = \iint_{S_C} \mathrm{i}\omega\mathbf{D} \cdot d\mathbf{s} + I_\mathrm{e}. \tag{A.39}$$

Here, I_e is the phasor of the current through the surface S_C.

2. Faraday's law

$$-\oint_C \mathbf{E} \cdot d\mathbf{c} = \iint_{S_C} \mathrm{i}\omega\mathbf{B} \cdot d\mathbf{s} + I_\mathrm{m}. \tag{A.40}$$

Here, I_m is the fictitious magnetic current through the surface S_C.

3. Gauss law of electricity

$$\oiint_S \mathbf{D} \cdot d\mathbf{s} = \iiint_{V_S} \rho_\mathrm{i} dv = Q_\mathrm{e}. \tag{A.41}$$

Here, Q_e is the phasor of the electric charge enclosed by S.

4. Gauss law of magnetism

$$\oiint_S \mathbf{B} \cdot d\mathbf{s} = \iiint_{V_S} m_\mathrm{i} dv = Q_\mathrm{m}. \tag{A.42}$$

Here, Q_m is the fictitious magnetic charge enclosed by S (measured in $\mathrm{V} \cdot \mathrm{s}$).

Finally, we restate the Lorenz force equation in its phasor form written for the force per unit volume:

$$\mathbf{F}' = \rho_\mathrm{e}\mathbf{E} + \mathbf{J} \times \mathbf{B}. \tag{A.43}$$

Table A.1 Dual Electromagnetic Equations

Electric Sources ($\mathbf{J} \neq 0$, $\mathbf{M} = 0$)	Magnetic Sources ($\mathbf{J} = 0$, $\mathbf{M} \neq 0$)
$\nabla \times \mathbf{E} = -\mathrm{i}\omega\mu\mathbf{H}$	$\nabla \times \mathbf{H} = \mathrm{i}\omega\varepsilon\mathbf{E}$
$\nabla \times \mathbf{H} = \mathrm{i}\omega\varepsilon\mathbf{E} + \mathbf{J}$	$\nabla \times \mathbf{E} = -\mathrm{i}\omega\mu\mathbf{H} - \mathbf{M}$
$\nabla \cdot \mathbf{D} = \rho$	$\nabla \cdot \mathbf{B} = m$
$\nabla \cdot \mathbf{J} = -\mathrm{i}\omega\rho$	$\nabla \cdot \mathbf{M} = -\mathrm{i}\omega m$

Here, ρ_e and $\mathbf{J}$ are the phasors of the electric charge and current densities, respectively, that must include all impressed, convection, and conduction terms.

A.3 Electromagnetic Duality

Duality in electromagnetism means that the EM field is described by two sets of quantities, which correspond to each other in such a manner that substituting the quantities from one set with the respective quantities from the other set in any given equation produces a valid equation (the dual of the given one). We deduce the dual sets by a simple comparison of the equations describing two dual fields: the field of electric sources and the field of magnetic sources. These equations are summarized in Table A.1. It must be emphasized that duality is valid regardless of whether sources are present in the region of interest or not. Table A.2 summarizes the duality correspondences of the EM quantities.

Table A.2 Dual Electromagnetic Quantities

given	$\mathbf{E}$	$\mathbf{H}$	$\mathbf{J}$	$\mathbf{M}$	ρ	m	ε	μ
dual	$\mathbf{H}$	$-\mathbf{E}$	$\mathbf{M}$	$-\mathbf{J}$	m	$-\rho$	μ	ε

Appendix B Electromagnetic Vector Wave and Helmholtz Equations

In Appendix A, we showed that, with the constitutive relations substituted in Maxwell's curl equations, we obtain a system of two first-order partial differential equations in two vector unknowns. For example, we stated Eq. (A.34) and Eq. (A.35) as the frequency-domain model for the case of a linear isotropic medium. Further, such a system can be reduced to a single second-order partial differential equation either for the electric or for the magnetic field vector. These are referred to as the vector wave equations in time-domain analysis, while in the frequency domain they are referred to as the vector Helmholtz equations.

B.1 Vector Wave Equations in a Uniform Medium

Here, we consider the case of a uniform, isotropic, and dispersion-free medium, which allows for simple closed-form formulations and serves well as an illustration of the impact of the various constitutive parameters.

The vector wave equation for the electric field $\mathcal{E}$ is obtained by taking the following steps. First, substitute the constitutive relation Eq. (A.9) into Eq. (A.1) to express the magnetic induction $\mathcal{B}$ in terms of $\mathcal{H}$. Then, take the curl of both sides of the resulting equation and make use of Eq. (A.3) to substitute $\nabla \times \mathcal{H}$ with the sum of all current densities: the displacement term $\frac{\partial \mathcal{D}}{\partial t}$, the conduction term $\sigma_{\rm e}\mathcal{E}$, and the source term $\mathcal{J}_{\rm i}$. In the final step, make use of the constitutive relation $\mathcal{D} = \epsilon\mathcal{E}$ and the assumption that the constitutive parameters are constant in space (homogeneity) and time (no dispersion) to rearrange the terms and obtain

$$\nabla \times \nabla \times \mathcal{E} + \mu\epsilon \frac{\partial^2 \mathcal{E}}{\partial t^2} + (\mu\sigma_{\rm e} + \epsilon\sigma_{\rm m})\frac{\partial \mathcal{E}}{\partial t} + \sigma_{\rm e}\sigma_{\rm m}\mathcal{E} = -\left(\mu\frac{\partial \mathcal{J}_{\rm i}}{\partial t} + \sigma_{\rm m}\mathcal{J}_{\rm i}\right) - \nabla \times \mathcal{M}_{\rm i}. \quad \text{(B.1)}$$

Note that the third and fourth terms in the left side of Eq. (B.1) are responsible for the wave damping in space and time.

Since microwave (MW) imaging rarely involves magnetic materials, it is often the case that magnetic loss is not a factor (i.e., $\sigma_{\rm m} = 0$), and the fourth term is identically zero.

The vector wave equation Eq. (B.1) leads to a system of three *coupled* scalar equations for the three components of the vector $\mathcal{E}$ regardless of what type of coordinate system is used. This is due to the spatial differential operator $\nabla \times \nabla \times$. If, on the other hand, we could reduce that operator to a Laplacian ∇^2, then we could exploit the fact that the Laplacian decouples the vector components in a Cartesian coordinate system and would reduce Eq. (B.1) to a system of three decoupled scalar equations for the x, y, and z components of $\mathcal{E}$.

This indeed can be done by using the identity $\nabla \times \nabla \times = \nabla\nabla \cdot -\nabla^2$. To find $\nabla \cdot \mathcal{E}$, we take the divergence of both sides of Eq. (A.11) and apply the conservation of charge, $\nabla \cdot \mathcal{J}_\mathrm{i} = -\frac{\partial \rho_\mathrm{i}}{\partial t}$, to obtain

$$\left(\epsilon \frac{\partial}{\partial t} + \sigma_\mathrm{e}\right) \nabla \cdot \mathcal{E} = \frac{\partial \rho_\mathrm{i}}{\partial t}. \tag{B.2}$$

Eq. (B.2) shows an explicit relation between $\nabla \cdot \mathcal{E}$ and the electric charge density ρ_i of the source. We can rewrite it formally as

$$\nabla \cdot \mathcal{E} = \left(\epsilon \frac{\partial}{\partial t} + \sigma_\mathrm{e}\right)^{-1} \frac{\partial \rho_\mathrm{i}}{\partial t}. \tag{B.3}$$

Substituting Eq. (B.3) in Eq. (B.1) results in

$$\begin{aligned}\nabla^2 \mathcal{E} - \epsilon\mu \frac{\partial^2 \mathcal{E}}{\partial t^2} - (\mu\sigma_\mathrm{e} + \epsilon\sigma_\mathrm{m}) \frac{\partial \mathcal{E}}{\partial t} - \sigma_\mathrm{e}\sigma_\mathrm{m}\mathcal{E} \\ = \left(\mu \frac{\partial \mathcal{J}_\mathrm{i}}{\partial t} + \sigma_\mathrm{m}\mathcal{J}_\mathrm{i}\right) + \nabla \times \mathcal{M}_\mathrm{i} + \left(\epsilon \frac{\partial}{\partial t} + \sigma_\mathrm{e}\right)^{-1} \nabla \left(\frac{\partial \rho_\mathrm{i}}{\partial t}\right).\end{aligned} \tag{B.4}$$

In numerical modeling, solving Eq. (B.2) or, equivalently, Eq. (B.3) for $\nabla \cdot \mathcal{E}$ is straightforward. However, the use of the inverse operator $\left(\sigma_\mathrm{e} + \epsilon \frac{\partial}{\partial t}\right)^{-1}$ in the right side of Eq. (B.4) obscures the physical meaning of this source term. This is why, Eq. (B.4) is often written in terms of the *total* electric charge density $\rho_\mathrm{e} = \rho_\mathrm{i} + \rho_\mathrm{c}$, where ρ_c is the charge density associated with the time-varying conduction current, $\frac{\partial \rho_\mathrm{c}}{\partial t} = -\nabla \cdot \mathcal{J}_\mathrm{c} = -\sigma_\mathrm{e}\nabla \cdot \mathcal{E}$. ρ_c is the loss term of the charge density just like $\mathcal{J}_\mathrm{c}$ is the loss term of the current density. Substituting the latter relation into Eq. (B.2) and integrating in time, we arrive at the equation of charge relaxation,

$$\frac{\partial \rho_\mathrm{c}}{\partial t} + \frac{\sigma_\mathrm{e}}{\epsilon}\rho_\mathrm{c} = -\frac{\sigma_\mathrm{e}}{\epsilon}\rho_\mathrm{i}, \tag{B.5}$$

which relates the conduction charge to the impressed one. Now $\nabla \cdot \mathcal{E}$ is expressed as

$$\nabla \cdot \mathcal{E} = \frac{\rho_\mathrm{e}}{\epsilon}, \tag{B.6}$$

and Eq. (B.1) is written as

$$\begin{aligned}\nabla^2 \mathcal{E} - \mu\epsilon \frac{\partial^2 \mathcal{E}}{\partial t^2} - (\mu\sigma_\mathrm{e} + \epsilon\sigma_\mathrm{m}) \frac{\partial \mathcal{E}}{\partial t} - \sigma_\mathrm{e}\sigma_\mathrm{m}\mathcal{E} \\ = \left(\mu \frac{\partial \mathcal{J}_\mathrm{i}}{\partial t} + \sigma_\mathrm{m}\mathcal{J}_\mathrm{i}\right) + \nabla \times \mathcal{M}_\mathrm{i} + \frac{1}{\epsilon}\nabla\rho_\mathrm{e}.\end{aligned} \tag{B.7}$$

Note that the right side contains the total charge ρ_e, which is an implicit function of $\mathcal{E}$ through ρ_c. That is why Eq. (B.7) is solved simultaneously with Eq. (B.5). This is equivalent to solving Eq. (B.4) together with Eq. (B.3). We also observe that in both Eq. (B.4) and Eq. (B.7), the right sides vanish if sources are not present.

Following an analogous procedure, we obtain the vector wave equation for the magnetic field in a uniform dispersion-free medium as

$$\nabla \times \nabla \times \mathcal{H} + \mu\epsilon \frac{\partial^2 \mathcal{H}}{\partial t^2} + (\mu\sigma_e + \epsilon\sigma_m)\frac{\partial \mathcal{H}}{\partial t} + \sigma_e\sigma_m \mathcal{H} = -\left(\epsilon \frac{\partial \mathcal{M}_i}{\partial t} + \sigma_e \mathcal{M}_i\right) + \nabla \times \mathcal{J}_i. \quad \text{(B.8)}$$

As in the case of the electric field, the continuity law for the magnetic source, $\nabla \cdot \mathcal{M}_i = -\frac{\partial m_i}{\partial t}$, is used in conjunction with the equation of charge relaxation to obtain $\nabla \cdot \mathcal{H}$ as

$$\nabla \cdot \mathcal{H} = \frac{m}{\mu} \quad \text{(B.9)}$$

where $m = m_i + m_c$ is the total (fictitious) magnetic charge density and m_c is the magnetic charge loss term. Now, Eq. (B.8) can be stated using the Laplacian operator as

$$\nabla^2 \mathcal{H} - \mu\epsilon \frac{\partial^2 \mathcal{H}}{\partial t^2} - (\mu\sigma_e + \epsilon\sigma_m)\frac{\partial \mathcal{H}}{\partial t} - \sigma_e\sigma_m \mathcal{H} = \left(\epsilon \frac{\partial \mathcal{M}_i}{\partial t} + \sigma_e \mathcal{M}_i\right) - \nabla \times \mathcal{J}_i + \frac{1}{\mu}\nabla m. \quad \text{(B.10)}$$

Comparing the wave equations governing the electric and magnetic field components, we observe that their manner of propagation is identical. Mathematically, this is stated by the identical second-order linear operators in the left-hand sides of Eq. (B.4) and Eq. (B.10). Physically, this is an expected result since the electric and magnetic field vectors are two manifestations of the same phenomenon. This is why the wave-equation model is applied either to the electric field or to the magnetic field, but (usually) not to both. If the electric field is known, then the magnetic field can be found from it and vice versa. Finally, we reiterate that the solution of the wave equation (Eq. (B.4) or Eq. (B.10) is simplest in a Cartesian coordinate system where it reduces to three decoupled scalar-wave equations as shown in Eq. (1.1) in Chapter 1.

The time-domain model presented here is applicable in the case of a uniform dispersion-free medium and is often used in representing the incident field.

B.2 Vector Helmholtz Equations in a Uniform Medium

Vector Helmholtz equations in a uniform medium are derived from the frequency-domain Maxwell equations Eq. (A.22) and Eq. (A.23) through a procedure analogous to the one used to obtain the vector wave equations. The frequency-domain analog of

Eq. (B.1) is obtained as

$$\nabla \times \nabla \times \mathbf{E} - \omega^2 \mu \varepsilon \mathbf{E}(\mathbf{r}) = -\mathrm{i}\omega\mu \mathbf{J}_{\mathrm{i}} - \nabla \times \mathbf{M}_{\mathrm{i}}, \tag{B.11}$$

where ε and μ are the *complex* permittivity and permeability, respectively. The vector wave equation does not require the medium to be dispersion free (see Section B.1) since the frequency-domain model by definition applies to harmonic waves only where ε and μ are specified at each frequency.

To be able to solve this vector equation one component at a time in rectangular coordinates, the double-curl operator needs to be reduced to the Laplace operator. We use Maxwell's divergence equations, Eq. (A.36) and Eq. (A.37), and the vector Helmholtz equation for the electric field vector is obtained as

$$\nabla^2 \mathbf{E} + \omega^2 \mu \varepsilon \mathbf{E} = \mathrm{i}\omega\mu \mathbf{J}_{\mathrm{i}} + \frac{1}{\varepsilon}\nabla \rho_{\mathrm{i}} + \nabla \times \mathbf{M}_{\mathrm{i}}, \tag{B.12}$$

whereas that for the magnetic field is

$$\nabla^2 \mathbf{H} + \omega^2 \mu \varepsilon \mathbf{H} = \mathrm{i}\omega\varepsilon \mathbf{M}_{\mathrm{i}} + \frac{1}{\mu}\nabla m_{\mathrm{i}} - \nabla \times \mathbf{J}_{\mathrm{i}}. \tag{B.13}$$

The impressed current and charge phasors must satisfy the continuity laws, e.g., $\nabla \cdot \mathbf{J}_{\mathrm{i}} = -\mathrm{i}\omega\rho_{\mathrm{i}}$.

In a Cartesian coordinate system, Eq. (B.12) and Eq. (B.13) can each be conveniently reduced to a set of three decoupled scalar Helmholtz equations for the x, y, and z components of the field vectors:

$$\nabla^2 F_{\xi} + \omega^2 \mu \varepsilon F_{\xi} = \mathcal{S}_{\xi}, \ \xi \equiv x, y, z, \tag{B.14}$$

where $F \equiv E, H$ denotes the field, the subscript ξ indicates the vector component, and $\mathcal{S}$ denotes the respective source. For example, if $F \equiv E$, then $\mathcal{S}_{\xi} = \mathrm{i}\omega\mu J_{\mathrm{i},\xi} + \frac{1}{\varepsilon}\frac{\partial \rho_{\mathrm{i}}}{\partial \xi}$ in the absence of magnetic sources.

Comparing Eq. (1.22) and Eq. (B.14), we see that the relationship between the propagation constant γ and the complex permittivity and permeability is

$$\gamma^2 = -\omega^2 \mu \varepsilon. \tag{B.15}$$

Explicitly, the relation to the real and imaginary parts of the complex permittivity $\varepsilon = \varepsilon' - \mathrm{i}\varepsilon''$ and permeability $\mu = \mu' - \mathrm{i}\mu''$ (see Appendix A) is

$$\gamma^2 = -\omega^2 \left[\mu'\varepsilon' - \mu''\varepsilon'' - \mathrm{i}\left(\varepsilon'\mu'' + \varepsilon''\mu'\right)\right]. \tag{B.16}$$

If one prefers to express the losses in terms of the effective conductivities σ_{e} and σ_{m} instead of ε'' and μ'', then Eq. (B.16) can be written as

$$\gamma^2 = -\omega^2 \mu'\varepsilon' + \sigma_{\mathrm{e}}\sigma_{\mathrm{m}} + \mathrm{i}\omega(\varepsilon'\sigma_{\mathrm{m}} + \mu'\sigma_{\mathrm{e}}), \tag{B.17}$$

which is identical with Eq. (1.23) in Section 1.3.

B.3 Vector Wave and Helmholtz Equations in a Nonuniform Medium

The uniform medium assumption is often adequate when modeling the incident field, i.e., the field that exists in the scatterer-free background. However, the total-field models must account for the scattering due to differences in the permittivity and/or the permeability of the inspected object under test (OUT). Thus the case of nonuniform regions is important in imaging as well. In inverse scattering, the OUT does not contain *impressed* sources. This is why we focus here on nonuniform isotropic source-free regions. Using a procedure analogous to the one in the case of a uniform medium, the vector Helmholtz equations for the **E** and **H** vectors are obtained as

$$\mu\nabla \times \frac{1}{\mu}\nabla \times \mathbf{E} - \omega^2\mu\varepsilon\mathbf{E} = 0, \tag{B.18}$$

$$\varepsilon\nabla \times \frac{1}{\varepsilon}\nabla \times \mathbf{H} - \omega^2\mu\varepsilon\mathbf{H} = 0. \tag{B.19}$$

We notice that in a nonmagnetic medium—a common case in MW imaging—the **E** field equation Eq. (B.18) reduces to

$$\nabla \times \nabla \times \mathbf{E} - \omega^2\mu\varepsilon\mathbf{E} = 0, \tag{B.20}$$

which is identical with the source-free form of Eq. (B.11), except that here μ and ε can depend on position.

For completeness, we also state that an analogous result is obtained for the time-domain vector wave equations of $\boldsymbol{\mathcal{E}}$ and $\boldsymbol{\mathcal{H}}$ in a nonuniform medium. The $\boldsymbol{\mathcal{E}}$ equation is

$$\mu\nabla \times \frac{1}{\mu}\nabla \times \boldsymbol{\mathcal{E}} + \mu\epsilon\frac{\partial^2\boldsymbol{\mathcal{E}}}{\partial t^2} + (\mu\sigma_{\mathrm{e}} + \epsilon\sigma_{\mathrm{m}})\frac{\partial\boldsymbol{\mathcal{E}}}{\partial t} + \sigma_{\mathrm{e}}\sigma_{\mathrm{m}}\boldsymbol{\mathcal{E}} = 0, \tag{B.21}$$

whereas the $\boldsymbol{\mathcal{H}}$ equation is the dual to Eq. (B.21):

$$\epsilon\nabla \times \frac{1}{\epsilon}\nabla \times \boldsymbol{\mathcal{H}} + \mu\epsilon\frac{\partial^2\boldsymbol{\mathcal{H}}}{\partial t^2} + (\mu\sigma_{\mathrm{e}} + \epsilon\sigma_{\mathrm{m}})\frac{\partial\boldsymbol{\mathcal{H}}}{\partial t} + \sigma_{\mathrm{e}}\sigma_{\mathrm{m}}\boldsymbol{\mathcal{H}} = 0. \tag{B.22}$$

Appendix C Scalarized Electromagnetic Models

The representation of the vectorial electromagnetic (EM) field in terms of scalar functions is a fascinating subject that, on one hand, provides a different view of the nature of the field and, on the other, may lead to computationally advantageous forward models. Such scalar functions are generally referred to as *wave potentials* or *wave functions*, and they are inherently related to the well-known magnetic vector potential **A**, electric scalar potential Φ, electric vector potential **F**, and magnetic scalar potential Ψ. Introductory-level treatment of this subject can be found in [1, 5, 265] where the simple case of uniform isotropic linear medium is explained. Numerous advanced applications have been proposed to treat heterogeneous isotropic media [266, 267, 268], heterogeneous bianisotropic media [269–272], anisotropic uniaxial media [273], gyrotropic media [274], and the scalarization of EM sources [275].

The objective of a scalar EM model is to reduce the classic vectorial model, which requires at least six scalar functions of space-time (three components for each of the electric and magnetic field vectors), to two and sometimes even one scalar-wave function. The concept dates back to 1904, when E. T. Whittaker proved that "only two functions are actually necessary (in place of four)"—he meant the 4-vector potential (**A**, Φ)—to describe the field associated with any configuration of moving or static charges [276].

Here, we focus on the wave-potential model in the simplest case of a uniform isotropic linear medium because this case is relevant in the modeling of the incident field in microwave imaging. In Appendix B, the vector wave (or Helmholtz) equations have been stated for this kind of medium in Cartesian coordinates as three decoupled scalar equations. Effectively, this is also a scalarization strategy. However, the advantages of the wave-potential model are that: (i) it can be applied in other (orthogonal) coordinate systems and (ii) the field is described by two (instead of three) scalar functions.

C.1 Vector Potentials of Constant Polarization and Field Scalarization

As shown in [267, 268], in the case of a linear dispersion-free isotropic (but heterogeneous) medium, the EM field can be analyzed in terms of the magnetic and electric vector potentials (instead of the electric and magnetic field vectors), with the resulting model being a system of two coupled wave equations:

$$\left(\nabla^2 - \mathcal{T}^2_{\mu\epsilon}\right)\boldsymbol{\mathcal{A}}_\mu + (\nabla\mathcal{T}_\epsilon)\,\varphi + (\nabla\mathcal{T}_\epsilon)\times\boldsymbol{\mathcal{F}}_\epsilon = -\boldsymbol{\mathcal{J}}_{\mathrm{i}}, \tag{C.1}$$

$$\left(\nabla^2 - \mathcal{T}^2_{\mu\epsilon}\right)\boldsymbol{\mathcal{F}}_\epsilon + \left(\nabla\mathcal{T}_\mu\right)\psi - \left(\nabla\mathcal{T}_\mu\right)\times\boldsymbol{\mathcal{A}}_\mu = -\boldsymbol{\mathcal{M}}_{\mathrm{i}}. \tag{C.2}$$

Here, $\boldsymbol{\mathcal{A}}_\mu$ (measured in *amperes*, A) and $\boldsymbol{\mathcal{F}}_\epsilon$ (measured in *volts*, V) are modified vector potential functions related to the conventional vector potentials $\boldsymbol{\mathcal{A}}$ (measured in *webers per meter*, Wb·m^{-1}) and $\boldsymbol{\mathcal{F}}$ (measured in *coulombs per meter*, C·m^{-1}) as

$$\boldsymbol{\mathcal{A}}_\mu = \boldsymbol{\mathcal{A}}/\mu \tag{C.3}$$

$$\boldsymbol{\mathcal{F}}_\epsilon = \boldsymbol{\mathcal{F}}/\epsilon. \tag{C.4}$$

$\boldsymbol{\mathcal{J}}_{\mathrm{i}}$ and $\boldsymbol{\mathcal{M}}_{\mathrm{i}}$ are the electric and magnetic current densities, respectively. The differential time-domain operators $\mathcal{T}_\epsilon$, $\mathcal{T}_\mu$ and $\mathcal{T}^2_{\mu\epsilon}$ are

$$\mathcal{T}_\epsilon = \epsilon\frac{\partial}{\partial t} + \sigma_{\mathrm{e}}, \tag{C.5}$$

$$\mathcal{T}_\mu = \mu\frac{\partial}{\partial t} + \sigma_{\mathrm{m}}, \tag{C.6}$$

and

$$\mathcal{T}^2_{\mu\epsilon} = \mathcal{T}_\mu\mathcal{T}_\epsilon = \mu\epsilon\frac{\partial^2}{\partial t^2} + (\epsilon\sigma_{\mathrm{m}} + \mu\sigma_{\mathrm{e}})\frac{\partial}{\partial t} + \sigma_{\mathrm{e}}\sigma_{\mathrm{m}}. \tag{C.7}$$

These operators are handy because they allow for easy conversion of the time-dependent equations into the frequency domain by simply replacing $\mathcal{T}_\epsilon$ with $\mathrm{i}\omega\boldsymbol{\varepsilon}$, $\mathcal{T}_\mu$ with $\mathrm{i}\omega\boldsymbol{\mu}$, and $\mathcal{T}^2_{\mu\epsilon}$ with $-\omega^2\boldsymbol{\mu}\boldsymbol{\varepsilon}$. Note that $\boldsymbol{\varepsilon}$ and $\boldsymbol{\mu}$ are the *complex* permittivity and permeability, respectively.

The operators $\nabla\mathcal{T}_\epsilon$ and $\nabla\mathcal{T}_\mu$ are defined through the gradients of the constitutive parameters as

$$\nabla\mathcal{T}_\epsilon = (\nabla\epsilon)\frac{\partial}{\partial t} + \nabla\sigma_{\mathrm{e}}, \tag{C.8}$$

$$\nabla\mathcal{T}_\mu = (\nabla\mu)\frac{\partial}{\partial t} + \nabla\sigma_{\mathrm{m}}. \tag{C.9}$$

In Eq. (C.1) and Eq. (C.2), the scalar potentials φ and ψ are related to their respective vector counterparts via the Lorenz gauge:

$$\nabla\cdot\boldsymbol{\mathcal{A}}_\mu = -\mathcal{T}_\epsilon\varphi \tag{C.10}$$

$$\nabla\cdot\boldsymbol{\mathcal{F}}_\epsilon = -\mathcal{T}_\mu\psi. \tag{C.11}$$

These are the conventional scalar potentials, i.e., φ is the electric scalar potential measured in *volts* while ψ is the magnetic scalar potential measured in *amperes*.

Using Eq. (C.1) and Eq. (C.2) and Maxwell's equations, two relationships are established between the field vectors and the modified vector potentials:

$$\mathcal{E} = -\mathcal{T}_\mu \boldsymbol{\mathcal{A}}_\mu - \nabla\varphi - \nabla \times \boldsymbol{\mathcal{F}}_\epsilon, \tag{C.12}$$

$$\mathcal{H} = -\mathcal{T}_\epsilon \boldsymbol{\mathcal{F}}_\epsilon - \nabla\psi + \nabla \times \boldsymbol{\mathcal{A}}_\mu, \tag{C.13}$$

and

$$\mathcal{T}_\epsilon \mathcal{E} = \nabla \times \left(\nabla \times \boldsymbol{\mathcal{A}}_\mu - \mathcal{T}_\epsilon \boldsymbol{\mathcal{F}}_\epsilon\right) - \mathcal{J}_\text{i}, \tag{C.14}$$

$$\mathcal{T}_\mu \mathcal{H} = \nabla \times \left(\nabla \times \boldsymbol{\mathcal{F}}_\epsilon + \mathcal{T}_\mu \boldsymbol{\mathcal{A}}_\mu\right) - \mathcal{M}_\text{i}. \tag{C.15}$$

The two relationships are equivalent. Thus, if the potential functions are known, the field vectors can always be obtained from them.

At first glance, it is not clear how the vector-potential model in Eq. (C.1) and Eq. (C.2) can be more advantageous that the system of Maxwell's equations. In fact, in general, it is not. It is more involved because it requires the solution of two coupled second-order partial differential equations for two vector functions as opposed to the Maxwell system where we have two coupled first-order equations for two vector functions. Its advantage comes only after we use this model to introduce vector potential functions of *constant* known (preferred) direction given by the unit vector $\hat{\mathbf{u}}$:

$$\boldsymbol{\mathcal{A}}_\mu = \mathcal{A}\hat{\mathbf{u}}, \;\; \boldsymbol{\mathcal{F}}_\epsilon = \mathcal{F}\hat{\mathbf{u}}. \tag{C.16}$$

Then, the EM field is described in terms of the two *scalar-wave functions* $\mathcal{A}$ and $\mathcal{F}$ and it is said to be *scalarized.*

Scalarization, however, is not always straightforward. One limitation is the presence of sources. Sources of arbitrary polarization preclude the possibility of fixing the direction of the vector potentials as in Eq. (C.16). It is possible to replace the actual sources of arbitrary polarizations with equivalent source distributions polarized only along $\hat{\mathbf{u}}$ as described in the next section. However, such source transformation is not trivial and is rarely used in the forward models of imaging. Thus, here, we will simply assume that field scalarization is practical only in source-free regions.

Scalarization is most advantageous in the case of source-free homogeneous regions or layered regions. In the former case, there is no preferred direction of $\hat{\mathbf{u}}$. In waveguides and transmission-line analyses, $\hat{\mathbf{u}}$ is usually chosen along the direction of propagation [1]. In the latter case, $\hat{\mathbf{u}}$ is chosen to be orthogonal to the layers' interfaces [267]. In both cases, the advantage is that the wave functions $\mathcal{A}$ and $\mathcal{F}$ are decoupled, i.e., they can be found from their respective equations independently. Mathematically, this means that the EM problem has been reduced to two decoupled scalar Helmholtz equations. This translates into a significant reduction in computational effort (memory and CPU time) in numerical implementations.

Let us consider the simple case of a source-free uniform medium. Eq. (C.1) and Eq. (C.2) now reduce to two decoupled equations

$$\nabla^2 \boldsymbol{\mathcal{A}}_\mu - \mathcal{T}_{\mu\epsilon}^2 \boldsymbol{\mathcal{A}}_\mu = 0, \tag{C.17}$$

$$\nabla^2 \boldsymbol{\mathcal{F}}_\epsilon - \mathcal{T}_{\mu\epsilon}^2 \boldsymbol{\mathcal{F}}_\epsilon = 0, \tag{C.18}$$

because the operators $\nabla\mathcal{T}_\epsilon$ and $\nabla\mathcal{T}_\mu$ are now zero. Further, setting the vector potentials as

$$\boldsymbol{\mathcal{A}}_\mu = \mathcal{A}\hat{\mathbf{u}},\ \ \boldsymbol{\mathcal{F}}_\epsilon = \mathcal{F}\hat{\mathbf{u}}, \tag{C.19}$$

where $\hat{\mathbf{u}}$ is constant, reduces Eq. (C.17) and Eq. (C.18) to two decoupled scalar equations:

$$\nabla^2 U - \mathcal{T}_{\mu\epsilon}^2 U = 0, U \equiv \mathcal{A},\ \mathcal{F}. \tag{C.20}$$

Further, let us apply the above theory in a rectangular coordinate system where the setting $\hat{\mathbf{u}} = \hat{\mathbf{z}}$ is used:

$$\boldsymbol{\mathcal{A}}_\mu = \mathcal{A}\hat{\mathbf{z}},\ \ \boldsymbol{\mathcal{F}}_\epsilon = \mathcal{F}\hat{\mathbf{z}}. \tag{C.21}$$

The frequency-domain wave functions are then governed (independently of each other) by the Helmholtz equation:

$$\nabla^2 U + \omega^2\mu\varepsilon U = 0, U \equiv \mathcal{A}, \mathcal{F}. \tag{C.22}$$

Once the wave functions are found, the field vectors **E** and **H** can be computed using Eq. (C.14) and Eq. (C.15) with the sources set to zero:

$$\mathrm{i}\omega\varepsilon\mathbf{E} = \nabla\times\nabla\times(\hat{\mathbf{z}}\mathcal{A}) - \mathrm{i}\omega\varepsilon\nabla\times(\hat{\mathbf{z}}\mathcal{F}), \tag{C.23}$$

$$\mathrm{i}\omega\mu\mathbf{H} = \nabla\times\nabla\times(\hat{\mathbf{z}}\mathcal{F}) + \mathrm{i}\omega\mu\nabla\times(\hat{\mathbf{z}}\mathcal{A}). \tag{C.24}$$

Let us write these relations explicitly in a component-wise form:

$$\begin{aligned} E_x &= \frac{1}{\mathrm{i}\omega\varepsilon}\frac{\partial^2\mathcal{A}}{\partial x\partial z} - \frac{\partial\mathcal{F}}{\partial y} & H_x &= \frac{1}{\mathrm{i}\omega\mu}\frac{\partial^2\mathcal{F}}{\partial x\partial z} + \frac{\partial\mathcal{A}}{\partial y} \\ E_y &= \frac{1}{\mathrm{i}\omega\varepsilon}\frac{\partial^2\mathcal{A}}{\partial y\partial z} + \frac{\partial\mathcal{F}}{\partial x} & H_y &= \frac{1}{\mathrm{i}\omega\mu}\frac{\partial^2\mathcal{F}}{\partial y\partial z} - \frac{\partial\mathcal{A}}{\partial x} \\ E_z &= -\frac{1}{\mathrm{i}\omega\varepsilon}\left(\frac{\partial^2\mathcal{A}}{\partial x^2} + \frac{\partial^2\mathcal{A}}{\partial y^2}\right) & H_z &= -\frac{1}{\mathrm{i}\omega\mu}\left(\frac{\partial^2\mathcal{F}}{\partial x^2} + \frac{\partial^2\mathcal{F}}{\partial y^2}\right). \end{aligned} \tag{C.25}$$

Bearing in mind that the wave functions satisfy Helmholtz's equation Eq. (C.22), E_z and H_z can also be expressed as

$$E_z = \frac{1}{\mathrm{i}\omega\varepsilon}\left(\frac{\partial^2\mathcal{A}}{\partial z^2} + \omega^2\mu\varepsilon\mathcal{A}\right) \qquad H_z = \frac{1}{\mathrm{i}\omega\mu}\left(\frac{\partial^2\mathcal{A}}{\partial z^2} + \omega^2\mu\varepsilon\mathcal{F}\right). \tag{C.26}$$

We observe that Eqs. (C.23) and (C.24) or, equivalently, Eqs. (C.25) and (C.26) describe the field as a superposition of two independent field terms. The field with vectors

$$\mathbf{E}_F = -\nabla\times(\hat{\mathbf{z}}\mathcal{F}), \tag{C.27}$$

$$\mathbf{H}_F = \frac{1}{\mathrm{i}\omega\mu}\nabla\times\nabla\times(\hat{\mathbf{z}}\mathcal{F}), \tag{C.28}$$

depends entirely and only on the wave function $\mathcal{F}$, and it features an electric field vector $\mathbf{E}_F$, which is perpendicular to $\hat{\mathbf{z}}$, $\mathbf{E}_F \perp \hat{\mathbf{z}}$. Thus, with rectangular components, $\mathbf{E}_F$ is described by two components only (along x and y). This is a type of field termed *transverse electric with respect to z*, or, for short, a TE$_z$ field. The magnetic field vector $\mathbf{H}_F$ may in general have three nonzero components.

Analogously, $\mathcal{A}$ determines the portion of the field,

$$\mathbf{E}_A = \frac{1}{i\omega\varepsilon}\nabla \times \nabla \times (\hat{\mathbf{z}}\mathcal{A}), \tag{C.29}$$

$$\mathbf{H}_A = \nabla \times (\hat{\mathbf{z}}\mathcal{A}), \tag{C.30}$$

which is *transverse magnetic with respect to z*, or, for short, TM$_z$, since $\mathbf{H}_A \perp \hat{\mathbf{z}}$. In this field mode, it is $\mathbf{H}_A$ that has only two components (x and y), whereas $\mathbf{E}_A$ may in general have three nonzero components.

The total field expressed in terms of the wave functions $\mathcal{A}$ and $\mathcal{F}$ is thus a superposition of a TM$_z$ and a TE$_z$ field. This is why the process of field scalarization is often referred to as *TE/TM decomposition*. A word of caution: since superposition applies only to linear media, the method of TE/TM field decomposition is limited to such media.

Note that in a uniform source-free medium, $\hat{\mathbf{u}}$ can be chosen along any of the three base unit vectors of a rectangular coordinate system ($\hat{\mathbf{x}}$, $\hat{\mathbf{y}}$, or $\hat{\mathbf{z}}$). The choice is usually dictated by convenience in implementing the boundary conditions for the wave functions that complement Helmholtz's equation Eq. (C.22). For example, a metallic planar surface of unit normal along $\pm\hat{\mathbf{z}}$ requires vanishing x and y $\mathbf{E}$-field components. This requirement is easily fulfilled by the vector potentials $\hat{\mathbf{z}}\mathcal{A}$ and $\hat{\mathbf{z}}\mathcal{F}$ when the wave-function boundary conditions are set as $\frac{\partial \mathcal{A}}{\partial z} = 0$ and $\mathcal{F} = 0$, respectively.

A planar dielectric interface in a layered medium with a unit normal along $\hat{\mathbf{z}}$ is also easily accommodated by the vector potentials $\hat{\mathbf{z}}\mathcal{A}$ and $\hat{\mathbf{z}}\mathcal{F}$. On such interfaces, the tangential $\mathbf{E}$-field components (E_x and E_y), as well as all three $\mathbf{H}$-field components, are continuous. The normal $\mathbf{E}$-field component E_z is discontinuous, obeying $\varepsilon_1 E_z^{(1)} = \varepsilon_2 E_z^{(2)}$, where ε_i $(i = 1, 2)$ is the permittivity in the i-the region and $E_z^{(i)}$ is the field normal component at the interface on the side of the i-th region [12]. From the field boundary conditions at the interface, those for $\mathcal{A}$ and $\mathcal{F}$ follow directly from Eq. (C.25) as

$$\begin{aligned} &\mathcal{A}^{(1)} = \mathcal{A}^{(2)}, \\ &\frac{1}{\varepsilon_1}\frac{\partial \mathcal{A}^{(1)}}{\partial z} = \frac{1}{\varepsilon_2}\frac{\partial \mathcal{A}^{(2)}}{\partial z}, \\ &\mathcal{F}^{(1)} = \mathcal{F}^{(2)}, \\ &\frac{\partial \mathcal{F}^{(1)}}{\partial z} = \frac{\partial \mathcal{F}^{(2)}}{\partial z}. \end{aligned} \tag{C.31}$$

It is worth noting that the wave potentials are *always* continuous across material interfaces. In contrast, the normal components of the $\mathbf{E}$ and $\mathbf{H}$ field vectors are discontinuous across dielectric and magnetic interfaces, respectively.

Reduction of Maxwell's equations to a scalar model is also possible in cylindrical coordinates if $\hat{\mathbf{u}}$ is chosen as $\hat{\mathbf{z}}$ [1]. Note that $\hat{\mathbf{u}}$ *cannot* be along $\hat{\rho}$ or $\hat{\phi}$. This is because $\hat{\rho}$ or $\hat{\phi}$ are *not* constant in space, but rather depend on the angular position of the observation point. The same limitation arises in a spherical coordinate system where all three base unit vectors depend on the angular position of the observation point. However, in Section C.4 it is shown that it is possible to perform TE/TM field decomposition in spherical coordinates by making use of modified wave functions with the preferred direction being along $\hat{\mathbf{r}}$; see also [1].

C.2 TE/TM Decomposition (Scalarization) of the Electromagnetic Sources

The impact of the impressed sources on the scalarization problem is addressed next. For simplicity, let us revisit the case of a uniform medium, where the frequency-domain Eqs. (C.1) and (C.2) reduce to

$$\nabla^2 \mathcal{A} + \omega^2 \mu\varepsilon \mathcal{A} = -\mathbf{J}_\mathrm{i}, \tag{C.32}$$

$$\nabla^2 \mathcal{F} + \omega^2 \mu\varepsilon \mathcal{F} = -\mathbf{M}_\mathrm{i}. \tag{C.33}$$

Let us consider only Eq. (C.32) since analogous arguments apply for Eq. (C.33). Clearly, we cannot reduce Eq. (C.32) to a single scalar equation with the preferred direction being $\hat{\mathbf{z}}$, i.e., set $\mathcal{A} = \hat{\mathbf{z}}A$, unless the electric density source $\mathbf{J}_\mathrm{i}$ is aligned with the preferred axis $\hat{\mathbf{z}}$ everywhere. That would be a linearly polarized radiator, which is a very limited case.

However, using the equivalent source transformations [275], $\mathbf{J}_\mathrm{i}$ can be transformed into a pair of equivalent electric and magnetic current density distributions, $\hat{\mathbf{z}}J_\mathrm{i}^\mathrm{e}$ and $\hat{\mathbf{z}}M_\mathrm{i}^\mathrm{e}$, both of which are along $\hat{\mathbf{z}}$.[1] According to [275], these are found as

$$J_\mathrm{i}^\mathrm{e} = J_{\mathrm{i},z} + \frac{\partial P}{\partial z}, \tag{C.34}$$

$$M_\mathrm{i}^\mathrm{e} = \mathrm{i}\omega\mu Q, \tag{C.35}$$

where $J_{\mathrm{i},z} = \hat{\mathbf{z}} \cdot \mathbf{J}_\mathrm{i}$ is the z-component of the original impressed source. This component does not need to be transformed because it is already aligned with $\hat{\mathbf{z}}$. P and Q are auxiliary functions computed from

$$\nabla_\perp^2 Q = -\hat{\mathbf{z}} \cdot \nabla_\tau \times \mathbf{J}_\tau, \tag{C.36}$$

$$\nabla_\perp^2 P = -\nabla_\tau \cdot \mathbf{J}_\tau. \tag{C.37}$$

Here, $\mathbf{J}_\tau = \mathbf{J}_\mathrm{i} - \hat{\mathbf{z}}J_{\mathrm{i},z}$ is the component of the original source that is transverse to $\hat{\mathbf{z}}$; therefore, it needs to be transformed. Finally, $\nabla_\tau \times$ and $\nabla_\perp^2$ are the 2D curl and Laplacian operators in the plane transverse to $\hat{\mathbf{z}}$.

The main point here is that the current density sources can be subjected to scalarization just like the field itself. As a result, the field representation in terms of two scalar-wave function is possible not only in source-free regions, which is well known [1, 5], but also at points where sources exist.

It is worth emphasizing that we do not have to associate the preferred direction $\hat{\mathbf{u}}$ with the polarization of the sources. This is obvious in the case of a source-free region where

[1] A source is said to be equivalent to another source if it produces the same field outside of its own domain. In frequency-domain analysis, the source's domain is the 3D volume it occupies. In time-domain analysis, it is the 3D volume it occupies plus its temporal extent.

the choice of $\hat{\mathbf{u}}$ is dictated mostly by the boundary conditions. For example, to analyze the modes in a hollow waveguide, $\hat{\mathbf{u}}$ is chosen along the direction of propagation. At the same time, the excitation of waveguides is done through short monopoles or loops [13], which are mathematically described by electric/magnetic current densities transverse to the direction of propagation. Harrington writes in [1], p. 130, "As a word of caution, *do not* make the mistake of thinking of **A** as due to **J** and **F** as due to **M**." Indeed, the polarization of the vector potentials can be disassociated from that of the currents. Then, the scalar-wave functions associated with the preferred direction $\hat{\mathbf{u}}$ become a purely mathematical construct, and no physical meaning is attached to them.

C.3 Implications of the Source Equivalence for the Uniqueness of the Inverse Source Problem

There is an important implication of source equivalence [275], namely, the fact that different sources can generate the same field. Thus, the inverse problem of finding the source from measurements of the field it generates is inherently nonunique. This nonuniqueness has nothing to do with insufficient information (insufficient sampling) or the noise and uncertainties of the measurement. It is rooted in Maxwell's equations.

The reader should be also aware of the existence of *nonradiating sources*—sources whose field is identically zero outside of their support. Their support is the space and time (or frequency) where and when the source distribution is not zero. Since the non-radiating sources have zero field outside of their support, it is impossible to detect them remotely. The possibility of a nonradiating component existing within a source that we are reconstructing adds a degree of nonuniqueness to the solution, which cannot be eliminated unless we have some a priori information.

As an example, consider a current-density source, which can be represented as the gradient of a scalar function in 3-D space. This is a nonradiating source [40]. This implies that the field we measure outside of a source is due to its rotational component only. So, if we expect the source to be rotational (or divergence free) we can require that the respective constraint is imposed, which would reduce the number of possible solutions.

The subject of nonradiating sources is examined in depth in [26] in the context of the inverse-source problem and the inverse-scattering problem. The mathematical types of nonradiating EM sources are systematically derived and summarized in [40]. Note that the subject of nonradiating sources has a long history. It dates back to publications by Ehrenfest (1910) [277] and Schott (1933) [278] describing motions of charge that involve acceleration but do not radiate. It is of great interest in physics, in particular to our understanding of the role of the EM field in the formation of matter. The physics literature on this subject is abundant.

C.4 Field Scalarization through Spherical-Wave Functions

Section C.1 describes how the EM field can be decomposed into a TE_u mode and a TM_u mode, each described by a wave function satisfying the scalar-wave (or Helmholtz) equation. The underlying requirement there is that the magnetic and the electric vector potentials—their values being represented by the wave functions—are of constant polarization in space given by the unit vector $\hat{\mathbf{u}}$.

When constructing forward models in spherical coordinates, however, we have an alternative choice, which may be advantageous, especially when solving unbounded problems. We can choose the polarization of the magnetic and electric vector potentials along the radial direction, $\hat{\mathbf{u}} = \hat{\mathbf{r}}$. The radial unit vector $\hat{\mathbf{r}}$ is not constant in space; however, with the proper choice of gauge (i.e., the relation between the scalar and the vector potentials), decomposition into TM_r and TE_r modes is possible. The details of this development can be found in [1, 5]. Here, a summary is presented along with the final result.

The vector potentials are chosen as

$$\boldsymbol{\mathcal{A}}_\mu = \mathcal{A}_r \hat{\mathbf{r}}, \ \boldsymbol{\mathcal{F}}_\epsilon = \mathcal{F}_r \hat{\mathbf{r}}, \tag{C.38}$$

where $\mathcal{A}_r$ and $\mathcal{F}_r$ are the spherical wave functions. Exploiting the freedom in choosing the corresponding electric (φ_r) and magnetic (ψ_r) scalar potentials, they are set as

$$-\mathcal{T}_\epsilon \varphi_r = \frac{\partial \mathcal{A}_r}{\partial r}, \tag{C.39}$$

$$-\mathcal{T}_\mu \psi_r = \frac{\partial \mathcal{F}_r}{\partial r}. \tag{C.40}$$

Note that the relations in Eqs. (C.39) and (C.40) are *not* in the form of the Lorenz gauge as in Eq. (C.10) and Eq. (C.11). However, they allow the reduction of each of the vector wave equations for $\boldsymbol{\mathcal{A}}_\mu$ and $\boldsymbol{\mathcal{F}}_\epsilon$ to a single scalar equation in $\mathcal{A}_r$ and $\mathcal{F}_r$, respectively:

$$\left(\nabla^2 - \mathcal{T}_{\mu\epsilon}^2\right) \frac{\mathcal{A}_r}{r} = 0, \tag{C.41}$$

$$\left(\nabla^2 - \mathcal{T}_{\mu\epsilon}^2\right) \frac{\mathcal{F}_r}{r} = 0. \tag{C.42}$$

The functions $\Psi_A = \frac{\mathcal{A}_r}{r}$ and $\Psi_F = \frac{\mathcal{F}_r}{r}$ are solutions to the scalar-wave equation in the time domain (or the Helmholtz equation in the frequency domain) where the ∇^2 operator can be applied in the spherical coordinate system.

Eq. (C.41) and Eq. (C.42) can be solved in *any* coordinate system, although usually the $\mathrm{TM}_r/\mathrm{TE}_r$ decomposition goes hand in hand with a solution in spherical coordinates.

Once the spherical wave functions Ψ_A and Ψ_F are found, the field can be computed by any of the two equivalent field-potential relationships; see Eqs. (C.12) and (C.13) or Eqs. (C.14) and (C.15). For example,

$$\mathcal{T}_\epsilon \boldsymbol{\mathcal{E}} = \nabla \times \nabla \times (\mathbf{r}\Psi_A) - \nabla \times (\mathbf{r}\mathcal{T}_\epsilon \Psi_F), \tag{C.43}$$

$$\mathcal{T}_\mu \boldsymbol{\mathcal{H}} = \nabla \times \nabla \times (\mathbf{r}\Psi_F) + \nabla \times (\mathbf{r}\mathcal{T}_\mu \Psi_A), \tag{C.44}$$

where $\mathbf{r} = \hat{\mathbf{r}}r$ is the position vector of the point at which the field is calculated.

Appendix D Causal, Acausal, and Adjoint Solutions to the Wave Equation

D.1 Causal, Acausal, and Adjoint Green's Functions

The relationship between causal and acausal Green's functions Eq. (1.91) is derived here. Let Green's causal function g_+ satisfy the generalized wave equation, the partial differential operator $\mathcal{L}$ of which is given as

$$\mathcal{L} = \nabla^2 - \frac{1}{\upsilon^2}\frac{\partial^2}{\partial t^2} - a\frac{\partial}{\partial t} - b. \tag{D.1}$$

This operator accommodates losses through the last two terms. We then assume that an acausal counterpart g_- exists such that it satisfies an auxialiary wave equation, the operator of which is given as

$$\mathcal{L}^\dagger = \nabla^2 - \frac{1}{\upsilon^2}\frac{\partial^2}{\partial t^2} + a\frac{\partial}{\partial t} - b. \tag{D.2}$$

Let the impulse excitations for g_+ and g_- have space-time coordinates $(\mathbf{r}_1, t_1)$ and $(\mathbf{r}_2, t_2)$, respectively. The resulting equations are then

$$\left(\nabla^2 - \frac{1}{\upsilon^2}\frac{\partial^2}{\partial t^2} - a\frac{\partial}{\partial t} - b\right) g_+(\mathbf{r}, t; \mathbf{r}_1, t_1) = \delta(\mathbf{r} - \mathbf{r}_1, t - t_1), \tag{D.3}$$

$$\left(\nabla^2 - \frac{1}{\upsilon^2}\frac{\partial^2}{\partial t^2} + a\frac{\partial}{\partial t} - b\right) g_-(\mathbf{r}, t; \mathbf{r}_2, t_2) = \delta(\mathbf{r} - \mathbf{r}_2, t - t_2). \tag{D.4}$$

Note that g_+ is by definition zero in all space for $t < t_1$ while g_- vanishes everywhere for $t > t_2$.

Multiply Eq. (D.3) by g_-, multiply Eq. (D.4) by g_+ and subtract the two resulting equations. Then perform an integration over the region of interest V and over the time period of interest $[T_{\min}, T_{\max}]$. The result is

$$\begin{aligned}
&\iiiint_{T_{\min} V}^{T_{\max}} (g_-\nabla^2 g_+ - g_+\nabla^2 g_-)dvdt - \iiiint_{T_{\min} V}^{T_{\max}} \frac{1}{\upsilon^2}\left(g_-\frac{\partial^2 g_+}{\partial t^2} - g_+\frac{\partial^2 g_-}{\partial t^2}\right)dvdt \\
&- \iiiint_{T_{\min} V}^{T_{\max}} a\left(g_-\frac{\partial g_+}{\partial t} + g_+\frac{\partial g_-}{\partial t}\right)dvdt - \iiiint_{T_{\min} V}^{T_{\max}} b(g_-g_+ - g_+g_-)dvdt \\
&= g_-(\mathbf{r}_1, t_1; \mathbf{r}_2, t_2) - g_+(\mathbf{r}_2, t_2; \mathbf{r}_1, t_1).
\end{aligned} \tag{D.5}$$

Next, apply Green's integral theorem Eq. (1.46) to the first term in Eq. (D.5). Also, apply the identities

$$g_-\frac{\partial^2 g_+}{\partial t^2} - g_+\frac{\partial^2 g_-}{\partial t^2} = \frac{\partial}{\partial t}\left(g_-\frac{\partial g_+}{\partial t} - g_+\frac{\partial g_-}{\partial t}\right)$$

and

$$g_-\frac{\partial g_+}{\partial t} + g_+\frac{\partial g_-}{\partial t} = \frac{\partial}{\partial t}\left(g_+ g_-\right),$$

to the second and the third terms in Eq. (D.5), respectively. The fourth term is zero. The resulting equation is

$$\int_{T_{\min}}^{T_{\max}}\oiint_S \left(g_-\frac{\partial g_+}{\partial n} - g_+\frac{\partial g_-}{\partial n}\right) ds dt - \iiint_V \frac{1}{v^2}\left[g_-\frac{\partial g_+}{\partial t} - g_+\frac{\partial g_-}{\partial t}\right]_{t=T_{\min}}^{t=T_{\max}} dv$$
$$- \iiint_V a\left[g_+ g_-\right]_{t=T_{\min}}^{t=T_{\max}} dv = g_-(\mathbf{r}_1, t_1; \mathbf{r}_2, t_2) - g_+(\mathbf{r}_2, t_2; \mathbf{r}_1, t_1). \quad (D.6)$$

So far, no constraints have been imposed on g_- and g_+ with regard to the boundary conditions they satisfy on the surface S. Here, we require that the two functions satisfy the same boundary conditions. This makes the first term in Eq. (D.6) vanish.

Next, we require that the time interval $(T_{\min}, T_{\max})$ contains the excitation instances t_1 and t_2. This makes the second and third terms in Eq. (D.6) vanish. This is because at $t = T_{\min}$, Green's causal function $g_+(\mathbf{r}, t; \mathbf{r}_1, t_1)$ and its time derivative vanish, while at $t = T_{\max}$ it is Green's acausal function $g_-(\mathbf{r}, t; \mathbf{r}_2, t_2)$ and its time derivative that vanish. Thus we obtain that

$$g_-(\mathbf{r}_1, t_1; \mathbf{r}_2, t_2) = g_+(\mathbf{r}_2, t_2; \mathbf{r}_1, t_1), \quad (D.7)$$

which completes the derivation.

A note on mathematical terminology is needed, The relationship in Eq. (D.7) is well known in mathematics as the one between two mutually *adjoint Green's functions* [19]. These are Green's functions of two mutually *adjoint operators*.

The definition of an adjoint operator is as follows: given a linear functional operator L, the adjoint operator $L^\dagger$ is defined as the one satisfying

$$\langle L^\dagger u, v\rangle = \langle u, Lv\rangle, \quad (D.8)$$

where $\langle\cdot,\cdot\rangle$ indicates an inner product, i.e., integration over the domain $\mathcal{D}$ of u and v. Eq. (D.8) is in essence Green's generalized theorem where the contribution of the surface integral over $\partial\mathcal{D}$ is set to zero; see Eq. (1.56).

If $g(\mathbf{x}, \mathbf{x}')$ is Green's function for L and $g^\dagger(\mathbf{x}, \mathbf{x}')$ is Green's function for $L^\dagger$ it can be shown [19] that

$$g(\mathbf{x}, \mathbf{x}') = g^\dagger(\mathbf{x}', \mathbf{x}), \quad (D.9)$$

provided the two functions satisfy the same boundary conditions. The proof follows similar lines as the ones that we used in deriving this relationship in the case of d'Alembert's operator; see Eq. (D.7). Eq. (D.9) shows a relationship between Green's original and adjoint functions that is the same as that between Green's causal and acausal functions in Eq. (D.7). Therefore, g_- is the adjoint of g_+ and vice versa (since the adjoint of an adjoint operator is the operator itself).

D.2 Relationship between Causal and Acausal Wave Solutions

The fact that for every causal Green's function there exists an acausal Green's function satisfying Eq. (D.7) implies that for every causal solution $u_+(\mathbf{r}, t)$ there exists an acausal solution $u_-(\mathbf{r}, t)$, and vice versa. This is indeed the case, and the acausal solution is such that

$$u_-(\mathbf{r}, \tau) = u_+(\mathbf{r}, t), \tag{D.10}$$

where $t \in [0, T_{\max}]$, $\tau \in [0, T_{\max}]$, and $\tau = T_{\max} - t$. In case $t \in (-\infty, +\infty)$, $\tau = -t$. The proof uses Eq. (D.7) and is provided next.

Consider the causal contribution of the source function $s(\mathbf{r}, t)$ to the interior solution [see Eq. (1.99) in Section 1.6],

$$u_+(\mathbf{r}, t) = \int_0^{T_{\max}} \iiint_V g_+(\mathbf{r}, t; \mathbf{r}', t') s(\mathbf{r}', t') dv(\mathbf{r}') dt', \tag{D.11}$$

where $t \in [0, T_{\max}]$ is the time interval of a nontrivial solution. $T_{\max}$ is sufficiently large to allow for both sources and responses to decay to negligible values everywhere in the region of interest. Therefore, the turn-off time of the source is $T_s << T_{\max}$. As is customary for the setup of a causal solution, both sources and field must be zero at negative time, i.e., $u_+(\mathbf{r}, t) = 0$ and $s(\mathbf{r}, t) = 0$ for $\forall \mathbf{r}$ and $\forall t < 0$.

The respective acausal wave is given by

$$u_-(\mathbf{r}, t) = \int_0^{T_{\max}} \iiint_V g_-(\mathbf{r}, t; \mathbf{r}', t') s(\mathbf{r}', t') dv(\mathbf{r}') dt'. \tag{D.12}$$

Note that this wave is nonzero in the temporal period $(-\infty, T_s]$. Using Eq. (D.7), Eq. (D.12) is rewritten as

$$u_-(\mathbf{r}, t) = \int_0^{T_{\max}} \iiint_V g_+(\mathbf{r}', t'; \mathbf{r}, t) s(\mathbf{r}', t') dv(\mathbf{r}') dt'. \tag{D.13}$$

The reciprocity of the wave-equation Green's function further leads to

$$u_-(\mathbf{r}, -t) = \int_0^{T_{\max}} \iiint_V g_+(\mathbf{r}, -t; \mathbf{r},' -t') s(\mathbf{r}', -t') dv(\mathbf{r}') dt'. \tag{D.14}$$

Next, the time variables τ and τ' are introduced in Eq. (D.14) through $\tau = T_{\max} - t$ and $\tau' = T_{\max} - t'$, leading to

$$u_-(\mathbf{r}, \tau) = \int_0^{T_{\max}} \iiint_V g_+(\mathbf{r}, \tau; \mathbf{r},' \tau') s^\dagger(\mathbf{r}', \tau') dv(\mathbf{r}') d\tau', \tag{D.15}$$

where $s^\dagger(\mathbf{r}', \tau') = s(\mathbf{r}', T_{\max} - t')$. The comparison between Eq. (D.15) and Eq. (D.11) proves the statement made in Eq. (D.10).

In a more abstract scenario, the upper limit of the observation period may be set at infinity, i.e., $T_{\max} \to \infty$. In this case, the reverse time variables τ and τ' cannot be introduced as $T_{\max} - t$ and $T_{\max} - t'$, respectively. This problem, however, is easily resolved by first writing the causal solution as

$$u_+(\mathbf{r}, t) = \int_{-\infty}^{+\infty} \iiint_V g_+(\mathbf{r}, t; \mathbf{r}', t') s(\mathbf{r}', t') dv(\mathbf{r}') dt'. \tag{D.16}$$

Notice that this representation does not change the causal solution compared to the lower and upper limits of the time integral being set to 0 and $+\infty$, respectively, because: (i) the source is nonzero only for $t \in [0, T_s]$ and (ii) Green's causal function g_+ vanishes for $\forall t < 0$ (observation instances preceding the excitation). Following the same steps of transformation as earlier, only this time introducing the reversed time variables as $\tau = -t$ and $\tau' = -t'$, we arrive at

$$u_-(\mathbf{r}, \tau) = \int_{-\infty}^{+\infty} \iiint_V g_+(\mathbf{r}, \tau; \mathbf{r},' \tau') s^\dagger(\mathbf{r}', \tau') dv(\mathbf{r}') d\tau', \tag{D.17}$$

where $s^\dagger(\mathbf{r}', \tau') = s(\mathbf{r}', -t')$. The comparison between Eq. (D.17) and Eq. (D.16) again proves the statement made in Eq. (D.10).

Fig. D.1 illustrates a 1D causal wave propagating along $+z$, together with its acausal counterpart when the reverse time variable is defined as $\tau = T_{\max} - t$. The two waveforms are "recorded" at locations z_1, z_2, and $z_{\max}$, which are increasingly distant from the source location z_0, i.e., $z_{\max} > z_2 > z_1 > z_0$.

Notice that the acausal wave can be viewed as a wave propagating toward the past (in reversed time) due to a fictitious source function $s^\dagger(\mathbf{r}', \tau')$, which has "mirror" temporal symmetry with respect to the original source $s(\mathbf{r}', t')$, just like the "mirror" symmetry between $u_-(\mathbf{r}, \tau)$ and $u_+(\mathbf{r}, t)$. Fig. D.2 illustrates the relationship $s^\dagger(\mathbf{r}', T_{\max} - t') = s(\mathbf{r}', t')$ between the original and fictitious sources. This fictitious source is referred to as the *adjoint source* [279]. This is because it exists in the domain $\mathcal{D}^\dagger$ of d'Alembert's adjoint operator, i.e., in reversed time. For the same reason, $u_-(\mathbf{r}, \tau)$ is also referred to as the *adjoint solution*. The important point here is that if the adjoint source $s^\dagger(\mathbf{r}, \tau)$ appears in reversed time in the same form as the original source in forward time, $s^\dagger(\mathbf{r}, T_{\max} - t) = s(\mathbf{r}, t)$, then the adjoint solution appears in reverse time in the same form as the original one, $u_-(\mathbf{r}, \tau) = u_-(\mathbf{r}, T_{\max} - t) = u_+(\mathbf{r}, t)$.

To complete the illustration of the acausal adjoint wave, its spatial behaviour is illustrated in Fig. D.3 through the 1D wave already shown in Fig. D.1 and Fig. D.2. Notice

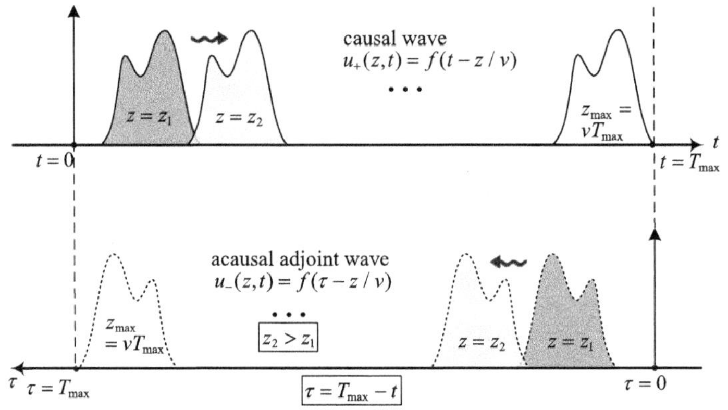

Figure D.1 Illustration of the temporal dependence of a 1D causal wave propagating along $+z$ together with its acausal (adjoint) counterpart when the reverse time variable is defined as $\tau = T_{max} - t$. The waveforms are "recorded" at z_1, z_2, and $z_{max} = vT_{max}$, where z_1 is the closest to the source location z_0 (not shown) and z_{max} is the farthest.

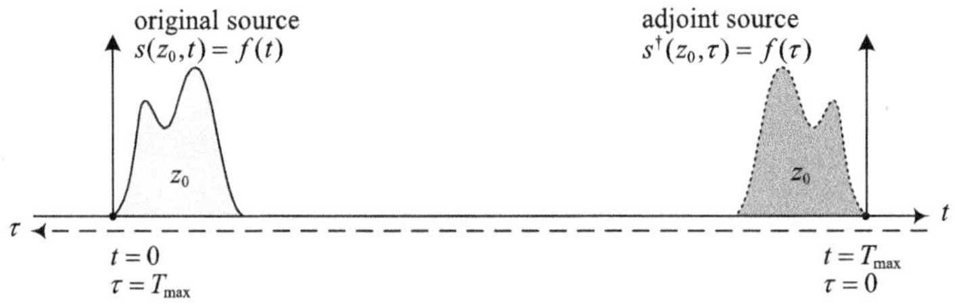

Figure D.2 Illustration of the temporal dependence of a 1D source function in forward time and its adjoint counterpart when the reverse time variable is defined as $\tau = T_{max} - t$.

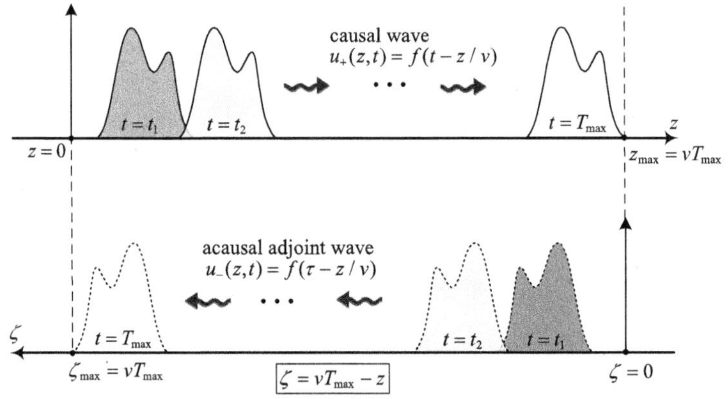

Figure D.3 Illustration of the spatial dependence of a 1D causal wave propagating along $+z$ together with its acausal (adjoint) counterpart. The wave "snapshots" are taken at t_1, t_2 ($t_2 > t_1$), and T_{max} ($T_{max} > t_2$).

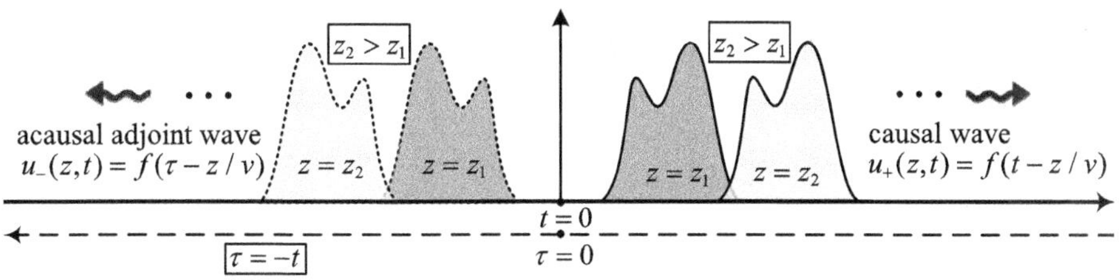

Figure D.4 Illustration of the temporal dependence of a 1D causal wave propagating along $+z$ together with its acausal (adjoint) counterpart when the reverse time variable is defined as $\tau = -t$. The waveforms are "recorded" at z_1 and z_2, where $z_2 > z_1$.

that the spatial behavior of the acausal adjoint wave with respect to the auxiliary spatial variable $\zeta = z_{\text{max}} - z$, $z_{\text{max}} = vT_{\text{max}}$, is identical to that of the causal wave with respect to z.

To illustrate the difference between an adjoint solution with $\tau = T_{\text{max}} - t$ and that with $\tau = -t$, Fig. D.4 shows again the 1D causal wave, which propagates along $+z$, this time together with the acausal adjoint counterpart where $\tau = -t$. We notice that the adjoint solution with $\tau = T_{\text{max}} - t$ shown in Fig. D.1 is simply a translated version (by T_{max}) of that in Fig. D.4. The same observation holds for the adjoint source here, $s^\dagger(\tau = -t)$, as compared to the adjoint source $s^\dagger(\tau = T_{\text{max}} - t)$ shown in Fig. D.2. The adjoint source $s^\dagger(z_0, \tau = -t)$ is illustrated in Fig. D.5 together with the respective original (causal) 1D source function $s(z_0, t)$.

In summary, *a causal wave is a disturbance that moves away from the source as time advances*. In contrast, *its acausal (or adjoint) counterpart is the same disturbance that moves toward the source as time advances*.

It is also important to remember that the acausal counterpart to a causal solution must satisfy the same boundary conditions. If the boundary conditions are not the same (in their respective domains $\mathcal{D}$ and $\mathcal{D}^\dagger$), Eq. (D.10) would not hold.

The interior boundary-value solution to the wave equation (see the term $\mathcal{I}_{\partial\mathcal{D}}$ in Eq. (1.100) in Section 1.6) allows for an acausal solution in a manner analogous to that considered earlier. The causal solution to the interior boundary-value problem is

$$u_+(\mathbf{r}, t) = \int_0^{T_{\text{max}}} \oint\!\!\!\oint_S \left(u_+(\mathbf{r}', t') \frac{\partial g_+(\mathbf{r}, t; \mathbf{r}', t')}{\partial n} - g_+(\mathbf{r}, t; \mathbf{r}', t') \frac{\partial u_+(\mathbf{r}', t')}{\partial n} \right) d\sigma(\mathbf{r}') dt'. \tag{D.18}$$

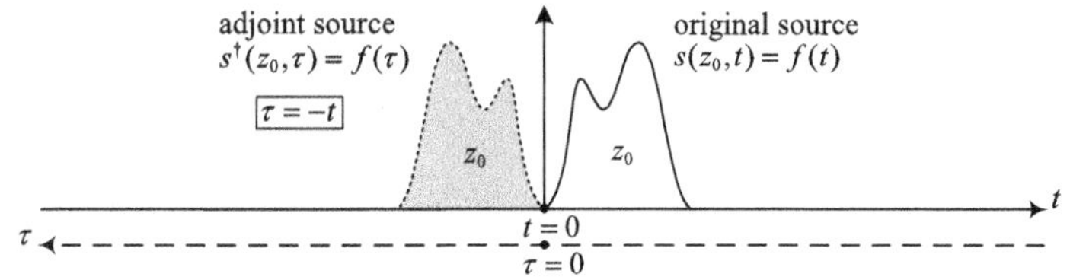

Figure D.5 Illustration of the temporal dependence of a 1D source function in forward time and its adjoint counterpart when the reverse time variable is defined as $\tau = -t$.

The only difference in the source-contribution integral in Eq. (D.11) is that here we are dealing with a surface integral instead of a volume integral. Thus all transformations regarding the time dependencies are the same as before, and they are not repeated here. The final result is that Eq. (D.10) holds for the boundary contribution as well.

Finally, the proof below shows that Eq. (D.10) holds for the Cauchy initial-value problem as well; see Section 1.6.5. In this case, the causal solution is given by

$$u_+(\mathbf{r},t) = \iiint_V \frac{1}{\upsilon^2}\left[u_+(\mathbf{r}',T_0)\frac{\partial g_+(\mathbf{r},t;\mathbf{r}',T_0)}{\partial t'} - g_+(\mathbf{r},t;\mathbf{r}',T_0)\frac{\partial u_+(\mathbf{r}',T_0)}{\partial t'}\right]dv(\mathbf{r}'),\, t \geq T_0, \tag{D.19}$$

which is simply a restatement of Eq. (1.124) in Section 1.6.5. Here, T_0 is the initial moment at which the field initial values are known throughout space. Thus, Eq. (D.19) is applicable only if $t \geq T_0$.

An acausal solution $u_-(\mathbf{r},t)$ exists only for $t \leq T_0$, and it is obtained using Green's acausal function and the field values at T_0, which are now its *terminal* values:

$$u_-(\mathbf{r},t) = \iiint_V \frac{1}{\upsilon^2}\left[u_+(\mathbf{r}',T_0)\frac{\partial g_-(\mathbf{r},t;\mathbf{r}',T_0)}{\partial t'} - g_-(\mathbf{r},t;\mathbf{r}',T_0)\frac{\partial u_+(\mathbf{r}',T_0)}{\partial t'}\right]dv(\mathbf{r}'),\, t \leq T_0\,. \tag{D.20}$$

Making use of Eq. (D.7) and Green's function reciprocity, we arrive at

$$u_-(\mathbf{r},-t) = \iiint_V \frac{1}{\upsilon^2}\left[u_+(\mathbf{r}',-T_0)\frac{\partial g_+(\mathbf{r},-t;\mathbf{r}',-T_0)}{\partial(-t')} - g_+(\mathbf{r},-t;\mathbf{r}',-T_0)\frac{\partial u_+(\mathbf{r}',-T_0)}{\partial(-t')}\right]dv(\mathbf{r}'). \tag{D.21}$$

We now translate all time dependencies by $2T_0$ (or shift the origin of the time axis to the left by $2T_0$) to obtain

$$u_-(\mathbf{r},\tau) = \iiint_V \frac{1}{\upsilon^2}\left[u_+(\mathbf{r}',T_0)\frac{\partial g_+(\mathbf{r},\tau;\mathbf{r}',T_0)}{\partial\tau'} - g_+(\mathbf{r},\tau;\mathbf{r}',T_0)\frac{\partial u_+(\mathbf{r}',T_0)}{\partial\tau'}\right]dv(\mathbf{r}'), \tag{D.22}$$

where $\tau = 2T_0 - t$. Eq. (D.22) is nonzero only for $\tau \geq T_0$ due to the causality of g_+ but it is a time-shifted copy of the solution in Eq. (D.20) .

Comparing Eq. (D.22) with Eq. (D.19) reveals that the causal initial-value solution and its adjoint acausal counterpart relate through

$$u_-(\mathbf{r},\tau) = u_+(\mathbf{r},t), \tag{D.23}$$

which is the same as Eq. (D.10), only here $\tau = 2T_0 - t$ instead of $\tau = T_{\max} - t$. If $T_0 = 0$, which is the usual setting for initial-value problems, then $\tau = -t$.

References

[1] R. F. Harrington, *Time-Harmonic Electromagnetic Fields*. McGraw-Hill, 1961.

[2] R. S. Elliott, *Electromagnetics: History, Theory and Applications*. IEEE Press, 1993.

[3] J. A. Stratton, *Electromagnetic Theory*. IEEE Press, 2007.

[4] J. V. Bladel, *Electromagnetic Fields*, 2nd ed. IEEE Press, 2007.

[5] C. A. Balanis, *Advanced Engineering Electromagnetics*, 2nd ed. John Wiley & Sons, 2012.

[6] F. Trèves, *Basic Linear Partial Differential Equations*. Academic Press, 1975.

[7] H. Bateman, *Partial Differential Equations of Mathematical Physics*. Cambridge University Press, 1932.

[8] J. Hadamard, *Lectures on Cauchy's Problem in Linear Partial Differential Equations*. Dover Publications, 2003.

[9] W. S. Weiglhofer and A. Lakhtakia, *Introduction to Complex Mediums for Optics and Electromagnetics*. SPIE Press, 2003.

[10] "MEFiSTo-3D Nova ver. 6.0 for Windows, Faustus Scientific Corporation," http://www.faustcorp.com/, 2014.

[11] K. F. Riley, M. P. Hobson, and S. J. Bence, *Mathematical Methods for Physics and Engineering*. Cambridge University Press, 1998.

[12] W. H. Hayt and J. A. Buck, *Engineering Electromagnetics*, 8th ed. McGraw-Hill, 2011.

[13] D. M. Pozar, *Microwave Engineering*, 4th ed. John Wiley & Sons, 2012.

[14] M. Born and E. Wolf, *Principles of Optics*, 7th ed. Cambridge University Press, 1999.

[15] C. A. Balanis, *Antenna Theory: Analysis and Design*, 3rd ed. John Wiley & Sons, 2005.

[16] L. D. Landau and E. M. Lifshitz, *Fluid Mechanics*, 3rd ed., ser. Course of Theoretical Physics. Pergamon Press, 1966, vol. 6.

[17] A. T. de Hoop, "A modification of Cagniard's method for solving seismic pulse problems," *Applied Scientific Research, Section B*, vol. 8, no. 1, pp. 349–356, 1960. Available: http://dx.doi.org/10.1007/BF02920068

[18] I. Stakgold and M. J. Holst, *Green's Functions and Boundary Value Problems*, 3rd ed. Wiley, 2011.

[19] S. I. Hayek, *Advanced Mathematical Methods in Science and Engineering*. Marcel Dekker, 2001.

[20] P. M. Morse and H. Feshbach, *Methods of Theoretical Physics - Part I*. McGraw-Hill, 1953.

[21] W. J. R. Hoefer, *MEFiSTo-3D Pro User Guide and Operating Manual*, 11th ed. Faustus Scientific Corporation, 2008.

[22] C. Christopoulos, *The Transmission–Line Modeling Method TLM*. Wiley – IEEE Press, 1995.

[23] P. Johns and R. Beurle, "Numerical solution of 2-dimensional scattering problems using a transmission Line matrix," *Proc. of the IEE*, vol. 118, no. 9, pp. 1203–1208, Sep. 1971.

[24] W. Hoefer, "The transmission-line matrix method - theory and applications," *IEEE Trans. Microw. Theory Tech.*, vol. 33, no. 10, pp. 882–893, Oct. 1985.
[25] C. R. Vogel, *Computational Methods for Inverse Problems*. SIAM, 2002.
[26] A. J. Devaney, *Mathematical Foundations of Imaging, Tomography, and Wavefield Inversion*. Cambridge University Press, 2012.
[27] D. M. Sullivan, *Electromagnetic Simulation Using the FDTD Method*, 2nd ed. John Wiley & Sons, 2013.
[28] J. Jin, *The Finite Element Method in Electromagnetics*, 2nd ed. John Wiley & Sons, 2002.
[29] M. N. Sadiku, *Numerical Techniques in Electromagnetics with MATLAB*, 3rd ed. CRC Press, 2009.
[30] R. F. Harrington, "Formulation of boundary integral equations by the equivalent source method," in *Boundary Element Technology VII - Proceedings of the Seventh International Conference on Boundary Element Technology (Albuquerque, June 1992)*. Springer, 1992, pp. 293–303.
[31] W. C. Chew, *Waves and Fields in Inhomogeneous Media*. IEEE Press, 1990.
[32] D. Poljak and C. A. Brebbia, *Boundary Element Methods for Electrical Engineers*. WIT Press, 2005.
[33] R. F. Harrington, *Field Computation by Moment Methods*. Wiley–IEEE Press, 1993.
[34] W. C. Chew, M. S. Tong, and B. Hu, *Integral Equation Methods for Electromagnetic and Elastic Waves*. Morgan & Claypool, 2009.
[35] F. D. Gakhov, *Boundary Value Problems*. Dover Publications, 1990 (translation from Russian by I.N. Sneddon).
[36] D. Colton and R. Kress, *Integral Equation Methods in Scattering Theory*. John Wiley & Sons, 1983.
[37] W. H. Press, S. A. Teukolsky, W. T. Vetterling, and B. P. Flannery, *Numerical Recipes in C: The Art of Scientific Computing*. Cambridge University Press, 1992.
[38] A. Devaney, "Nonuniqueness in the inverse scattering problem," *J. Mathematical Phys.*, vol. 19, no. 7, pp. 1526–1531, 2008.
[39] A. Devaney and E. Wolf, "Radiating and nonradiating classical current distributions and the fields they generate," *Phys. Rev. D*, vol. 8, no. 4, pp. 1044–1047, 1973.
[40] N. K. Nikolova and Y. S. Rickard, "Nonradiating electromagnetic sources in a nonuniform medium," *Phys. Rev. E*, vol. 71, no. 1, p. 016617, 2005.
[41] I. V. Lindell, *Methods for Electromagnetic Field Analysis*. New York, NY (United States); Oxford Univ. Press, 1992.
[42] A. Shlivinski, E. Heyman, and R. Kastner, "Antenna characterization in the time domain," *IEEE Trans. Antennas Propag.*, vol. 45, no. 7, pp. 1140–1149, 1997.
[43] E. W. Weisstein. (2014, Feb.) Delta function. http://mathworld.wolfram.com/DeltaFunction.html
[44] A. T. de Hoop, *Handbook of Radiation and Scattering of Waves*. Academic Press, 1995.
[45] A. C. Pipkin, *A Course on Integral Equations*, ser. Texts in Applied Mathematics. Springer-Verlag, 1991, no. 9.
[46] M. Pastorino, *Microwave Imaging*, ser. Wiley Series in Microwave and Optical Engineering. Wiley, 2010.
[47] F. Natterer, "An error bound for the Born approximation," *Inverse Problems*, vol. 20, no. 2, pp. 447–452, 2004.
[48] M. Slaney, A. C. Kak, and L. E. Larsen, "Limitations of imaging with first-order diffraction tomography," *IEEE Trans. Microw. Theory Tech.*, vol. 32, no. 8, pp. 860–874, 1984.

[49] T. M. Habashy, R. W. Groom, and B. R. Spies, "Beyond the Born and Rytov approximations: A nonlinear approach to electromagnetic scattering," *J. Geophysical Res.: Solid Earth (1978–2012)*, vol. 98, no. B2, pp. 1759–1775, 1993.

[50] B. Chen and J. J. Stamnes, "Validity of diffraction tomography based on the first Born and the first Rytov approximations," *Appl. Optics*, vol. 37, no. 14, pp. 2996–3006, 1998.

[51] C. K. Avinash and M. Slaney, *Principles of Computerized Tomographic Imaging*. Society for Industrial and Applied Mathematics, 2001.

[52] J. C. Kolecki, "An introduction to tensors for students of physics and engineering," *NASA Scientific and Technical Information (STI) Program Office*, vol. NASA/TM-2002-211716, September 2002. https://www.grc.nasa.gov

[53] R. Mittra, *Computer Techniques for Electromagnetics: International Series of Monographs in Electrical Engineering*. Elsevier, 2013, vol. 7.

[54] J. Mautz and R. Harrington, "E-field, H-field, and combined field solutions for conducting bodies of revolution," *Arch. Electron. Übertragungstech*, vol. 32, no. 4, pp. 159–164, April 1978.

[55] N. K. Nikolova, "Electromagnetic boundary conditions and uniqueness revisited," *IEEE Antennas Propag. Mag.*, vol. 46, no. 5, pp. 141–149, 2004.

[56] A. Love, "The integration of the equations of propagation of electric waves," *Phyl. Trans. Royal Soc. London, Ser. A*, vol. 198, pp. 1–45, 1901.

[57] G. S. Smith, *An Introduction to Classical Electromagnetic Radiation*. Cambridge University Press, 1997.

[58] P.-H. Tsao, "Derivation and implications of the symmetry property of the permittivity tensor," *Am. J. Phys.*, vol. 61, no. 9, pp. 823–825, Sep. 1993.

[59] M. Haynes and M. Moghaddam, "Vector Green's function for S-parameter measurements of the electromagnetic volume integral equation," *IEEE Trans. Antennas Propag.*, vol. 60, no. 3, pp. 1400–1413, March 2012.

[60] J. L. Volakis, Ed., *Antenna Engineering Handbook*, 4th ed. McGraw–Hill, 2007.

[61] J. D. Kraus, *Antennas*. McGraw–Hill, 1988.

[62] D. M. Sheen, D. L. McMakin, and T. E. Hall, "Three-dimensional millimeter-wave imaging for concealed weapon detection," *IEEE Trans. Microw. Theory Tech.*, vol. 49, no. 9, pp. 1581–1592, Sep. 2001.

[63] D. Sheen, D. McMakin, and T. Hall, "Near-field three-dimensional radar imaging techniques and applications," *Applied Optics*, vol. 49, no. 19, pp. E83–E93, 2010.

[64] R. K. Amineh, M. Ravan, A. Khalatpour, and N. K. Nikolova, "Three-dimensional near–field microwave holography using reflected and transmitted signals," *IEEE Trans. Antennas Propag.*, vol. 59, no. 12, pp. 4777–4789, 2011.

[65] M. Ravan, R. K. Amineh, and N. K. Nikolova, "Two-dimensional near–field microwave holography," *Inverse Problems*, vol. 26, no. 5, p. 055011, 2010.

[66] R. K. Amineh, J. McCombe, and N. K. Nikolova, "Microwave holographic imaging using the antenna phaseless radiation pattern," *IEEE Antennas Wireless Propag. Lett.*, vol. 11, pp. 1529–1532, 2012.

[67] D. G. Swanson and W. J. Hoefer, *Microwave Circuit Modeling Using Electromagnetic Field Simulation*. Artech House, 2003.

[68] K. S. Kunz and R. J. Luebbers, *The Finite Difference Time Domain Method for Electromagnetics*. CRC Press, 1993.

[69] A. Taflove and S. C. Hagness, *Computational Electrodynamics*, 3rd ed. Artech House, 2004.

[70] J.-F. Lee, R. Lee, and A. Cangellaris, "Time-domain finite-element methods," *IEEE Trans. Antennas Propag.*, vol. 45, no. 3, pp. 430–442, 1997.

[71] J.-M. Jin, *The Finite Element Method in Electromagnetics*, 3rd ed. John Wiley & Sons, 2014.

[72] D. Jiao, J.-M. Jin, E. Michielssen, and D. J. Riley, "Time-domain finite-element simulation of three-dimensional scattering and radiation problems using perfectly matched layers," *IEEE Trans. Antennas Propag.*, vol. 51, no. 2, pp. 296–305, Feb. 2003.

[73] J. Li, P. Monk, and D. Weile, "Time domain integral equation methods in computational electromagnetism," in *Computational Electromagnetism*. Springer, 2015, pp. 111–189.

[74] Z. Mei, Y. Zhang, T. K. Sarkar, B. H. Jung, A. García-Lampérez, and M. Salazar-Palma, "An improved marching-on-in-degree method using a new temporal basis," *IEEE Trans. Antennas Propag.*, vol. 59, no. 12, pp. 4643–4650, 2011.

[75] R. Mittra, "Integral equation methods for transient scattering," in *Transient Electromagnetic Fields*. Springer, 1976, pp. 73–128.

[76] L. B. Felsen, "Transient electromagnetic fields," in *Topics in Applied Physics*, vol. 10, 1976.

[77] K. Kurokawa, "Power waves and the scattering matrix," *IEEE Trans. Microw. Theory Tech.*, vol. 13, no. 2, pp. 194–202, Apr. 1965.

[78] P. Penfield, "Noise in negative–resistance amplifiers," *IRE Trans. Circuit Theory*, vol. 7, no. 2, pp. 166–170, June 1960.

[79] MATLAB, *ver. 9.0.0.341360 (R2016a)*. MathWorks Inc., 2016.

[80] A. H. Golnabi, P. M. Meaney, S. Geimer, and K. D. Paulsen, "Microwave imaging for breast cancer detection and therapy monitoring," in *Biomedical Wireless Technologies, Networks, and Sensing Systems (BioWireleSS), 2011 IEEE Topical Conference on*. IEEE, 2011, pp. 59–62.

[81] C. Gilmore, P. Mojabi, A. Zakaria, M. Ostadrahimi, C. Kaye, S. Noghanian, L. Shafai, S. Pistorius, and J. LoVetri, "A wideband microwave tomography system with a novel frequency selection procedure," *IEEE Trans. Biomedical Engineering*, vol. 57, no. 4, pp. 894–904, 2010.

[82] M. Klemm, D. Gibbins, J. Leendertz, T. Horseman, A. Preece, R. Benjamin, and I. Craddock, "Development and testing of a 60-element UWB conformal array for breast cancer imaging," in *Proc. 5th European Conf. Antennas & Propagation (EuCAP)*. IEEE, 2011, pp. 3077–3079.

[83] M. Hiebel, *Fundamentals of Vector Network Analysis*, 5th ed. Rohde&Schwarz, 2011.

[84] J. P. Dunsmore, *Handbook of Microwave Component Measurements: with Advanced VNA Techniques*. John Wiley & Sons, 2012.

[85] D. G. Cacuci, *Sensitivity and Uncertainty Analysis, Volume I: Theory*. CRC Press, 2003, vol. 1.

[86] D. G. Cacuci, M. Ionescu-Bujor, and I. M. Navon, *Sensitivity and Uncertainty Analysis, Volume II: Applications to Large Scale Systems*. CRC Press, 2005, vol. 2.

[87] A. Saltelli, S. Tarantola, F. Campolongo, and M. Ratto, *Sensitivity Analysis in Practice: A Guide to Assessing Scientific Models*. John Wiley & Sons, 2004.

[88] V. Komkov, K. K. Choi, and E. J. Haug, *Design Sensitivity Analysis of Structural Systems*. Academic Press, 1986, vol. 177.

[89] A. D. Belegundu and T. R. Chandrupatla, *Optimization Concepts and Applications in Engineering*. Cambridge University Press, 2011.

[90] K. K. Choi and N.-H. Kim, *Structural Sensitivity Analysis and Optimization 1: Linear Systems*, ser. Mechanical Engineering Series. Springer Science & Business Media, 2005.

[91] ——, *Structural Sensitivity Analysis and Optimization 2: Nonlinear Systems and Applications*, ser. Mechanical Engineering Series. Springer Science & Business Media, 2005.

[92] M. Bakr, *Nonlinear Optimization in Electrical Engineering with Applications in Matlab™*. Institution of Engineering and Technology, 2013.

[93] K. D. Paulsen, P. M. Meaney, and L. C. Gilman, *Alternative Breast Imaging: Four Model Based Approaches*. Springer Science & Business Media, 2004.

[94] Y. Song and N. K. Nikolova, "Memory-efficient method for wideband self-adjoint sensitivity analysis," *IEEE Trans. Microw. Theory Tech.*, vol. 56, no. 8, pp. 1917–1927, 2008.

[95] L. Liu, A. Trehan, and N. Nikolova, "Near-field detection at microwave frequencies based on self-adjoint response sensitivity analysis," *Inverse Problems*, vol. 26, no. 10, p. 105001, 2010.

[96] Y. Zhang, S. Tu, R. K. Amineh, and N. K. Nikolova, "Resolution and robustness to noise of the sensitivity-based method for microwave imaging with data acquired on cylindrical surfaces," *Inverse Problems*, vol. 28, no. 11, p. 115006, 2012.

[97] S. Tu, J. J. McCombe, D. S. Shumakov, and N. K. Nikolova, "Fast quantitative microwave imaging with resolvent kernel extracted from measurements," *Inverse Problems*, vol. 31, no. 4, p. 045007, 2015.

[98] V. Monaco and P. Tiberio, "On linear network scattering matrix sensitivity," *Alta Frequenza*, vol. 39, no. 2, pp. 193–195, 1970.

[99] G. Iuculano, V. Monaco, and P. Tiberio, "Network sensitivities in terms of scattering parameters," *Electronics Lett.*, vol. 2, no. 7, pp. 53–55, 1971.

[100] J. W. Bandler and R. E. Seviora, "Wave sensitivities of networks," *IEEE Trans. Microw. Theory Tech.*, vol. 20, no. 2, pp. 138–147, 1972.

[101] K. Gupta, R. Garg, and R. Chadha, *Computer-Aided Design of Microwave Circuits*. Artech House, 1981.

[102] J. W. Bandler, Q.-J. Zhang, and R. Biernacki, "A unified theory for frequency-domain simulation and sensitivity analysis of linear and nonlinear circuits," *IEEE Trans. Microw. Theory Tech.*, vol. 36, no. 12, pp. 1661–1669, 1988.

[103] N. K. Nikolova, J. W. Bandler, and M. H. Bakr, "Adjoint techniques for sensitivity analysis in high-frequency structure CAD," *IEEE Trans. Microw. Theory Tech.*, vol. 52, no. 1, pp. 403–419, 2004.

[104] H. Akel and J. Webb, "Design sensitivities for scattering-matrix calculation with tetrahedral edge elements," *IEEE Trans. Magnetics*, vol. 36, no. 4, pp. 1043–1046, 2000.

[105] Y.-S. Chung, J. Ryu, C. Cheon, I.-H. Park, and S.-Y. Hahn, "Optimal design method for microwave device using time domain method and design sensitivity analysis – part I: FETD case," *IEEE Trans. Magnetics*, vol. 37, no. 5, pp. 3289–3293, 2001.

[106] N. K. Nikolova, J. Zhu, D. Li, M. H. Bakr, and J. W. Bandler, "Sensitivity analysis of network parameters with electromagnetic frequency-domain simulators," *IEEE Trans. Microw. Theory Tech.*, vol. 54, no. 2, pp. 670–681, 2006.

[107] N. K. Nikolova, Y. Li, Y. Li, and M. H. Bakr, "Sensitivity analysis of scattering parameters with electromagnetic time-domain simulators," *IEEE Trans. Microw. Theory Tech.*, vol. 54, no. 4, pp. 1598–1610, 2006.

[108] N. K. Nikolova, X. Zhu, Y. Song, A. Hasib, and M. H. Bakr, "S-parameter sensitivities for electromagnetic optimization based on volume field solutions," *IEEE Trans. Microw. Theory Tech.*, vol. 57, no. 6, pp. 1526–1538, 2009.

[109] Q. Fang, P. M. Meaney, S. D. Geimer, A. V. Streltsov, and K. D. Paulsen, "Microwave image reconstruction from 3-D fields coupled to 2-D parameter estimation," *IEEE Trans. Medical Imaging*, vol. 23, no. 4, pp. 475–484, 2004.

[110] Q. Fang, P. M. Meaney, and K. D. Paulsen, "Singular value analysis of the Jacobian matrix in microwave image reconstruction," *IEEE Trans. Antennas Propag.*, vol. 54, no. 8, pp. 2371–2380, 2006.

[111] A. Fhager and M. Persson, "Using a priori data to improve the reconstruction of small objects in microwave tomography," *IEEE Trans. Microw. Theory Tech.*, vol. 55, no. 11, pp. 2454–2462, 2007.

[112] A. Zakaria, C. Gilmore, and J. LoVetri, "Finite-element contrast source inversion method for microwave imaging," *Inverse Problems*, vol. 26, no. 11, p. 115010, 2010.

[113] M. S. Dadash, N. K. Nikolova, and J. W. Bandler, "Analytical adjoint sensitivity formula for the scattering parameters of metallic structures," *IEEE Trans. Microw. Theory Tech.*, vol. 60, no. 9, pp. 2713–2722, 2012.

[114] M. S. Dadash and N. K. Nikolova, "Analytical S-parameter sensitivity formula for the shape parameters of dielectric objects," *IEEE Microw. Wireless Comp. Lett.*, vol. 24, no. 5, pp. 291–293, 2014.

[115] A. S. Beaverstone, D. S. Shumakov, and N. K. Nikolova, "Integral equations of scattering for scalar frequency–domain responses," *IEEE Trans. Microw. Theory Tech.*, vol. 64, no. 4, pp. 1120–1132, 2017.

[116] R. K. Amineh, M. Ravan, A. Khalatpour, and N. K. Nikolova, "Three-dimensional near-field microwave holography using reflected and transmitted signals," *IEEE Trans. Antennas Propag.*, vol. 59, no. 12, pp. 4777–4789, 2011.

[117] M. Ostadrahimi, P. Mojabi, C. Gilmore, A. Zakaria, S. Noghanian, S. Pistorius, and J. LoVetri, "Analysis of incident field modeling and incident/scattered field calibration techniques in microwave tomography," *IEEE Antennas Wireless Propag. Lett.*, vol. 10, pp. 900–903, 2011.

[118] R. K. Amineh, J. J. McCombe, A. Khalatpour, and N. K. Nikolova, "Microwave holography using point–spread functions measured with calibration objects," *IEEE Trans. Instrumentation Measurement*, vol. 64, no. 2, pp. 403–417, 2015.

[119] C. J. Bouwkamp, "Diffraction theory," *Reports on Progress in Physics*, vol. 17, no. 1, pp. 35–100, 1954.

[120] P. C. Clemmow, *The Plane Wave Spectrum Representation of Electromagnetic Fields (Reissue 1996 with Additions)*. Oxford University Press in association with IEEE Press, 1996.

[121] L. Novotny. (2013, Feb.) Lecture notes on electromagnetic fields and waves. ETH Zürich, Photonics Laboratory. https://www.photonics.ethz.ch/fileadmin/user_upload/Courses/EM_FieldsAndWaves/CompleteNotes.pdf

[122] FEKO, *Suite 6.2*. 32 Techno Avenue, Technopark, Stellenbosch, 7600, South Africa: EM Software & Systems-S.A. (Pty) Ltd, 2012.

[123] G. H. Golub and C. F. Van Loan, *Matrix Computations*, 4th ed. Johns Hopkins University Press, 2013.

[124] H. Eyal and R. Daisy, "Three dimensional real time ultra-wide band whole body imaging system," in *2012 IEEE Radar Conference*. IEEE, 2012, pp. 0022–0027.

[125] R. K. Amineh, A. Trehan, and N. K. Nikolova, "Tem horn antenna for ultra-wide band microwave breast imaging," *Progress in Electromagnetics Research B*, vol. 13, pp. 59–74, 2009.

[126] J. Sachs, *Handbook of Ultra-Wideband Short-Range Sensing*. Wiley–VCH Verlag & Co. KGaA, 2012.

[127] J. D. Taylor, *Ultrawideband Radar: Applications and Design*. CRC Press, 2016.

[128] D. J. Daniels, *Ground Penetrating Radar*. IET, 2004, vol. 1.

[129] H. M. Jol, *Ground Penetrating Radar Theory and Applications*. Elsevier, 2008.

[130] R. Persico, *Introduction to Ground Penetrating Radar*. IEEE Press, 2014.
[131] M. G. Amin, Ed., *Through-the-Wall Radar Imaging*. CRC Press, 2016.
[132] Ø. Aardal and J. Hammerstad. (2010, Apr.) "Medical radar literature overview." http://www.ffi.no/no/Rapporter/10-00958.pdf
[133] X. Li, E. J. Bond, B. D. Van Veen, and S. C. Hagness, "An overview of ultra-wideband microwave imaging via space-time beamforming for early-stage breast-cancer detection," *IEEE Antennas Propag. Mag.*, vol. 47, no. 1, pp. 19–34, 2005.
[134] N. Nikolova, "Microwave imaging for breast cancer," *IEEE Microw. Mag.*, vol. 7, no. 12, pp. 78–94, 2011.
[135] N. K. Nikolova, "Microwave biomedical imaging," *Wiley Encyclopedia of Electrical and Electronics Engineering*, 2014.
[136] M. Klemm, I. Craddock, J. Leendertz, A. Preece, D. Gibbins, M. Shere, and R. Benjamin, "Clinical trials of a UWB imaging radar for breast cancer," in *Proc. of the Fourth European Conf. on Antennas and Propagation*, 2010.
[137] E. Fear, J. Bourqui, C. Curtis, D. Mew, B. Docktor, and C. Romano, "Microwave breast imaging with a monostatic radar-based system: A study of application to patients," *IEEE Trans. Microw. Theory Tech.*, vol. 61, no. 5, pp. 2119–2128, 2013.
[138] E. Porter, E. Kirshin, A. Santorelli, M. Coates, and M. Popovi, "Time-domain multistatic radar system for microwave breast screening," *IEEE Antennas Wireless Propag. Lett.*, vol. 12, pp. 229–232, 2013.
[139] F. Sabath, D. V. Giri, F. Rachidi-Haeri, and A. Kaelin, Eds., *Ultra-Wideband, Short Pulse Electromagnetics 9*. Springer, 2010.
[140] T. Savelyev and A. Yarovoy, "Fast imaging by 3D deconvolution in short-range UWB radar for concealed weapon detection," in *9th European Radar Conference (EuRAD 2012)*. IEEE, 2012, pp. 26–29.
[141] "openEMS, Laboratory for General and Theoretical Electrical Engineering (ATE), University of Duisburg-Essen," http://openems.de/start/index.php.
[142] M. Fink, "Time reversal of ultrasonic fields - Part I: basic principles," *IEEE Trans. Ultrasonics, Ferroelectrics, and Frequency Control*, vol. 39, no. 5, pp. 555–566, 1992.
[143] F. Wu, J.-L. Thomas, and M. Fink, "Time reversal of ultrasonic fields - Part II: experimental results," *IEEE Trans. Ultrasonics, Ferroelectrics, and Frequency Control*, vol. 39, no. 5, pp. 567–578, 1992.
[144] D. Cassereau and M. Fink, "Time-reversal of ultrasonic fields - Part III: theory of the closed time-reversal cavity," *IEEE Trans. Ultrasonics, Ferroelectrics, and Frequency Control*, vol. 39, no. 5, pp. 579–592, 1992.
[145] M. Fink, "Time-reversal mirrors," *J. Phys. D: Applied Physics*, vol. 26, no. 9, pp. 1333–1350, 1993.
[146] D. Liu, G. Kang, L. Li, Y. Chen, S. Vasudevan, W. Joines, Q. H. Liu, J. Krolik, and L. Carin, "Electromagnetic time-reversal imaging of a target in a cluttered environment," *IEEE Trans. Antennas Propag.*, vol. 53, no. 9, pp. 3058–3066, 2005.
[147] D. Liu, S. Vasudevan, J. Krolik, G. Bal, and L. Carin, "Electromagnetic time-reversal source localization in changing media: Experiment and analysis," *IEEE Trans. Antennas Propag.*, vol. 55, no. 2, pp. 344–354, 2007.
[148] G. Lerosey, J. De Rosny, A. Tourin, A. Derode, and M. Fink, "Time reversal of wideband microwaves," *Applied Phys. Lett.*, vol. 88, no. 15, pp. 154 101–154 101, 2006.
[149] G. Lerosey, J. De Rosny, A. Tourin, and M. Fink, "Focusing beyond the diffraction limit with far-field time reversal," *Science*, vol. 315, no. 5815, pp. 1120–1122, 2007.

[150] J. M. Moura and Y. Jin, "Detection by time reversal: Single antenna," *IEEE Trans. Signal Processing*, vol. 55, no. 1, pp. 187–201, 2007.

[151] L. Bellomo, S. Pioch, M. Saillard, and E. Spano, "Time reversal experiments in the microwave range: Description of the radar and results," *Progress in Electromagnetics Research*, vol. 104, pp. 427–448, 2010.

[152] C. Prada and M. Fink, "Eigenmodes of the time reversal operator: A solution to selective focusing in multiple–target media," *Wave Motion*, vol. 20, no. 2, pp. 151–163, 1994.

[153] ——, "Selective focusing through inhomogeneous media: The DORT method," in *Proc. IEEE Ultrasonics Symposium 1995*, vol. 2. IEEE, 1995, pp. 1449–1453.

[154] H. Lev-Ari and A. J. Devaney, "The time-reversal technique re-interpreted: Subspace-based signal processing for multi-static target location," in *Proc. 2000 IEEE Sensor Array and Multichannel Signal Processing Workshop*. IEEE, 2000, pp. 509–513.

[155] A. J. Devaney, "Super-resolution processing of multi-static data using time reversal and music," 2000.

[156] ——, "Time reversal imaging of obscured targets from multistatic data," *IEEE Trans. Antennas Propag.*, vol. 53, no. 5, pp. 1600–1610, 2005.

[157] C. J. Leuschen and R. G. Plumb, "A matched-filter-based reverse-time migration algorithm for ground-penetrating radar data," *IEEE Trans. Geoscience Remote Sensing*, vol. 39, no. 5, pp. 929–936, 2001.

[158] N. Maaref, P. Millot, X. Ferrières, C. Pichot, and O. Picon, "Electromagnetic imaging method based on time reversal processing applied to through-the-wall target localization," *Progress In Electromagnetics Research M*, vol. 1, pp. 59–67, 2008.

[159] W. J. Hoefer, "Computational time reversal: A frontier in electromagnetic structure synthesis and design," *IEEE Trans. Microw. Theory Tech.*, vol. 63, no. 1, pp. 3–10, 2015.

[160] W. J. Hoefer and P. P. So, "Reconstruction of non-simultaneous impulsive sources with superresolution in tlm by computational time reversal," in *2015 IEEE MTT-S Int. Microw. Symp. (IMS 2015)*. IEEE, 2015, pp. 1–3.

[161] W. J. Hoefer, "Tracking impulsive sources in space-time by TLM time reversal," in *2016 IEEE MTT-S Int. Microw. Symp. (IMS 2016)*. IEEE, 2016, pp. 1–3.

[162] P. Kosmas and C. M. Rappaport, "A matched-filter FDTD-based time reversal approach for microwave breast cancer detection," *IEEE Trans. Antennas Propag.*, vol. 54, no. 4, pp. 1257–1264, 2006.

[163] ——, "FDTD-based time reversal for microwave breast cancer detection — localization in three dimensions," *IEEE Trans. Microw. Theory Tech.*, vol. 54, no. 4, pp. 1921–1927, 2006.

[164] J. Detlefsen, A. Dallinger, S. Huber, and S. Schelkshorn, "Effective reconstruction approaches to millimeter–wave imaging of humans," in *Proc. 28th General Assembly Int. Union Radio Science*, 2005, pp. 23–29.

[165] M. C. Kemp, "Millimetre wave and terahertz technology for the detection of concealed threats: A review," in *Optics/Photonics in Security and Defence*. International Society for Optics and Photonics, 2006, pp. 64 020D–64 020D.

[166] H. D. Griffiths and C. J. Baker, "Radar imaging for combatting terrorism," in *Imaging for Detection and Identification*. Springer, 2007, pp. 29–48.

[167] R. Zoughi, *Microwave Non-destructive Testing and Evaluation Principles*. Springer Science & Business Media, 2012, vol. 4.

[168] S. Kharkovsky and R. Zoughi, "Microwave and millimeter wave nondestructive testing and evaluation – overview and recent advances," *IEEE Instrumentation Measurement Mag.*, vol. 10, no. 2, pp. 26–38, April 2007.

[169] M. G. Guler and E. B. Joy, "High resolution spherical microwave holography," *IEEE Trans. Antennas Propag.*, vol. 43, no. 5, pp. 464–472, 1995.

[170] B. Gao, H. Zhang, W. L. Woo, G. Y. Tian, L. Bai, and A. Yin, "Smooth nonnegative matrix factorization for defect detection using microwave nondestructive testing and evaluation," *IEEE Trans. Instrumentation Measurement*, vol. 63, no. 4, pp. 923–934, 2014.

[171] O. Mudanyali, S. Yildiz, A. Yapar, I. Akduman *et al.*, "A microwave tomographic approach for nondestructive testing of dielectric coated metallic surfaces," *IEEE Geoscience Remote Sensing Lett.*, vol. 5, no. 2, pp. 180–184, 2008.

[172] K. P. Gaikovich, P. K. Gaikovich, Y. S. Maksimovitch, and V. A. Badeev, "Subsurface near-field microwave holography," *IEEE J. Selected Topics Applied Earth Observations Remote Sensing*, vol. 9, no. 1, pp. 74–82, 2016.

[173] D. Smith, O. Yurduseven, B. Livingstone, and V. Schejbal, "Microwave imaging using indirect holographic techniques," *IEEE Antennas Propag. Mag.*, vol. 56, no. 1, pp. 104–117, 2014.

[174] E. J. Feleppa, "Holography and medicine," *IEEE Trans. Biomedical Engineering*, no. 3, pp. 194–205, 1972.

[175] N. Farhat, "Microwave holography and coherent tomography," *Medical Applications of Microwave Imaging*, pp. 66–81, 1986.

[176] M. Elsdon, D. Smith, M. Leach, and S. Foti, "Experimental investigation of breast tumor imaging using indirect microwave holography," *Microw. Optical Technology Lett.*, vol. 48, no. 3, pp. 480–482, 2006.

[177] R. K. Amineh, A. Khalatpour, H. Xu, Y. Baskharoun, and N. K. Nikolova, "Three-dimensional near-field microwave holography for tissue imaging," *J. Biomedical Imaging*, vol. 2012, p. 5, 2012.

[178] L. Wang, R. Simpkin, and A. Al-Jumaily, "Holographic microwave imaging for medical applications," *J. Biomedical Science Engineering*, vol. 2013, 2013.

[179] L. Wang, A. M. Al-Jumaily, and R. Simpkin, "Investigation of antenna array configurations using far-field holographic microwave imaging technique," *Progress in Electromagnetics Research M*, vol. 42, pp. 1–11, 2015.

[180] D. J. Rochblatt and B. L. Seidel, "Microwave antenna holography," *IEEE Trans. Microw. Theory Tech.*, vol. 40, no. 6, pp. 1294–1300, 1992.

[181] D. Smith, M. Leach, M. Elsdon, and S. Foti, "Indirect holographic techniques for determining antenna radiation characteristics and imaging aperture fields," *IEEE Antennas Propag. Mag.*, vol. 49, no. 1, pp. 54–67, 2007.

[182] D. Gabor *et al.*, "A new microscopic principle," *Nature*, vol. 161, no. 4098, pp. 777–778, 1948.

[183] D. Gabor, "Microscopy by reconstructed wave-fronts," in *Proc. Royal Soc. London A: Mathematical, Physical and Engineering Sciences*, vol. 197, no. 105, pp. 454–487, 1949.

[184] E. N. Leith and J. Upatnieks, "Reconstructed wavefronts and communication theory," *J. Opt. Soc. Amer.*, vol. 52, no. 10, pp. 1123–1130, 1962.

[185] ——, "Wavefront reconstruction with continuous–tone objects," *J. Opt. Soc. Amer.*, vol. 53, no. 12, pp. 1377–1381, 1963.

[186] H. J. Caulfield, *Handbook of Optical Holography*. Elsevier, 2012.

[187] B. P. Hildebrand and B. B. B, *An Introduction to Acoustical Holography*. Plenum Press, 1972.

[188] E. G. Williams, *Fourier Acoustics: Sound Radiation and Nearfield Acoustical Holography*. Academic Press, 1999.

[189] S. I. Hayek, "Nearfield acoustical holography," in *Handbook of Signal Processing in Acoustics*. Springer, 2008, pp. 1129–1139.

[190] R. Dooley, "X-band holography," *Proc. of the IEEE*, vol. 53, no. 11, pp. 1733–1735, 1965.

[191] D. Duffy, "Optical reconstruction from microwave holograms," *J. Optical Soc. Amer.*, vol. 56, no. 6, pp. 832–832, 1966.

[192] G. Tricoles and E. L. Rope, "Reconstructions of visible images from reduced–scale replicas of microwave holograms," *J. Opt. Soc. Amer.*, vol. 57, no. 1, pp. 97–99.

[193] W. Kock, "Stationary coherent (hologram) radar and sonar," *Proc. of the IEEE*, vol. 56, no. 12, pp. 2180–2181, 1968.

[194] W. E. Kock, "Microwave holography," in *Engineering Applications of Lasers and Holography*. Springer, 1975, pp. 179–223.

[195] E. N. Leith, "Quasi-holographic techniques in the microwave region," *Proc. of the IEEE*, vol. 59, no. 9, pp. 1305–s–1318, Sept 1971.

[196] G. Tricoles and N. H. Farhat, "Microwave holography – applications and techniques," vol. 65, pp. 108–121, 1977.

[197] A. Anderson, "Microwave holography," *Proc. of the IEEE*, vol. 124, no. 11, pp. 946–962, 1977.

[198] N. H. Farhat, "Microwave holography and coherent tomography," in *Medical Applications of Microwave Imaging*.

[199] D. Smith, O. Yurduseven, B. Livingstone, and V. Schejbal, "Microwave imaging using indirect holographic techniques," *IEEE Antennas Propag. Mag.*, vol. 56, no. 1, pp. 104–117, 2014.

[200] W. M. Brown and L. J. Porcello, "Reconstructions of visible images from reduced-scale replicas of microwave holograms," *IEEE Spectrum*, vol. 6, no. 9, pp. 52–62.

[201] E. Abbe, "Ueber einen neuen beleuchtungsapparat am mikroskop," *Archiv für mikroskopische Anatomie*, vol. 9, no. 1, pp. 469–480, 1873.

[202] M. Bertero and P. Boccacci, *Introduction to Inverse Problems in Imaging*. CRC Press, 1998.

[203] F. Natterer and F. Wübbeling, *Mathematical Methods in Image Reconstruction*, ser. Monographs on Mathematical Modeling and Computation. SIAM, 2001.

[204] W. H. Press, *Numerical Recipes: The Art of Scientific Computing*, 3rd ed. Cambridge University Press, 2007.

[205] Emerson and Cuming Microwave Products, a unit of Laird Technologies. http://www.eccosorb.com/products-overview.htm

[206] Y. Rahmat-Samii and J. Lemanczyk, "Application of spherical near–field measurements to microwave holographic diagnosis of antennas," *IEEE Trans. Antennas Propag.*, vol. 36, no. 6, pp. 869–878, 1988.

[207] L. Wang, A. M. Al-Jumaily, and R. Simpkin, "Imaging of 3-d dielectric objects using far-field holographic microwave imaging technique," *Progress In Electromagnetics Research B*, vol. 61, pp. 135–147, 2014.

[208] ——, "Investigation of antenna array configurations using far-field holographic microwave imaging technique," *Progress In Electromagnetics Research M*, vol. 42, pp. 1–11, 2015.

[209] M. Leach, M. Elsdon, S. Foti, and D. Smith, "Imaging dielectric objects using a novel synthetic off–axis holographic technique," *Microwave Optical Tech. Lett.*, vol. 48, no. 10, pp. 1957–1961, 2006.

[210] M. Elsdon, D. Smith, M. Leach, and S. Foti, "Experimental investigation of breast tumor imaging using indirect microwave holography," *Microwave Optical Tech. Lett.*, vol. 48, no. 3, pp. 480–482, 2006.

[211] E. Wolf, "Three-dimensional structure determination of semi-transparent objects from holographic data," *Optics Communications*, vol. 1, no. 4, pp. 153–156, 1969.

[212] R. K. Mueller, M. Kaveh, and G. Wade, "Reconstructive tomography and applications to ultrasonics," *Proc. of the IEEE*, vol. 67, no. 4, pp. 567–587, 1979.

[213] R. Mueller, M. Kaveh, and R. Iverson, "A new approach to acoustic tomography using diffraction techniques," in *Acoustical Imaging*. Springer, 1980, pp. 615–628.

[214] A. Devaney, "A filtered backpropagation algorithm for diffraction tomography," *Ultrasonic Imaging*, vol. 4, no. 4, pp. 336–350, 1982.

[215] ——, "A computer simulation study of diffraction tomography," *IEEE Trans. Biomed. Eng.*, no. 7, pp. 377–386, 1983.

[216] S. Pan and A. Kak, "A computational study of reconstruction algorithms for diffraction tomography: Interpolation versus filtered-back-propagation," *IEEE Trans. Acoustics, Speech, Signal Proc.*, vol. 31, no. 5, pp. 1262–1275, 1983.

[217] K. Iwata and R. Nagata, "Calculation of refractive index distribution from interferograms using the Born and Rytov's approximation," *Japanese Journal of Applied Physics*, vol. 14, no. S1, p. 379, 1975.

[218] A. Devaney, "Inverse-scattering theory within the Rytov approximation," *Optics Lett.*, vol. 6, no. 8, pp. 374–376, 1981.

[219] M. F. Adams and A. P. Anderson, "Synthetic aperture tomographic (SAT) imaging for microwave diagnostics," *IEEE Proc. H (Microwaves, Optics and Antennas)*, vol. 129, no. 2, pp. 83–88, 1982.

[220] M. Baribaud, F. Dubois, R. Floyrac, M. Kom, and S. Wang, "Tomographic image reconstitution of biological objects from coherent microwave diffraction data," *IEEE Proc. H (Microwaves, Optics and Antennas)*, vol. 129, no. 6, pp. 356–359, 1982.

[221] J. C. Bolomey, A. Izadnegahdar, L. Jofre, C. Pichot, G. Peronnet, and M. Solaimani, "Microwave diffraction tomography for biomedical applications," *IEEE Trans. Microw. Theory Tech.*, vol. 30, no. 11, pp. 1998–2000, Nov. 1982.

[222] C. Pichot, L. Jofre, G. Peronnet, and J. Bolomey, "Active microwave imaging of inhomogeneous bodies," *IEEE Trans. Antennas Propag.*, vol. 33, no. 4, pp. 416–425, 1985.

[223] F. J. Paoloni, "Implementation of microwave diffraction tomography for measurement of dielectric constant distribution," vol. 134, no. 1, pp. 25–29, 1987.

[224] A. C. Kak and M. Slaney, *Principles of Computerized Tomographic Imaging*. IEEE Press, 1988.

[225] O. K. Ersoy, *Diffraction, Fourier Optics and Imaging*. John Wiley & Sons, 2006, vol. 30.

[226] K. Knaell and G. Cardillo, "Radar tomography for the generation of three-dimensional images," *IEEE Proc. Radar, Sonar and Navigation*, vol. 142, no. 2, pp. 54–60, 1995.

[227] L. Jofre, M. S. Hawley, A. Broquetas, E. de Los Reyes, M. Ferrando, and A. R. Elias-Fuste, "Medical imaging with a microwave tomographic scanner," *IEEE Trans. Biomed. Eng.*, vol. 37, no. 3, pp. 303–312, 1990.

[228] J. V. Candy and C. Pichot, "Active microwave imaging: A model-based approach," *IEEE Trans. Antennas Propag.*, vol. 39, no. 3, pp. 285–290, 1991.

[229] L. Chommeloux, C. Pichot, and J.-C. Bolomey, "Electromagnetic modeling for microwave imaging of cylindrical buried inhomogeneities," *IEEE Transactions on Microwave Theory and techniques*, vol. 34, no. 10, pp. 1064–1076, 1986.

[230] S. Y. Semenov, R. H. Svenson, A. E. Boulyshev, A. E. Souvorov, V. Y. Borisov, Y. Sizov, A. N. Starostin, K. R. Dezern, G. P. Tatsis, and V. Y. Baranov, "Microwave tomography: Two-dimensional system for biological imaging," *IEEE Trans. Biomed. Eng.*, vol. 43, no. 9, pp. 869–877, 1996.

[231] J.-C. Bolomey and C. Pichot, "Microwave tomography: From theory to practical imaging systems," *Int. J. Imaging Systems Tech.*, vol. 2, no. 2, pp. 144–156, 1990.

[232] M. Miyakawa, "Tomographic measurement of temperature change in phantoms of the human body by chirp radar-type microwave computed tomography," *Medical & Biological Engineering & Computing*, vol. 31, no. 1, pp. S31–S36, 1993.

[233] M. Miyakawa and J. C. Bolomey, "Microwave imaging - 1: microwave computed tomography," in *Non-Invasive Thermometry of Human Body*, 1996, pp. 105–126.

[234] M. Bertero, M. Miyakawa, P. Boccacci, F. Conte, K. Orikasa, and M. Furutani, "Image restoration in chirp-pulse microwave ct (cp-mct)," *IEEE Trans. Biomed. Eng.*, vol. 47, no. 5, pp. 690–699, 2000.

[235] M. Miyakawa, T. Yokoo, N. Ishii, and M. Bertero, "Visualization of human arms and legs by CP-MCT," in *38th European Microwave Conference (EuMC 2008)*. IEEE, 2008, pp. 412–415.

[236] A. Massone, M. Miyakawa, M. Piana, F. Conte, and M. Bertero, "A linear model for chirp-pulse microwave computerized tomography: Applicability conditions," *Inverse Problems*, vol. 22, no. 6, pp. 2209–2218, 2006.

[237] T. Takahashi, M. Miyakawa, M. Tamura, and T. Ogawa, "High fidelity breast phantom and its microwave imaging by CP-MCT," in *Asia-Pacific Microwave Conference 2011*. IEEE, 2011, pp. 1490–1493.

[238] R. F. Wagner and D. G. Brown, "Unified SNR analysis of medical imaging systems," *Physics in Medicine and Biology*, vol. 30, no. 6, p. 489, 1985.

[239] R. F. Wagner, "Low contrast sensitivity of radiologic, CT, nuclear medicine, and ultrasound medical imaging systems," *IEEE Trans. Medical Imaging*, vol. 2, no. 3, pp. 105–121, 1983.

[240] M. Tapiovaara and R. Wagner, "SNR and noise measurements for medical imaging: I. A practical approach based on statistical decision theory," *Physics in Medicine and Biology*, vol. 38, no. 1, p. 71, 1993.

[241] K. W. Bowyer, "Validation of medical image analysis techniques," in *Handbook of Medical Imaging*, vol. 2. Bellingham, WA: SPIE, 2000, pp. 567–607.

[242] A. Khalatpour, R. K. Amineh, H. Xu, Y. Baskharoun, and N. K. Nikolova, "Image quality enhancement in the microwave raster scanning method," in *2011 IEEE MTT-S Int. Microwave Symp. Digest*. IEEE, 2011, pp. 1–4.

[243] I. Walterscheid, A. Brenner, and J. Ender, "Results on bistatic synthetic aperture radar," *Electronics Lett.*, vol. 40, no. 19, pp. 1224–1225, 2004.

[244] R. K. Amineh, M. Ravan, J. McCombe, and N. K. Nikolova, "Three-dimensional microwave holographic imaging employing forward-scattered waves only," *Int. J. Antennas Propag.*, vol. 2013, 2013.

[245] R. F. Wagner, M. F. Insana, and D. G. Brown, "Statistical properties of radio-frequency and envelope-detected signals with applications to medical ultrasound," *JOSA A*, vol. 4, no. 5, pp. 910–922, 1987.

[246] Keysight Technologies: Understanding and Improving Network Analyzer Dynamic Range. http://literature.cdn.keysight.com/litweb/pdf/5980-2778EN.pdf

[247] Agilent: Fundamentals of RF and Microwave Noise Figure Measurements, Application Note 57-1. http://cp.literature.agilent.com/litweb/pdf/5952-8255E.pdf

[248] L. Kaufman, D. M. Kramer, L. E. Crooks, and D. A. Ortendahl, "Measuring signal-to-noise ratios in MR imaging," *Radiology*, vol. 173, no. 1, pp. 265–267, 1989.

[249] M. Firbank, A. Coulthard, R. Harrison, and E. Williams, "A comparison of two methods for measuring the signal to noise ratio on MR images," *Physics in Medicine and Biology*, vol. 44, no. 12, p. N261, 1999.

[250] M. Firbank, R. Harrison, E. Williams, and A. Coulthard, "Quality assurance for MRI: Practical experience." *British Journal of Radiology*, vol. 73, no. 868, pp. 376–383, 2000.

[251] O. Dietrich, J. G. Raya, S. B. Reeder, M. F. Reiser, and S. O. Schoenberg, "Measurement of signal-to-noise ratios in MR images: Influence of multichannel coils, parallel imaging, and reconstruction filters," *Journal of Magnetic Resonance Imaging*, vol. 26, no. 2, pp. 375–385, 2007.

[252] X. Song, B. W. Pogue, S. Jiang, M. M. Doyley, H. Dehghani, T. D. Tosteson, and K. D. Paulsen, "Automated region detection based on the contrast-to-noise ratio in near-infrared tomography," *Applied Optics*, vol. 43, no. 5, pp. 1053–1062, 2004.

[253] J. J. McCombe and N. K. Nikolova, "SNR assessment of microwave imaging systems," in *2014 IEEE AP-S/URSI Int. Symp. on Antennas and Propagation*, 2014.

[254] J. J. McCombe, "Noise reduction and clutter suppression in microwave imaging and detection," Master's thesis, McMaster University, Canada, 2014.

[255] X. Zeng, A. Fhager, M. Persson, P. Linner, and H. Zirath, "Accuracy evaluation of ultrawideband time domain systems for microwave imaging," *IEEE Trans. Antennas Propag.*, vol. 59, no. 11, pp. 4279–4285, 2011.

[256] K. Moussakhani, J. J. McCombe, and N. K. Nikolova, "Sensitivity of microwave imaging systems employing scattering-parameter measurements," *IEEE Trans. Microwave Theory Tech.*, vol. 62, no. 10, pp. 2447–2455, 2014.

[257] R. C. Conceição, J. J. Mohr, and M. O'Halloran, *An Introduction to Microwave Imaging for Breast Cancer Detection*, ser. Biological and Medical Physics. Springer, 2016.

[258] R. Fletcher, *Practical Methods of Optimization*. John Wiley & Sons, 2013.

[259] S. S. Rao, *Engineering Optimization: Theory and Practice*. John Wiley & Sons, 2009.

[260] H. Blok and M. Oristaglio, "Wavefield imaging and inversion in electromagnetics and acoustics," *Delft University of Technology. Report Number Et/EM*, vol. 21, 1995.

[261] R. F. Remis and P. Van den Berg, "On the equivalence of the Newton-Kantorovich and distorted Born methods," *Inverse Problems*, vol. 16, no. 1, pp. L1–L4, 2000.

[262] J. D. Jackson, *Classical Electrodynamics*, 3rd ed. Wiley, 1999.

[263] T. G. Mackay and A. Lakhtakia, *Electromagnetic Anisotropy and Bianisotropy: A Field Guide*. World Scientific, 2010.

[264] A. Taflove and S. Hagness, *Computational Electrodynamics: The Finite-Difference Time-Domain Method*, 3rd ed. Artech House, 2005.

[265] N. K. Georgieva and H. W. Tam, "Potential formalisms in electromagnetic field analysis," *IEEE Trans. Microw. Theory Tech.*, vol. 51, no. 4, pp. 1330–1338, 2003.

[266] S.-G. Pan and I. Wolff, "Scalarization of dyadic spectral Green's functions and network formalism for three-dimensional full-wave analysis of planar lines and antennas," *IEEE Trans. Microw. Theory Tech.*, vol. 42, no. 11, pp. 2118–2127, 1994.

[267] N. K. Georgieva, "Construction of solutions to electromagnetic problems in terms of two collinear vector potentials," *IEEE Trans. Microw. Theory Tech.*, vol. 50, no. 8, pp. 1950–1959, 2002.

[268] N. K. Georgieva and W. S. Weiglhofer, "Electromagnetic vector potentials in isotropic nonhomogeneous materials: Mode equivalence and scalarization," *IEE Proc. H (Microwaves, Antennas and Propagation)*, vol. 150, no. 3, pp. 164–170, 2003.

[269] W. S. Weiglhofer, "Scalarisation of Maxwell's equations in general inhomogeneous bianisotropic media," *IEE Proc. H (Microwaves, Antennas and Propagation)*, vol. 134, no. 4, pp. 357–360, 1987.

[270] W. Weiglhofer, "Scalar Hertz potentials for linear bianisotropic mediums," in *Electromagnetic Fields in Unconventional Materials and Structures*, pp. 1–37, 2000.

[271] N. Zhuck, "Electromagnetic theory of arbitrarily anisotropic layered media. Part I: scalarization of field quantities," *Int. J. Electronics Theoretical and Experimental*, vol. 75, no. 1, pp. 141–148, 1993.

[272] T. G. Mackay and A. Lakhtakia, "Electromagnetic fields in linear bianisotropic mediums," *Progress in Optics*, vol. 51, pp. 121–209, 2008.

[273] W. S. Weiglhofer, "Dyadic Green's functions for general uniaxial media," *IEE Proc. H (Microwaves, Antennas and Propagation)*, vol. 137, no. 1, pp. 5–10, Feb 1990.

[274] S. Przeździecki and W. Laprus, "On the representation of electromagnetic fields in gyrotropic media in terms of scalar Hertz potentials," *J. Mathematical Physics*, vol. 23, no. 9, pp. 1708–1712, 1982.

[275] N. K. Georgieva and W. S. Weiglhofer, "Electromagnetic vector potentials and the scalarization of sources in a nonhomogeneous medium," *Phys. Rev. E*, vol. 66, no. 4, p. 046614, 2002.

[276] E. T. Whittaker, *A History of the Theories of Aether & Electricity*. Harper & Brothers, 1960 reprint.

[277] P. Ehrenfest, "Ungleichförmige elektrizitätsbewegungen ohne magnet- und strahlungsfeld," *Physikalische Zeitschrift*, vol. 11, pp. 708–709, 1910.

[278] G. A. Schott, "The electromagnetic field of a moving uniformly and rigidly electrified sphere and its radiationless orbits," *Phil. Mag. S. 7*, vol. 15, no. 100, pp. 752–761, 1933.

[279] N. K. Nikolova, Y. Li, Y. Li, and M. H. Bakr, "Sensitivity analysis of scattering parameters with electromagnetic time-domain simulators," *IEEE Trans. Microw. Theory Tech.*, vol. 54, no. 4, pp. 1598–1610, 2006.

Index